Approximate Models of Mechanics of Composites

Approximate Models of Mechanics of Composites: An Asymptotic Approach is an essential guide to constructing asymptotic models and mathematical methods to correctly identify the mechanical behavior of composites. It provides methodology for predicting and evaluating composite behavior in various structures, leading to accurate mathematical and physical assessments.

The book estimates the error of approximations through comparing asymptotic solutions with the results of numerical and analytical solutions to gain a holistic view of the data. The authors have developed asymptotic models based on mathematical and physical rigorous approaches, which include three-phase models of fibrous composites, a modernized three-phase composite model with cylindrical inclusions, and models of two-dimensional composites of hexagonal structure. Also covered are two-phase models of composites related to the Maxwell formula and a percolation transition model for elastic problems based on the self-consistency method and Padé approximations. By obtaining analytical expressions to effectively characterize composite materials, their physical and geometric parameters can be accurately assessed.

This book suits engineers and students working in material science, mechanical engineering, physics, and mathematics, as well as composite materials in industries such as construction, transport, aerospace, and chemical engineering.

Approximate Models of Mechanics of Composites

An Asymptotic Approach

Igor V. Andrianov
Jan Awrejcewicz
Galina A. Starushenko

CRC Press
Taylor & Francis Group
Boca Raton London New York

CRC Press is an imprint of the
Taylor & Francis Group, an **informa** business

Cover image: I.V. Andrianov, J. Awrejcewicz and G.A. Starushenko

First edition published 2023
by CRC Press
6000 Broken Sound Parkway NW, Suite 300, Boca Raton, FL 33487-2742

and by CRC Press
4 Park Square, Milton Park, Abingdon, Oxon, OX14 4RN

CRC Press is an imprint of Taylor & Francis Group, LLC

Library of Congress Cataloging-in-Publication Data

Names: Andrianov, I. V. (Igor' Vasil'evich), 1948- author. | Awrejcewicz,
   J. (Jan), author. | Starushenko, G. A., author.
Title: Approximate models of mechanics of composites : an asymptotic
   approach / I.V. Andrianov, J. Awrejcewicz, G.A. Starushenko.
Description: First edition. | Boca Raton : CRC Press, [2023] | Includes
   bibliographical references and index.
Identifiers: LCCN 2022056927 | ISBN 9781032488301 (hbk) | ISBN
   9781032488349 (pbk) | ISBN 9781003391029 (ebk)
Subjects: LCSH: Composite materials--Mechanical properties--Mathematical
   models. | Asymptotic expansions.
Classification: LCC TA418.9.C6 A455 2023 | DDC
   620.1/18015118--dc23/eng/20230111
LC record available at https://lccn.loc.gov/2022056927

ISBN: 978-1-032-48830-1 (hbk)
ISBN: 978-1-032-48834-9 (pbk)
ISBN: 978-1-003-39102-9 (ebk)

DOI: 10.1201/9781003391029

Typeset in font Nimbus font
by KnowledgeWorks Global Ltd.

Publisher's note: This book has been prepared from camera-ready copy provided by the authors.

Contents

Preface

We live in the era of big data, artificial intelligence (AI), neural networks, and powerful commercial software packages. However, our ability to understand and manage what is happening is still based on simple models [23]. Fifty years ago it was written, "The purpose of computing is insight, not numbers" [107]. In our opinion, there is no question of "asymptotic or computer (numerical) simulations". However, it is well recognized that both of these scientific directions are equally important [9, 10].

Without a clear understanding of the asymptotic nature of any model of applied mathematics, it is impossible to correctly determine its place in the hierarchy of descriptions of various aspects of the studied processes [7]. In addition, the more powerful numerical algorithms become, the more important the role of a successful zero approximation. For multidimensional and multifactorial problems, it is important to know where in the vicinity we are building a solution. Otherwise, even the most advanced numerical algorithms can be unstable, the solutions may diverge, or we can get spurious solutions. Asymptotic solutions usually allow one to find correct seed solutions for numerical studies.

It is equally important to test any numerical solutions. From this point of view, asymptotic methods are not only useful for analysis, but also allow one to obtain important and useful benchmark solutions [103].

Recently, the assertion has been ubiquitous so that the possibilities of obtaining information have increased immeasurably. However, information is not a meaningless collection of data. It needs to be analyzed in order to identify patterns, create small-scale models, compactification, and aggregation, determine the most important singularities (for example, bifurcation points, areas of unstable and chaotic behavior), etc. Analytical, especially asymptotic, methods are indispensable here [23].

A few words about what is meant here by the term "analytical solution" are provided. Unfortunately, in recent times this term is very often used in vain [24]. A solution will be called analytical if it contains a finite number of elementary or special functions, arithmetic operations, integrals, derivatives, and series. This is how it is understood in our book.

Let's get to the heart of the matter. The text is devoted to the construction of asymptotic models that make it possible to calculate the effective characteristics of composite materials, which describe their internal structure.

Periodically inhomogeneous composite materials, consisting of several components with different physical properties, are widely used in aircraft, rocket, and shipbuilding, mechanical engineering, industrial and civil engineering, and other areas of modern industry. The choice of various materials and forms of inclusions and matrix allows obtaining materials with useful properties, i.e., having high strength and rigidity, low thermal conductivity, etc. [69, 208, 161, 224, 226].

When dealing with a composite material, we can be interested in both its global (effective) characteristics (for example, the thermal conductivity coefficient or the elastic modulus of a sufficiently large volume) and the local distribution of the desired fields. The latter are determined by the real (internal) structure of the composite.

As a rule, the size of a typical repeating element of a composite is significantly smaller than the size of the entire structure, which allows us to consider their ratio as a small parameter and use the asymptotic homogenization method of partial differential equations with rapidly oscillating or periodically discontinuous coefficients [38, 44, 145, 207]. This method can be described as follows. First, some periodically repeating boundary value problem ("problem on a cell" and "local problem") is singled out and its solution is sought under the boundary conditions of periodic continuation. In this case (explicitly or implicitly) local coordinates are introduced ("fast variables" if the multiple scale asymptotic method is applied). Next, the averaging over local (fast) variables is performed. The efficiency of the homogenization method depends on whether it is possible to solve problems on a cell quite simply, preferably analytically. The book shows that this can indeed be done by successively applying the asymptotic methods of regular and singular perturbations, the alternating Schwarz method [160], as well as the Padé summation [8, 30, 31] and Padé interpolation [8, 152, 215] methods. In the studies carried out, not only analytical solutions of problems are obtained, but also the limits of applicability of the theories on the basis of which they were derived are indicated, and, where necessary, refined relations are proposed.

The following asymptotic models are constructed, mathematically described, and physically substantiated in the text:

(i) three-phase model of fibrous composite with circle and square inclusions;
(ii) modified three phase for composite model with cylindrical inclusions;
(iii) model of two-dimensional composites of hexagonal structure;
(iv) models of two-phase fibrous composites for inclusions of various shapes and the "inclusion-matrix" contact;
(v) two-phase models of composites with circle and square inclusions for constructing generalizing relations of the Maxwell formula;
(vi) a percolation transition model for elastic problems based on the self-consistency method and Padé approximations.

In the mechanics of composites, it is conditionally possible to distinguish four main areas of research: mathematical, physical, computational, and asymptotic approaches [12]. There is no clear line between them because they are not competing and not mutually exclusive approaches. In contrast, each of them has its own strengths and specific areas of application; therefore, they should be considered in an inseparable unit with emphasis put on complementary aspects, allowing a comprehensive description of the object of study from various points of view.

Mathematical theories of composite materials enable a rigorous mathematical justification of the methods, algorithms, and calculation schemes used [1, 32, 33, 34, 35, 36, 37, 38, 94, 191, 192, 193]. On the other hand, physical theories are aimed, first of all, at describing the physical essence of the ongoing processes full-stop, *then* (upper case T) they allow a deeper understanding of these processes and assess the adequacy of the mathematical apparatus used to describe them [53, 59, 69, 139, 213, 219]. Finally, computational theories are used both in engineering calculations and in scientific research of a fundamental and applied nature [107].

The trend of modern research in the age of AI, big data, and digital technologies is aimed at the use of tools, i.e., mathematical packages and computational numerical methods of analysis. However, in our opinion, a qualitative breakthrough in numerical research, associated with the advent of powerful PCs and the latest software, does not diminish but rather increases, the significance of the asymptotic approach [11, 12, 23]. Asymptotic approaches allow one to obtain important information about the properties of the object under study, which is difficult to expect from numerical algorithms. If analytical formulas describe one or another characteristic of the object or process under study, then their asymptotics allow us to identify the most general patterns and essential components of solutions that, with a slight perturbation of the system, lead to significant changes in its behavior.

The need to use asymptotic information in the development of computational or experimental schemes was pointed out by a number of prominent researchers [53, 77, 103, 107]. They emphasized that without such information it is not possible to identify the critical values parameters of the processes under consideration.

Asymptotic methods are effective in the range of limiting values of the structural and physical parameters of composites, precisely where the use of pure computational algorithms is problematic. Moreover, as experience shows, often asymptotic solutions can be used far beyond their nominal range of applicability [77]. It is even more important that they can be treated as first approximations of some processes, which make it possible within the framework of a unified approach to construct subsequent approximations that refine the solutions of zero approximation [11, 12, 23].

In this regard, it seems relevant to build asymptotic models and develop adequate methods for their mathematical description, all the more so in the light of modern trends in the creation of new composite materials, "smart" composites with desired properties, and predictable behavior under operating conditions.

In terms of the development of applied aspects of the mechanics of composites, research related to the construction of various asymptotic models of composite materials that take into account the features of their internal structure seems promising. The use of such an integrated approach is expedient in the study of many practically important applied problems that arise in modern engineering and still do not have a satisfactory solution.

In the book, much attention is paid to estimating the error of the main approximations. In real problems, the "small parameter" is finite. Asymptotic character of solution does little to estimate the actual error, and, as a rule, it is not possible to prove convergence (in those rare cases when this can be done, the corresponding estimates are too pessimistic). Therefore, in practice, it seems more consistent to compare the asymptotic solutions with the results of numerical solutions or analytical estimates. In the text, these comparisons are made carefully for all available data, and reliable verification of the results is provided.

This book summarizes the results of many years of research by the authors in the field of mechanics of composite materials [5, 6, 13, 14, 15, 16, 17, 18, 19, 20, 21, 22, 25, 26, 27, 28, 29, 117, 118, 113].

The first two parts of the book provide basic information about composite materials, outline the most important stages in the development of composite mechanics as a science, and offer a brief overview of research and the most significant results in this area.

Chapter 3 discusses the three-phase model of the composite. The possibilities of generalizing the three-phase model based on the perturbation position of the boundary and using Padé approximants are analyzed. A modernized three-phase model of the composite is constructed, and its asymptotic analysis is carried out. Possible approaches to generalizations of the three-phase model to the case of 3D composite structures and to the problems of elasticity theory are shown.

Chapter 4 is devoted to the analysis of the lubrication approach on the basis of which solutions for composites with different structure and shape of inclusions are obtained. A technique for applying two- and three-point Padé approximants to determine two-sided estimations of the effective thermal conductivity is described.

In the fifth chapter, to study the limit states of composite structures the apparatus of asymptotically equivalent functions is used. Models are constructed, and asymptotic representations are obtained for fibrous composites of various structures. Using nonsmooth periodic coordinates, the concept of physical equivalence of composite structures is analyzed. The question of the influence of a thin interface at the phase boundary of a composite on its effective characteristics is studied. The issue of percolation transition in problems of elasticity theory is touched upon.

Chapter 6 considers the problem of constructing higher approximations of the Maxwell formula. An asymptotic analysis of the relations constructed by the alternating Schwarz method based on the two-phase model of the composite is carried out. Analytical expressions are obtained that refine the Maxwell formula for composite materials reinforced with fibres of square and round cross sections. Generalizing relations for the Maxwell formula are constructed using the Schwarz–Padé expansion, and both numerical and asymptotic analysis of their range of applicability are performed.

In conclusion, the obtained results are briefly summarized, their place among other studies is indicated, and directions for further research are formulated.

We would like to acknowledge V.V. Danishevskyy, S. Gluzman, A.L. Kalamkarov, V.I. Malyi, L.I. Manevitch, V.V. Mityushev, S. Tokarzewski, and D. Weichert for the helpful and inspiring exchanges of ideas, fruitful collaboration, and stimulating discussions.

To read this book, it is enough to know the standard courses of higher mathematics, strength of materials, and the theory of elasticity of a technical university.

For those wishing to get acquainted with the basics of the mechanics of composites, we highly recommend the book by Christensen [71].

The text will be interesting and useful to engineers, mechanical scientists, physicists, and mathematicians whose interests lie in the field of composite materials research.

Abbreviations

2PhM	two-phase model
AP	Padé approximants
EMA	effective media approximation
LA	lubrication approach
MF	Maxwell formula
PPB	perturbation position of the boundary
MThPhM	modified three-phase model
ThPhM	three-phase model
q_{ASM}	alternating Schwarz method
$q_{asf}^{(\infty)}$	asymptotic formula
q_{ass}	asymptotic solution
$q_{ci}^{(\infty)}$	circular inclusion
$q_{ciint}^{(\infty)}$	circular inclusion with interlayer
q_{cm}	contact of matrixes
q_{cr}	critical
$q_{cri}^{(\infty)}$	curvilinear rhombic inclusion
$q_{criint}^{(\infty)}$	curvilinear rhombic inclusion with interlayer
q_{cd}	domain of contact
q_{ef}	effective value
q_f	formula
$\underline{q}_{HS}$	Hashin-Shtrikman
$q_{m.incl.}$	inclusion of middle size
λ_i	inclusion parameter
λ_{int}	interface parameter
$q_{LA}^{(0)}$	lubrication approach
λ_m	matrix parameter
q_{MF}	Maxwell formula
q_{MThPhM}	modified three-phase model
q_{num}	numerical solution
q_{AP}	Padé approximants
q_{PPB}	perturbation position of the boundary
q_{cp}	point of contact
q_{ASM-AP}	Schwarz–Padé
q_{ThPhM}	three-phase model
q_{ThPhM}^{hex}	three-phase model for hexagonal lattice
q_{2PhM}	two-phase model

1 Introduction

This chapter serves as an introduction to the mechanics of composite materials. The composite definition is given, and its importance of both theoretical and application aspects is outlined. In particular, emphasis is paid to the description of fibrous composites and classification of inhomogeneous structures, as well as materials of composite and fillers. The advantage of analytical approaches to understand the behavior of composites is finally addressed.

1.1 REVIEW OF COMPOSITE MATERIAL DEVELOPMENT

The mechanics of composite materials is a rapidly developing branch of knowledge in all aspects, whether it be theoretical research, experimental study, or practical applications. This is due to the modern requirements of scientific and technological progress with regard to the use of complex composite structures in all areas of technology, which must combine lightness with high strength, the use of innovative production technologies, and the creation of new composite materials.

In production, composite materials of various properties are employed, and they take a variety of shapes. However, initially a material rarely has a combination of properties that exactly match the requirements of a particular application. Therefore, from the point of view of structural feasibility, it is reasonable to use composite materials, the most important advantage of which is the possibility of constructing structures from them with predetermined properties that would correspond as fully as possible to the nature and conditions of the required work.

In a broad sense, composites can be defined as complex materials consisting of two or more components (a binder matrix and discrete inclusions placed in it) and having specific properties that are different from the properties of their constituent elements.

The most common binders are polymers, polyesters, phenols, epoxy compounds, silicones, alkyds, melamines, polyamides, fluorocarbon compounds, polypropylene, polyethylene, polystyrene, etc. [151, 224].

Binders are also classified into thermoplastics, which are capable of softening and hardening with temperature changes; thermosetting polymers, in which, when heated, irreversible structural and chemical transformations occur.

Discrete elements can play a passive or active role: in the first case, they serve as a filler; in the second, they are used as reinforcing elements. When creating composite materials and structures, inert fillers are used, as a rule, to fill the volume and reduce the cost of the composite. Active fillers are employed to improve the strength characteristics of composites, reduce the weight of structures, and optimize their physical and mechanical, structural, or functional properties.

DOI: 10.1201/9781003391029-1

The reinforcing components may be fibers, microspheres, crystals, powders, "whiskers" of organic or inorganic compounds, metallic materials or ceramics (whiskers are very short fibers, usually from single crystals [151]).

The creation of new composite materials implies selection of materials with the necessary properties for each specific area of their practical use. And the scope of composites is unusually wide and affects almost all areas of production: from the manufacture of the simplest household items to the creation of structural elements for nuclear reactors and spacecraft. Thus, the use of new composite materials has raised the aircraft industry to a new qualitative level, contributed to the creation of rocket and space technology of a fundamentally higher level.

Important factors in the creation of new composite materials, in addition to their mechanical properties, are also their cost, the availability of appropriate technologies and the environmental safety of production and operation.

The variety of components of composite materials include fibers and matrix filler, as well as reinforcement schemes, and allow one to develop technological processes for creating a composite with needed physical characteristics like strength, rigidity, stability, operating temperature level, etc., achieved by varying its composition, as well as by changing the ratio of the components and macrostructure.

Practice shows that by combining materials, it is often possible to get a favorable combination of their properties. The explanation of such empirical results and accidental discoveries is one of the tasks of applied science. However, its more important purpose is the creation of new materials based on the fundamental study of heterogeneous media. Ultimately, only the exact knowledge of the properties of heterogeneous media can provide the key to the optimal use of structural materials.

Let us give two typical examples illustrating the directions of research in the field of creating composite materials.

Many homogeneous polymers are glassy and brittle. By trial and error, technologists have found that the uniform dispersion of spherical rubber particles in the polymer can significantly reduce its sensitivity to impact. Most of the glassy polymers currently produced contain a rubber "strengthening" additive as a performance enhancer.

The second example is fiber composites. In many cases it is expedient to use stiff fibers combined into a monolithic material with the help of a soft component, i.e. a matrix (often a polymer). The examples given are typical in terms of research directions on the purposeful creation and use of heterogeneous materials.

The variable concept of "fibrous composite" unites a very wide class of heterogeneous structures. For their classification, several approaches based on various features [47, 94, 113, 133, 158, 164] can be used:

1. Materials science-based classification yields of the following large groups of composites:
   1.1. Based on the matrix material choice [224]: polymer matrix (plastics); metal matrix (metal composites); ceramic matrix; carbon matrix.
   1.2. Based on the nature of the reinforcing fibers [123], such composites on a polymer matrix: fiberglass; carbon fiber; boroplasty; organoplasty, etc.

Figure 1.1 Chaotically reinforced composites: a) with long fibers; b) with short fibers; c) with identical chaotic inclusions; d) with chaotic inclusions of various geometries.

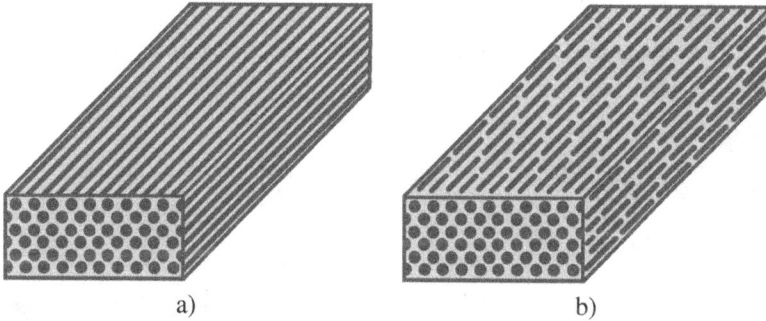

Figure 1.2 One-dimensionally reinforced composites: a) with unidirectional long fibers; b) with unidirectional short fibers.

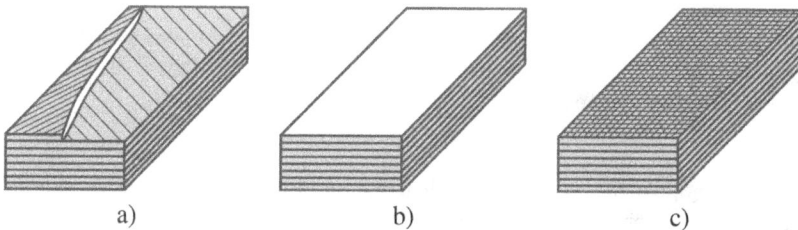

Figure 1.3 Two-dimensionally reinforced composites: a) with continuous filaments; b) with continuous parallel layers; c) with fabrics.

There are similarly named composites on other matrices.

2. According to the type of reinforcement and its orientation in the matrix [230]: formed from layers reinforced with continuous parallel fibers (their properties are determined mainly by the properties of the unidirectional layer); reinforced with fabrics (textolites), chaotic reinforcement and spatial reinforcement.

3. According to the method of manufacture and processing into products.

Figures 1.1–1.4 illustrate various types of laminated composite materials classified based on their design characteristic features.

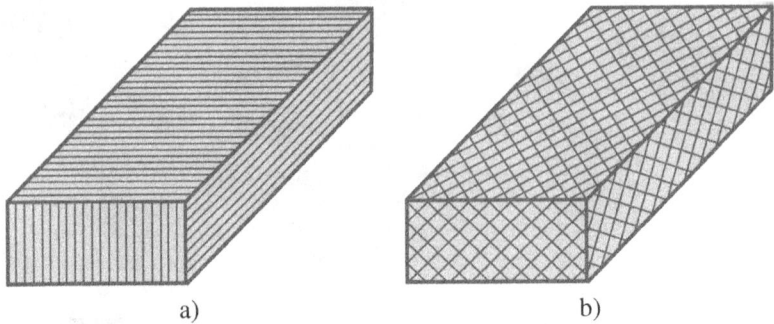

Figure 1.4 Spatially reinforced composites: a) with three families of threads; b) with many families of threads.

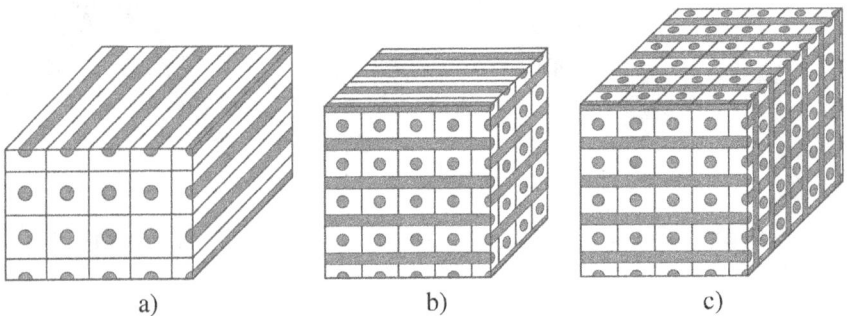

Figure 1.5 Composite materials of periodic structure: a) 1D-reinforced periodic structure composite; b) 2D orthogonally reinforced periodic structure composite; c) 3D orthogonally reinforced periodic structure composite.

Fibrous composites with a periodic distribution of reinforcing fibers are classified based on the reinforcement structure, and they include unidirectionally reinforced (1D); reinforced with fibers in two planes (2D); spatially reinforced (3D).

1D-reinforced and 2D- and 3D-orthogonally reinforced composite materials of a periodic structure are shown in Fig. 1.5, respectively.

The internal structure of composite materials, especially of natural origin, is characterized, as a rule, by a complex hierarchical structure. The structure of polydisperse composite media presents a complex statistical combination of macro- and microelements being different in their physicochemical properties, particle-size distribution, diverse in geometric shape and size, distributed in the volume of a certain continuum, and interacting with each other. Based on an extended interpretation of the concept of structure, the following approaches to the classification of inhomogeneous structures were proposed in [53]:

(i) by chemical and mineralogical phase composition, including properties that do not depend on geometrical parameters;

(ii) by geometric structure, including a set of properties that characterize the geometric structure and the relative position of structural elements;

(iii) according to the structure of links between separate elements;

(iv) according to the characteristics of the elements that determine the properties of the material.

From a historical perspective, the creation of the first high-strength composite materials should be attributed to the early 1940s. The impetus for the development of research in this direction was the political situation in the world at that time, which foreshadowed the approach of war. By the beginning of the war in the United States, the industrial production of parts from a composite, which was a cotton cloth impregnated with phenolic resin, had already been established.

A few years later, composite materials reinforced with glass fibers appeared. Later, honeycomb and sandwich structures made of fiberglass composites were widely used. In terms of structure, a sandwich structure is a panel, the outer surfaces of which are thin high-strength plates, between which a light aggregate (cellular material, foam plastic, honeycomb structure, etc.) is placed. With a small weight, this design exhibits highly demanded mechanical characteristics.

In the mid-1950s, the US Air Force used reinforced composite materials in the aircraft industry [151]. The practical production of boron and carbon fibers led to the creation of novel composites. The research and technology programs were curated by the National Aeronautics and Space Administration (NACA) and the US Air Force. The implementation of these programs from the beginning of the 1970s allowed one to put into practice the results of scientific developments and to begin the widespread use of composite materials for the production of aircraft.

However, a characteristic feature of the initial stage of the production of composite materials was rather high prices for raw materials and insufficient productivity of technological processes. Currently, composite materials and structures are widely used in almost all areas of technology: aircraft and rocket manufacturing, construction, instrumentation, in the production of household appliances, in medicine, sports, etc. Moreover, there is a tendency to outpace the growth rate of demand for composites and new technologies for their production in comparison with the development of experimental and theoretical methods for studying such structures.

In the manufacture of composite materials, the most common reinforcing components include glass, polyamide, asbestos, and cellulose fibers; natural fibers from cotton, sisal, jute, etc., are also used; the scope of application of carbon, graphite, boron, steel fibers, and "whiskers" is expanding [123].

In addition to fibers, other fillers are also used, like powders based on aluminum, iron oxide; granites, silicon, and tungsten carbides; metals; natural materials, such as cotton, etc.

In addition to polymers, metals and ceramics serve also as the matrices material.

The main physicochemical processes of filling polymers that contribute to the creation of a set of their desired properties, the effect of fillers on the mechanical and rheological properties of filled systems, and the mechanisms of the reinforcing effect of fillers in polymers are discussed in the book [146].

It is important to take into account one more aspect of the problem, i.e. the combination of materials in composites, which improves their individual properties (in some cases accompanied by a deterioration in other properties). Therefore, in practice, one should take into account all the defining characteristics of the composite and choose the optimal solution. Concerning the issue of fundamental scientific research in the field of mechanics of composites, it is necessary, first of all, to dwell on the problem of substantiation and a rigorous mathematical description of the relationship between the effective properties of inhomogeneous media of different types with the characteristics of the components, as well as the analysis of the behavior of composites due to their heterogeneity. Considering composite structures from the standpoint of continuum mechanics, it is simply unthinkable to take into account all the effects inherent in composites that include a large number of discrete phases.

This problem can be approached from different standpoints and with varying degrees of generality. One of the possible directions of research is to collect extensive experimental data of mechanical tests, carefully analyze them, compare them with each other, interpret and, ultimately, use them in the form of graphs drawn in dimensionless coordinates. Or one can take the next step and, on the basis of empirical results, construct analytical expressions that described experimental data. Such approaches are used in solving engineering problems – and are very useful and convenient – but their scope is limited by the conditions under which the empirical data are obtained. In addition, such methods do not work to describe the behavior of a material outside of laboratory experience.

A more fundamental approach suitable for calculating and predicting the behavior of heterogeneous media is based on the development of a rigorous theories containing mathematically justified analytical solutions. At its core, such a theory should contain some assumptions or hypotheses corresponding to the physical meaning of the process under consideration, which determine the limits of applicability of the results obtained. A well-constructed and mathematically substantiated theoretical base makes it possible to adequately model the physical processes occurring in a composite medium.

The development of a theoretical basis for studying the mechanical behavior of inhomogeneous media is based, as a rule, on the use of idealized geometric models of composite structures, which yield analytical estimates of the macroscopic properties of composites through the geometric and physical characteristics of their constituent phases. The advantage of this approach is that it gives better understanding at the physical level and designate the behavioral features in a single composite array of various combinations of structural materials.

In addition, the use of idealized models makes it possible to describe from a single point of view and find relationships between many results obtained for various problems of composite mechanics. In other words, to create a theoretical basis for a reasonable qualitative and quantitative analysis of the physical state of heterogeneous media.

At the same time, it should be noted that the fundamental differences in research approaches lie on the formulation stage of solving the problem. For regular structures with a sufficiently large number of inhomogeneities, the behavior of the material can be described by macroscopic constitutive equations, and it is only necessary to find a reasonable way to average the statistical variations of the material.

In many cases, however, the researcher is not just dealing with the issue of effective averaging of structural parameters since changes occur on a fairly small scale, and it is necessary to consider local phenomena and describe their physical essence (percolation, superconductivity, etc.). These provisions illustrate various aspects of the study of the problem, which require the construction of appropriate models and the use of adequate solution methods.

A significant structural difference of composite materials (both in terms of their physical properties and geometric structure), the widest range of their practical application, and features of production technologies do not exhaust a complete list of factors that determine the diversity and complexity of scientific problems, arising in the mechanics of composites. The study of these problems requires a deep understanding of the physical processes, a fundamental theoretical justification, and a rigorous mathematical solution.

2 Mathematical Approaches Used in the Theory of Composite Materials

In this brief review, we only indicate the main mathematical methods used in the theory of composites. Further sections of the book also provide brief reviews of the literature on topics related to the problems considered in these sections. The following short reviews of different branches of theory of composite materials are addressed: early stage studies, numerical approaches, exact results, Einstein's work, Rayleigh's method, complex analysis methods, Schwartz alternating method, functional approaches, virial expansions, lubrication approach, Padé approximations, self-consistent approach, three-phase and composite sphere models, variational approaches, percolation theory, and homogenization approach.

2.1 EARLY STAGE

The development of the theory of composites as a branch of science has a long history [140, 141, 144]. Poisson back in 1826 [200] built the theory of induced magnetism, which considered a structure consisting of conducting spheres embedded in a non-conducting matrix. Apparently, the concept of effective properties was introduced for the first time in this work. To a certain extent, also the classical works of Poisson [200], Maxwell [154], Rayleigh [203, 202], Voigt [233], Reuss [204] stand as the forerunners of modern asymptotic homogenization theory. In 1839, Faraday [85] proposed a model for dielectric materials, consisting of metal balls separated by an insulating substance. Mossotti in 1850 [180] and Clausius in 1879 [72] used the Poisson's method to study inhomogeneous dielectrics. In 1873, Maxwell [154] considered the problem of the conductivity of a suspension, which is a conductive inclusion-sphere placed in a conductive matrix. L. Lorenz [147, 148] and H.A. Lorentz [150, 149] studied the refractive index of media depending on the polarizability and concentration of particles. L. Lorenz published the first papers in this area in 1875 [147] in Danish, but only his paper, written in 1880 in the language of science of that time, German, gained fame. Rayleigh [202] in 1892 constructed a system of linear equations, the solution of which gives the effective conductivity of a structure with square cylindrical or spherical inclusions forming a cubic lattice. Informative historical reviews covering the development of composite mechanics as a science are given in [53, 212, 213]. Extensive bibliographic material, related to the development of various areas of composite mechanics, is given in the monograph by Milton [161].

It should be noted that formulas that are essentially the same in essence have been rediscovered in different fields of science and have different names. For example, the

DOI: 10.1201/9781003391029-2

Maxwell formula is also referred to as the Clausius-Mossotti approximation [56], Maxwell Garnett [155], Maxwell-Odelevskii [188], Lorenz-Lorentz [147, 148, 150, 149], Landauer [140, 141] and Wiener-Wagner formula [234].

Since a review of all advances in the field of composite mechanics is unnecessary here, we will further focus on composites with a regular structure. We will explore the theoretical studies leading to analytical expressions of the effective properties of the considered materials and structures.

2.2 NUMERICAL APPROACHES

The application of numerical methods in the theory of composites deserves a separate study. The advent of personal computers and information technologies of a new generation contributed to a qualitatively higher level of development of numerical methods and, accordingly, the accuracy of numerical solutions. One of the (not always observed) requirements of the application of numerical methods is a thorough check of the convergence of the process.

We note the solution of a number of problems in the theory of composites for piecewise inhomogeneous media by the grid method [53, 106, 228, 229].

Bourgat [60] used finite element methods (FEM) and evaluated the accuracy of various approximate schemes for obtaining effective characteristics. The use of computer technologies made it possible to provide the necessary accuracy of the solution by the Rayleigh method [203, 202] of the problem of determining the effective dielectric constant [157, 158, 194].

Research by Manevitch and his collaborators have been dedicated to homogenization problems for layered, fibrous and particulate-filled composites. The Bakhvalov technique [38] was used, to obtain analytically the solution of cell problems for layered composites. However, in other cases considered, the finite difference method was used. The matrix was considered elastic or elastoplastic. Both the case of perfect contact and the influence of the interface between inclusions and the matrix on the effective characteristics of the composite were studied. The calculation of elastic fields in composites filled with elongated rectangular parallelepipeds was also performed [53, 130].

Efficient methods have also been developed for the numerical solution of integral equations obtained in the theory of composites [117]. The solution of the problem is reduced to an integral equation, the numerical solution of which allows for estimating numerically the effective coefficients of a two-dimensional two-phase structure with a high degree of accuracy. The same method was used to study a two-phase composite with anisotropic phases, when solving a plane problem for a composite with two isotropic phases, and in the case of a three-dimensional elastic problem [161]. In addition. the boundary element method is also widely used [226].

In [134], the numerical homogenization of the characteristics of composites was carried out using the wavelet transform. In works [88, 89], using FEM and finite difference approach, periodic problems of the mechanics of composites with dispersed inclusions, short fibers and lamellar particles were solved. FEM was used to determine elastic moduli and analyze stress distribution in orthogonally reinforced fiber composites. In the monograph [219], on the basis of the FEM, the method of local approximations was developed, which yielded the thickness of the transition layer surrounding the inclusion.

2.3 EXACT RESULTS IN THE THEORY OF COMPOSITES

Exact results in the theory of composites are rare and therefore very valuable. In particular, exact results can be used as benchmark examples for approximate methods. Interesting results were obtained for self-dual media, i.e. structures whose components are geometrically the same. Namely, they occupy the same volume and do not change when the matrix is replaced by an inclusion and vice versa (a chessboard structure on a plane is an example of such a medium). For the first time, duality relations for such composites were found by Keller [127]. Under more general assumptions, the duality formula was derived by Matheron [153]. Mendelson obtained a formula for the effective conductivity in the two-dimensional case [159]. Dykhne showed that the effective conductivity of a self-dual media is equal to the geometric mean of the characteristics of its components [49, 80, 217]. For problems of elasticity theory, it is shown [236] that the effective Young's modulus of the composite for which Poisson's ratios of components are the same is equal to the mean geometric value of Young's moduli for both components.

Complex analysis procedures allows for some two-dimensional structures to accurately determine local fields in two-dimensional composites. In these cases, the Riemann problem of linear conjugation can be solved exactly. Berdichevsky [48] considered the problem of heat propagation in composites with square checkerboard field. Obnosov [190] considered rectangular checkerboard field, Craster and Obnosov [74, 76] solved problem for three- and four-phase checkerboard composites. Some additional information concerning formulas of this type can be found in the book by Milton [161].

The rigorous link between the conductivity and elastic moduli of fiber-reinforced composite materials [161, 226, 210] can also be referred to the number of useful exact results.

2.4 EINSTEIN'S WORK AND ITS GENERALIZATIONS

An instructive example from the field of composite materials is the determination of the effective viscosity of a suspension, which refers to a liquid with particles suspended in it. For the first time, the solution of this problem was obtained (in the approximation of a low concentration of suspended particles) by Einstein in his doctoral dissertation, published in [81]. Considering suspended particles as solid uniformly distributed spheres of the same radius, Einstein obtained the following formula for the ratio of the effective viscosity of a suspension μ_e to the viscosity of the liquid μ_l

$$\mu_e/\mu_l = 1 + c, \tag{2.1}$$

where c stands for the volume fraction of suspended particles.

Later, it turned out that the formula (2.1) is not confirmed by the experimental results of J.B. Perrin. L. Hopf made corrections to the work [81], as a result, the Einstein formula acquired the form [81]

$$\mu_e/\mu_l = 1 + 2.5c. \tag{2.2}$$

The same result is obtained for effective shear modulus for absolutely rigid spherical inclusions in incompressible matrix. Formula (2.2) gives a good approximation only for low concentrations. The construction of the next approximation required a lot of effort [71], as a result of which was obtained

$$\mu_e/\mu_l = 1 + 2.5c + 5.01c^2. \tag{2.3}$$

However, formula (2.3) does not provide a significant improvement in the results compared to the experimental data [136]. If we construct the Padé approximant of expression (2.3), then the resulting expression of the form

$$\mu_e/\mu_l = 0.5(2+c)/(1-2c) \tag{2.4}$$

confirms the experimental results well [71].

Interestingly, in Christensen's book [71], the expression (2.4) is treated as an empirical one. Applying the Padé approximant to the Einstein formula yields

$$\mu_e/\mu_l = 1/(1-2.5c), \tag{2.5}$$

which also significantly improves the agreement between theoretical and experimental data. Remarkably, self-consistent approach for effective shear modulus for absolutely rigid spherical inclusions in incompressible matrix [62] gives expression (2.5).

Here we come to an important conclusion: using various summation methods, one can significantly increase the real accuracy of the solution without increasing the number of terms in the asymptotic expansion.

2.5 RAYLEIGH'S METHOD

In 1892, Rayleigh [202] proposed a method based on considering a separate inclusion in an external field and finding a solution in the form of an expansion in the interaction coefficients, for the determination of which an infinite system of linear algebraic equations is obtained. In other words, this approach consists in representing the desired potentials as expansions in multipoles.

McPhedran and McKenzie [157] calculated the effective conductivity of periodic array of cylinders, having used multipole expansions for the fields in and around each inclusion and generalized an identity deduced by Rayleigh to provide a set of linear algebraic equations for the multipole coefficients. A pair of nearly touching cylinders of high conductivity was considered, and truncation errors occurring in the matrix solution of the corresponding transport problem for the square array of cylinders were estimated. These investigations were developed and summarized by

Movchan et al. [181]. Further development of Rayleigh's idea allowed to achieve many interesting results [39, 40]. A detailed overview of the use of this technique is given in [181]. The approach proposed by Mityushev [164] allows one to justify the Rayleigh method for rectangular arrays of cylinders. It also shows that in order to improve the Clausius-Mossotti approximation for a rectangular array, it is necessary to calculate the Rayleigh sum [79, 171]. The multipole expansion method was used also to calculate perforated media in [177, 178].

It should be noted that since the use of the Rayleigh method leads to the need to solve infinite systems of linear algebraic equations, the method works well for inclusions of low and medium concentration, but for densely packed inclusions computational problems arise [79, 161].

Here it is necessary to make the following important remark. The obtained infinite system of linear algebraic equations can be used for construction of justified approximate solutions. Let the regularity of such a system be proved, i.e., the possibility of its reduction with controlled accuracy to a finite system of equations (for regularity criteria for infinite systems of linear algebraic equations, see [125]). Then it is possible to estimate how many equations should be left in the system in order to obtain a solution with a given accuracy for a practically important range of parameters (as a rule, an engineer is quite satisfied with an accuracy of 5%, which he can easily cover with a safety factor). In particular, one can use the first few equations to obtain an approximate analytical expression. Unfortunately, in recent years, a number of authors have begun to pass off the reduction of the original problem to an infinite system of linear algebraic equations as an "exact solution" or "a closed-form solution". Often this appears in a veiled form, for example, an analytical expression is written out, which includes a coefficient determined by the exact solution (!) of the infinite system. A detailed analysis of these works is contained in the article [24] , which shows the inconsistency of such claims.

2.6 METHODS BASED ON COMPLEX ANALYSIS

In the survey paper [171], the approaches to problems of two-dimensional stationary conductivity and elasticity for a doubly periodic array of circular inclusions are analyzed. The seminal works from their historical perspective are related to the results of the famous scientists like Eisenstein, Weierstrass, Rayleigh, Natanzon (Natanson), and Filshtinsky (Fil'shtinskii). L.A. Filshtinskii was educated and started his career as an aeronautical engineer. When studying the strength of aircraft structures, one has to deal with composite materials and periodically inhomogeneous elements, e.g. ribbed, corrugated, perforated, etc. Filshtinskii, having a good mathematical background, quickly realized that complex analysis, and especially the apparatus of Weierstrass doubly periodic functions, is adequate for these problems. In this way, Filshtinskii obtained a number of important results [98, 99, 100, 101, 102], summarized in detail in [171].

The ideas of Filshtinskii were widely used by Balagurov et al. (method of series in elliptic functions) [39, 40]. Pobedrya [198] and Mol'kov and Pobedrya [179] wrote more convenient formulas for effective constants. For the first time, the effective constants were written in terms of the local fields by Natanson and Fil-shtinskii. Pobedrya successfully presented the problem from the point of view of homogenization and tensor theory through a well-defined problem for the cell. Unfortunately, these works do not always contain correct references to Filshtinskii's original works.

The complex analysis was also used in [171], followed by reduction to integral equations. When solving periodic and doubly periodic problems, special systems of functions (harmonic, spherical, Legendre polynomials) are employed [49, 171, 212].

In fundamental research [228, 229], using the complex analysis, solutions were obtained for a number of problems of tension, bending, and thermal conductivity for composites of various structures (with cylindrical, spherical, ellipsoidal inclusions). The desired solutions were presented in the form of series, the coefficients of which are determined from the conjugation conditions. This approach allows one to calculate the effective elastic parameters of structures with periodic inclusions, taking into account their sizes and stiffness's.

2.7 SCHWARZ ALTERNATING METHOD

In a number of works, the generalized Schwarz alternating method [160] has been successfully applied. A modification of the Schwarz alternating method was proposed in [171, 174] in order to obtain a convergent algorithm for any geometry. This result was extended to periodic problems by Mityushev [169]. This approach is equivalent to the contrast expansion (the contrast parameter $c = (\lambda - 1)/(\lambda + 1)$ was introduced by Bergman [50]).

Realization of the generalized Schwarz alternating method for circular inclusions is reduced to the method of functional equations and its solution by a method of successive approximations. Its convergence for $|c| \leq 1$ was proved in [79].

2.8 FUNCTIONAL EQUATIONS APPROACH AND CLUSTER METHOD

Mityushev and his co-authors obtained [79, 92, 93, 94, 95, 96, 163, 164, 165, 166, 167, 168, 169, 170, 171, 172, 173, 174, 175, 176] a number of interesting results for composites of various structures, including regular and random distribution of inclusions. The use of a rigorous mathematical apparatus, in combination with the technique of specially constructed Padé approximants and numerical simulation using the computational capabilities of the mathematical packages Mathematica, Maple, MATLAB, yielded high-precision analytical solutions of problems that are in good agreement with the known numerical and asymptotic results.

In book [79], using the method of functional equations, the asymptotic formula for composites with absolutely conducting circular cylindrical inclusions was obtained. Also the asymptotic relations are constructed, respectively, for densely packed high contrast composites with circular cylindrical inclusions and for a material weakened by an array of cylindrical cavities, circular in cross section.

The work [164] is devoted to the problem of conductivity of doubly periodic composite materials with circular inclusions; an exact formula for the effective conductivity tensor is obtained, which in a particular case reduces to the Maxwell formula.

In a number of publications, the Maxwell approach has been developed in combination with the cluster method [170]. Thus, the generalization in the article [176] of the Maxwell approach for a single inclusion to problems for n inclusions leads to cluster methods employed to calculate the effective transport properties of composites. In the article [172], the cluster method was generalized for two-dimensional elastic composites.

2.9 VIRIAL EXPANSIONS

One of the assumptions that makes it possible to analytically describe the distribution of elastic fields in composites is the condition of sufficient large distance between inclusions of a spherical or ellipsoidal shape [212]. The most adequate method for solving the problem for such media is the virial expansion method, which is based on the expansion of the effective tensors of the elastic moduli and elastic compliance in a series in terms of the volume fraction of inclusions [212]. The problem of deformation of an unbounded elastic medium with an elastic inclusion of a spherical shape was solved in [83] under the assumption that the components are isotropic and the mechanical fields are homogeneous at a sufficient distance from the inclusion.

An analytical method for calculating the elastic characteristics of a composite structure, based on the expansion in a series in concentrations, was proposed by Pobedrya in the works [198, 199]. The problem is solved under the assumption of a low concentration of inclusions; it is shown that in this case it is possible to confine ourselves to consider a single fiber in an infinite matrix.

2.10 LUBRICATION APPROACH

An analytical description of the effective characteristics of the composite is possible for a high concentration of inclusions. The solution methods in this case are based on the analysis of deformation of thin layers of the matrix located between adjacent rigid inclusion. In particular, the expressions for the effective elastic moduli, obtained on the basis of this model, are given in [54].

Bakhvalov and Panasenko [38] and Panasenko [191] studied the case of large cylindrical voids using the asymptotic method and obtained analytic expressions for the effective parameters of a fairly simple form.

One of the important areas of research is the study of the effective conductivity of planar and spatial composite structures in the limiting cases of the values of their physical and geometric parameters, namely, for the limiting concentrations of inclusions of extremely high conductivity [58, 126, 128].

Good results in this case are given by the lubrication approach [71]. Using this approach, in [71] asymptotic relations were obtained for the case of composites of various structures with inclusions of extremely possible large sizes and extremely high conductivity.

For regular structures, the relationship between the effective thermal conductivity for small and large conductivities of inclusions is established by the Keller's theorem [127]. With dense packing, the conductivity is determined by the neighbouring inclusions (i.e., by the asymptotic behavior of two bodies in an infinite matrix).

2.11 PADÉ APPROXIMANTS

There exist many approaches to solve this problem [8, 11, 12]. The most common are the approaches based on transformation of the power series to a fractional rational function, especially Padé approximants (AP) [30, 31]. AP performs meromorphic continuation of the function given in the form of the power series, and for this reason, it allows one to achieve success in cases where analytic continuation cannot be applied. If the AP sequence converges to a given function, then the roots of its denominators tend to singular points. The range of applicability of AP, as a rule, is much wider than the range of applicability of the original power series, which leads to the widespread use of AP in the analytical mechanics of composite materials [12, 90, 92, 209].

An analysis of numerous examples confirms that, as a rule, a kind of "principle of complementarity" is implemented: if $\varepsilon \to 0$, it is possible to construct a physically meaningful asymptotics, then there exists a nontrivial asymptotics for $\varepsilon \to \infty$. The most difficult for analysis, from the point of view of the asymptotic approach, is the intermediate case, $\varepsilon \sim 1$. Numerical methods usually work well in this area; however, if the task is to investigate the solution depending on the parameter ε, it is inconvenient to use different solutions in different areas. The construction of a solution which can be used for all values of asymptotic parameter is a non-trivial problem, which in many cases can be solved by a two-point Padé approximants (AP) [8, 19, 20, 215].

We note an important application of the AP in the theory of composites. The method of bounds is widely used in the theory of composite materials. Suffice it to mention the Wiener [235], Hill [115, 116] or Hashin-Shtrikman [110, 111] bounds. These estimates can be obtained from the properties of Padé approximants [30, 31]. Mentioned estimates are of a very general nature. The bounds become increasingly accurate when more information on geometrical properties of the medium is known. For two component isotropic composites, the bounds based on the AP for the effective transport constants already exist [161]. These bounds are usually obtained in the form of continued fractions on the basis of the analytic properties of the effective conductivity $\lambda_e(\lambda_1, \lambda_2)$, where λ_i are the conductivities of composite phases. Bergman [50] proved that $\lambda_e(\lambda_1, \lambda_2)/\lambda_1 = \lambda_e(1, \lambda_2/\lambda_1)$ is a Stieltjes function of λ_2/λ_1, analytic everywhere except on the negative real axes, satisfying $\lambda_e(\lambda_1, \lambda_2)/\lambda_1 > 0$ when $\lambda_2/\lambda_1 > 0$. The Stieltjes functions have been extensively studied in the mathematics literature, and their AP and continued fractions representations are well known [30, 31].

On the contrary, the analytic properties of two-point AP generated by two different power expansions of Stieltjes function have not been examined as deeply as the one-point AP. Tokarzewski [225] investigated the two-point AP for a non-equal, finite number of terms of two power expansions of the Stieltjes functions at zero and infinity. Under some assumptions, he proved that the diagonal two-point AP form sequences of lower and upper bounds uniformly converging to the Stieltjes function.

Quite often, one of the limiting solutions is not expressed by a Taylor or Laurent series. For example, in a problem with a periodic array of spherical inclusions, one of the limit solution contains logarithms [226]. In this case, one can use the two-point quasifractional approximant [152] or asymptotically equivalent functions [215] method.

2.12 SELF-CONSISTENT APPROACH

One of the most common and frequently used models for describing macroscopically inhomogeneous media is the self-consistent approach. For the first time, self-consistent effective medium approach was proposed for problems of conductivity and elasticity theory by Bruggeman [61] back in 1935–1937.

Later, the self-consistent approach was independently rediscovered by Landauer [141], which is why the theory of the self-consistent approach is sometimes called the Bruggeman-Landauer theory. The same method was also rediscovered by Odelevskii [188].

The self-consistent approach was used in 1954–1958 by Hershey [114] and Kröner [137] to the description of the effective properties of polycrystalline materials. Polycrystals as such are single-phase, but due to their random (or partially random) orientation, when crossing the boundaries of crystals, they experience abrupt changes in properties, which is a defining feature of an inhomogeneous medium.

When applying the self-consistent approach, a single anisotropic crystallite is considered as a spherical or ellipsoidal inclusion placed in a medium with unknown isotropic properties. At a large distance from switching on, uniform stresses or deformations act on the system. The self-consistent procedure – which gave the method its name – consists in setting the average stress or strain in the inclusion equal to the value of the applied stress or strain.

In 1965, the self-consistent approach was generalized by Hill [112] to describe the elastic characteristics of a two-phase composite with spherical inclusions. The modern interpretation and self-consistent approaches are divided into two directions [53, 208]. Within the framework of the first direction (the theory of self-consistent media), the selected particle is considered to be immersed in a medium with effective elastic moduli. The second direction (the theory of self-consistent field) involves the introduction to the matrix of a fictitious mechanical field, which depends on stresses or strains at infinity.

Self-consistent media models, in turn, can be divided into two groups according to the type of equations obtained, i.e. differential or algebraic. Differential models are obtained as a result of a small change in the concentration of inclusions by additional introduction of inclusion particles into a medium with effective properties. After the passage to the limit, the effective modules are determined from the system of ODEs. The characteristics of the matrix act as the initial conditions of the system of ODEs. Various implementations of this approach for the case of absolutely rigid particles and pores are described in [63, 208, 212]. In self-consistent media theories leading to an algebraic system of equations, an inclusion particle isolated in a material with effective properties is used to analyze mechanical fields in the dispersed phase ([62]).

As well as in the self-consistent theories, since components have equal status, this theory can be also successfully employed to percolation problems [216, 218].

2.13 THREE-PHASE MODEL

In 1956–58 Kerner [129] and van der Poel [227] proposed the so-called three-phase model of the composite. The essence of this model is that the entire periodic structure, with the exception of one cell, is replaced by a homogeneous averaged medium with unknown reduced characteristics. Further, the desired effective parameters determined from the energy principle: the energies stored in the composite and in an equivalent homogeneous medium are equal. The three-phase model of a medium with cylindrical inclusions was subsequently used in the works by Christensen and Lo [70] and Christensen [71].

2.14 COMPOSITE SPHERES MODEL

The development and description of the composite spheres model belongs to the 60s: Hashin [108] in 1962 proposed it for spherical inclusions and Hashin, Rosen [106] in 1964 – for circular cylindrical ones. Within the framework of this model, a continuous medium with inclusions of various sizes is considered, and each inclusion of radius r is encased in a sphere of matrix material of radius R. Absolute radii values r and R different, but their ratio r/R remains unchanged. The particle sizes are reduced to infinitesimal and completely fill the entire medium. The advantages and disadvantages of composite spheres model are thoroughly analyzed in monograph [71].

2.15 VARIATIONAL APPROACHES BOUND FOR EFFECTIVE PROPERTIES

Since the beginning of the 60s of the last century, variational methods extensively used for the study of inhomogeneous media are of a fairly general form. Particularly significant were the transformations of variational problems introduced by Hashin-Shtrikman [110, 111]. The Hashin-Shtrikman bounds, which are not unreasonably interpreted in the literature as one of the key results of the theory of composites, are the best of those that can be obtained without specifying the geometry of the composite phases. The accuracy of Hashin-Shtrikman bounds has been analysed using special structures introduced by Hashin [108].

Subsequently, this area of research was supplemented by a number of results based on rather complex heuristic methods that were used by mechanicians long before the creation of the mathematical homogenization theory. Hashin [108] extended the variational principles to inhomogeneous elastic bodies (not only composites).

Further generalization of these variational principles was carried out for composites with complex moduli and for nonlinear media [158].

The main results of such studies and literature references are given in the books Christensen [71] and Shermergor [212]. Of great importance in the theory of composites is the direction of research associated with the determination of estimates of the effective characteristics of composite structures focused on estimation of the upper and lower bounds of the effective parameters. The history of the issue originates from the time of Voigt [233], who proposed to calculate the parameters of a polycrystal by averaging the corresponding values over the volume and orientations. Reuss [204] used the averaging of the inverse tensor components for this purpose. Lichtenecker [144] based on the analysis of experimental data proposed to consider the logarithms of conductivities as additive.

From a mathematical point of view, Voigt averaging (direct averaging) is the arithmetic mean, and Reuss averaging [204] is the harmonic mean of the original tensors. In mechanics, the Voigt averaging corresponds to the additive stiffness averaging, while the Reuss averaging corresponds to the additive compliance averaging. It was further shown by Hill [115, 116] that averaging over Voigt and Reuss gives, respectively, upper and lower bounds for the corresponding parameters. For conductivity problems, this was previously shown by Wiener [235]. Each of these estimates may, depending on the nature of the inhomogeneity of the medium, turn out to be more or less close to the true value of the effective parameters.

However, in many cases, the difference between these values is too large [71, 224] and these bounds do not provide the necessary information to the real values of the effective parameters. Therefore, these ratios are of little use for practice.

It should be noted that the Hashin-Shtrikman bounds give good estimates only when the physical characteristics of the composite phases are of the same order. In the limiting cases of physical properties of inclusions one of the Hashin-Shtrikman bounds becomes trivial. For example, in problems of heat conduction in limiting cases, we have non-conductive or absolutely conductive inclusions. For non-conductive inclusions lower Hashin-Shtrikman bound is zero; for absolutely conducting inclusions, the upper Hashin-Shtrikman bound gives infinite value. In such cases, obviously, it is not possible to draw reasonable conclusions about the real values of the effective parameters.

Taking into account the shape and nature of the distribution of inhomogeneities in the composite structure, the Hashin-Shtrikman bounds can be improved (this issue is discussed in detail in monograph [161]). For example, narrow bounds were obtained for composites with spherical [108] or cylindrical [109] inclusions. Comparison with the results of numerical calculations showed a very good estimate of the effective conductivity.

2.16 PERCOLATION THEORY

Since the 1990s, percolation theory has been intensively developed, which makes it possible to adequately describe various systems that undergo a geometric phase transition. The main property of the environment studied by means of percolation theory is the degree of connectivity or clustering of certain elements of the system or fields associated with them. In the latter case, the degree of connectivity depends both on the concentration of field sources and on the radius of the sphere of influence.

A number of papers are devoted to the application of percolation theory to the problems of mechanics of composites [43, 51, 92, 95, 205, 216, 218, 226]. In particular, percolation theory with FEM was used to determine the elastic properties of composites of random structure [226]. In the article [216], a relation is constructed for a qualitative description of the percolation threshold in a macroscopically disordered medium; the connection of the found approximation with the AP for Maxwell formula is analyzed.

Manevitch and his collaborators [88, 89] proposed a method for modeling the elastic properties of composites of a random structure, which consists in replacement, a two- or three-dimensional space with a regular lattice, randomly filling its cells with the material of one of the phases, having probabilities equal to their volum fractions. Then, for calculating the fields of displacements, strains, and stresses arising in the system when external loads are applied to it, FEM is used. The materials of neighbouring cells were considered to be perfectly connected. A comparison of the elastic properties of composites with a regular structure and inhomogeneous materials with a random geometry of phase yielded a prediction of a significant qualitative difference in the concentration dependences of the effective moduli of the corresponding systems. Numerical results sufficiently well describe experimental data. Padé approximants and self-similar power series transforms were used in [92, 94, 95].

2.17   HOMOGENIZATION APPROACH

The most common composites in practice are described by PDEs with periodic and rapidly varying coefficients. Before the computer revolution and the advent of FEM software packages, an engineer needed simple calculation formulas (however, they are still important, especially in optimal design). The standard method for obtaining such formulas at that time was the reduction in the original problem for periodically inhomogeneous media to homogeneous ones ("structurally-orthotropic theory" in Russian terminology or "smeared theory" in Western terminology). The main problem here was the determination of parameters of homogeneous system, equivalent in some sense to the original one. Now it can be done using the asymptotic homogenization theory. It is characteristic that the approximate schemes used by engineers turned out to be the first approximations of asymptotic processes of a mathematically rigorously justified homogenization theory [12, 156].

Beginning in 1974, Bakhvalov and his school developed the method of homogenization of ODEs and PDEs with rapidly varying coefficients [32, 33, 34, 35, 36, 37, 38, 191, 193]. Bakhvalov built the solution by introducing some operator (Bakhvalov's ansatz). These results have been developed by Oleynik and her school [238], as well as Pobedrya and his school [97, 198]. In parallel, the averaging method was developed by Lions et al. [44, 145], Sanchez-Palencia [207] based on the multiple scale perturbation methods. The variant of the homogenization theory proposed by L. Manevitch and his school is based on asymptotic methods for PDEs with rapidly oscillating right-hand sides and boundary conditions. The important features of works of this group are the analytical solution of cell problems and the analytical description of the boundary layers [155].

The homogenization approach is based on the idea of separating the fast and slow components of the solution, but its implementation may be different, while some techniques are convenient for some problems and inconvenient for others. For our purposes, the most suitable method is based on multiple scale expansions.

The homogenization approach makes it possible to reduce the original problem for a multiply connected region to boundary value problems for a periodically repeating singly connected cell and region with some effective characteristics. In most mathematical works, the solution of the cell problem is not considered. It is believed that it can be obtained by one or another numerical method. Mathematically oriented authors usually confine themselves to proving existence and uniqueness theorems. An exception are spatially one-dimensional problems, for example, for layered media. Here we can say that, where problems end for a mathematician, for a mechanics researcher they are on the begining stage (especially, if the latter are focused on obtaining analytical results [11, 12, 155]).

In the framework of the two-scale asymptotic method [182, 183], the transformation of variables is performed in such a way that the "slow" variables describe the macroscopic properties of the composite, while the "fast" ones are local in nature and vary within structural cell of a heterogeneous array.

An essential advantage of the homogenization method is the possibility of generalization: if a solution of a cell problem for a given equation (or system of equations) is found, then the solution of both the initial problem and the eigenvalue problem is determined without fundamental difficulties. For the theory of composites, this means that by solving one cell problem, one can thereby obtain a solution to a whole class of problems including linear and quasi-linear, static, and dynamic. However, the efficiency of the homogenization method essentially depends on whether it is possible to obtain a solution to the problem on a cell quite simply.

As follows from the analysis of a number of works, the analytical solution of the cell problem requires using of additional small or large parameters. The combined use of homogenization theory and asymptotic solutions of cell problems has a wide area of practical application and leads to simple and clear formulas, especially useful in optimal design.

3 Three-Phase Composite Model

This chapter offers a brief introduction to the history of three-phase composite model (ThPhM) and its advantages over other mathematical models. It is viewed from the perturbation and homogenization methods perspective, and then the transformation of local problem in the framework of ThPhM is discussed (zeroth- and first-order approximations of the local problem are derived, and the results are compared with the Hashin-Shtrikman bounfds). Numerous results are presented for estimation of the thermal conductivity coefficient based on modified ThPhM approach, asymptotic solutions, Padé approximations, and direct numerical integration.

It is shown how the ThPhM can be generalized with the help of Padé approximations. Then special cases of the ThPhM for square cylindrical inclusions are studied in detail (Section 3.4), and next (Section 3.5), the modified ThPhM is developed to overcome the limitation of the standard ThPhM.

The effective characteristics of 3D composite structures with cubic inclusions are derived and discussed in the final section 3.6.

3.1 THREE-PHASE MODEL IN COMPOSITE MECHANICS: ESSENCE, MODIFICATIONS, AND APPLICATIONS

The first ThPhM of the composite was used in 1956–58 (see Kerner [129] and van der Poel [227]). The essence of this model is that the entire periodic structure, with the exception of one cell, is replaced by a homogeneous medium with unknown characteristics. Further, the unknown effective parameters are determined from the ratios obtained on the basis of the following energy principle: the energies stored in the composite and the equivalent homogeneous medium are equal.

ThPhM was used by Christensen and Lo [70] and Christensen [71] to solve problems for composites with cylindrical and spherical inclusions. Classical ThPhM is also described in the monographs by Milton [161], Torquato [226], and in an article by Buryachenko [63]. An asymptotic study of the relations obtained on the basis of ThPhM was carried out in [12]. Various aspects of using ThPhM in combination with the asymptotic homogenization method were analyzed and used to model periodic inhomogeneous materials in [13, 22]. In [27], ThPhM and the Padé approximants (AP) were employed to determine the effective thermal conductivity of a two-phase composite with square lattice array of a square inclusoins. In the latter case, the idealization of ThPhM from a mathematical point of view leads to the problem of conjugation of the matrix with an equivalent homogeneous medium, in which the temperature and heat flux distribution functions decay at a considerable distance from the inclusion.

DOI: 10.1201/9781003391029-3

The advantages of ThPhM include

(i) the possibility of applying to the study of composites of various shapes and structures: fibrous with inclusions of various profiles, spatially heterogeneous composite arrays with spherical and cubic inclusions, etc.;

(ii) independence of the general scheme of application of the model from the geometric and physical characteristics of the matrix and inclusions of the composite;

(iii) possibility of application and generalization for a wide class of problems: polydisperse medium,; periodically inhomogeneous mediums with inclusions of different profiles; structures with a more complex geometry and with an arbitrary (non-periodic) nature of the arrangement of inclusions, and structures with inclusions of various sizes, etc.

In this chapter, we consider the problem of determining the effective thermal conductivity of a composite with cylindrical inclusions of a circle and square profile periodically arranged in a square lattice.

We employ the following approaches: homogenization theory based on the method of multiple scale expansions [44, 145]; ThPhM [71] for solving the cell problem; boundary shape perturbation method [11, 105, 182, 183].

3.2   THREE-PHASE COMPOSITE MODEL WITH THE PERTURBATION POSITION OF THE BOUNDARY METHOD

3.2.1   BASIC PROVISIONS OF THE HOMOGENIZATION METHOD

Let us consider a two-phase micro-inhomogeneous material consisting of a continuous matrix and cylindrical inclusions periodically located in it. The size of the structure in the direction of the length of the fibers significantly exceed its size in plan, $\hat{L} \gg \hat{\ell}$. We assume that the structure is doubly periodic with the same period in both directions, and the inclusions are arranged in a square lattice (Fig. 3.1).

When studying the thermal conductivity of such a structure, the governing relation is the Poisson's equation with conjugation conditions at the phase boundary and boundary conditions at the outer contour of the entire composite. It should be noticed that such a mathematical model can have several different physical interpretations [41]. Therefore, all reasoning and mathematical relationships that relate to solving the problem of the effective thermal conductivity remain valid also in the case of other physical interpretations of the problem, when the electrical conductivity, dielectric constant, magnetic permeability, and other physical characteristics of the composite material are sought.

In what follows we present the mathematical formulation of the problem. Let us denote the area in the composite section $\Omega = \sum\limits_{i=1}^{n} \Omega_i^+ \bigcup \Omega_i^-$, where Ω_i^+ and Ω_i^- are the areas of the matrix and the inclusion with a characteristic size of $2\hat{a}$ (Fig. 3.2), respectively.

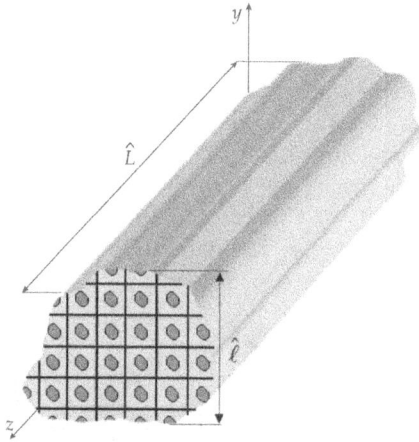

Figure 3.1   Composite structure with cylindrical inclusions.

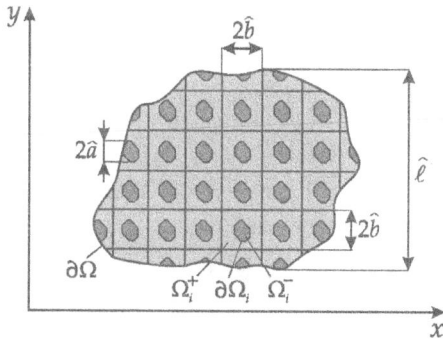

Figure 3.2   Cross section of a composite with periodically arranged cylindrical inclusions.

The phases of the composite array have different thermal conductivities, i.e. λ^+ and λ^- in the matrix and inclusions, respectively, and $\frac{\lambda^-}{\lambda^+} = \lambda$.

We assume that the structure period $2\hat{b}$ is small in comparison with the characteristic size of region Ω in plan, i.e. $\varepsilon = \frac{2\hat{b}}{\hat{\ell}} \ll 1$. Material behaviors in the areas Ω_i^+ and inclusions Ω_i^- are governed by the following Poisson's equations:

$$\lambda^+ \Delta u^+ = F \quad \text{in} \quad \Omega^+, \tag{3.1}$$

$$\lambda^- \Delta u^- = F \quad \text{in} \quad \Omega^-, \tag{3.2}$$

where u^+, u^- are the temperature distribution functions in the matrix and inclusions, respectively, and F is the heat source density.

At the junction of the matrix and inclusions, the conjugation conditions read

$$u^+ = u^- \quad \text{on} \quad \partial \Omega_i, \tag{3.3}$$

$$\lambda^+ \frac{\partial u^+}{\partial \mathbf{n}} = \lambda^- \frac{\partial u^-}{\partial \mathbf{n}} \quad \text{on} \quad \partial \Omega_i, \tag{3.4}$$

where $\mathbf{n}$ stands for an outer normal to the inclusion contour.

The presence in the structure of a natural small parameter (the period of the composite) encourages us to employ the technique of two scale expansions. In accordance with the multiple scales approach [182, 183], instead of each of the variables x, y, we introduce two: "slow" and "fast". "Slow" variables (let's keep the original notation of them, x, y) have the same scale as the original ones; "fast" variables have a scale within the period of the structure and both of them are defined as follows

$$\xi = \frac{x}{\varepsilon}, \quad \eta = \frac{y}{\varepsilon}. \tag{3.5}$$

We represent the solution of the boundary value problem (3.1)-(3.4) as asymptotic series in powers of a small parameter ε:

$$
\begin{aligned}
u^+ &= u_0^+(x,y,\xi,\eta) + \varepsilon u_1^+(x,y,\xi,\eta) + \varepsilon^2 u_2^+(x,y,\xi,\eta) + \dots \\
u^- &= u_0^-(x,y,\xi,\eta) + \varepsilon u_1^-(x,y,\xi,\eta) + \varepsilon^2 u_2^-(x,y,\xi,\eta) + \dots
\end{aligned}
\tag{3.6}
$$

Taking into account expressions (3.5), relations for the derivatives

$$\frac{\partial}{\partial x} = \frac{\partial}{\partial x} + \frac{1}{\varepsilon}\frac{\partial}{\partial \xi}; \quad \frac{\partial}{\partial y} = \frac{\partial}{\partial y} + \frac{1}{\varepsilon}\frac{\partial}{\partial \eta} \tag{3.7}$$

and expansions (3.6), the original equations (3.1)–(3.2), and the boundary conditions (3.3)–(3.4) split with regard to ε into the following infinite systems:

$$\varepsilon^{-2}: \quad \begin{aligned} \frac{\partial^2 u_0^+}{\partial \xi^2} + \frac{\partial^2 u_0^+}{\partial \eta^2} &= 0, \\ \frac{\partial^2 u_0^-}{\partial \xi^2} + \frac{\partial^2 u_0^-}{\partial \eta^2} &= 0, \end{aligned} \tag{3.8}$$

$$\varepsilon^{-1}: \quad \begin{aligned} 2\frac{\partial^2 u_0^+}{\partial x \partial \xi} + 2\frac{\partial^2 u_0^+}{\partial y \partial \eta} + \frac{\partial^2 u_1^+}{\partial \xi^2} + \frac{\partial^2 u_1^+}{\partial \eta^2} &= 0, \\ 2\frac{\partial^2 u_0^-}{\partial x \partial \xi} + 2\frac{\partial^2 u_0^-}{\partial y \partial \eta} + \frac{\partial^2 u_1^-}{\partial \xi^2} + \frac{\partial^2 u_1^-}{\partial \eta^2} &= 0, \end{aligned} \tag{3.9}$$

$$\varepsilon^0: \quad \lambda \begin{aligned} \frac{\partial^2 u_0^+}{\partial x^2} + \frac{\partial^2 u_0^+}{\partial y^2} + 2\frac{\partial^2 u_1^+}{\partial x \partial \xi} + 2\frac{\partial^2 u_1^+}{\partial y \partial \eta} + \frac{\partial^2 u_2^+}{\partial \xi^2} + \frac{\partial^2 u_2^+}{\partial \eta^2} &= F, \\ \left(\frac{\partial^2 u_0^-}{\partial x^2} + \frac{\partial^2 u_0^-}{\partial y^2} + 2\frac{\partial^2 u_1^-}{\partial x \partial \xi} + 2\frac{\partial^2 u_1^-}{\partial y \partial \eta} + \frac{\partial^2 u_2^-}{\partial \xi^2} + \frac{\partial^2 u_2^-}{\partial \eta^2} \right) &= F, \end{aligned} \tag{3.10}$$

$$\dots\dots$$

$$\varepsilon^0: \quad u_0^+ = u_0^-, \tag{3.11}$$

$$\varepsilon^1: \quad u_1^+ = u_1^-, \tag{3.12}$$

$$\varepsilon^2: \quad u_2^+ = u_2^-, \tag{3.13}$$

$$\dots\dots$$

$$\varepsilon^{-1}: \quad \frac{\partial u_0^+}{\partial \bar{\mathbf{n}}} = \lambda \frac{\partial u_0^-}{\partial \bar{\mathbf{n}}}, \tag{3.14}$$

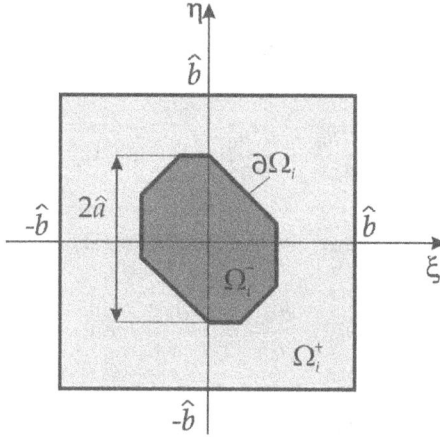

Figure 3.3 Characteristic structural cell of the composite.

$$\varepsilon^0 : \quad \frac{\partial u_1^+}{\partial \bar{n}} + \frac{\partial u_0^+}{\partial n} = \lambda \left(\frac{\partial u_1^-}{\partial \bar{n}} + \frac{\partial u_0^-}{\partial n} \right), \tag{3.15}$$

$$\varepsilon^1 : \quad \frac{\partial u_2^+}{\partial \bar{n}} + \frac{\partial u_1^+}{\partial n} = \lambda \left(\frac{\partial u_2^-}{\partial \bar{n}} + \frac{\partial u_1^-}{\partial n} \right), \tag{3.16}$$

$$\cdots \cdots$$

where $\frac{\partial}{\partial \bar{n}}$, $\frac{\partial}{\partial n}$ are the derivatives with respect to the outer normal to the inclusion contour, written, respectively, in "fast" and "slow" variables:

$$
\begin{aligned}
\frac{\partial}{\partial \bar{n}} &= \frac{\partial}{\partial \xi} \cos \alpha + \frac{\partial}{\partial \eta} \cos \beta \\
\frac{\partial}{\partial n} &= \frac{\partial}{\partial x} \cos \alpha + \frac{\partial}{\partial y} \cos \beta
\end{aligned}
\tag{3.17}
$$

and α, β are angles between the normal and the coordinate axes ξ, η.

The application of the homogenization method allows one to find the solution of the original boundary value problem for a multiply connected domain in two steps.

The first step is the solution of a local (or a cell) problem. In this case, the problem is considered in the domain of a periodically repeating characteristic cell of the composite (Fig. 3.3), on the opposite faces of which the conditions of periodic continuation (translational symmetry) must be satisfied:

$$
\begin{aligned}
&\left. u_i^+ \right|_{\xi=\hat{b}} = \left. u_i^+ \right|_{\xi=-\hat{b}}, \quad \left. u_i^+ \right|_{\eta=\hat{b}} = \left. u_i^+ \right|_{\eta=-\hat{b}}, \\
&\left. \frac{\partial u_i^+}{\partial \xi} \right|_{\xi=\hat{b}} = \left. \frac{\partial u_i^+}{\partial \xi} \right|_{\xi=-\hat{b}}, \quad \left. \frac{\partial u_i^+}{\partial \eta} \right|_{\eta=\hat{b}} = \left. \frac{\partial u_i^+}{\partial \eta} \right|_{\eta=-\hat{b}}, \quad i = 0, 1, 2, \dots .
\end{aligned}
\tag{3.18}
$$

Thus, as follows from (3.8)-(3.9), (3.11)-(3.12), (3.14)-(3.15), (3.18), we obtain the following sequence of local problems:

1)

$$\frac{\partial^2 u_0^+}{\partial \xi^2} + \frac{\partial^2 u_0^+}{\partial \eta^2} = 0 \quad \text{in} \quad \Omega_i^+,$$

$$\frac{\partial^2 u_0^-}{\partial \xi^2} + \frac{\partial^2 u_0^-}{\partial \eta^2} = 0 \quad \text{in} \quad \Omega_i^-,$$

$$u_0^+ = u_0^-, \quad \frac{\partial u_0^+}{\partial \bar{n}} = \lambda \frac{\partial u_0^-}{\partial \bar{n}} \quad \text{on} \quad \partial \Omega_i, \qquad (3.19)$$

$$u_0^+ \Big|_{\xi = \hat{b}} = u_0^+ \Big|_{\xi = -\hat{b}}, \quad u_0^+ \Big|_{\eta = \hat{b}} = u_0^+ \Big|_{\eta = -\hat{b}},$$

$$\frac{\partial u_0^+}{\partial \xi} \Big|_{\xi = \hat{b}} = \frac{\partial u_0^+}{\partial \xi} \Big|_{\xi = -\hat{b}}, \quad \frac{\partial u_0^+}{\partial \eta} \Big|_{\eta = \hat{b}} = \frac{\partial u_0^+}{\partial \eta} \Big|_{\eta = -\hat{b}},$$

2)

$$\frac{\partial^2 u_1^+}{\partial \xi^2} + \frac{\partial^2 u_1^+}{\partial \eta^2} + 2 \frac{\partial^2 u_0^+}{\partial x \partial \xi} + 2 \frac{\partial^2 u_0^+}{\partial y \partial \eta} = 0 \quad \text{in} \quad \Omega_i^+,$$

$$\frac{\partial^2 u_1^-}{\partial \xi^2} + \frac{\partial^2 u_1^-}{\partial \eta^2} + 2 \frac{\partial^2 u_0^-}{\partial x \partial \xi} + 2 \frac{\partial^2 u_0^-}{\partial y \partial \eta} = 0 \quad \text{in} \quad \Omega_i^-,$$

$$u_1^+ = u_1^-; \quad \frac{\partial u_1^+}{\partial \bar{n}} + \frac{\partial u_0^+}{\partial \mathbf{n}} = \lambda \left(\frac{\partial u_1^-}{\partial \bar{n}} + \frac{\partial u_0^-}{\partial \mathbf{n}} \right) \quad \text{at} \quad \partial \Omega_i, \qquad (3.20)$$

$$u_1^+ \Big|_{\xi = \hat{b}} = u_1^+ \Big|_{\xi = -\hat{b}}, \quad u_1^+ \Big|_{\eta = \hat{b}} = u_1^+ \Big|_{\eta = -\hat{b}},$$

$$\frac{\partial u_1^+}{\partial \xi} \Big|_{\xi = \hat{b}} = \frac{\partial u_1^+}{\partial \xi} \Big|_{\xi = -\hat{b}}, \quad \frac{\partial u_1^+}{\partial \eta} \Big|_{\eta = \hat{b}} = \frac{\partial u_1^+}{\partial \eta} \Big|_{\eta = -\hat{b}}.$$

Tthe boundary value problem (3.19) has only a trivial solution. It means that

$$u_0^+ = u_0^+ (x, y), \quad u_0^- = u_0^- (x, y),$$

i.e., the functions u_0^+, u_0^- depend only on "slow" variables and represent the average part of the desired solution.

Thus, in relations (3.6), the expansions begin with a term of order ε, and therefore, they have the following form:

$$u^+ = u_0(x,y) + \varepsilon u_1^+ (x,y,\xi,\eta) + \varepsilon^2 u_2^+ (x,y,\xi,\eta) + \dots$$
$$u^- = u_0(x,y) + \varepsilon u_1^- (x,y,\xi,\eta) + \varepsilon^2 u_2^- (x,y,\xi,\eta) + \dots \qquad (3.21)$$

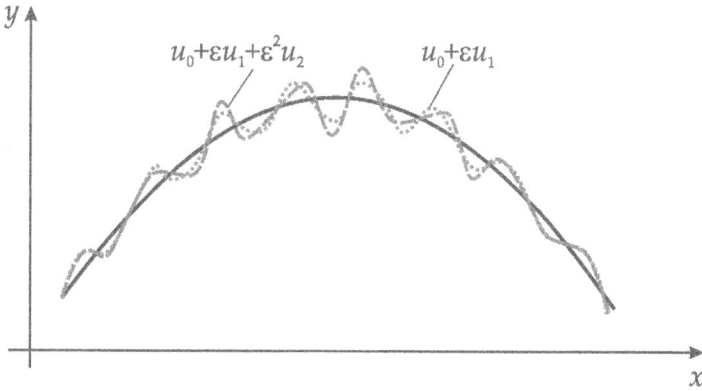

Figure 3.4 Principal form of solving the problem by the homogenization method.

The other components u_i^+, u_i^- $(i = 1, 2, ...)$ of expansions (3.21) are periodic, together with their derivatives, and are determined by solving the corresponding local problems.

To find the functions u_1^+, u_1^-, we have the following boundary value problem:

$$
\frac{\partial^2 u_1^+}{\partial \xi^2} + \frac{\partial^2 u_1^+}{\partial \eta^2} = 0 \text{ in } \Omega_i^+,
$$
$$
\frac{\partial^2 u_1^-}{\partial \xi^2} + \frac{\partial^2 u_1^-}{\partial \eta^2} = 0 \text{ in } \Omega_i^-,
$$
(3.22)

$$
u_1^+ = u_1^-, \quad \frac{\partial u_1^+}{\partial \bar{n}} - \lambda \frac{\partial u_1^-}{\partial \bar{n}} = (\lambda - 1) \frac{\partial u_0}{\partial n} \quad \text{at} \quad \partial \Omega_i,
$$
(3.23)

$$
u_1^+ \Big|_{\xi = \hat{b}} = u_1^+ \Big|_{\xi = -\hat{b}} \; ; \; u_1^+ \Big|_{\eta = \hat{b}} = u_1^+ \Big|_{\eta = -\hat{b}}
$$
$$
\frac{\partial u_1^+}{\partial \xi} \Big|_{\xi = \hat{b}} = \frac{\partial u_1^+}{\partial \xi} \Big|_{\xi = -\hat{b}} \; ; \; \frac{\partial u_1^+}{\partial \eta} \Big|_{\eta = \hat{b}} = \frac{\partial u_1^+}{\partial \eta} \Big|_{\eta = -\hat{b}}
$$
(3.24)

Thus, the solution of the original boundary value problem is represented as the sum of some slowly changed function and small rapidly oscillating correctors of the order ε, ε^2, ε^3, ... (Fig. 3.4) [23, 177, 178].

Further, at the second step of solving the problem by the homogenization method, we determine the main part of the desired solution, i.e. the function $u_0(x,y)$ from the homogenized equation.

To construct the homogenized equation, we apply to relation (3.10) the averaging operator, which is defined as

$$
\bar{\Phi}(x,y) = \frac{1}{|\Omega_i^*|} \left[\iint\limits_{\Omega_i^+} \Phi^+(x,y,\xi,\eta)\, d\xi d\eta + \lambda \iint\limits_{\Omega_i^-} \Phi^-(x,y,\xi,\eta)\, d\xi d\eta \right],
$$
(3.25)

where $|\Omega_i^*| = |\Omega_i^+ \cup \Omega_i^-|$ stands for the cell area; in this case $|\Omega_i^*| = 4\hat{b}^2$.

Using the Green's theorem, due to the boundary condition (3.16), we have

$$
\int\limits_{\partial\Omega_i} \left[\left(\frac{\partial u_2^+}{\partial \bar{n}} + \frac{\partial u_1^+}{\partial \mathbf{n}} \right) - \lambda \left(\frac{\partial u_2^-}{\partial \bar{n}} + \frac{\partial u_1^-}{\partial \mathbf{n}} \right) \right] d\ell = 0.
$$
(3.26)

Thus, taking into account relation (3.26), we obtain the following homogenized equation

$$
\frac{1}{|\Omega_i^*|} \left[\left(\frac{\partial^2 u_0}{\partial x^2} + \frac{\partial^2 u_0}{\partial y^2} \right) \cdot \left(|\Omega_i^+| + \lambda\,|\Omega_i^-| \right) + \right.
$$
$$
\left. + \iint\limits_{\Omega_i^+} \left(\frac{\partial^2 u_1^+}{\partial x \partial \xi} + \frac{\partial^2 u_1^+}{\partial y \partial \eta} \right) d\xi d\eta + \lambda \iint\limits_{\Omega_i^-} \left(\frac{\partial^2 u_1^-}{\partial x \partial \xi} + \frac{\partial^2 u_1^-}{\partial y \partial \eta} \right) d\xi d\eta \right] = F.
$$
(3.27)

After integration over "fast" variables in the domains Ω_i^+ and Ω_i^- relation (3.27) reduces to the Poisson's equation

$$
q\,\Delta u_0 = F,
$$
(3.28)

where the effective conductivity q is defined as follows:

$$
q = 1 + (\lambda - 1)\frac{|\Omega_i^-|}{4\hat{b}^2} + \frac{1}{4\hat{b}^2} \left(\iint\limits_{\Omega_i^+} \frac{\partial u_{1\xi}^+}{\partial \xi}\, d\xi d\eta + \lambda \iint\limits_{\Omega_i^-} \frac{\partial u_{1\xi}^-}{\partial \xi}\, d\xi d\eta \right)
$$
(3.29)

$$
u_1^\pm = \frac{\partial u_0}{\partial x} u_{1\xi}^\pm(\xi,\eta) + \frac{\partial u_0}{\partial y} u_{1\eta}^\pm(\xi,\eta)
$$

$$
u_{1\xi}^\pm(\xi,\eta) = u_{1\eta}^\pm(\xi,\eta) \quad (\xi \leftrightarrow \eta).
$$

Adding to (3.28) and (3.29), the averaged boundary conditions for function $u_0(x,y)$ on the outer contour of region $\partial\Omega$, we obtain boundary value problem for determining the main ("slow") part of the solution.

Since (3.28) is a Poisson's equation and is considered in a simply connected domain Ω^* with effective thermal conductivity $q = const$, the resulting boundary value problem is much simpler than the original one and can be solved by standard methods.

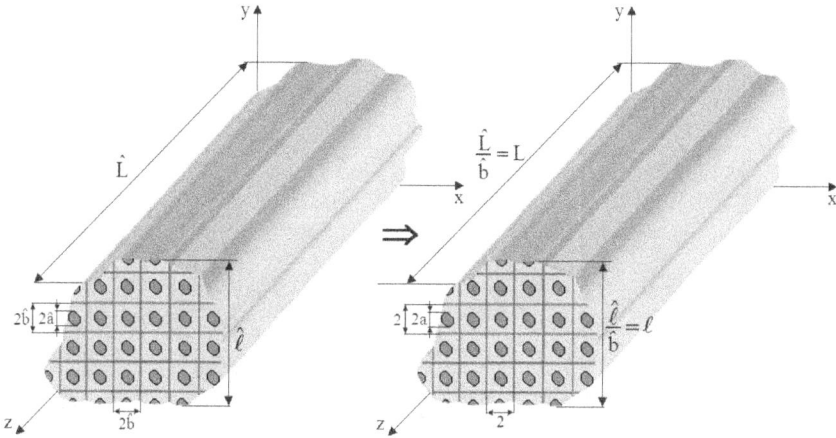

Figure 3.5 Geometric parameters of the composite structure in dimensionless quantities.

Thus, the main difficulties lie precisely in solving a local problem. In this regard, let us further consider a method for solving the cell problem (3.22)–(3.24), based on the ThPhM. In this case, without loss of generality, we accept natural normalization, in which the value of the geometric parameter a is defined as the ratio of the characteristic size of the inclusion $2\hat{a}$ to the characteristic size of the unit cell $2\hat{b}$.

In particular, for a square lattice (Fig. 3.5), we have $\hat{b} = 1$; $a = \frac{\hat{a}}{\hat{b}}$ characterizes the inclusion size in dimensionless units, $0 \leq a \leq 1$; volume fraction of inclusions $c = \frac{|\Omega_i^-(a)|}{4}$.

3.2.2 TRANSFORMATION OF A LOCAL PROBLEM IN THE FRAMEWORK OF THPHM

Consider the two-phase 2D periodic composite structure with cylindrical inclusions, circular in cross section (Fig. 3.6).

In this case, the characteristic structural cell of the composite is shown in Fig. 3.7, and the cell problem can be written as follows:

$$\frac{\partial^2 u_1^+}{\partial \xi^2} + \frac{\partial^2 u_1^+}{\partial \eta^2} = 0 \quad \text{in} \quad \Omega_i^+,$$

$$\frac{\partial^2 u_1^-}{\partial \xi^2} + \frac{\partial^2 u_1^-}{\partial \eta^2} = 0 \quad \text{in} \quad \Omega_i^-, \tag{3.30}$$

$$u_1^+ = u_1^-, \quad \frac{\partial u_1^+}{\partial \bar{n}} - \lambda \frac{\partial u_1^-}{\partial \bar{n}} = (\lambda - 1) \frac{\partial u_0}{\partial \mathbf{n}} \quad \text{on} \quad \partial \Omega_i, \tag{3.31}$$

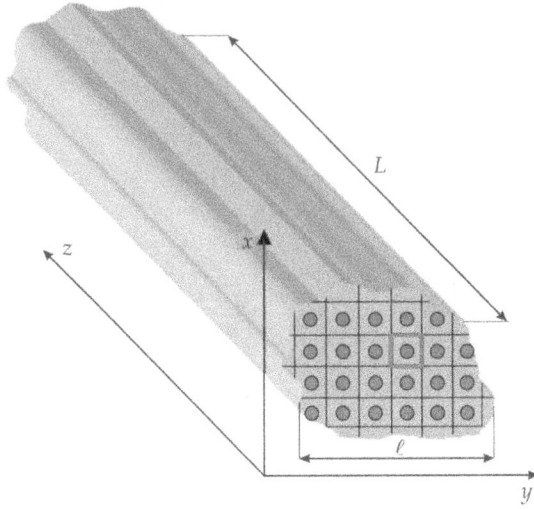

Figure 3.6 Two-phase 2D periodic composite with circular cylindrical inclusions.

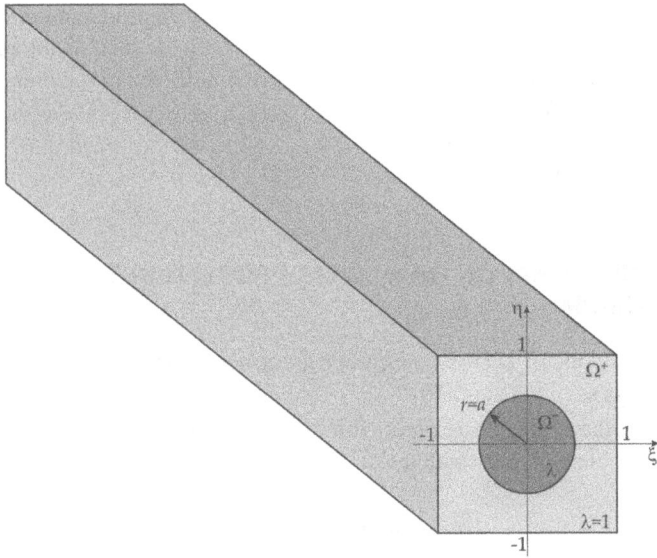

Figure 3.7 Structural cell of a composite with circular cylindrical inclusions.

$$
u_1^+\Big|_{\xi=1} = u_1^+\Big|_{\xi=-1}, \qquad u_1^+\Big|_{\eta=1} = u_1^+\Big|_{\eta=-1},
$$

$$
\frac{\partial u_1^+}{\partial \xi}\Big|_{\xi=1} = \frac{\partial u_1^+}{\partial \xi}\Big|_{\xi=-1}, \qquad \frac{\partial u_1^+}{\partial \eta}\Big|_{\eta=1} = \frac{\partial u_1^+}{\partial \eta}\Big|_{\eta=-1}.
$$

(3.32)

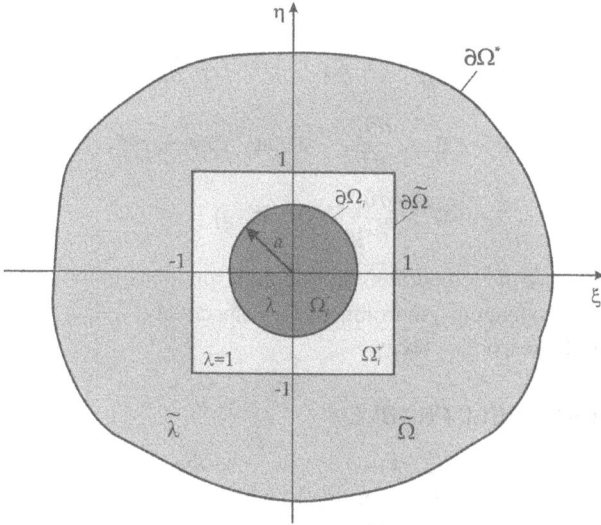

Figure 3.8 ThPhM of composite.

In accordance with the ThPhM of the composite, we replace the entire considered periodic structure, with the exception of one cell, by a homogeneous medium with an unknown thermal conductivity $\tilde{\lambda}$ (Fig. 3.8).

Mathematically, this idealization leads to:

(i) the replacement of the periodicity conditions (3.32) by the conditions of conjugation of the matrix with an equivalent homogeneous medium;

(ii) the decaying conditions for the functions of temperature and heat flux in an averaged medium at a considerable distance from the inclusion.

Consequently, using ThPhM, relations (3.21) must be supplemented by an expansion into an asymptotic series in powers of the small parameter ε of the function $\tilde{u}(x,y,\xi,\eta)$, defined in a homogeneous medium $\tilde{\Omega}$ with the conductivity $\tilde{\lambda}$:

$$\tilde{u} = \tilde{u}_0(x,y) + \varepsilon\,\tilde{u}_1(x,y,\xi,\eta) + \varepsilon^2\tilde{u}_2(x,y,\xi,\eta) + \dots \qquad (3.33)$$

and the local problem (3.30)–(3.32) is transformed to the following form

$$\frac{\partial^2 u_1^+}{\partial \xi^2} + \frac{\partial^2 u_1^+}{\partial \eta^2} = 0 \quad \text{in} \quad \Omega_i^+,$$

$$\frac{\partial^2 u_1^-}{\partial \xi^2} + \frac{\partial^2 u_1^-}{\partial \eta^2} = 0 \quad \text{in} \quad \Omega_i^-, \qquad (3.34)$$

$$\frac{\partial^2 \tilde{u}_1}{\partial \xi^2} + \frac{\partial^2 \tilde{u}_1}{\partial \eta^2} = 0 \quad \text{in} \quad \tilde{\Omega},$$

$$u_1^+ = u_1^-, \quad \frac{\partial u_1^+}{\partial \bar{\mathbf{n}}} - \lambda\frac{\partial u_1^-}{\partial \bar{\mathbf{n}}} = (\lambda - 1)\frac{\partial u_0}{\partial \mathbf{n}} \quad \text{at} \quad \partial\Omega_i, \qquad (3.35)$$

$$u_1^+ = \tilde{u}_1, \quad \frac{\partial u_1^+}{\partial \bar{\mathbf{n}}} - \tilde{\lambda} \frac{\partial \tilde{u}_1}{\partial \bar{\mathbf{n}}} = \left(\tilde{\lambda} - 1 \right) \frac{\partial u_0}{\partial \mathbf{n}} \quad \text{at} \quad \partial \Omega, \qquad (3.36)$$

$$\tilde{u}_1 \to 0, \quad \frac{\partial \tilde{u}_1}{\partial \xi} \to 0 \quad \text{at} \quad \xi = \to \pm\infty,$$

$$\tilde{u}_1 \to 0, \quad \frac{\partial \tilde{u}_1}{\partial \eta} \to 0 \quad \text{at} \quad \eta = \to \pm\infty,$$

$$(3.37)$$

where u_1^+, u_1^-, $\tilde{u}_1$ are the temperature distribution functions in the matrix Ω_i^+, inclusions Ω_i^- and homogeneous environment $\tilde{\Omega}$, respectively; λ^+, λ^-, $\tilde{\lambda}$ stand for the conductivity of the respective areas.

3.2.3 SOLVING A CELL PROBLEM

To solve the local problem (3.34)–(3.37), we use the perturbation position of the boundary (PPB) method [105], according to which the square contour of cell $\partial \tilde{\Omega}$ in polar coordinates r, θ is described by an equation of the following form

$$r = r_0 + \varepsilon_1 f(\theta), \qquad (3.38)$$

where $r_0 = const > 0$; $f(\theta)$ is the differentiable function characterizing the geometric shape of the considered contour; ε_1 is the small parameter ($|\varepsilon_1| << 1$).

In the case under consideration, we replace the square boundary of unit cell $\partial \tilde{\Omega}$ with a smooth contour in the form of a square with rounded corners. Cosequently, a function is used that implements a conformal mapping of an infinite plane with a circular inclusion into an infinite plane. The inclusion is in the form of a square with rounded corners [184]. This function can be written as follows:

$$z = \xi + i\eta = \zeta + \varepsilon_1 \zeta^{-3}, \quad \zeta = r \exp(i\theta), \quad i = \sqrt{-1}.$$

The contour equation in this case takes the form:

$$f(\zeta) = r_0 \left[\zeta + \varepsilon_1 \zeta^{-3} \right].$$

Neglecting terms of the higher order of smallness, the unit cell boundary can be approximated by the following expressions [184]

$$\xi = r_0 \left(\cos \theta + \varepsilon_1 \cos 3\theta \right), \quad \eta = r_0 \left(\sin \theta - \varepsilon_1 \sin 3\theta \right), \qquad (3.39)$$

$$0 \leq \theta \leq 2\pi, \quad \frac{1}{9} \leq |\varepsilon_1| \leq \frac{1}{6}.$$

The unit cell area described by equation (3.39) differs slightly from a square with rounded corners. Note that the singularities at the corner points of the original contour are not taken into account in this approximation. However, this does not significantly affect the global characteristics, including the effective coefficients. To determine the local characteristics, it is necessary to use additional techniques to take these features into account [11].

In our case, relation (3.38) can be represented as follows:

$$r^+ = r_0^+ \left(1 + \varepsilon_1 \cos 4\theta + \ldots \right). \tag{3.40}$$

The mathematical meaning of this approximation lies in the fact that in the zero approximation the square contour of the cell is replaced by a circle. In this case, the radius of the circle r_0^+ is chosen in such a way that the equality of the areas of the original and transformed areas of the cell is preserved:

$$\pi \left(r_0^+\right)^2 = 4, \quad \text{i.e.} \quad r_0^+ = \frac{2}{\sqrt{\pi}}. \tag{3.41}$$

Let us represent the functions u_1^+, u_1^-, $\tilde{u}_1$, respectively, in the matrix Ω_i^+, inclusions Ω_i^- and homogeneous medium $\tilde{\Omega}$ in the form of asymptotic series in powers of a small parameter ε_1:

$$u_1^+ = u_{10}^+ + \varepsilon_1 u_{11}^+ + \ldots ,$$
$$u_1^- = u_{10}^- + \varepsilon_1 u_{11}^- + \ldots , \tag{3.42}$$
$$\tilde{u}_1 = \tilde{u}_{10} + \varepsilon_1 \tilde{u}_{11} + \ldots .$$

In a similar way, we represent parameter

$$\tilde{\lambda} = \tilde{\lambda}_0 + \varepsilon_1 \tilde{\lambda}_1 + \ldots . \tag{3.43}$$

The value of the small parameter ε_1 for a square contour, as a rule, varies within [184]: $\frac{1}{9} \leq |\varepsilon_1| \leq \frac{1}{6}$. Here we suppose that

$$\varepsilon_1 = \frac{1 - \frac{2}{\sqrt{\pi}}}{1} = 1 - \frac{2}{\sqrt{\pi}} \approx -0.1284.$$

Next, we use the expansion, taking into account the size of the region. In the first approximation, under the assumption of a small inclusion size, we consider the inclusion problem in an infinite domain. Then the constitutive relations for the function $\tilde{u}_1$ in the polar coordinate system r, θ can be written as follows:

$$\Delta_r \tilde{u}_{11} = 0, \quad \Delta_r = \frac{\partial^2}{\partial r^2} + r^{-1} \frac{\partial}{\partial r} + r^{-2} \frac{\partial^2}{\partial \theta^2},$$

$$\tilde{u}_{1,r} = \tilde{u}_{11,r} + \frac{4}{R} \tilde{u}_{11,\theta} \, \varepsilon_1 \sin 4\theta + \tilde{u}_{10,x} \cos \theta + \tilde{u}_{10,y} \sin \theta -$$
$$- 4 \left(\dot{u}_{10,x} \sin \theta - \tilde{u}_{10,y} \cos \theta\right) \varepsilon_1 \sin 4\theta, \quad \tilde{u}_{11} = 0 \quad \text{for} \quad r \to \infty.$$

In accordance with the PPB [105, 184], the boundary condition on contour $\partial \tilde{\Omega}$ should be transformed into the condition on circle $r = r_0^+$ using the Taylor series expansion:

$$f\left(r_0^+ \left(1 + \varepsilon_1 \cos 4\theta\right), \, \theta\right) = f\left(r_0^+, \, \theta\right) + r_0^+ \, \varepsilon_1 f_1 \left(r_0^+, \, \theta\right) \cos 4\theta + \ldots,$$

where $f(r, \theta)$ is the differentiable function; $f_1\left(r_0^+, \theta\right) = \left. \frac{\partial f(r, \theta)}{\partial r} \right|_{r=r_0^+}$.

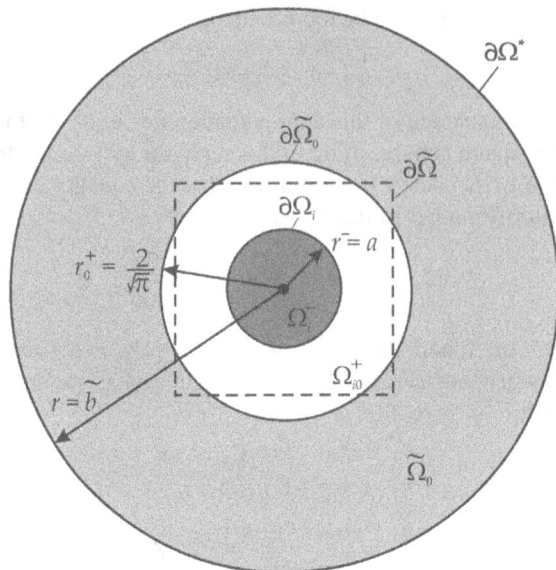

Figure 3.9 Calculation model of a three-phase structure in the zero approximation.

3.2.4 THE LOCAL PROBLEM IN THE ZEROTH APPROXIMATION

Let us obtain the solution of the cell problem in the zero approximation, the computational model is shown in Fig. 3.9.

Taking into account relations (3.40)–(3.43), the local problem (3.34)–(3.37) in "fast" polar coordinates r, θ can be written in the zero approximation for functions u_{10}^+, u_{10}^-, $\tilde{u}_{10}$ as follows:

$$\frac{\partial^2 u_{10}^+}{\partial r^2} + \frac{1}{r} \cdot \frac{\partial u_{10}^+}{\partial r} + \frac{1}{r^2} \cdot \frac{\partial^2 u_{10}^+}{\partial \theta^2} = 0 \quad \text{in} \quad \Omega_{i0}^+,$$

$$\frac{\partial^2 u_{10}^-}{\partial r^2} + \frac{1}{r} \cdot \frac{\partial u_{10}^-}{\partial r} + \frac{1}{r^2} \cdot \frac{\partial^2 u_{10}^-}{\partial \theta^2} = 0 \quad \text{in} \quad \Omega_i^-, \tag{3.44}$$

$$\frac{\partial^2 \tilde{u}_{10}}{\partial r^2} + \frac{1}{r} \cdot \frac{\partial \tilde{u}_{10}}{\partial r} + \frac{1}{r^2} \cdot \frac{\partial^2 \tilde{u}_{10}}{\partial \theta^2} = 0 \quad \text{in} \quad \tilde{\Omega}_0,$$

$$\text{at} \quad \partial \Omega_i : \ r = r^- = a;$$

$$u_{10}^+ = u_{10}^-,$$

$$\frac{\partial u_{10}^+}{\partial r} - \lambda \frac{\partial u_{10}^-}{\partial r} = (\lambda - 1) \left(\frac{\partial u_0}{\partial x} \cos \theta + \frac{\partial u_0}{\partial y} \sin \theta \right), \tag{3.45}$$

$$\text{at} \quad \partial \tilde{\Omega}_0 : \ r = r_0^+ = \frac{2}{\sqrt{\pi}};$$

$$u_{10}^+ = \tilde{u}_{10},$$

$$\frac{\partial u_{10}^+}{\partial r} - \tilde{\lambda}\frac{\partial \tilde{u}_{10}}{\partial r} = \left(\tilde{\lambda}_0 - 1\right)\left(\frac{\partial u_0}{\partial x}\cos\theta + \frac{\partial u_0}{\partial y}\sin\theta\right), \tag{3.46}$$

$$\tilde{u}_{10} \to 0, \quad \frac{\partial \tilde{u}_{10}}{\partial r} \to 0 \quad \text{at} \quad r \to \infty. \tag{3.47}$$

The solution of the boundary value problem (3.44)-(3.47) is sought in the following form

$$u_{10}^- = \overline{A}_{10} r\cos\theta + \overline{\overline{A}}_{10} r\sin\theta,$$

$$u_{10}^+ = \left(\overline{B}_{10} r + \frac{\overline{C}_{10}}{r}\right)\cos\theta + \left(\overline{\overline{B}}_{10} r + \frac{\overline{\overline{C}}_{10}}{r}\right)\sin\theta, \tag{3.48}$$

$$\tilde{u}_{10} = \frac{\overline{D}_{10}}{r}\cos\theta + \frac{\overline{\overline{D}}_{10}}{r}\sin\theta,$$

where $\overline{A}_{10}, \overline{\overline{A}}_{10}, \overline{B}_{10}, \overline{\overline{B}}_{10}, \overline{C}_{10}, \overline{\overline{C}}_{10}, \overline{D}_{10}, \overline{\overline{D}}_{10}$ are the arbitrary constants.

Note that in expressions (3.48), the representation of the function u_{10}^- is written taking into account the boundedness of the temperature distribution function and its derivative $\frac{\partial u_{10}^-}{\partial r}$ for $r = 0$, and the expression of function $\tilde{u}_{10}$ satisfies the decaying conditions at $r \to \infty$ (3.47).

Relations (3.48) include eight arbitrary constants, four for each of the basis functions $\cos\theta$ and $\sin\theta$ (respectively, $\overline{A}_{10}, \overline{B}_{10}, \overline{C}_{10}, \overline{D}_{10}$ and $\overline{\overline{A}}_{10}, \overline{\overline{B}}_{10}, \overline{\overline{C}}_{10}, \overline{\overline{D}}_{10}$), which are determined from the conditions (3.45), (3.46). Systems of equations for integration constants $\overline{A}_{10}, \overline{B}_{10}, \overline{C}_{10}, \overline{D}_{10}$ and $\overline{\overline{A}}_{10}, \overline{\overline{B}}_{10}, \overline{\overline{C}}_{10}, \overline{\overline{D}}_{10}$ are completely identical, so we give a solution to only one of them:

$$\overline{A}_{10} = A_{10}^*\frac{\partial u_0}{\partial x}, \quad \overline{B}_{10} = B_{10}^*\frac{\partial u_0}{\partial x},$$

$$\overline{C}_{10} = C_{10}^*\frac{\partial u_0}{\partial x}, \quad \overline{D}_{10} = D_{10}^*\frac{\partial u_0}{\partial x}, \tag{3.49}$$

$$A_{10}^* = -1 + 4\tilde{\lambda}_0\Delta_0,$$

$$B_{10}^* = -1 + 2(\lambda + 1)\tilde{\lambda}_0\Delta_0,$$

$$C_{10}^* = -2(\lambda - 1)\tilde{\lambda}_0 a^2\Delta_0, \tag{3.50}$$

$$D_{10}^* = \frac{4}{\pi}\left(1 - 2\left((\lambda - 1)\frac{\pi a^2}{4} + (\lambda + 1)\right)\Delta_0\right),$$

where

$$\Delta_0 = \left(\left(\tilde{\lambda}_0 + 1\right)(\lambda + 1) - \left(\tilde{\lambda}_0 - 1\right)(\lambda - 1)\frac{\pi a^2}{4}\right)^{-1}. \tag{3.51}$$

Obviously, for arbitrary constants $\overline{\overline{A}}_{10}$, $\overline{\overline{B}}_{10}$, $\overline{\overline{C}}_{10}$, $\overline{\overline{D}}_{10}$, we have

$$
\begin{aligned}
\overline{\overline{A}}_{10} &= \overline{A}_{10}, & \overline{\overline{B}}_{10} &= \overline{B}_{10}, \\
\overline{\overline{C}}_{10} &= \overline{C}_{10}, & \overline{\overline{D}}_{10} &= \overline{D}_{10} \quad \left(\frac{\partial u_0}{\partial x} \to \frac{\partial u_0}{\partial y} \right).
\end{aligned}
\tag{3.52}
$$

The further solution of the problem is reduced to the construction of homogenized relations. In this case, it should be taken into account that averaging should be carried out over the region $\Omega_0^* = \Omega_{i0}^+ \bigcup \Omega_i^- \bigcup \tilde{\Omega}_0$. Thus, the equation to be averaged is

$$
\begin{aligned}
&\frac{\partial^2 u_0}{\partial x^2} + \frac{\partial^2 u_0}{\partial y^2} + 2\frac{\partial^2 u_{10}^+}{\partial x \partial \xi} + 2\frac{\partial^2 u_{10}^+}{\partial y \partial \eta} + \frac{\partial^2 u_{20}^+}{\partial \xi^2} + \frac{\partial^2 u_{20}^+}{\partial \eta^2} + \\
&+ \lambda \left(\frac{\partial^2 u_0}{\partial x^2} + \frac{\partial^2 u_0}{\partial y^2} + 2\frac{\partial^2 u_{10}^-}{\partial x \partial \xi} + 2\frac{\partial^2 u_{10}^-}{\partial y \partial \eta} + \frac{\partial^2 u_{20}^-}{\partial \xi^2} + \frac{\partial^2 u_{20}^-}{\partial \eta^2} \right) + \\
&+ \tilde{\lambda}_0 \left(\frac{\partial^2 u_0}{\partial x^2} + \frac{\partial^2 u_0}{\partial y^2} + 2\frac{\partial^2 \tilde{u}_{10}}{\partial x \partial \xi} + 2\frac{\partial^2 \tilde{u}_{10}}{\partial y \partial \eta} + \frac{\partial^2 \tilde{u}_{20}}{\partial \xi^2} + \frac{\partial^2 \tilde{u}_{20}}{\partial \eta^2} \right) = F.
\end{aligned}
\tag{3.53}
$$

The averaging operator is generalized in this case as follows:

$$
\overline{(...)} = \frac{1}{|\Omega_0^*|} \left(\iint_{\Omega_{i0}^+} (...) \, d\xi \, d\eta + \lambda \iint_{\Omega_i^-} (...) \, d\xi \, d\eta + \tilde{\lambda}_0 \iint_{\tilde{\Omega}_0} (...) \, d\xi \, d\eta \right),
\tag{3.54}
$$

where $|\Omega_0^*| = |\Omega_{i0}^+ \bigcup \Omega_i^- \bigcup \tilde{\Omega}_0|$ stands for the measure of a region, the boundary of which can formally be considered a circle of infinitely large radius $\tilde{b} \to \infty$.

Applying the averaging operator (3.54) to relation (3.53), we obtain the homogenized equation in the following form

$$
\frac{1}{|\Omega_0^*|} \left[\iint_{\Omega_{i0}^+} \left(\frac{\partial^2 u_0}{\partial x^2} + \frac{\partial^2 u_0}{\partial y^2} + \frac{\partial^2 u_{10}^+}{\partial x \partial \xi} + \frac{\partial^2 u_{10}^+}{\partial y \partial \eta} \right) d\xi \, d\eta + \right.
$$

$$
+ \int_{\partial \Omega_i} \left(\frac{\partial u_{20}^+}{\partial \bar{n}} + \frac{\partial u_{10}^+}{\partial n} \right) d\ell + \int_{\partial \tilde{\Omega}_0} \left(\frac{\partial u_{20}^+}{\partial \bar{n}} + \frac{\partial u_{10}^+}{\partial n} \right) d\ell +
$$

$$
+ \lambda \iint_{\Omega_i^-} \left(\frac{\partial^2 u_0}{\partial x^2} + \frac{\partial^2 u_0}{\partial y^2} + \frac{\partial^2 u_{10}^-}{\partial x \partial \xi} + \frac{\partial^2 u_{10}^-}{\partial y \partial \eta} \right) d\xi \, d\eta +
\tag{3.55}
$$

$$+\lambda \int\limits_{\partial\Omega_i} \left(\frac{\partial u_{20}^-}{\partial \bar{n}} + \frac{\partial u_{10}^-}{\partial n} \right) d\ell + \tilde{\lambda}_0 \iint\limits_{\tilde{\Omega}_0} \left(\frac{\partial^2 u_0}{\partial x^2} + \frac{\partial^2 u_0}{\partial y^2} + \frac{\partial^2 \tilde{u}_{10}}{\partial x \partial \xi} + \frac{\partial^2 \tilde{u}_{10}}{\partial y \partial \eta} \right) d\xi\, d\eta +$$

$$\left. +\tilde{\lambda}_0 \int\limits_{\partial\tilde{\Omega}_0} \left(\frac{\partial \tilde{u}_{20}}{\partial \bar{n}} + \frac{\partial \tilde{u}_{10}}{\partial n} \right) d\ell + \tilde{\lambda}_0 \int\limits_{\partial\Omega^*} \left(\frac{\partial \tilde{u}_{20}}{\partial \bar{n}} + \frac{\partial \tilde{u}_{10}}{\partial n} \right) d\ell \right] = F.$$

Taking into account the expressions for u_{10}^+, u_{10}^-, $\tilde{u}_{10}$ (3.48)-(3.52), the homogenized equation (3.55) is transformed to the following form:

$$\frac{1}{|\Omega_0^*|} \left[|\Omega_{i0}^+| \, (B_{10}^* + 1) + \lambda \, |\Omega_i^-| \, (A_{10}^* + 1) + \tilde{\lambda}_0 \, |\tilde{\Omega}_0| \right] \Delta u_0 = F, \qquad (3.56)$$

where the effective parameter

$$q = \frac{1}{|\Omega_0^*|} \left[|\Omega_{i0}^+| \, (B_{10}^* + 1) + \lambda \, |\Omega_i^-| \, (A_{10}^* + 1) + \tilde{\lambda}_0 \, |\tilde{\Omega}_0| \right]. \qquad (3.57)$$

It is obvious that the coefficient q, determined by expression (3.57), is exactly the parameter $\tilde{\lambda}_0$. Therefore, equating $\tilde{\lambda}_0$ to expression (3.57), we arrive at a linear algebraic equation with respect to the coefficient $\tilde{\lambda}_0$. The solution in the zeroth approximation yields the following analytical expression for the effective thermal conductivity coefficient

$$q_{ThPhM} = \tilde{\lambda}_0 = \frac{1 - \frac{\pi a^2}{4} + \lambda \left(1 + \frac{\pi a^2}{4} \right)}{1 + \frac{\pi a^2}{4} + \lambda \left(1 - \frac{\pi a^2}{4} \right)}. \qquad (3.58)$$

3.2.5 EVALUATION OF THE RESULTS OF THE PPB SOLUTION

In the zero approximation, ThPhM describes the effective thermal conductivity by expression (3.58), which coincides with the Maxwell approximation (see Chapter 6). Let us estimate the limits of applicability of the obtained relation.

1) As is known, by virtue of the Keller's theorem [127], the effective thermal conductivity parameters of composites with inclusions that are symmetric in the cell with respect to the straight line $\eta = \xi$ are coupled by the relation

$$q(\lambda) = \frac{1}{q\left(\frac{1}{\lambda}\right)}. \qquad (3.59)$$

Expression (3.58) exactly satisfies the Keller's theorem for all values of the conductivity λ, including the limiting cases: $\lambda \to 0$ and $\lambda \to \infty$.

2) If $\lambda \sim 1$, i.e., the thermal conductivity of the matrix and inclusions of the same order, then for arbitrary values of the inclusion size a we find that

$$q = 1 + \frac{\pi a^2}{4}(\lambda - 1) - \frac{\pi a^2}{8}\left(1 - \frac{\pi a^2}{4}\right)(\lambda - 1)^2 + O\left((\lambda - 1)^3\right).$$

(3.60)

In the limit $\lambda \to 1$, we obtain a homogeneous structure, $q = 1$.

3) For $\lambda \ll 1$, we have the case of low conductivity of inclusions:

$$q = \frac{1 - \frac{\pi a^2}{4}}{1 + \frac{\pi a^2}{4}} + \left(1 - \frac{\left(1 - \frac{\pi a^2}{4}\right)^2}{\left(1 + \frac{\pi a^2}{4}\right)^2}\right)\lambda - \frac{4\left(1 - \frac{\pi a^2}{4}\right)\frac{\pi a^2}{4}}{\left(1 + \frac{\pi a^2}{4}\right)^3}\lambda^2 + O\left(\lambda^3\right).$$

(3.61)

In the limit $\lambda \to 0$, we have a composite with heat insulating inclusions; then from (3.61), we find

$$q = \frac{1 - \frac{\pi a^2}{4}}{1 + \frac{\pi a^2}{4}}.$$

(3.62)

4) Case $\lambda \gg 1$ corresponds to a composite with high conductivity inclusions:

$$q = \frac{1 + \frac{\pi a^2}{4}}{1 - \frac{\pi a^2}{4}} + \left(1 - \frac{\left(1 + \frac{\pi a^2}{4}\right)^2}{\left(1 - \frac{\pi a^2}{4}\right)^2}\right)\lambda^{-1} + \frac{4\left(1 + \frac{\pi a^2}{4}\right)\frac{\pi a^2}{4}}{\left(1 - \frac{\pi a^2}{4}\right)^3}\lambda^{-2} + O\left(\lambda^{-3}\right).$$

(3.63)

In limit $\lambda \to \infty$, we obtain a composite with absolutely conducting inclusions:

$$q = \frac{1 + \frac{\pi a^2}{4}}{1 - \frac{\pi a^2}{4}}.$$

(3.64)

Comparison of relations (3.62), (3.64) yields

$$q(\lambda)\Big|_{\lambda \to \infty} = \frac{1}{q(\lambda)}\Big|_{\lambda \to 0},$$

i.e., the relations obtained satisfy the Keller's theorem (3.59).

5) The sizes of inclusions are small: $a \ll 1$. Then, for an arbitrary value of the conductivity of the inclusions λ, we obtain

$$q = 1 + \frac{\lambda - 1}{\lambda + 1}\frac{\pi a^2}{2} + O\left(a^4\right).$$

(3.65)

In the limit $a \to 1$, we get a homogeneous structure, $q = 1$.

6) Expressions (3.62) for $a \to 0$ and (3.65) for $\lambda \to 0$ yield the following asymptotic relation

$$q = 1 - \frac{\pi a^2}{2},$$

(3.66)

coinciding with the result given in [21] and obtained as a solution to the problem of torsion of a perforated rod.

7) Obviously, in the case of inclusions of small sizes ($a \ll 1$) with absolute conductivity ($\lambda \to \infty$) from (3.64) for $a \to 0$ and from (3.65) for $\lambda \to \infty$, we have

$$q = 1 + \frac{\pi a^2}{2}. \qquad (3.67)$$

8) It should also be noted that the ThPhM cannot be used for large inclusion sizes ($a \to 1$) with extremely large ($\lambda \to \infty$) or extremely small ($\lambda \to 0$) conductivity. As shown in [158], the asymptotic relation for $\lambda \to \infty$, $a \to 1$ looks like

$$q_{as}^{(\infty)} = \frac{\pi}{\sqrt{1 - a^2}} - \pi + 1. \qquad (3.68)$$

From the comparison of expressions (3.58) and (3.68), it follows that in this case, ThPhM does not describe even qualitatively the effective conductivity of the composite.

Thus, the analysis of ThPhM in the zero approximation showed:

   (i)   for small sizes of inclusions, ThPhM reliably describes the effective thermal conductivity parameter at any values of their conductivity;

   (ii) for large sizes of inclusions, the model gives acceptable results only in the case of conductivity $\lambda \sim 1$;

   (iii) ThPhM cannot be used, even qualitatively, for large inclusion sizes ($a \to 1$) and extremely large ($\lambda \to \infty$) or extremely small ($\lambda \to 0$) conductivity.

In this regard, it seems appropriate to refine relation (3.58) by constructing the first approximation in the solution of the local problem using PPB.

3.2.6 MATHEMATICAL STRUCTURE OF THE SOLUTION OF THE FIRST APPROXIMATION

The local problem in the first approximation of PPB is written in "fast" polar coordinates r, θ as follows:

$$\frac{\partial^2 u_{11}^+}{\partial r^2} + \frac{1}{r} \cdot \frac{\partial u_{11}^+}{\partial r} + \frac{1}{r^2} \cdot \frac{\partial^2 u_{11}^+}{\partial \theta^2} = 0 \quad \text{in} \quad \Omega_{i1}^+,$$

$$\frac{\partial^2 u_{11}^-}{\partial r^2} + \frac{1}{r} \cdot \frac{\partial u_{11}^-}{\partial r} + \frac{1}{r^2} \cdot \frac{\partial^2 u_{11}^-}{\partial \theta^2} = 0 \quad \text{in} \quad \Omega_i^-, \qquad (3.69)$$

$$\frac{\partial^2 \tilde{u}_{11}}{\partial r^2} + \frac{1}{r} \cdot \frac{\partial \tilde{u}_{11}}{\partial r} + \frac{1}{r^2} \cdot \frac{\partial^2 \tilde{u}_{11}}{\partial \theta^2} = 0 \quad \text{in} \quad \tilde{\Omega}_1,$$

$$u_{11}^+ = u_{11}^- \quad \text{at} \quad r = r_0^- = a,$$

$$\frac{\partial u_{11}^+}{\partial r} = \lambda \frac{\partial u_{11}^-}{\partial r} \quad \text{at} \quad r = r_0^- = a, \qquad (3.70)$$

$$u_{11}^+ + \frac{\partial u_{10}^+}{\partial r} r_0^+ \cos 4\theta = \tilde{u}_{11} + \frac{\partial \tilde{u}_{10}}{\partial r} r_0^+ \cos 4\theta \quad \text{at} \quad r = r_0^+ = \frac{2}{\sqrt{\pi}},$$

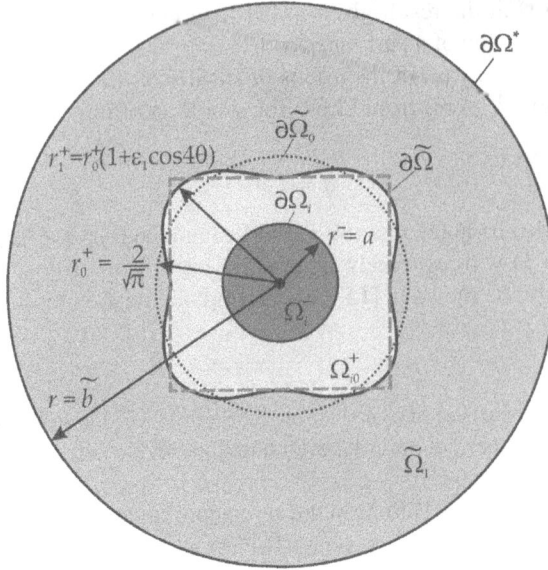

Figure 3.10 Model of a three-phase structure in the first approximation.

$$\frac{\partial u_{11}^+}{\partial r} + \frac{\partial^2 u_{10}^+}{\partial r^2} r_0^+ \cos 4\theta + \frac{4}{r_0^+} \cdot \frac{\partial u_{10}^+}{\partial \theta} \sin 4\theta - 4 \frac{\partial u_0}{\partial x} \sin \theta \sin 4\theta +$$

$$+ 4 \frac{\partial u_0}{\partial y} \cos \theta \sin 4\theta = \tilde{\lambda}_0 \left(\frac{\partial \tilde{u}_{11}}{\partial r} + \frac{\partial^2 \tilde{u}_{10}}{\partial r^2} r_0^+ \cos 4\theta + \right.$$

$$+ \frac{4}{r_0^+} \cdot \frac{\partial \tilde{u}_{10}}{\partial \theta} \sin 4\theta - 4 \frac{\partial u_0}{\partial x} \sin \theta \sin 4\theta + 4 \frac{\partial u_0}{\partial y} \cos \theta \sin 4\theta \right) + \tag{3.71}$$

$$+ \tilde{\lambda}_1 \frac{\partial \tilde{u}_{10}}{\partial r} \quad \text{at} \quad r = r_0^+ = \frac{2}{\sqrt{\pi}},$$

$$\tilde{u}_{11} \to 0, \quad \frac{\partial \tilde{u}_{11}}{\partial r} \to 0 \quad \text{at} \quad r \to \infty. \tag{3.72}$$

The geometric model of the structure in the first approximation is shown in Fig. 3.10.

Taking into account the effective thermal conductivity coefficient of the zeroth approximation, the expressions for the constants A_{10}^*, B_{10}^*, C_{10}^*, D_{10}^* in the solution of the zeroth approximation

$$u_{10}^- = A_{10}^* r \left(\frac{\partial u_0}{\partial x} \cos \theta + \frac{\partial u_0}{\partial x} \sin \theta \right),$$

$$u_{10}^+ = \left(B_{10}^* r + \frac{C_{10}^*}{r} \right) \left(\frac{\partial u_0}{\partial x} \cos \theta + \frac{\partial u_0}{\partial x} \sin \theta \right), \tag{3.73}$$

$$\tilde{u}_{10} = \frac{D_{10}^*}{r} \left(\frac{\partial u_0}{\partial x} \cos \theta + \frac{\partial u_0}{\partial x} \sin \theta \right)$$

convert to the form

$$A_{10}^* = -(\lambda - 1)\left(1 - \frac{\pi a^2}{4}\right)\Delta, \quad B_{10}^* = \frac{(\lambda - 1)\pi a^2 \Delta}{4},$$

$$C_{10}^* = -\frac{(\lambda - 1)\pi a^2 \Delta}{4}, \quad D_{10}^* = 0,$$

(3.74)

where

$$\Delta = \left((\lambda + 1) - (\lambda - 1)\frac{\pi a^2}{4}\right)^{-1}.$$

Taking into account the conditions of boundedness of the temperature distribution function u_{11}^- and its derivative $\frac{\partial u_{11}^-}{\partial r}$ for $r = 0$, as well as decaying conditions (3.72), we obtain that the solution of the problem (3.69)-(3.72) can be represented as follows:

$$u_{11}^- = \overline{A}_{11} r \cos\theta + \overline{A}_{21} r^3 \cos 3\theta + \overline{A}_{31} r^5 \cos 5\theta +$$
$$+ \overline{\overline{A}}_{11} r \sin\theta + \overline{\overline{A}}_{21} r^3 \sin 3\theta + \overline{\overline{A}}_{31} r^5 \sin 5\theta,$$

$$u_{11}^+ = \left(\overline{B}_{11} r + \frac{\overline{C}_{11}}{r}\right)\cos\theta + \left(\overline{B}_{21} r^3 + \frac{\overline{C}_{21}}{r^3}\right)\cos 3\theta + \left(\overline{B}_{31} r^5 + \frac{\overline{C}_{31}}{r^5}\right)\cos 5\theta +$$
$$+ \left(\overline{\overline{B}}_{11} r + \frac{\overline{\overline{C}}_{11}}{r}\right)\sin\theta + \left(\overline{\overline{B}}_{21} r^3 + \frac{\overline{\overline{C}}_{21}}{r^3}\right)\sin 3\theta + \left(\overline{\overline{B}}_{31} r^5 + \frac{\overline{\overline{C}}_{31}}{r^5}\right)\sin 5\theta,$$

$$\tilde{u}_{11} = \frac{\overline{D}_{11}}{r}\cos\theta + \frac{\overline{D}_{21}}{r^3}\cos 3\theta + \frac{\overline{D}_{31}}{r^5}\cos 5\theta +$$
$$+ \frac{\overline{\overline{D}}_{11}}{r}\sin\theta + \frac{\overline{\overline{D}}_{21}}{r^3}\sin 3\theta + \frac{\overline{\overline{D}}_{31}}{r^5}\sin 5\theta,$$

(3.75)

where $\overline{A}_{i1}$, $\overline{\overline{A}}_{i1}$, $\overline{B}_{i1}$, $\overline{\overline{B}}_{i1}$, $\overline{C}_{i1}$, $\overline{\overline{C}}_{i1}$, $\overline{D}_{i1}$, $\overline{\overline{D}}_{i1}$ $(i = 1, 2, 3)$ are the arbitrary constants.

Relations (3.75) include 24 arbitrary constants, which are determined from the conditions (3.70), (3.71). Since the systems of equations for finding the constants of integration $\overline{A}_{i1}$, $\overline{B}_{i1}$, $\overline{C}_{i1}$, $\overline{D}_{i1}$ and $\overline{\overline{A}}_{i1}$, $\overline{\overline{B}}_{i1}$, $\overline{\overline{C}}_{i1}$, $\overline{\overline{D}}_{i1}$ $(i = 1, 2, 3)$ are completely identical, we present the only the following relations for $\cos\theta$:

$$\begin{cases} -\overline{A}_{11} r_0^- + \overline{B}_{11} r_0^- + \dfrac{\overline{C}_{11}}{r_0^-} = 0, \\[2mm] -\lambda \overline{A}_{11} + \overline{B}_{11} - \dfrac{\overline{C}_{11}}{r_0^{-2}} = 0, \\[2mm] \overline{B}_{11} r_0^+ + \dfrac{\overline{C}_{11}}{r_0^+} - \dfrac{\overline{D}_{11}}{r_0^+} = 0, \\[2mm] \overline{B}_{11} - \dfrac{\overline{C}_{11}}{r_0^{+2}} + \tilde{\lambda}_0 \dfrac{\overline{D}_{11}}{r_0^{+2}} = -\tilde{\lambda}_1 \dfrac{\overline{D}_{10}}{r_0^{+2}}, \end{cases}$$

(3.76)

for cos 3θ:

$$
\begin{cases}
-\overline{A}_{21}\, r_0^{-3} + \overline{B}_{21}\, r_0^{-3} + \dfrac{\overline{C}_{21}}{r_0^{-3}} = 0, \\[2mm]
-\lambda\overline{A}_{21}\, r_0^{-2} + \overline{B}_{21}\, r_0^{-2} - \dfrac{\overline{C}_{21}}{r_0^{-4}} = 0, \\[2mm]
\overline{B}_{21}\, r_0^{+3} + \dfrac{\overline{C}_{21}}{r_0^{+3}} - \dfrac{\overline{D}_{21}}{r_0^{+3}} = \dfrac{r_0^+}{2}\left(-B_{10} - \dfrac{\overline{C}_{10}}{r_0^{+2}} - \dfrac{\overline{D}_{10}}{r_0^{+2}}\right), \\[2mm]
\overline{B}_{21}\, r_0^{+2} - \dfrac{\overline{C}_{21}}{r_0^{+4}} + \tilde{\lambda}_0\dfrac{\overline{D}_{21}}{r_0^{+4}} = \dfrac{1}{3}\left(2\overline{B}_{10} + \dfrac{\overline{C}_{10}}{r_0^{+2}}\right) - \dfrac{2}{3}\left(\tilde{\lambda}_0 - 1\right)\dfrac{\partial u_0}{\partial x},
\end{cases}
\tag{3.77}
$$

for cos 5θ:

$$
\begin{cases}
-\overline{A}_{31}\, r_0^{-5} + \overline{B}_{31}\, r_0^{-5} + \dfrac{\overline{C}_{31}}{r_0^{-5}} = 0, \\[2mm]
-\lambda\overline{A}_{31}\, r_0^{-4} + \overline{B}_{31}\, r_0^{-4} - \dfrac{\overline{C}_{31}}{r_0^{-6}} = 0, \\[2mm]
\overline{B}_{31}\, r_0^{+5} + \dfrac{\overline{C}_{31}}{r_0^{+5}} - \dfrac{\overline{D}_{31}}{r_0^{+5}} = \dfrac{r_0^+}{2}\left(-\overline{B}_{10} + \dfrac{\overline{C}_{10}}{r_0^{+2}} - \dfrac{\overline{D}_{10}}{r_0^{+2}}\right), \\[2mm]
\overline{B}_{31}\, r_0^{+4} - \dfrac{\overline{C}_{31}}{r_0^{+6}} + \tilde{\lambda}_0\dfrac{\overline{D}_{31}}{r_0^{+6}} = \dfrac{1}{5}\left(-2\overline{B}_{10} - 3\dfrac{\overline{C}_{10}}{r_0^{+2}} + 3\tilde{\lambda}_0\dfrac{\overline{D}_{10}}{r_0^{+2}}\right) + \\[2mm]
\qquad\qquad\qquad +\dfrac{2}{5}\left(\tilde{\lambda}_0 - 1\right)\dfrac{\partial u_0}{\partial x}.
\end{cases}
\tag{3.78}
$$

Taking into account the expression for D_{10}^* in (3.74), from the system of equations (3.76), we directly find that

$$
\overline{A}_{11} = \overline{B}_{11} = \overline{C}_{11} = \overline{D}_{11} = 0,
\tag{3.79}
$$

and, hence

$$
\overline{\overline{A}}_{11} = \overline{\overline{B}}_{11} = \overline{\overline{C}}_{11} = \overline{\overline{D}}_{11} = 0.
\tag{3.80}
$$

Taking into account relations (3.74), after algebraic simplifications, the systems of equations (3.77), (3.78) are transformed to the following form
for cos 3θ:

$$
\begin{cases}
-\overline{A}_{21}\, r_0^{-6} + \overline{B}_{21}\, r_0^{-6} + \overline{C}_{21} = 0, \\[2mm]
-\lambda\overline{A}_{21}\, r_0^{-6} + \overline{B}_{21}\, r_0^{-6} - \overline{C}_{21} = 0, \\[2mm]
\overline{B}_{21}\, r_0^{+6} + \overline{C}_{21} - \overline{D}_{21} = \overline{C}_{10}\, r_0^{+2}, \\[2mm]
\overline{B}_{21}\, r_0^{+6} - \overline{C}_{21} + \tilde{\lambda}_0\overline{D}_{21} = \overline{C}_{10}\, r_0^{+2},
\end{cases}
\tag{3.81}
$$

for cos 5θ:

$$\begin{cases} -\overline{A}_{31}\, r_0^{-10} + \overline{B}_{31}\, r_0^{-10} + \overline{C}_{31} = 0, \\ -\lambda\overline{A}_{31}\, r_0^{-10} + \overline{B}_{31}\, r_0^{-10} - \overline{C}_{31} = 0, \\ \overline{B}_{31}\, r_0^{+10} + \overline{C}_{31} - \overline{D}_{31} = \overline{C}_{10}\, r_0^{+4}, \\ \overline{B}_{31}\, r_0^{+10} - \overline{C}_{31} + \tilde{\lambda}_0 \overline{D}_{31} = -\overline{C}_{10}\, r_0^{+4}. \end{cases} \tag{3.82}$$

Solution of systems of equations (3.81), (3.82) follows

$$\overline{A}_{21} = A_{21}^{*}\,\frac{\partial u_0}{\partial x} = -\left(\lambda^2 - 1\right)\frac{\pi a^2}{2}\,\Delta\Delta_1\,\frac{\partial u_0}{\partial x},$$

$$\overline{B}_{21} = B_{21}^{*}\,\frac{\partial u_0}{\partial x} = -\left(\lambda^2 - 1\right)\left(\lambda + 1\right)\frac{\pi a^2}{4}\,\Delta\Delta_1\,\frac{\partial u_0}{\partial x},$$

$$\overline{C}_{21} = C_{21}^{*}\,\frac{\partial u_0}{\partial x} = \left(\lambda^2 - 1\right)\left(\lambda - 1\right)\left(\frac{\pi a^2}{4}\right)^4\,\Delta\Delta_1\,\frac{\partial u_0}{\partial x}, \tag{3.83}$$

$$\overline{D}_{21} = D_{21}^{*}\,\frac{\partial u_0}{\partial x} = \left(\lambda^2 - 1\right)\left(\frac{\pi a^2}{4}\right)^4\,\Delta_1\,\frac{\partial u_0}{\partial x},$$

where

$$\Delta = \left((\lambda + 1) - (\lambda - 1)\left(\frac{\pi a^2}{4}\right)^2\right)^{-1},$$

$$\Delta_1 = \left((\lambda + 1)^2 - (\lambda - 1)^2\left(\frac{\pi a^2}{4}\right)^4\right)^{-1},$$

$$\overline{A}_{31} = A_{31}^{*}\,\frac{\partial u_0}{\partial x} = -2\left(\lambda - 1\right)^2\left(\frac{\pi a^2}{4}\right)^2\,\Delta\Delta_2\,\frac{\partial u_0}{\partial x},$$

$$\overline{B}_{31} = B_{31}^{*}\,\frac{\partial u_0}{\partial x} = -\left(\lambda - 1\right)^2\left(\lambda - 1\right)\left(\frac{\pi a^2}{4}\right)^2\,\Delta\Delta_2\,\frac{\partial u_0}{\partial x}, \tag{3.84}$$

$$\overline{C}_{31} = C_{31}^{*}\,\frac{\partial u_0}{\partial x} = \left(\lambda - 1\right)^3\left(\frac{\pi a^2}{4}\right)^7\,\Delta\Delta_2\,\frac{\partial u_0}{\partial x},$$

$$\overline{D}_{31} = D_{31}^{*}\,\frac{\partial u_0}{\partial x} = \left(\lambda^2 - 1\right)\frac{\pi a^2}{4}\,\Delta_2\,\frac{\partial u_0}{\partial x},$$

and

$$\Delta_2 = \left((\lambda + 1)^2 - (\lambda - 1)^2\left(\frac{\pi a^2}{4}\right)^6\right)^{-1}.$$

For the integration constants $\overline{\overline{A}}_{i1}$, $\overline{\overline{B}}_{i1}$, $\overline{\overline{C}}_{i1}$, $\overline{\overline{D}}_{i1}$ ($i = 2, 3$), we have

$$\overline{\overline{A}}_{i1} = \overline{A}_{i1}, \quad \overline{\overline{B}}_{i1} = \overline{B}_{i1}, \quad \overline{\overline{C}}_{i1} = \overline{C}_{i1}, \quad \overline{\overline{D}}_{i1} = \overline{D}_{i1} \quad \left(\frac{\partial u_0}{\partial x} \rightarrow \frac{\partial u_0}{\partial y}\right). \tag{3.85}$$

Homogenized equation of the first approximation is transformed to the following form

$$
\frac{1}{|\Omega_1^*|}\left\{ \iint\limits_{\Omega_{i0}^+} \left(\frac{\partial^2 u_0}{\partial x^2} + \frac{\partial^2 u_0}{\partial y^2} + \frac{\partial^2 u_{10}^+}{\partial x \partial \xi} + \frac{\partial^2 u_{10}^+}{\partial y \partial \eta} \right) d\xi\, d\eta + \right.
$$

$$
+\lambda \iint\limits_{\Omega_i^-} \left(\frac{\partial^2 u_0}{\partial x^2} + \frac{\partial^2 u_0}{\partial y^2} + \frac{\partial^2 u_{10}^-}{\partial x \partial \xi} + \frac{\partial^2 u_{10}^-}{\partial y \partial \eta} \right) d\xi\, d\eta +
$$

$$
+\tilde\lambda_0 \iint\limits_{\tilde\Omega_0} \left(\frac{\partial^2 u_0}{\partial x^2} + \frac{\partial^2 u_0}{\partial y^2} + \frac{\partial^2 \tilde u_{10}}{\partial x \partial \xi} + \frac{\partial^2 \tilde u_{10}}{\partial y \partial \eta} \right) d\xi\, d\eta +
$$

$$
+\varepsilon_1 \left[\iint\limits_{\Omega_{i0}^+} \left(\frac{\partial^2 u_{11}^+}{\partial x \partial \xi} + \frac{\partial^2 u_{11}^+}{\partial y \partial \eta} \right) d\xi\, d\eta + \lambda \iint\limits_{\Omega_i^-} \left(\frac{\partial^2 u_{11}^-}{\partial x \partial \xi} + \frac{\partial^2 u_{11}^-}{\partial y \partial \eta} \right) d\xi\, d\eta + \right.
$$

$$
+\tilde\lambda_0 \iint\limits_{\tilde\Omega_0} \left(\frac{\partial^2 \tilde u_{11}}{\partial x \partial \xi} + \frac{\partial^2 \tilde u_{11}}{\partial y \partial \eta} \right) d\xi\, d\eta + \tag{3.86}
$$

$$
+ \underline{\iint\limits_{\Omega_{i1}^+ \setminus \Omega_{i0}^+} \left(\frac{\partial^2 u_0}{\partial x^2} + \frac{\partial^2 u_0}{\partial y^2} + \frac{\partial^2 u_{10}^+}{\partial x \partial \xi} + \frac{\partial^2 u_{10}^+}{\partial y \partial \eta} \right) d\xi\, d\eta} +
$$

$$
+\underline{\tilde\lambda_0 \iint\limits_{\tilde\Omega_1 \setminus \tilde\Omega_0} \left(\frac{\partial^2 u_0}{\partial x^2} + \frac{\partial^2 u_0}{\partial y^2} + \frac{\partial^2 \tilde u_{10}}{\partial x \partial \xi} + \frac{\partial^2 \tilde u_{10}}{\partial y \partial \eta} \right) d\xi\, d\eta} +
$$

$$
\left. \left. + \tilde\lambda_1 \iint\limits_{\tilde\Omega_0} \left(\frac{\partial^2 u_0}{\partial x^2} + \frac{\partial^2 u_0}{\partial y^2} + \frac{\partial^2 \tilde u_{10}}{\partial x \partial \xi} + \frac{\partial^2 \tilde u_{10}}{\partial y \partial \eta} \right) d\xi\, d\eta \right] \right\} =
$$

$$
= \left(\frac{\partial^2 u_0}{\partial x^2} + \frac{\partial^2 u_0}{\partial y^2} \right) \left(\tilde\lambda_0 + \varepsilon_1 \tilde\lambda_1 \right),
$$

where $|\Omega_1^*| = |\Omega_{i1}^+ \cup \Omega_i^- \cup \tilde\Omega_1|$ stands for a measure of three-phase area.

In relation (3.86), the underlined terms determine the correction to the homogenized coefficient due to the perturbation of contour $r = r_0^+$, i.e., the correction determined by the order term ε_1. "Perturbed" regions of integration $\Omega_{i1}^+ \setminus \Omega_{i0}^+$, $\tilde\Omega_1 \setminus \tilde\Omega_0$ are shown in Fig. 3.11, while the signs "+" and "-" mark positive and negative additions to the corresponding areas.

Fig. 3.12 presents the contour of the cell in the zeroth approximation and the curves that determine the correction to it in the first approximation (correction of the order ε_1). Equating the expressions at ε_1 in (3.86), we obtain the relation for the first

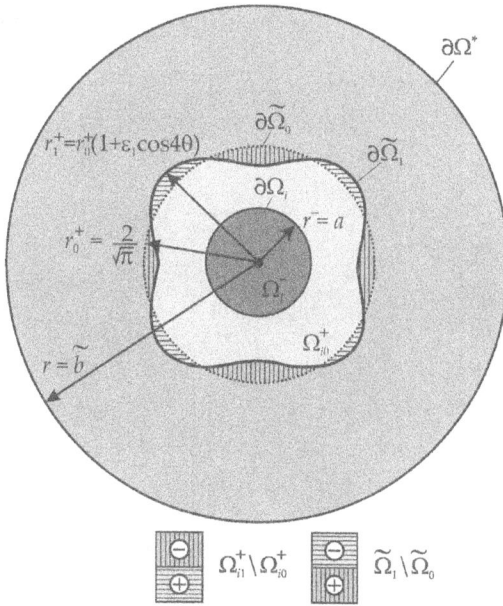

Figure 3.11 Modification of the cell area taking into account the first approximation.

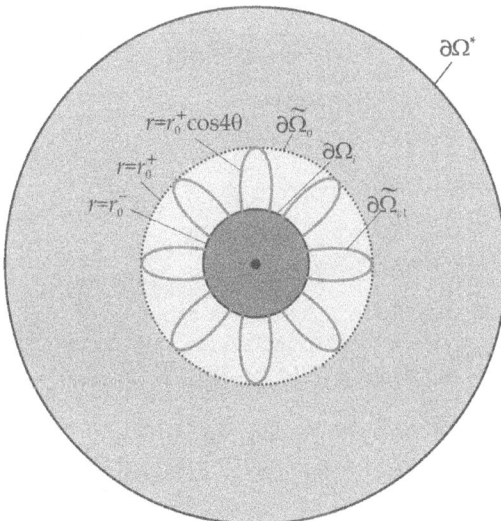

Figure 3.12 Cell boundary in the zeroth approximation and the first correction to it.

correction $\tilde{\lambda}_1$ to the effectivs thermal conductivity coefficient in the following form

$$\tilde{\lambda}_1 \Delta u_0 = \iint\limits_{\Omega_{i0}^+} \left(\frac{\partial^2 u_{11}^+}{\partial x \partial \xi} + \frac{\partial^2 u_{11}^+}{\partial y \partial \eta} \right) d\xi \, d\eta + \lambda \iint\limits_{\Omega_i^-} \left(\frac{\partial^2 u_{11}^-}{\partial x \partial \xi} + \frac{\partial^2 u_{11}^-}{\partial y \partial \eta} \right) d\xi \, d\eta +$$

$$+ \tilde{\lambda}_0 \iint\limits_{\tilde{\Omega}_0} \left(\frac{\partial^2 \tilde{u}_{11}}{\partial x \partial \xi} + \frac{\partial^2 \tilde{u}_{11}}{\partial y \partial \eta} \right) d\xi \, d\eta +$$

$$+ \iint\limits_{\Omega_{i1}^+ \backslash \Omega_{i0}^+} \left(\frac{\partial^2 u_0}{\partial x^2} + \frac{\partial^2 u_0}{\partial y^2} + \frac{\partial^2 u_{10}^+}{\partial x \partial \xi} + \frac{\partial^2 u_{10}^+}{\partial y \partial \eta} \right) d\xi \, d\eta +$$

$$+ \tilde{\lambda}_0 \iint\limits_{\tilde{\Omega}_1 \backslash \tilde{\Omega}_0} \left(\frac{\partial^2 u_0}{\partial x^2} + \frac{\partial^2 u_0}{\partial y^2} + \frac{\partial^2 \tilde{u}_{10}}{\partial x \partial \xi} + \frac{\partial^2 \tilde{u}_{10}}{\partial y \partial \eta} \right) d\xi \, d\eta +$$

$$+ \tilde{\lambda}_1 \iint\limits_{\tilde{\Omega}_0} \left(\frac{\partial^2 u_0}{\partial x^2} + \frac{\partial^2 u_0}{\partial y^2} + \frac{\partial^2 \tilde{u}_{10}}{\partial x \partial \xi} + \frac{\partial^2 \tilde{u}_{10}}{\partial y \partial \eta} \right) d\xi \, d\eta,$$

or after obvious transformations

$$\tilde{\lambda}_1 \Delta u_0 = \frac{1}{4b^2} \left[\iint\limits_{\Omega_{i1}^+ \backslash \Omega_{i0}^+} \Delta u_0 \, d\xi \, d\eta + \tilde{\lambda}_0 \iint\limits_{\tilde{\Omega}_1 \backslash \tilde{\Omega}_0} \Delta u_0 \, d\xi \, d\eta + \oint\limits_{\partial \tilde{\Omega}_{\varepsilon_1}} \frac{\partial u_{10}^+}{\partial n} \, d\ell + \right.$$

$$+ \tilde{\lambda}_0 \oint\limits_{\partial \tilde{\Omega}_{\varepsilon_1}} \frac{\partial \tilde{u}_{10}}{\partial n} \, d\ell + \oint\limits_{\partial \tilde{\Omega}_0} \frac{\partial u_{11}^+}{\partial n} \, d\ell + \oint\limits_{\partial \Omega_{i0}} \frac{\partial u_{11}^+}{\partial n} \, d\ell + \qquad (3.87)$$

$$\left. + \lambda \oint\limits_{\partial \Omega_{i0}} \frac{\partial u_{11}^-}{\partial n} \, d\ell + \tilde{\lambda}_0 \oint\limits_{\partial \tilde{\Omega}_0} \frac{\partial \tilde{u}_{11}}{\partial n} \, d\ell \right],$$

where $\frac{\partial(\ldots)}{\partial \mathbf{n}} = \frac{\partial(\ldots)}{\partial x} \cos(\mathbf{n}, \xi) + \frac{\partial(\ldots)}{\partial y} \cos(\mathbf{n}, \eta)$; $\mathbf{n}$ is the outer normal to the contour, written in "slow" variables x, y.

Performing integration in (3.87), and taking into account the expressions for u_{10}^+, u_{10}^-, $\tilde{u}_{10}$ (3.73), (3.74) and u_{11}^+, u_{11}^-, $\tilde{u}_{11}$ (3.75), (3.83)-(3.85), we have

$$\tilde{\lambda}_1 = -\frac{1}{4} \int\limits_0^{2\pi} d\theta \int\limits_{\frac{2}{\sqrt{\pi}}}^{\frac{2}{\sqrt{\pi}}(1+\varepsilon_1 \cos 4\theta)} \left(1 + B_{10}^* - \tilde{\lambda}_0 \right) r \, dr = -\frac{\varepsilon_1}{2} \left(1 + B_{10}^* - \tilde{\lambda}_0 \right),$$

i.e.

$$\tilde{\lambda}_1 = \frac{\varepsilon_1}{2} \cdot \frac{(\lambda - 1) \frac{\pi a^2}{4}}{1 + \frac{\pi a^2}{4} + \lambda \left(1 - \frac{\pi a^2}{4} \right)} \qquad (3.88)$$

or

$$\tilde{\lambda}_1 = \frac{\varepsilon_1}{4} \cdot \left(\tilde{\lambda}_0 - 1\right). \tag{3.89}$$

Thus, the perturbation of the cell contour of order ε_1 contributes to the homogenized coefficint of the order ε_1^2. This means that in the case of a composite with periodic cylindrical inclusions of a circular profile, the first corrector obtained on the basis of ThPhM is equal to zero up to terms of the order of ε_1:

$$\tilde{\lambda}_1 = 0.$$

Therefore, taking into account the solution of the first approximation, the effective thermal conductivity coefficient $q^{(1)}_{ThPhM}$ is determined by the relation:

$$q^{(1)}_{ThPhM} = \frac{1 - \frac{\pi a^2}{4} + \lambda \left(1 + \frac{\pi a^2}{4}\right)}{1 + \frac{\pi a^2}{4} + \lambda \left(1 - \frac{\pi a^2}{4}\right)} + \varepsilon_1^2 \frac{1}{2} \cdot \frac{(\lambda - 1)\frac{\pi a^2}{4}}{1 + \frac{\pi a^2}{4} + \lambda \left(1 - \frac{\pi a^2}{4}\right)} + O\left(\varepsilon_1^2\right) + ..., \tag{3.90}$$

or

$$q^{(1)}_{ThPhM} = q_{ThPhM} + \varepsilon_1^2 \frac{q_{ThPhM} - 1}{4} + O\left(\varepsilon_1^2\right) + ..., \tag{3.91}$$

where q_{ThPhM} is zero approximation solution (3.58) coinciding with the Maxwell formula.

Note that the solution (3.58), obtained on the basis of ThPhM in the zeroth approximation of PPB, identically coincides:

(i) with the lower Hashin-Shtrikman bound for $1 \leq \lambda < \infty$;
(ii) with the upper Hashin-Shtrikman bound for $0 \leq \lambda \leq 1$.

The Hashin-Shtrikman bounds $\underline{q}_{HS}$, $\bar{q}_{HS}$ for the problem under consideration are written as:

for $1 \leq \lambda < \infty$:

$$\frac{1 - \frac{\pi a^2}{4} + \lambda \left(1 + \frac{\pi a^2}{4}\right)}{1 + \frac{\pi a^2}{4} + \lambda \left(1 - \frac{\pi a^2}{4}\right)} = \underline{q}_{HS} \leq q \leq \bar{q}_{HS} = \lambda \frac{2 - \frac{\pi a^2}{4} + \lambda \frac{\pi a^2}{4}}{\frac{\pi a^2}{4} + \lambda \left(2 - \frac{\pi a^2}{4}\right)}, \tag{3.92}$$

for $0 \leq \lambda \leq 1$:

$$\lambda \frac{2 - \frac{\pi a^2}{4} + \lambda \frac{\pi a^2}{4}}{\frac{\pi a^2}{4} + \lambda \left(2 - \frac{\pi a^2}{4}\right)} = \underline{q}_{HS} \leq q \leq \bar{q}_{HS} = \frac{1 - \frac{\pi a^2}{4} + \lambda \left(1 + \frac{\pi a^2}{4}\right)}{1 + \frac{\pi a^2}{4} + \lambda \left(1 - \frac{\pi a^2}{4}\right)}. \tag{3.93}$$

Therefore, we have

$$\underline{q}_{HS} = q_{ThPhM} \leq \bar{q}_{HS} \quad for \quad 1 \leq \lambda < \infty,$$

$$\underline{q}_{HS} \leq q_{ThPhM} = \bar{q}_{HS} \quad for \quad 0 \leq \lambda \leq 1.$$

Thus, the solution of the first approximation of PPB in ThPhM falls into the Hashin-Shtrikman bounds for any values $0 \leq \lambda < \infty$:

$$\underline{q}_{HS} \leq q^{(1)}_{ThPhM} \leq \bar{q}_{HS} \quad for \quad 0 \leq \lambda < \infty.$$

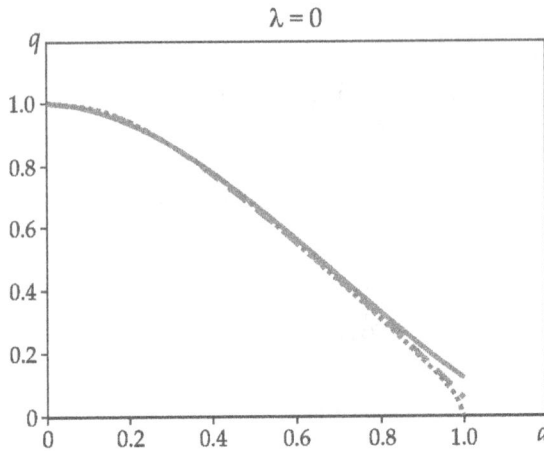

Figure 3.13 Effective coefficient of thermal conductivity for non-conductive inclusions ($\lambda = 0$) (PPB – solid line, MThPhM [13] – dotted line, asymptotic [165] – dashed line).

3.2.7 NUMERICAL ANALYSIS

Figures 3.13–3.18 show graphs of the effective thermal conductivity coefficient found in accordance with PPB for various values of the geometrical parameter of the composite. For comparison, the same graphs present the results of calculating the effective thermal conductivity parameter according to data obtained by other authors and based on other models:

   (i) modified ThPhM (MThPhM) [13];
  (ii) asymptotic solution for non-conducting inclusions [165];
 (iii) asymptotic solutions for inclusions of high conductivity and absolutely conducting inclusions [158];
 (iv) the expression for the effective thermal conductivity coefficient obtained using AP for absolutely conducting inclusions [119];
  (v) expression for the effective thermal conductivity coefficient of a composite with ideally conducting cylindrical fibers [119];
 (vi) numerical solutions [194].

Figures 3.13–3.17 show the dependence of the effective thermal conductivity coefficient q on the size of inclusions a at a fixed conductivity of inclusions λ, corresponding to:

   (i) $\lambda = 0$ – non-conductive inclusions (Fig. 3.13);
  (ii) $\lambda = 10^{-2}$, $\lambda = 10^{-1}$ – inclusions on low conductivity $\lambda \ll 1$ (Fig. 3.14);
 (iii) $\lambda = 0.5$, $\lambda = 2$ – the conductivity of the inclusions is close to the conductivity of the matrix $\lambda \sim 1$ (Fig. 3.15);
 (iv) $\lambda = 10$, $\lambda = 10^{2}$ – inclusion of high conductivity (Fig. 3.16);
  (v) $\lambda \to \infty$ – absolutely conducting inclusions (Fig. 3.17).

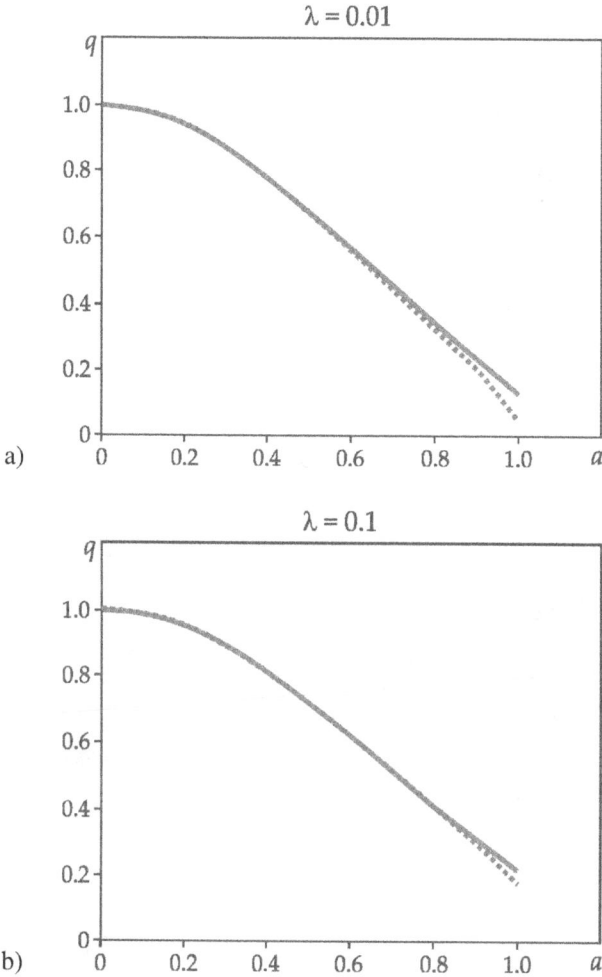

Figure 3.14 Effective thermal conductivity for inclusions of low conductivity ($\lambda = 10^{-2}$ (a), $\lambda = 10^{-1}$ (b)) (PPB – solid line, MThPhM [13] – dotted line).

Fig. 3.18 illustrates the dependence of the effective thermal conductivity coefficient q on the conductivity of inclusions λ for a fixed size of inclusions a, corresponding to:

(i) $a = 0.3568$ (concentration of inclusions: $c = \pi a^2/4 = 0.1$) – small inclusions (Fig. 3.18a);

(ii) $a = 0.5046$ ($c = 0.2$) – inclusions of medium size (Fig. 3.18b);

(iii) $a = 0.7979$ ($c = 0.5$) – large inclusions (Fig. 3.18c).

Comparison of the calculation results shown in Figs. 3.13–3.18 yields the following conclusions:

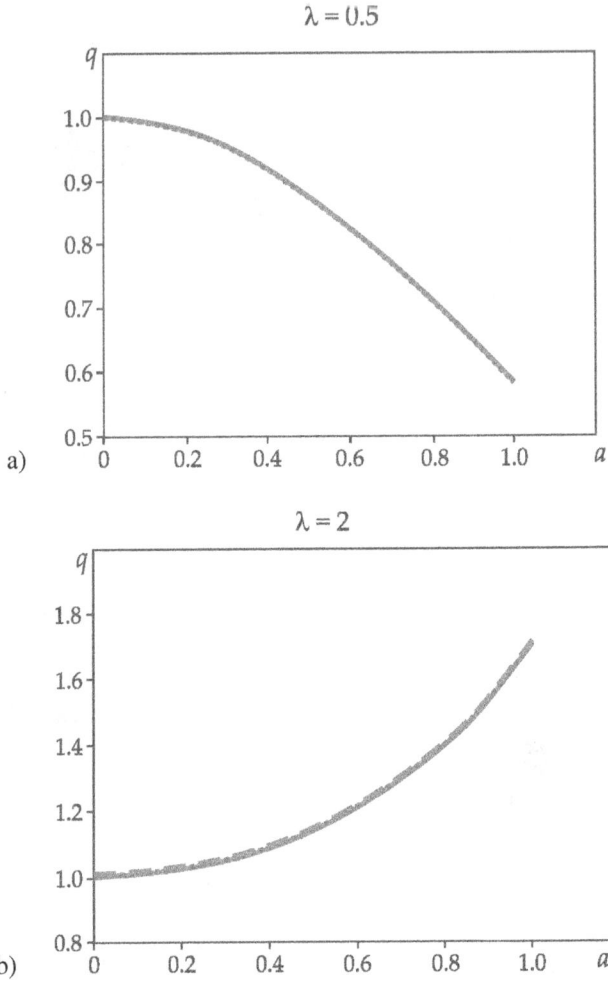

Figure 3.15 Effective thermal conductivity coefficient for the conductivity of inclusions close to the conductivity of the matrix ($\lambda = 0.5$ (a), $\lambda = 2$ (b)) (PPB – solid line, MThPhM [13] – dotted line, numerical solution [194] – dashed line).

1) Formula (3.90), obtained on the basis of ThPhM in the first-order PPB approximation, correctly describes the behavior of a homogenized medium:

   (i) in the case of conductivities of inclusions and a matrix of the same order ($\lambda \sim 1$) for inclusions of any size $0 \leq a \leq 1$;

  (ii) if the conductivity of the inclusions differs significantly from the conductivity of the matrix, i.e. if $0 \leq \lambda \ll 1$ or $1 \ll \lambda < \infty$, $\lambda \to \infty$, for inclusion sizes $0 \leq a \leq 0.8$.

2) For non-conductive fibers ($\lambda = 0$) and $a = 0.8$ relative error δ solution (3.90) in comparison with the results of [13] and [165] is $2.7\% - 5.9\%$. For

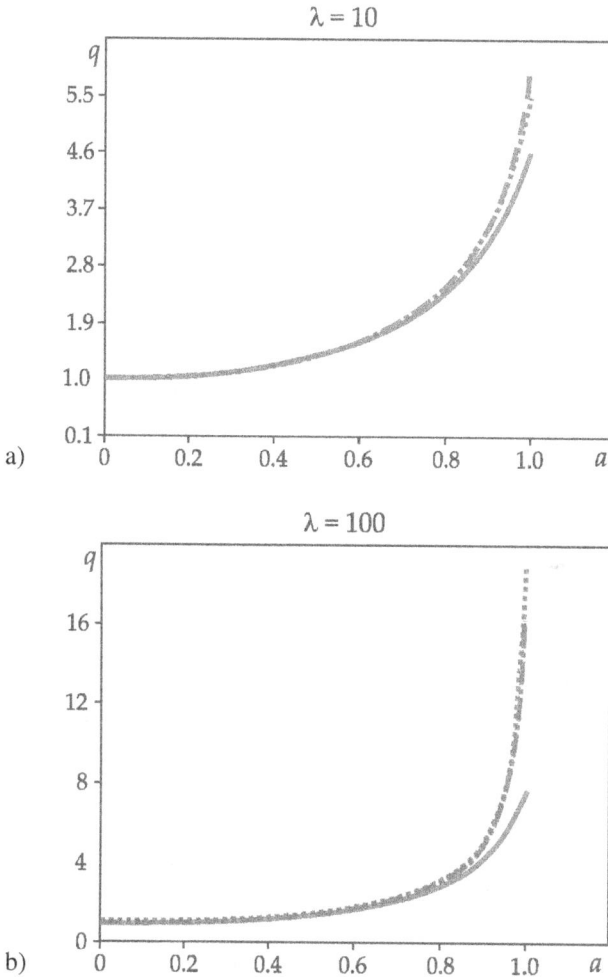

Figure 3.16 Effective thermal conductivity coefficient for highly conductive inclusions ($\lambda = 10$ (a), $\lambda = 100$ (b)) (PPB – solid line, MThPhM [13] – dotted line, numerical solution [194] – dashed line, asymptotics $\lambda \gg 1$ [158] – pointed line).

absolutely conductive fibers ($\lambda \to \infty$) compared to results [13, 119, 158, 194] $\delta = 2.6\%$-5.4%.

3) Relation (3.90) does not properly describe the properties of a composite with high conductivity of inclusions ($\lambda \to \infty$) and their large sizes ($a \to 1$):

  (i) from a physical point of view, this means that equation (3.90) does not give a quantitative description of the percolation threshold, or even a qualitative picture of the phenomena occurring in the composite (the formation of an infinite cluster);

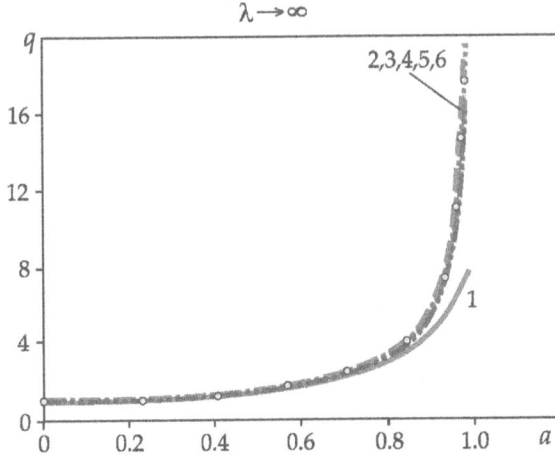

Figure 3.17 Effective thermal conductivity coefficient for absolutely conductive inclusions ($\lambda \to \infty$) (1- PPB, 2- MThPhM [13], 3- numerical solution [194], 4- asymptotics $\lambda \to \infty$ [158], 5- formula (3.13) [119], 6- PA [119]).

(ii)   mathematically, this leads to a discrepancy between the asymptotic relation obtained from equation (3.90) with $a \to 1$, $\lambda \to \infty$, and the expression (3.68) [158];

(iii) by virtue of the Keller's theorem [127], a similar conclusion is also applicable in the case of inclusions of large size ($a \to 1$) with very low conductivity ($\lambda \to 0$).

In order to illustrate the qualitative nature of the dependence on two parameters (inclusion size a and conductivity λ) of the effective thermal conductivity coefficient q obtained by PPB, as well as to visualize the limits of applicability of the obtained solution, in Fig. 3.19, 3D graphs are constructed: according to the formula (3.90) and to the MThPhM.

3.2.8   THPHM: CONCLUSIONS AND PHYSICAL INTERPRETATION OF RESULTS

1. In the case of a composite with cylindrical inclusions of a circular profile, the use of the ThPhM, PPB, and the homogenization theory gives in the zero approximation an expression for the effective thermal conductivity, which coincides with the Maxwell approximation.
   The relation obtained adequately describes the composite at small sizes of inclusions and coincides in the limiting case $a \to 0$, $\lambda \to 0$ with known result [26].
2. For large inclusions ($a \to 1$), ThPhM gives acceptable results only if the conductivity of the inclusions is close to that of the matrix: $\lambda \sim 1$.

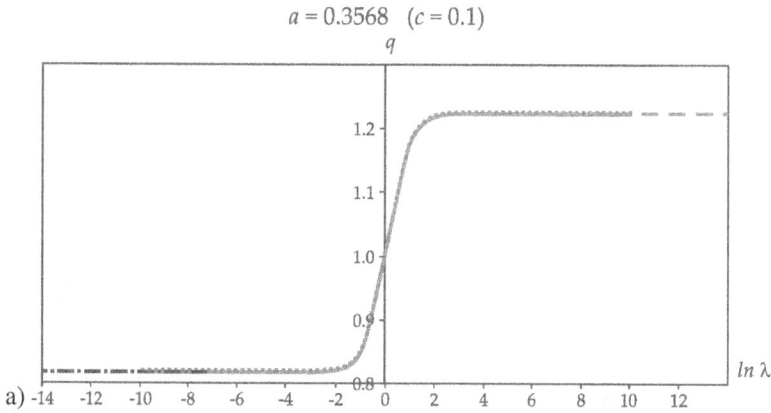

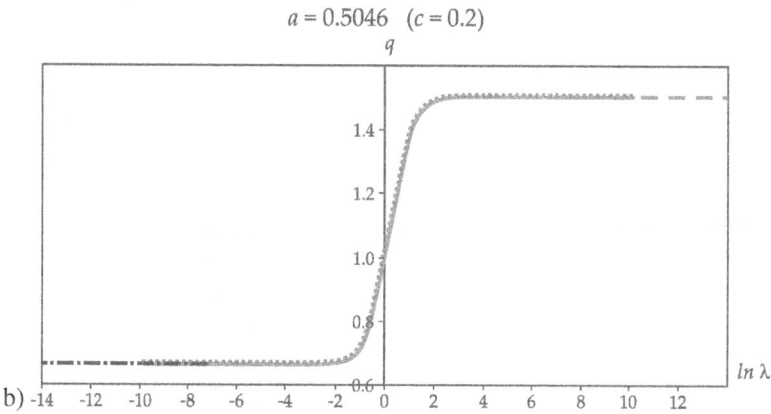

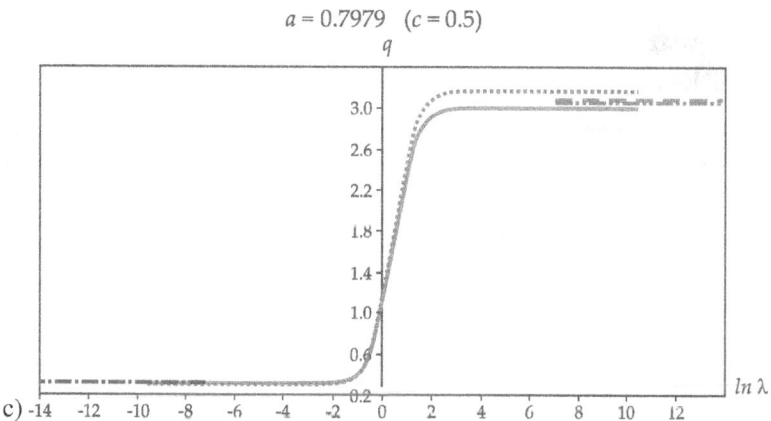

Figure 3.18 Effective coefficient of thermal conductivity in the case of: a) small sizes of inclusions ($a = 0.3568$), b) medium size of inclusions ($a = 0.5046$), c) large size of inclusions ($a = 0.7979$) (PPB – solid line, MThPhM [13] – dotted line, numerical solution $\lambda \to \infty$ [194] – dashed line, asymptotics $\lambda = 0$ [165] – pointed line, asymptotics $\lambda \to \infty$ [158] – two-pointed line).

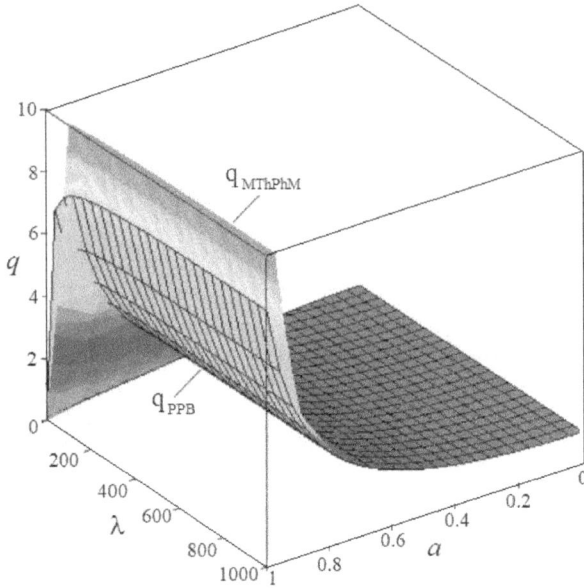

Figure 3.19 Solution $q_{PPB}(a, \lambda)$ (3.90) and illustration of its applicability.

3. ThPhM cannot be used, even qualitatively, for large inclusion sizes $(a \to 1)$ with extremely large $(\lambda \to \infty)$ or extremely small $(\lambda \to 0)$ conductivity.

A small perturbation of the cell contour of order ε_1 leads to corrections in the homogenized relations of the order ε_1^2.

3.3 GENERALIZATION OF A THREE-PHASE COMPOSITE MODEL USING PADÉ APPROXIMATIONS

We consider the use of AP to determine the effective thermal conductivity of a composite material with circular cylindrical inclusions in cases of extreme values of its physical and / or geometric characteristics.

If function $f(z)$ can be represented as a power series:

$$f(z) = \sum_{i=0}^{\infty} c_i z^i,\tag{3.94}$$

then AP [30, 31] is called a rational function of the form

$$f_{[L/M]}(z) = \frac{a_0 + a_1 z + \ldots + a_L z^L}{b_0 + b_1 z + \ldots + b_M z^M},$$

whose expansion into the Maclaurin series coincides with the expansion (3.94) as exact as it is possible.

3.3.1 AP FOR A THPHM SOLUTION

From formula (3.58), one obtains:

$$q^{(\infty)} = \frac{1 + \frac{\pi a^2}{4}}{1 - \frac{\pi a^2}{4}} \quad \text{at} \quad \lambda \to \infty, \tag{3.95}$$

$$q^{(0)} = \frac{1 - \frac{\pi a^2}{4}}{1 + \frac{\pi a^2}{4}} \quad \text{at} \quad \lambda \to 0. \tag{3.96}$$

1) Using expression (3.95), we construct the sequence of AP $q^{(\infty)}_{[0,2]}$, $q^{(\infty)}_{[0,6]}$, ..., $q^{(\infty)}_{[0,18]}$:

$$q^{(\infty)}_{[0/2]}(a) = \left(1 - \frac{\pi a^2}{2}\right)^{-1},$$

$$q^{(\infty)}_{[0/6]}(a) = \left\{\left[q_0^{(\infty)}_{[0/2]}(a)\right]^{-1} + 2\left(1 - \frac{\pi a^2}{4}\right)\cdot\left(\frac{\pi a^2}{4}\right)^2\right\}^{-1},$$

$$q^{(\infty)}_{[0/10]}(a) = \left\{\left[q_0^{(\infty)}_{[0/6]}(a)\right]^{-1} + 2\left(1 - \frac{\pi a^2}{4}\right)\cdot\left(\frac{\pi a^2}{4}\right)^4\right\}^{-1},$$

$$\tag{3.97}$$

$$q^{(\infty)}_{[0/14]}(a) = \left\{\left[q_0^{(\infty)}_{[0/10]}(a)\right]^{-1} + 2\left(1 - \frac{\pi a^2}{4}\right)\cdot\left(\frac{\pi a^2}{4}\right)^6\right\}^{-1},$$

$$q^{(\infty)}_{[0/18]}(a) = \left\{\left[q_0^{(\infty)}_{[0/14]}(a)\right]^{-1} + 2\left(1 - \frac{\pi a^2}{4}\right)\cdot\left(\frac{\pi a^2}{4}\right)^8\right\}^{-1}.$$

In Fig. 3.20 for the case of absolutely conductive inclusions graphs of the effective thermal conductivity coefficient are shown, calculated using:

(i) ThPhM in the first approximation of PPB, limited to terms of order ε_1 inclusive (expression (3.95));
(ii) the asymptotic representation obtained in [158] (formula (3.68));
(iii) sequences AP $q^{(\infty)}_{[0/2]}$, $q^{(\infty)}_{[0/6]}$,...., $q^{(\infty)}_{[0/18]}$ (ratios (3.97)).

Analysis of expressions (3.97) and Fig. 3.20 shows that for absolutely conductive inclusions, the sequence of APs can be treated as recursive representations of the effective thermal conductivity parameter, moreover:

(i) at small sizes of inclusions $q^{(\infty)}_{[0/M]}(a)$, $M = 2$, 6, ... , 18 coincide with the series expansion of expression (3.95) for $a \to 0$ up to order terms a^2, a^6, ... , a^{18} inclusive. For $q^{(\infty)}_{[0/18]}(a)$, we have

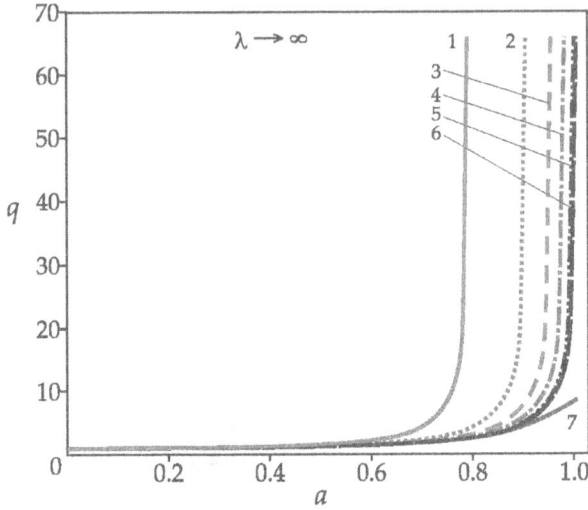

Figure 3.20 The effective thermal conductivity parameter for the ThPhM solution in the case of absolutely conducting inclusions $(1 - q^{(\infty)}_{[0/2]}(a), 2 - q^{(\infty)}_{[0/6]}(a), 3 - q^{(\infty)}_{[0/10]}(a), 4 - q^{(\infty)}_{[0/14]}(a), 5 - q^{(\infty)}_{[0/18]}(a), 6$ – Asymptotic [158], 7 – ThPhM).

$$q^{(\infty)}{}_{[0/18]}(a) \sim q^{(\infty)} \sim 1 + \frac{\pi a^2}{2} + \frac{\pi^2 a^4}{2^3} + \frac{\pi^3 a^6}{2^5} + \frac{\pi^4 a^8}{2^7} + \frac{\pi^5 a^{10}}{2^9} +$$

$$+ \frac{\pi^6 a^{12}}{2^{11}} + \frac{\pi^7 a^{14}}{2^{13}} + \frac{\pi^8 a^{16}}{2^{15}} + \frac{\pi^9 a^{18}}{2^{17}} + O\left(a^{20}\right) \quad \text{at} \quad a \to 0;$$

(ii) applicability of the ratios $q^{(\infty)}{}_{[0/M]}(a)$, $M = 2, 6, \ldots , 18$ consistently increase from "not very small" inclusions up to almost $a \to 1$;

(iii) AP $q^{(\infty)}{}_{[0/18]}(a)$ reliably describes the effective coefficient for small sizes of inclusions, and close to asymptotic estimates of the effective coefficient for the size of the inclusion near to 1.

It should be noted that a further increase in the order of AP is inexpedient, since starting from $q^{(\infty)}{}_{[0/22]}(a)$, the AP sequence diverges.

In a similar way, we reconstruct expression (3.95) in AP with respect to the inclusion size a and find the sequence AP of the form $q^{(0)}{}_{[2/0]}(a)$, $q^{(0)}{}_{[6/0]}(a)$, ... $q^{(0)}{}_{[18/0]}(a)$:

$$q^{(0)}{}_{[2/0]}(a) = 1 - \frac{\pi a^2}{2},$$

$$q^{(0)}{}_{[6/0]}(a) = q^{(0)}{}_{[2/0]}(a) + 2\left(1 - \frac{\pi a^2}{4}\right) \cdot \left(\frac{\pi a^2}{4}\right)^2,$$

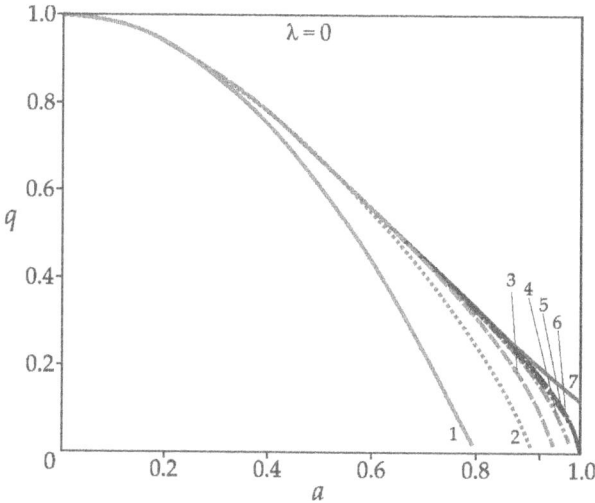

Figure 3.21 Effective thermal conductivity parameter for the ThPhM solution in the case of non-conductive inclusions $(1- q^{(\infty)}_{[2/0]}(a), 2 - q^{(\infty)}_{[6/0]}(a), 3 - q^{(\infty)}_{[10/0]}(a), 4 - q^{(\infty)}_{[14/0]}(a), 5 - q^{(\infty)}_{[18/0]}(a), 6 -$ Asymptotic [127, 158], 7 – ThPhM).

$$q^{(0)}_{[10/0]}(a) = q^{(0)}_{[6/0]}(a) + 2 \left(1 - \frac{\pi a^2}{4}\right) \cdot \left(\frac{\pi a^2}{4}\right)^4, \qquad (3.98)$$

$$q^{(0)}_{[14/0]}(a) = q^{(0)}_{[10/0]}(a) + 2 \left(1 - \frac{\pi a^2}{4}\right) \cdot \left(\frac{\pi a^2}{4}\right)^6,$$

$$q^{(0)}_{[18/0]}(a) = q^{(0)}_{[14/0]}(a) + 2 \left(1 - \frac{\pi a^2}{4}\right) \cdot \left(\frac{\pi a^2}{4}\right)^8.$$

Graphs in Fig. 3.21 illustrate the effective thermal conductivity in the case of non-conductive inclusions, computed using:

(i) ThPhM (expression (3.96));
(ii) the expression (3.68) and the Keller's theorem (3.59) [127], in the form:

$$q^{(0)}_{as} = \frac{\sqrt{1 - a^2}}{\pi + (1 - \pi)\sqrt{1 - a^2}} \quad \text{at} \quad \lambda \to 0, \quad a \to 1; \qquad (3.99)$$

(iii) sequences AP $q^{(0)}_{[2/0]}(a)$, $q^{(0)}_{[6/0]}(a)$,..., $q^{(0)}_{[18/0]}(a)$ (relations (3.98)).

Thus, in the case of non-conductive inclusions, the sequence of AP $q^{(0)}_{[L/0]}(a)$, $l = 2, 6, \dots, 18$, converges to:

(i) expression (3.96) for small sizes of inclusions. AP $q^{(0)}_{[18/0]}(a)$ coincides with the series expansion of expression (3.96) for $a \to 0$ up to terms a^{18}

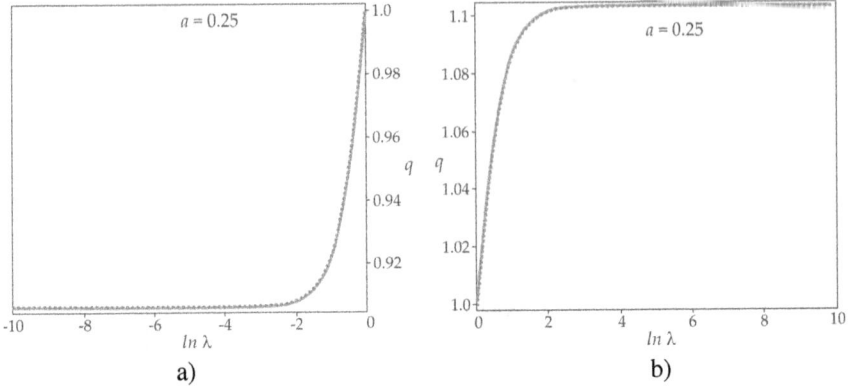

Figure 3.22 The effective thermal conductivity found for small inclusion sizes ($a = 0.25$) and: a) $0 < \lambda \leq 1$ ($q_{[18/0]}(a)$ – solid line, ThPhM – dotted line), b) $1 \leq \lambda < \infty$ ($q_{[0/18]}(a)$ – dotted line, ThPhM – solid line).

inclusive:

$$q^{(0)}{}_{[18/0]}(a) \sim q^{(0)} \sim 1 - \frac{\pi a^2}{2} + \frac{\pi^2 a^4}{2^3} - \frac{\pi^3 a^6}{2^5} + \frac{\pi^4 a^8}{2^7} - \frac{\pi^5 a^{10}}{2^9} +$$

$$+ \frac{\pi^6 a^{12}}{2^{11}} - \frac{\pi^7 a^{14}}{2^{13}} + \frac{\pi^8 a^{16}}{2^{15}} - \frac{\pi^9 a^{18}}{2^{17}} + O\left(a^{20}\right) \quad \text{for} \quad a \to 0;$$

(ii) asymptotic representation (3.99) for large inclusion sizes.

In a similar way, AP (3.58) can be constructed:

$$q_{[0/18]}(a) = \frac{(\lambda + 1)^9}{\sum\limits_{n=0}^{9} \frac{(-1)^n}{2^n} \left(\frac{\pi a^2}{2}\right)^n (\lambda + 1)^{9-n}(\lambda - 1)^n} \qquad \text{at} \quad 1 \leq \lambda < \infty, \qquad (3.100)$$

$$q_{[18/0]}(a) = \frac{\sum\limits_{n=0}^{9} \frac{1}{2^n} \left(\frac{\pi a^2}{2}\right)^n (\lambda + 1)^{9-n}(\lambda - 1)^n}{(\lambda + 1)^9} \qquad \text{at} \quad 0 \leq \lambda \leq 1. \qquad (3.101)$$

Note that in the limit $\lambda \to \infty$ or $\lambda \to 0$ expressions (3.100), (3.101) coincide with the expansions $q^{(\infty)}{}_{[0/18]}(a)$, $q^{(0)}{}_{[18/0]}(a)$ in (3.97), (3.98), obtained, respectively, for absolutely conducting and non-conducting inclusions.

Figs. 3.22–3.25 show graphs of the effective thermal conductivity as a function of the conductivity of inclusions λ for various cases of their sizes a, wherein

(i) $a = 0.25$ corresponds to small inclusions;
(ii) $a = 0.5$; $a = 0.75$ corresponds to inclusions of medium size;
(iii) $a = 0.995$ corresponds to the size of the inclusions is close to the limit value.

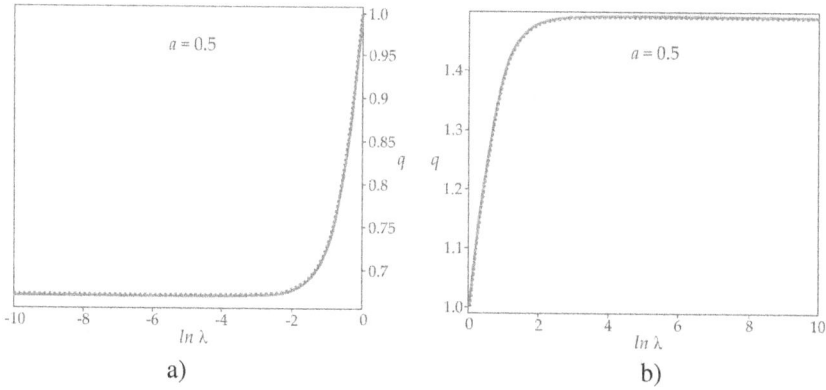

Figure 3.23   The effective thermal conductivity for medium-sized inclusions ($a = 0.5$) and: a) $0 < \lambda \leq 1$ ($q_{[18/0]}(a)$ – solid line, ThPhM – dotted line), b) $1 \leq \lambda < \infty$ ($q_{[0/18]}(a)$ – dotted line, ThPhM – solid line).

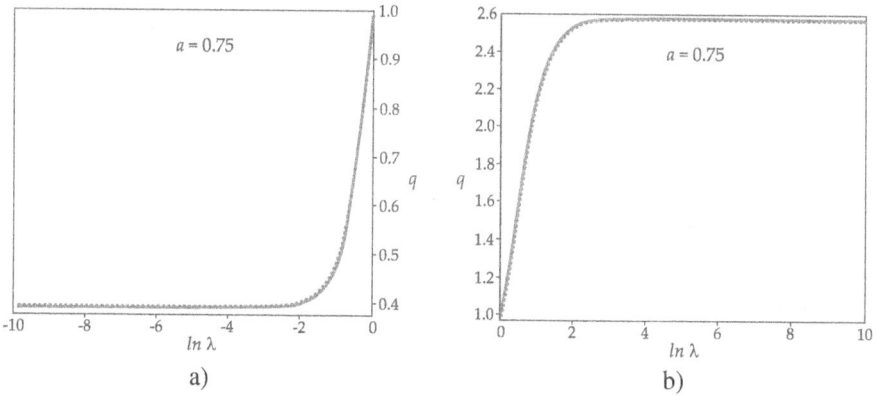

Figure 3.24   The effective thermal conductivity for the middle-sized of inclusions ($a = 0.75$) and: a) $0 < \lambda \leq 1$ ($q_{[18/0]}(a)$ – solid line, ThPhM – dotted line), b) $1 \leq \lambda < \infty$ ($q_{[0/18]}(a)$ – solid line, ThPhM – dotted line).

To compare the results, the effective parameter was calculated as a solution in terms of ThPhM (formula (3.58)) and in the form of AP (relations (3.100), (3.101)); when $a = 0.995$, the corresponding asymptotic representations (3.68) and (3.99) are also given.

3.3.2   ANALYSIS OF THE RESULTS OF NUMERICAL CALCULATIONS

Let us compare the obtained results with the known data of numerical and analytical calculations.

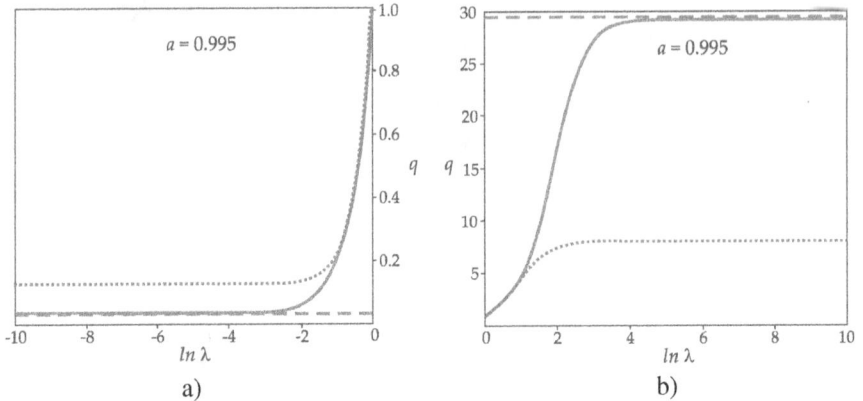

Figure 3.25 The effective thermal conductivity and asymptotic representations [126, 158] for extremely large inclusions ($a = 0.995$) and: a) $0 < \lambda \leq 1$ ($q_{[18/0]}(a)$ – solid line, ThPhM – dotted line, asymptotic – dashed line), b) $1 \leq \lambda < \infty$ ($q_{[0/18]}(a)$ – solid line, ThPhM – dotted line, asymptotic – dashed line).

1) Solutions (3.100), (3.101), found with the help of AP, fall into the Hashin-Shtrikman bounds (3.92), (3.93):

$$\underline{q}_{HS} \leq q_{[0/18]}(a) \leq \bar{q}_{HS} \quad \text{for} \quad 1 \leq \lambda < \infty,$$

$$\underline{q}_{HS} \leq q_{[18/0]}(a) \leq \bar{q}_{HS} \quad \text{for} \quad 0 \leq \lambda \leq 1.$$

Solutions $q_{[0/18]}(a)$, $q_{[18/0]}(a)$, and Hashin-Shtrikman bounds for different values of the conductivity of the inclusions λ are illustrated in Figs. 3.26–3.27.

2) For small inclusion sizes, an expression for the effective thermal conductivity was obtained using the Schwarz alternating method. Table 3.1 shows the calculation data of the effective coefficient by ThPhM (relation (3.58)), using AP (expressions (3.100), (3.101)) and by the Schwarz alternating method for various cases of inclusion conductivity λ and small sizes of inclusions.

Thus, for small sizes of inclusions:

(i) the values of the effective thermal conductivity calculated from ThPhM practically coincide with its values found using AP $q_{[18/0]}(a)$ and $q_{[0/18]}(a)$ for any conductivity of inclusions $0 \leq \lambda \leq 1$ and $1 \leq \lambda < \infty$, $\lambda \to \infty$, respectively;

(ii) the values of the effective thermal conductivity according to the Schwarz alternating method slightly differ from those found using ThPhM and AP: the maximum discrepancy between the results does not exceed 1.4%.

3) For inclusions of medium and large sizes, the papers [119, 194] present the values of the effective thermal conductivity coefficient calculated by various methods for absolutely conducting inclusions.

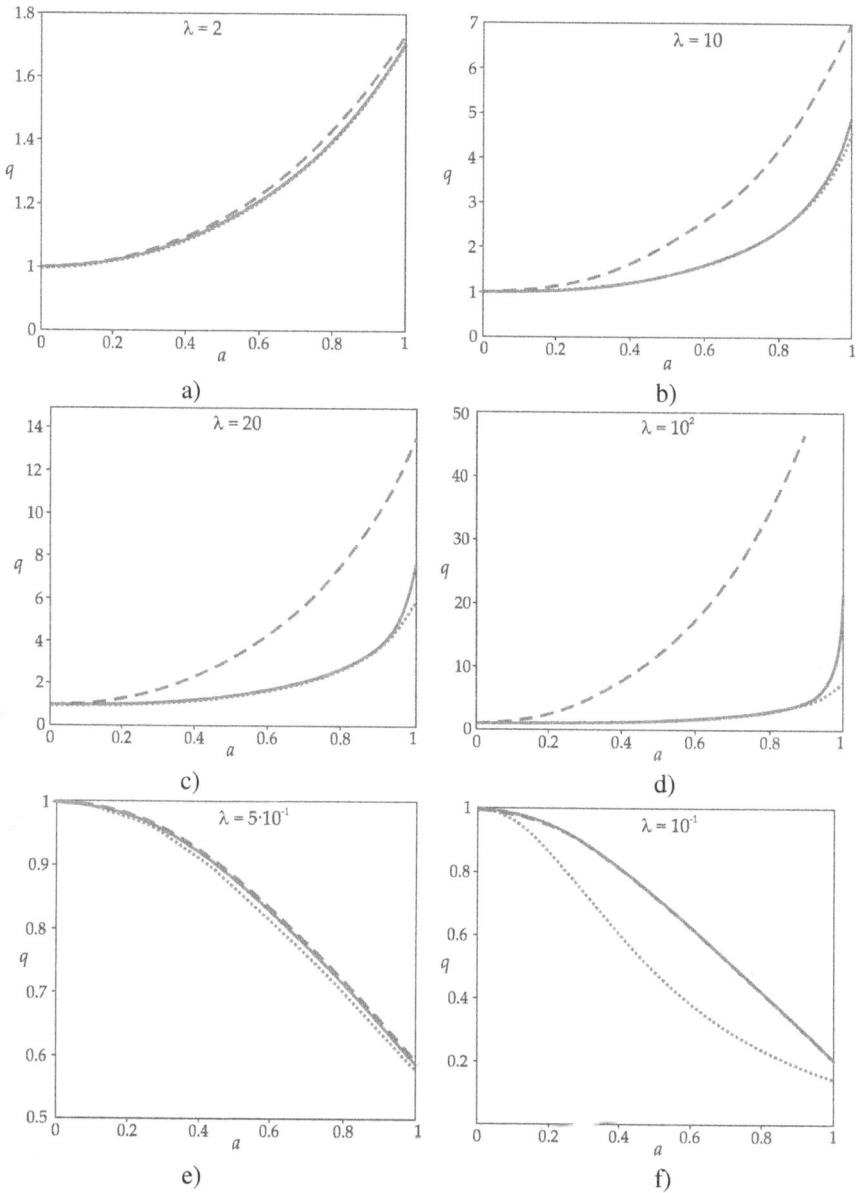

Figure 3.26 AP $q_{[0/18]}(a)$ and Hashin-Shtrikman bounds for the conductivity of inclusions: a) $\lambda = 2$; b) $\lambda = 10$; c) $\lambda = 20$, d) $\lambda = 10^2$, e) $\lambda = 5 \cdot 10^{-1}$, f) $\lambda = 10^{-1}$ ($q_{[0/18]}(a)$ – solid line, $\underline{q}_{HS}$ – dotted line, $\bar{q}_{HS}$ – dashed line).

Table 3.3 compares the calculation data of the effective coefficient found using AP $q^{(\infty)}_{[0/18]}$ (a), with known results.

Comparison of data in Table 3.3 indicates a good agreement between the results of calculations according to the AP $q^{(\infty)}_{[0/18]}$ (a) from (3.97) with

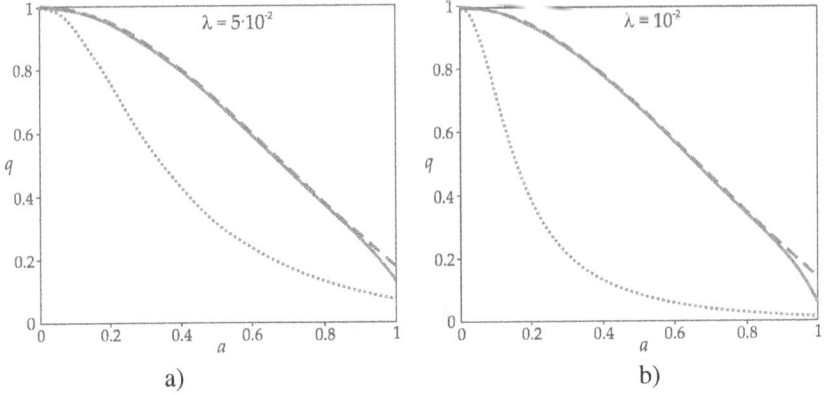

Figure 3.27 AP $q_{[18/0]}$ (a) and Hashin-Shtrikman bounds for the conductivity of inclusions $\lambda = 5 \cdot 10^{-2}$ (a) and $\lambda = 10^{-2}$ (b).

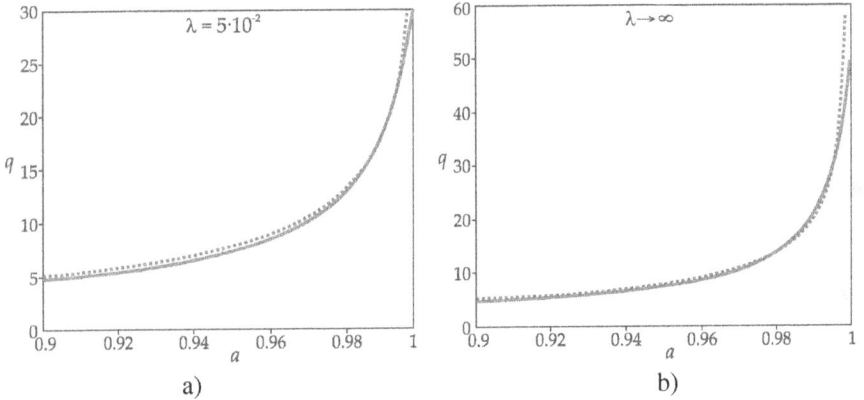

Figure 3.28 AP $q_{[0/18]}$ (a) (solid line) and the asymptotic solution [158] (dotted line) for the conductivity of inclusions: (a) $\lambda = 5 \cdot 10^2$; (b) $\lambda \to \infty$.

known results up to the inclusion size $a = 0.9966$ (concentration $c = 0.78$); in the latter case, the AP error $q^{(\infty)}_{[0/18]}$ (a) is about 7%.

4) For inclusions of large sizes, close to the limit possible ($a \to 1$) and high conductivity ($\lambda \gg 1$, including the case of absolutely conducting inclusions, $\lambda \to \infty$), in [158] an asymptotic expression for the effective thermal conductivity has been obtained.

Table 3.4 shows the calculation data of the effective coefficient according to the AP $q_{[0/18]}$ (a) in comparison with the asymptotic solution [158].

Fig. 3.28 presents graphs of the effective thermal conductivity coefficient, calculated, respectively, by the AP $q_{[0/18]}$ (a) (3.100), AP $q^{(\infty)}_{[0/18]}$ (a) (3.97), in comparison with the asymptotic solution [158].

Table 3.1

The results of calculations of the effective thermal conductivity for small sizes of inclusions.

Conductivity of inclusions $\lambda = 0$			Conductivity of inclusions $\lambda = 10^{-2}$		
Inclusion size a	AP $q_{[18/0]}$ (a)	ThPhM	Schwarz Method [26]	Inclusion size a	AP $q_{[18/0]}$ (a)	ThPhM	Schwarz Method [26]
0.05	0.9961	0.9961	0.9961	0.05	0.9962	0.9962	0.9962
0.1	0.9844	0.9844	0.9844	0.1	0.9847	0.9847	0.9847
0.15	0.9653	0.9653	0.9650	0.15	0.9659	0.9659	0.9657
0.2	0.9391	0.9391	0.9382	0.2	0.9403	0.9403	0.9394
0.25	0.9064	0.9064	0.9044	0.25	0.9082	0.9082	0.9063
0.3	0.8680	0.8680	0.8639	0.3	0.8704	0.8704	0.8666

Conductivity of inclusions $\lambda = 10^{-1}$			Conductivity of inclusions $\lambda = 5 \cdot 10^{-1}$		
Inclusion size a	AP $q_{[18/0]}$ (a)	ThPhM	Schwarz Method [26]	Inclusion size a	AP $q_{[18/0]}$ (a)	ThPhM	Schwarz Method [26]
0.05	0.9968	0.9968	0.9968	0.05	0.9987	0.9987	0.9987
0.1	0.9872	0.9872	0.9872	0.1	0.9948	0.9948	0.9948
0.15	0.9715	0.9715	0.9714	0.15	0.9883	0.9883	0.9883
0.2	0.9499	0.9499	0.9494	0.2	0.9793	0.9793	0.9794
0.25	0.9228	0.9228	0.9218	0.25	0.9678	0.9678	0.9681
0.3	0.8907	0.8907	0.8887	0.3	0.9540	0.9540	0.9546

Conductivity of inclusions $\lambda = 2$			Conductivity of inclusions $\lambda = 10$		
Inclusion size a	AP $q_{[18/0]}$ (a)	ThPhM	Schwarz Method [26]	Inclusion size a	AP $q_{[18/0]}$ (a)	ThPhM	Schwarz Method [26]
0.05	1.0013	1.0013	1.0013	0.05	1.0032	1.0032	1.0032
0.1	1.0052	1.0052	1.0052	0.1	1.0129	1.0129	1.0130
0.15	1.0119	1.0119	1.0118	0.15	1.0293	1.0293	1.0295
0.2	1.0212	1.0212	1.0210	0.2	1.0528	1.0528	1.0532
0.25	1.0333	1.0333	1.0329	0.25	1.0839	1.0839	1.0849
0.3	1.0483	1.0483	1.0475	0.3	1.1228	1.1228	1.1253

Conductivity of inclusions $\lambda = 10^2$			Conductivity of inclusions $\lambda \rightarrow \infty$		
Inclusion size a	AP $q_{[18/0]}$ (a)	ThPhM	Schwarz Method [26]	Inclusion size a	AP $q_{[18/0]}$ (a)	ThPhM	Schwarz Method [26]
0.05	1.0039	1.0039	1.0039	0.05	1.0039	1.0039	1.0039
0.1	1.0155	1.0155	1.0156	0.1	1.0158	1.0158	1.0159
0.15	1.0353	1.0353	1.0355	0.15	1.0360	1.0360	1.0363
0.2	1.0635	1.0635	1.0645	0.2	1.0649	1.0649	1.0659
0.25	1.1011	1.1011	1.1034	0.25	1.1032	1.1032	1.1057
0.3	1.1489	1.1489	1.1539	0.3	1.1521	1.1521	1.1575

Table 3.2

Comparison of the values of the effective thermal conductivity for the medium-sized inclusions.

Conductivity of inclusion a	Inclusion concentration c	The size of inclusions a	FEM [60]	AP $q_{[0/18]}$ (a)	ThPhM
$\lambda = 10$	0.25	0.5642	1.516	1.5143	1.5143

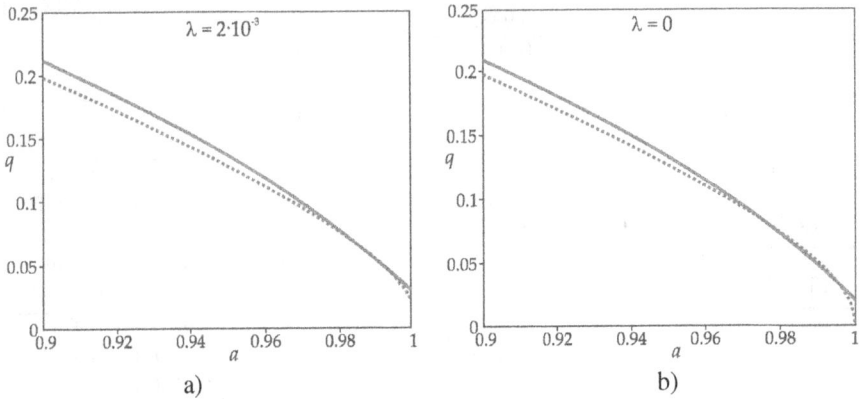

a)   b)

Figure 3.29 AP $q_{[18/0]}$ (a) (solid line) and the asymptotic solution [158] (dotted line) for the conductivity of inclusions: (a) $\lambda = 2 \cdot 10^{-3}$; (b) $\lambda = 0$.

As shown in Fig. 3.29, graphs of the thermal conductivity coefficient are calculated by the AP $q_{[18/0]}$ (a) (3.101) and AP $q^{(0)}_{[18/0]}$ (a) (3.98); the asymptotic relation was obtained from results of [158] using Keller's theorem.

3.3.3 USING THE AP TO IMPROVE THE THPHM

The analysis of the obtained results and their comparison with known data leads to the following conclusions:

1. The ThPhM solution can be transform to the AP as follows:
   (i)   AP $q_{[18/0]}$ (a) for the conductivity of inclusions $0 \leq \lambda \leq 1$;
   (ii) AP $q_{[0/18]}$ (a) for the conductivity $1 \leq \lambda < \infty$, including the limiting case $\lambda \to \infty$.
2. The obtained AP:
   (i)   reliably describe the effective coefficient of thermal conductivity for small sizes of inclusions;

Table 3.3

The results of calculations of the effective coefficient of thermal conductivity for absolutely conductive inclusions of medium and large size.

Concentration inclusions c	The size inclusions a	Formula (3.9) [119]	Formula (3.12) [119]	Formula (3.13) [119]	Numerical results [194]	AP $q_{[0/18]}^{(\infty)}(a)$
0.1	0.3568	1.210	1.247	1.223	1.222	1.2222
0.2	0.5046	1.470	1.544	1.506	1.500	1.5000
0.3	0.6180	1.811	1.918	1.879	1.860	1.8572
0.4	0.7136	2.306	2.417	2.395	2.351	2.3341
0.5	0.7979	3.270	3.145	3.172	3.080	3.0118
0.6	0.8740	7.106	4.386	4.517	4.342	4.1247
0.7	0.9441	73.92	7.409	7.769	7.433	6.9814
0.74	0.9707	–	10.91	11.46	11.01	10.7726
0.76	0.9837	–	15.29	15.99	15.44	15.7958
0.77	0.9901	–	20.18	21.04	20.43	21.2061
0.78	0.9966	–	35.01	36.60	35.93	33.4056

(ii) close to asymptotic estimations of the effective coefficient at a close to 1;

(iii) give practically acceptable results for calculating the effective thermal conductivity coefficient for inclusions of any conductivity ($0 \leq \lambda < \infty$, including the limiting case $\lambda \to \infty$) with sizes up to $a = 0.996$ inclusive.

3. Thus, using of AP to improve the ThPhM solution allows, sufficiently expand the area one to applicability of the latter, and also to describe the composite at large sizes of inclusions for relatively large $\lambda \to \infty$. Note that the proposed approach allows for the generalization of the case of inclusions of different geometry and different arrangements.

3.4 THREE-PHASE COMPOSITE MODEL FOR SQUARE CYLINDRICAL INCLUSIONS

3.4.1 USING OF THPHM WITH ASYMPTOTIC HOMOGENIZATION THEORY

Let us consider a two-phase microinhomogeneous material consisting of a continuous matrix and cylindrical inclusions of a square profile periodically located in it. We will assume that the structure is doubly periodic with the same period in both directions, and the inclusions are arranged in a square lattice (Fig. 3.30).

When studying the thermal conductivity of such a structure, the constitutive relations are the Poisson's equation with the corresponding conjugation conditions at the

Table 3.4

The results of calculations of the effective thermal conductivity coefficient for inclusions of large sizes with high conductivity.

Conductivity of inclusions $\lambda = 10^2$			Conductivity of inclusions $\lambda = 5 \cdot 10^2$		
Inclusion size a	Asymptotic solution [152]	AP $q_{[0/18]}(a)$	Inclusion size a	Asymptotic solution [152]	AP $q_{[0/18]}(a)$
0.9	4.9108	4.5270	0.9	5.0347	4.7283
0.91	5.2473	4.8118	0.91	5.3980	5.0517
0.92	5.6413	5.1482	0.92	5.8277	5.4404
0.93	6.1112	5.5546	0.93	6.3467	5.9197
0.94	6.6841	6.0584	0.94	6.9901	6.5296
0.95	7.4037	6.7043	0.95	7.8164	7.3380
0.96	8.3438	7.5684	0.96	8.9315	8.4687
0.97	9.6432	8.7920	0.97	10.5536	10.1733
0.98	11.5936	10.6703	0.98	13.2351	13.0554
0.99	14.8171	13.9399	0.99	19.0663	19.0093
0.991	15.2200	14.4058	0.991	20.1058	19.9743
0.992	15.6149	14.9093	0.992	21.3187	21.0549
0.993	15.9721	15.4549	0.993	22.7595	22.2732
0.994	16.2297	16.0484	0.994	24.5101	23.6571
0.995	16.2513	16.6962	0.995	26.7012	25.2431
0.996	–	17.4063	0.996	29.5560	27.0788
Conductivity of inclusions $\lambda = 10^3$			Conductivity of inclusions $\lambda \to \infty$		
Inclusion size a	Asymptotic solution [152]	AP $q_{[0/18]}(a)$	Inclusion size a	Asymptotic solution [152]	AP $q^\infty_{[0/18]}(a)$
0.9	5.0502	4.7552	0.9	5.0657	4.7826
0.91	5.4168	5.0841	0.91	5.4357	5.1171
0.92	5.8510	5.4803	0.92	5.8743	5.5210
0.93	6.3761	5.9701	0.93	6.4056	6.0218
0.94	7.0283	6.5958	0.94	7.0666	6.6639
0.95	7.8680	7.4290	0.95	7.9200	7.5230
0.96	9.0049	8.6017	0.96	9.0784	8.7403
0.97	10.6674	10.3864	0.97	10.7812	10.6109
0.98	13.4403	13.4495	0.98	13.6455	13.8728
0.99	19.5974	19.9647	0.99	20.1286	21.0343
0.991	20.7166	21.0452	0.991	21.3273	22.2524
0.992	22.0317	22.2634	0.992	22.7447	23.6364
0.993	23.6079	23.6475	0.993	24.4563	25.2227
0.994	25.5452	25.2337	0.994	26.5802	27.0590
0.995	28.0074	27.0699	0.995	29.3137	29.2098
0.996	31.2868	29.2203	0.996	33.0177	31.7631

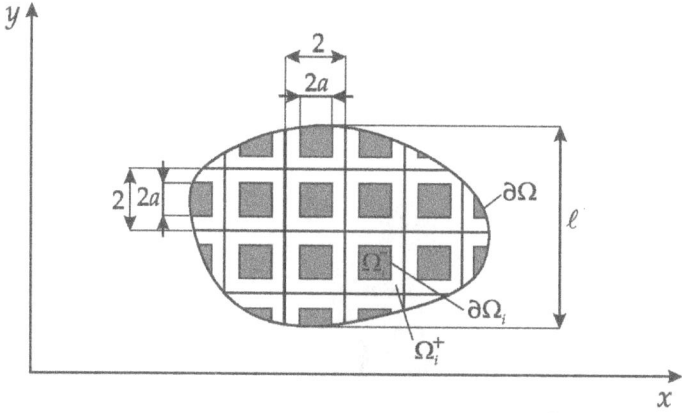

Figure 3.30 Cross section of a composite with cylindrical square inclusions.

phase boundaries and boundary conditions at the external contour of the composite:

$$
\lambda^+ \Delta u^+ = F \quad \text{in} \quad \Omega_i^+,
$$
$$
\lambda^- \Delta u^- = F \quad \text{in} \quad \Omega_i^-,
$$
(3.102)

$$
u^+ = u^- \quad \text{at} \quad \partial\Omega_i,
$$
(3.103)

$$
\lambda^+ \frac{\partial u^+}{\partial \mathbf{n}} = \lambda^- \frac{\partial u^-}{\partial \mathbf{n}} \quad \text{at} \quad \partial\Omega_i,
$$
(3.104)

where u^+, u^- are the temperature distribution functions, respectively, in the matrix Ω_i^+ and inclusions Ω_i^-; F is the heat source density; $\mathbf{n}$ is the outer normal to the inclusion contour.

Application of the homogenization method allows one to reduce the original problem (3.102)–(3.104) to the cell problem (Fig. 3.31).

In "fast" variables ξ, η the cell problem is written as follows:

$$
\frac{\partial^2 u_1^+}{\partial \xi^2} + \frac{\partial^2 u_1^+}{\partial \eta^2} = 0 \text{ in } \Omega_i^+;
$$
$$
\frac{\partial^2 u_1^-}{\partial \xi^2} + \frac{\partial^2 u_1^-}{\partial \eta^2} = 0 \text{ in } \Omega_i^-;
$$
(3.105)

$$
u_1^+ = u_1^- ; \quad \frac{\partial u_1^+}{\partial \xi} - \lambda \frac{\partial u_1^-}{\partial \xi} = (\lambda - 1) \frac{\partial u_0}{\partial \xi} \text{ at } \xi = \pm a;
$$
$$
u_1^+ = u_1^- ; \quad \frac{\partial u_1^+}{\partial \eta} - \lambda \frac{\partial u_1^-}{\partial \eta} = (\lambda - 1) \frac{\partial u_0}{\partial \eta} \text{ at } \eta = \pm a;
$$
(3.106)

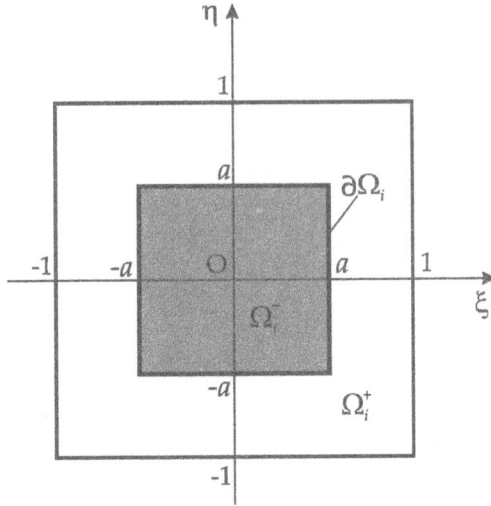

Figure 3.31 Typical cell of a composite under consideration.

$$\left. u_1^+ \right|_{\xi=1} = \left. u_1^+ \right|_{\xi=-1} , \qquad \left. u_1^+ \right|_{\eta=1} = \left. u_1^+ \right|_{\eta=-1} ,$$

$$\left. \frac{\partial u_1^+}{\partial \xi} \right|_{\xi=1} = \left. \frac{\partial u_1^+}{\partial \xi} \right|_{\xi=-1} , \qquad \left. \frac{\partial u_1^+}{\partial \eta} \right|_{\eta=1} = \left. \frac{\partial u_1^+}{\partial \eta} \right|_{\eta=-1} . \qquad (3.107)$$

We use ThPhM to solve the problem (3.105)–(3.107). The physical essence of the application of this model in this case is that the entire periodic structure, with the exception of one cell, is replaced by a homogeneous medium with an unknown thermal conductivity $\tilde{\lambda}$ (Fig. 3.32).

Mathematically, this idealization leads to:

1) the replacement of the periodicity conditions (3.107) by the conditions of conjugation of the matrix with an equivalent homogeneous medium;
2) the decaying conditions for the functions of temperature and heat flux in an averaged medium at a considerable distance from the inclusion.

Consequently, the local problem (3.105)–(3.107) is transformed to the following form

$$\frac{\partial^2 u_1^+}{\partial \xi^2} + \frac{\partial^2 u_1^+}{\partial \eta^2} = 0 \quad \text{in} \quad \Omega_i^+,$$

$$\frac{\partial^2 u_1^-}{\partial \xi^2} + \frac{\partial^2 u_1^-}{\partial \eta^2} = 0 \quad \text{in} \quad \Omega_i^-, \qquad (3.108)$$

$$\frac{\partial^2 \tilde{u}_1}{\partial \xi^2} + \frac{\partial^2 \tilde{u}_1}{\partial \eta^2} = 0 \quad \text{in} \quad \tilde{\Omega},$$

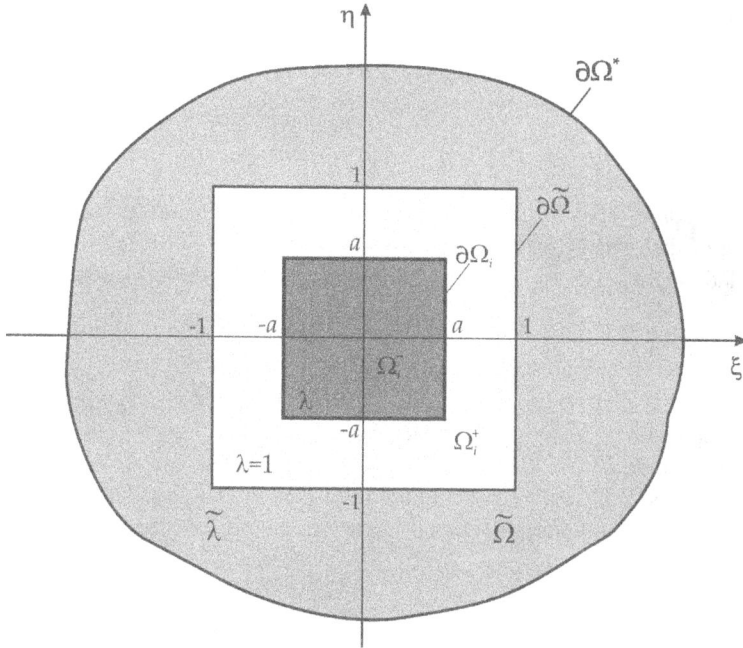

Figure 3.32 Model of a three-phase structure.

$$u_1^+ = u_1^-, \quad \frac{\partial u_1^+}{\partial \xi} - \lambda \frac{\partial u_1^-}{\partial \xi} = (\lambda - 1) \frac{\partial u_0}{\partial \xi} \quad \text{at} \quad \xi = \pm a,$$

$$u_1^+ = u_1^-, \quad \frac{\partial u_1^+}{\partial \eta} - \lambda \frac{\partial u_1^-}{\partial \eta} = (\lambda - 1) \frac{\partial u_0}{\partial \eta} \quad \text{at} \quad \eta = \pm a,$$

$$(3.109)$$

$$u_1^+ = \tilde{u}_1, \quad \frac{\partial u_1^+}{\partial \xi} - \tilde{\lambda} \frac{\partial \tilde{u}_1}{\partial \xi} = (\tilde{\lambda} - 1) \frac{\partial u_0}{\partial \xi} \quad \text{at} \quad \xi = \pm 1,$$

$$u_1^+ = \tilde{u}_1, \quad \frac{\partial u_1^+}{\partial \eta} - \tilde{\lambda} \frac{\partial \tilde{u}_1}{\partial \eta} = (\tilde{\lambda} - 1) \frac{\partial u_0}{\partial \eta} \quad \text{at} \quad \eta = +1,$$

$$(3.110)$$

$$\tilde{u}_1 \to 0, \quad \frac{\partial \tilde{u}_1}{\partial \xi} \to 0 \quad \text{at} \quad \xi = \to \pm\infty,$$

$$\tilde{u}_1 \to 0, \quad \frac{\partial \tilde{u}_1}{\partial \eta} \to 0 \quad \text{at} \quad \eta = \to \pm\infty,$$

$$(3.111)$$

where u_1^+, u_1^-, $\tilde{u}_1$ are the temperature distribution functions in the matrix Ω_i^+, inclusions Ω_i^- and homogeneous environment $\tilde{\Omega}$; λ^+, λ^-, $\tilde{\lambda}$ are the conductivity of the respective areas.

3.4.2 USING PPB

To solve the local problem (3.108)–(3.111), we use PPB, according to which the inclusion contours $\partial\Omega_i$ and cells $\partial\tilde\Omega$ in polar coordinates r, θ are described by equation

$$r = r_0 + \varepsilon_1 f(\theta),\tag{3.112}$$

where $r_0 = const > 0$; $f(\theta)$ is the differentiable function characterizing the geometric shape of the considered contour; ε_1 is the small parameter ($|\varepsilon_1| << 1$). In this case, relation (3.112) as applied to contours $\partial\Omega_i$ and $\partial\tilde\Omega$ can be written as follows

$$
\begin{aligned}
r^- &= r_0^- \left(1 + \varepsilon_1 \cos 4\theta + ...\right); \\
r^+ &= r_0^+ \left(1 + \varepsilon_1 \cos 4\theta + ...\right).
\end{aligned}
\tag{3.113}
$$

The mathematical meaning of this approximation lies in the fact that in the zero approximation, the square contours of the inclusion and cells are replaced by circles. In this case, the radii of the corresponding circles r_0^-, r_0^+ are chosen in such a way that the areas of the original and transformed areas of the inclusion and the cell remain equal:

$$\pi\left(r_0^-\right)^2 = 4a^2, \qquad \pi\left(r_0^+\right)^2 = 4,$$

i.e.

$$r_0^- = \frac{2a}{\sqrt{\pi}}, \qquad r_0^+ = \frac{2}{\sqrt{\pi}}.\tag{3.114}$$

Let us represent the temperature distribution functions in the matrix Ω_i^+, inclusions Ω_i^- and homogeneous environment $\tilde\Omega$ in the form of series in powers of a small parameter ε_1:

$$
\begin{aligned}
u_1^+ &= u_{10}^+ + \varepsilon_1 u_{11}^+ + ...; \\
u_1^- &= u_{10}^- + \varepsilon_1 u_{11}^- + ...; \\
\tilde u_1 &= \tilde u_{10} + \varepsilon_1 \tilde u_{11} + ...
\end{aligned}
\tag{3.115}
$$

In a similar way, we represent unknown parameter $\tilde\lambda$:

$$\tilde\lambda = \tilde\lambda_0 + \varepsilon_1 \tilde\lambda_1 + ...\tag{3.116}$$

The value of the small parameter ε_1 for a square contour, as a rule, varies within: $\frac{1}{9} \le |\varepsilon_1| \le \frac{1}{6}$. In this case, the value is taken as:

$$\varepsilon_1 = \frac{1 - \frac{2}{\sqrt{\pi}}}{\frac{2}{\sqrt{\pi}}} = \frac{a - \frac{2a}{\sqrt{\pi}}}{\frac{2a}{\sqrt{\pi}}} = \frac{\sqrt{\pi}}{2} - 1 \approx -0.1138.$$

3.4.3 SOLUTION OF A LOCAL PROBLEM IN THE ZEROTH APPROXIMATION

The computational model of the cell problem in the zeroth approximation is shown in Fig. 3.33.

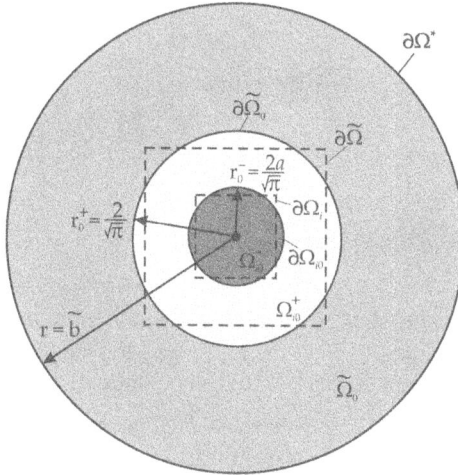

Figure 3.33 Calculation model in zero approximation.

Taking into account relations (3.113)–(3.116), the local problem (3.108)–(3.111) in "fast" polar coordinates r, θ can be written in the zero approximation for the functions u_{10}^+, u_{10}^-, $\tilde{u}_{10}$ in the following way:

$$\frac{\partial^2 u_{10}^+}{\partial r^2} + \frac{1}{r}\frac{\partial u_{10}^+}{\partial r} + \frac{1}{r^2}\frac{\partial^2 u_{10}^+}{\partial \theta^2} = 0 \quad \text{in} \quad \Omega_{i0}^+,$$

$$\frac{\partial^2 u_{10}^-}{\partial r^2} + \frac{1}{r}\frac{\partial u_{10}^-}{\partial r} + \frac{1}{r^2}\frac{\partial^2 u_{10}^-}{\partial \theta^2} = 0 \quad \text{in} \quad \Omega_{i0}^-, \qquad (3.117)$$

$$\frac{\partial^2 \tilde{u}_{10}}{\partial r^2} + \frac{1}{r}\frac{\partial \tilde{u}_{10}}{\partial r} + \frac{1}{r^2}\frac{\partial^2 \tilde{u}_{10}}{\partial \theta^2} = 0 \quad \text{in} \quad \tilde{\Omega}_0,$$

at $r = r_0^- = \frac{2a}{\sqrt{\pi}}$:

$$u_{10}^+ = u_{10}^-; \quad \frac{\partial u_{10}^+}{\partial r} - \lambda \frac{\partial u_{10}^-}{\partial r} = (\lambda - 1)\left(\frac{\partial u_0}{\partial x}\cos\theta + \frac{\partial u_0}{\partial y}\sin\theta\right), \qquad (3.118)$$

at $r = r_0^+ = \frac{2}{\sqrt{\pi}}$:

$$u_{10}^+ = \tilde{u}_{10}; \quad \frac{\partial u_{10}^+}{\partial r} - \tilde{\lambda}\frac{\partial \tilde{u}_{10}}{\partial r} = (\tilde{\lambda}_0 - 1)\left(\frac{\partial u_0}{\partial x}\cos\theta + \frac{\partial u_0}{\partial y}\sin\theta\right), \qquad (3.119)$$

$$\tilde{u}_{10} \to 0, \quad \frac{\partial \tilde{u}_{10}}{\partial r} \to 0 \quad \text{at} \quad r \to \infty. \qquad (3.120)$$

The solution of the boundary value problem (3.117)–(3.120) is sought in the form:

$$u_{10}^- = \bar{A}_{10}\, r\cos\theta + \bar{\bar{A}}_{10}\, r\sin\theta,$$

$$u_{10}^+ = \left(\bar{B}_{10}\, r + \frac{\bar{C}_{10}}{r}\right)\cos\theta + \left(\bar{\bar{B}}_{10}\, r + \frac{\bar{\bar{C}}_{10}}{r}\right)\sin\theta, \qquad (3.121)$$

$$\tilde{u}_{10} = \frac{\bar{D}_{10}}{r}\cos\theta + \frac{\bar{\bar{D}}_{10}}{r}\sin\theta,$$

where $\bar{A}_{10}, \bar{\bar{A}}_{10}, \bar{B}_{10}, \bar{\bar{B}}_{10}, \bar{C}_{10}, \bar{\bar{C}}_{10}, \bar{D}_{10}, \bar{\bar{D}}_{10}$ are the arbitrary constants.

Note that in expressions (3.121) the representation of the function u_{10}^- written taking into account the boundedness of the temperature distribution function and its derivative $\frac{\partial u_{10}^-}{\partial r}$ at $r = 0$, and the expression of function $\tilde{u}_{10}$ satisfies the decaying conditions for these functiond at $r \to \infty$ (3.120).

Relation (3.121) includes eight arbitrary constants, which are determined from the conjugation conditions (3.118), (3.119). Systems of equations for integration constants $\bar{A}_{10}, \bar{B}_{10}, \bar{C}_{10}, \bar{D}_{10}$ and $\bar{\bar{A}}_{10}, \bar{\bar{B}}_{10}, \bar{\bar{C}}_{10}, \bar{\bar{D}}_{10}$ are completely identical, so we give a solution to only one of them:

$$\bar{A}_{10} = A_{10}^* \frac{\partial u_0}{\partial x}, \quad \bar{B}_{10} = B_{10}^* \frac{\partial u_0}{\partial x}, \quad \bar{C}_{10} = C_{10}^* \frac{\partial u_0}{\partial x}, \quad \bar{D}_{10} = D_{10}^* \frac{\partial u_0}{\partial x}, \quad (3.122)$$

$$A_{10}^* = -1 + 4\tilde{\lambda}_0 \Delta_0, \qquad B_{10}^* = -1 + 2(\lambda + 1)\,\tilde{\lambda}_0 \Delta_0,$$

$$C_{10}^* = -\frac{8a^2}{\pi}(\lambda - 1)\,\tilde{\lambda}_0 \Delta_0, \qquad (3.123)$$

$$D_{10}^* = \frac{4}{\pi}\left(1 - 2\left(a^2(\lambda - 1) + (\lambda + 1)\right)\Delta_0\right),$$

where

$$\Delta_0 = \left(\left(\tilde{\lambda}_0 + 1\right)(\lambda + 1) - \left(\tilde{\lambda}_0 - 1\right)(\lambda - 1)\,a^2\right)^{-1}. \qquad (3.124)$$

Obviously, for arbitrary constants $\bar{\bar{A}}_{10}, \bar{\bar{B}}_{10}, \bar{\bar{C}}_{10}, \bar{\bar{D}}_{10}$, we have

$$\bar{\bar{A}}_{10} = \bar{A}_{10}, \quad \bar{\bar{B}}_{10} = \bar{B}_{10}, \quad \bar{\bar{C}}_{10} = \bar{C}_{10}, \quad \bar{\bar{D}}_{10} = \bar{D}_{10} \left(\frac{\partial u_0}{\partial x} \to \frac{\partial u_0}{\partial y}\right). \qquad (3.125)$$

Further solution of the problem is reduced to the construction of homogenized relations. In this case, it should be taken into account that the averaging must be carried out over the region $\Omega_0^* = \Omega_{i0}^+ \cup \Omega_{i0}^- \cup \tilde{\Omega}_0$ – matrix, inclusion, homogeneous environment. Thus, the equation to be averaged is

$$\frac{\partial^2 u_0}{\partial x^2} + \frac{\partial^2 u_0}{\partial y^2} + 2\frac{\partial^2 u_{10}^+}{\partial x \partial \xi} + 2\frac{\partial^2 u_{10}^+}{\partial y \partial \eta} + \frac{\partial^2 u_{20}^+}{\partial \xi^2} + \frac{\partial^2 u_{20}^+}{\partial \eta^2} +$$

$$+\lambda \left(\frac{\partial^2 u_0}{\partial x^2} + \frac{\partial^2 u_0}{\partial y^2} + 2\frac{\partial^2 u_{10}^-}{\partial x \partial \xi} + 2\frac{\partial^2 u_{10}^-}{\partial y \partial \eta} + \frac{\partial^2 u_{20}^-}{\partial \xi^2} + \frac{\partial^2 u_{20}^-}{\partial \eta^2} \right) + \tag{3.126}$$

$$+\tilde{\lambda}_0 \left(\frac{\partial^2 u_0}{\partial x^2} + \frac{\partial^2 u_0}{\partial y^2} + 2\frac{\partial^2 \tilde{u}_{10}}{\partial x \partial \xi} + 2\frac{\partial^2 \tilde{u}_{10}}{\partial y \partial \eta} + \frac{\partial^2 \tilde{u}_{20}}{\partial \xi^2} + \frac{\partial^2 \tilde{u}_{20}}{\partial \eta^2} \right) = F.$$

The averaging operator is generalized in this case and takes the following form

$$\tilde{(...)} = \frac{1}{|\Omega_0^*|} \left(\iint_{\Omega_{i0}^+} (...) \, d\xi \, d\eta + \lambda \iint_{\Omega_{i0}^-} (...) \, d\xi \, d\eta + \tilde{\lambda}_0 \iint_{\tilde{\Omega}_0} (...) \, d\xi \, d\eta \right),$$
$$\tag{3.127}$$

where $|\Omega_0^*| = |\Omega_{i0}^+ \cup \Omega_{i0}^- \cup \tilde{\Omega}_0|$ is the measure of a three phase region, the boundary of which can formally be considered a circle of infinitely large radius $\tilde{b} \to \infty$.

Applying the averaging operator (3.127) to relation (3.126), we obtain the homogenized equation in the following form

$$\frac{1}{|\Omega_0^*|} \left[\iint_{\Omega_{i0}^+} \left(\frac{\partial^2 u_0}{\partial x^2} + \frac{\partial^2 u_0}{\partial y^2} + \frac{\partial^2 u_{10}^+}{\partial x \partial \xi} + \frac{\partial^2 u_{10}^+}{\partial y \partial \eta} \right) d\xi \, d\eta + \right.$$

$$+ \int_{\partial \Omega_{i0}} \left(\frac{\partial u_{20}^+}{\partial \bar{n}} + \frac{\partial u_{10}^+}{\partial n} \right) d\ell + \int_{\partial \tilde{\Omega}_0} \left(\frac{\partial u_{20}^+}{\partial \bar{n}} + \frac{\partial u_{10}^+}{\partial n} \right) d\ell +$$

$$+\lambda \iint_{\Omega_{i0}^-} \left(\frac{\partial^2 u_0}{\partial x^2} + \frac{\partial^2 u_0}{\partial y^2} + \frac{\partial^2 u_{10}^-}{\partial x \partial \xi} + \frac{\partial^2 u_{10}^-}{\partial y \partial \eta} \right) d\xi \, d\eta +$$

$$+\lambda \int_{\partial \Omega_{i0}} \left(\frac{\partial u_{20}^-}{\partial \bar{n}} + \frac{\partial u_{10}^-}{\partial n} \right) d\ell + \tag{3.128}$$

$$+\tilde{\lambda}_0 \iint_{\tilde{\Omega}_0} \left(\frac{\partial^2 u_0}{\partial x^2} + \frac{\partial^2 u_0}{\partial y^2} + \frac{\partial^2 \tilde{u}_{10}}{\partial x \partial \xi} + \frac{\partial^2 \tilde{u}_{10}}{\partial y \partial \eta} \right) d\xi \, d\eta +$$

$$+\tilde{\lambda}_0 \int_{\partial \tilde{\Omega}_0} \left(\frac{\partial \tilde{u}_{20}}{\partial \bar{n}} + \frac{\partial \tilde{u}_{10}}{\partial n} \right) d\ell + \tilde{\lambda}_0 \int_{\partial \Omega^*} \left(\frac{\partial \tilde{u}_{20}}{\partial \bar{n}} + \frac{\partial \tilde{u}_{10}}{\partial n} \right) d\ell \right] = F,$$

where $\bar{n}$ is the outer normal to the contour; n is the outer normal to the contour, written in "slow" variables x, y; $\frac{\partial(...)}{\partial n} = \frac{\partial(...)}{\partial x} \cos(n, \xi) + \frac{\partial(...)}{\partial y} \cos(n, \eta)$.

Taking into account the expressions (3.121)–(3.125), the homogenized equation (3.128) is transformed to the following form

$$\frac{1}{|\Omega_0^*|} \left[|\Omega_{i0}^+| (B_{10}^* + 1) + \lambda |\Omega_{i0}^-| (A_{10}^* + 1) + \tilde{\lambda}_0 |\tilde{\Omega}_0| \right] \Delta u_0 = F, \tag{3.129}$$

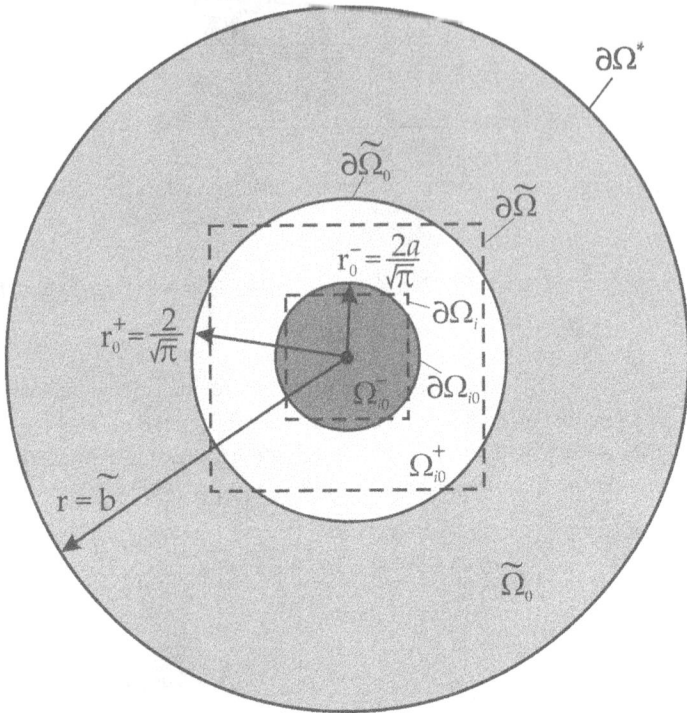

Figure 3.34 Model of the first approximation.

where q is the effective parameter:

$$q = \frac{1}{|\Omega_0^*|} \left[|\Omega_{i0}^+| \, (B_{10}^* + 1) + \lambda \, |\Omega_{i0}^-| \, (A_{10}^* + 1) + \tilde{\lambda}_0 \, |\tilde{\Omega}_0| \right]. \tag{3.130}$$

It is obvious that the coefficient q, determined by expression (3.130), is the unknown parameter $\tilde{\lambda}_0$. Therefore, equating $\tilde{\lambda}_0$ to expression (3.130), we arrive at a linear algebraic equation with respect to the coefficient $q = \tilde{\lambda}_0$.

Thus, the solution of the resulting equation gives, in the zeroth approximation, the following analytical expression for the effective thermal conductivity:

$$q = \tilde{\lambda}_0 = \frac{1 - a^2 + \lambda \, (1 + a^2)}{1 + a^2 + \lambda \, (1 - a^2)}. \tag{3.131}$$

3.4.4 CONSTRUCTION OF THE FIRST PPB APPROXIMATION

Next, we construct the first approximation of the solution of the local problem (3.108)–(3.111) using the PPB. The model of the structure in the first approximation is shown in Fig. 3.34.

In "fast" polar coordinates r, θ local problem (3.108)–(3.111) for functions of the first approximation u_{11}^+, u_{11}^-, $\tilde{u}_{11}$ written as follows:

$$\frac{\partial^2 u_{11}^+}{\partial r^2} + \frac{1}{r}\frac{\partial u_{11}^+}{\partial r} + \frac{1}{r^2}\frac{\partial^2 u_{11}^+}{\partial \theta^2} = 0 \quad \text{in} \quad \Omega_{i1}^+,$$

$$\frac{\partial^2 u_{11}^-}{\partial r^2} + \frac{1}{r}\frac{\partial u_{11}^-}{\partial r} + \frac{1}{r^2}\frac{\partial^2 u_{11}^-}{\partial \theta^2} = 0 \quad \text{in} \quad \Omega_{i1}^-, \qquad (3.132)$$

$$\frac{\partial^2 \tilde{u}_{11}}{\partial r^2} + \frac{1}{r}\frac{\partial \tilde{u}_{11}}{\partial r} + \frac{1}{r^2}\frac{\partial^2 \tilde{u}_{11}}{\partial \theta^2} = 0 \quad \text{in} \quad \tilde{\Omega}_1,$$

$$u_{11}^+ + \frac{\partial u_{10}^+}{\partial r}r_0^- \cos 4\theta = u_{11}^- + \frac{\partial u_{10}^-}{\partial r}r_0^- \cos 4\theta \quad \text{at} \quad r = r_0^- = \frac{2a}{\sqrt{\pi}},$$

$$\frac{\partial u_{11}^+}{\partial r} + \frac{\partial^2 u_{10}^+}{\partial r^2}r_0^- \cos 4\theta + \frac{4}{r_0^-}\frac{\partial u_{10}^+}{\partial \theta}\sin 4\theta - 4\frac{\partial u_0}{\partial x}\sin \theta \sin 4\theta +$$

$$+4\frac{\partial u_0}{\partial y}\cos \theta \sin 4\theta = \lambda \left(\frac{\partial u_{11}^-}{\partial r} + \frac{\partial^2 u_{10}^-}{\partial r^2}r_0^- \cos 4\theta + \frac{4}{r_0^-}\frac{\partial u_{10}^-}{\partial \theta}\sin 4\theta - \right. \qquad (3.133)$$

$$\left. -4\frac{\partial u_0}{\partial x}\sin \theta \sin 4\theta + 4\frac{\partial u_0}{\partial y}\cos \theta \sin 4\theta \right) \quad \text{at} \quad r = r_0^- = \frac{2a}{\sqrt{\pi}},$$

$$u_{11}^+ + \frac{\partial u_{10}^+}{\partial r}r_0^+ \cos 4\theta = \tilde{u}_{11} + \frac{\partial \tilde{u}_{10}}{\partial r}r_0^+ \cos 4\theta \quad \text{at} \quad r = r_0^+ = \frac{2}{\sqrt{\pi}},$$

$$\frac{\partial u_{11}^+}{\partial r} + \frac{\partial^2 u_{10}^+}{\partial r^2}r_0^+ \cos 4\theta + \frac{4}{r_0^+}\frac{\partial u_{10}^+}{\partial \theta}\sin 4\theta - 4\frac{\partial u_0}{\partial x}\sin \theta \sin 4\theta +$$

$$+4\frac{\partial u_0}{\partial y}\cos \theta \sin 4\theta = \tilde{\lambda}_0 \left(\frac{\partial \tilde{u}_{11}}{\partial r} + \frac{\partial^2 \tilde{u}_{10}}{\partial r^2}r_0^+ \cos 4\theta + \frac{4}{r_0^+} \cdot \frac{\partial \tilde{u}_{10}}{\partial \theta}\sin 4\theta - \right. \qquad (3.134)$$

$$\left. -4\frac{\partial u_0}{\partial x}\sin \theta \sin 4\theta + 4\frac{\partial u_0}{\partial y}\cos \theta \sin 4\theta \right) + \tilde{\lambda}_1 \frac{\partial \tilde{u}_{10}}{\partial r} \quad \text{at} \quad r = r_0^+ = \frac{2}{\sqrt{\pi}},$$

$$\tilde{u}_{11} \to 0, \quad \frac{\partial \tilde{u}_{11}}{\partial r} \to 0 \quad \text{at} \quad r \to \infty. \qquad (3.135)$$

The expressions for the constants (3.123)–(3.125) are transformed to the following form

$$A_{10}^* = -(\lambda - 1)(1 - a^2)\Delta, \qquad B_{10}^* = (\lambda - 1)a^2\Delta,$$

$$C_{10}^* = -\frac{4}{\pi}(\lambda - 1)a^2\Delta, \qquad D_{10}^* = 0, \qquad (3.136)$$

where

$$\Delta = \left((\lambda + 1) - (\lambda - 1)a^2 \right)^{-1}.$$

Taking into account the conditions of boundedness of the function u_{11}^- and its derivative $\frac{\partial u_{11}^-}{\partial r}$ at $r = 0$, as well as decaying conditions (3.135), we obtain the solution of the problem (3.132)–(3.135):

$$u_{11}^- = \overline{A}_{11} r \cos\theta + \overline{A}_{21} r^3 \cos 3\theta + \overline{A}_{31} r^5 \cos 5\theta +$$

$$+ \overline{\overline{A}}_{11} r \sin\theta + \overline{\overline{A}}_{21} r^3 \sin 3\theta + \overline{\overline{A}}_{31} r^5 \sin 5\theta,$$

$$u_{11}^+ = \left(\overline{B}_{11} r + \frac{\overline{C}_{11}}{r} \right) \cos\theta + \left(\overline{B}_{21} r^3 + \frac{\overline{C}_{21}}{r^3} \right) \cos 3\theta +$$

$$+ \left(\overline{B}_{31} r^5 + \frac{\overline{C}_{31}}{r^5} \right) \cos 5\theta + \left(\overline{\overline{B}}_{11} r + \frac{\overline{\overline{C}}_{11}}{r} \right) \sin\theta + \qquad (3.137)$$

$$+ \left(\overline{\overline{B}}_{21} r^3 + \frac{\overline{\overline{C}}_{21}}{r^3} \right) \sin 3\theta + \left(\overline{\overline{B}}_{31} r^5 + \frac{\overline{\overline{C}}_{31}}{r^5} \right) \sin 5\theta,$$

$$\tilde{u}_{11} = \frac{\overline{D}_{11}}{r} \cos\theta + \frac{\overline{D}_{21}}{r^3} \cos 3\theta + \frac{\overline{D}_{31}}{r^5} \cos 5\theta +$$

$$+ \frac{\overline{\overline{D}}_{11}}{r} \sin\theta + \frac{\overline{\overline{D}}_{21}}{r^3} \sin 3\theta + \frac{\overline{\overline{D}}_{31}}{r^5} \sin 5\theta,$$

where $\overline{A}_{i1}$, $\overline{\overline{A}}_{i1}$, $\overline{B}_{i1}$, $\overline{\overline{B}}_{i1}$, $\overline{C}_{i1}$, $\overline{\overline{C}}_{i1}$, $\overline{D}_{i1}$, $\overline{\overline{D}}_{i1}$ $(i = 1, 2, 3)$ are the arbitrary constants.

Relation (3.137) includes 24 arbitrary constants, which are determined from the conditions (3.133), (3.134). Since the systems of equations for the constants $\overline{A}_{i1}$, $\overline{B}_{i1}$, $\overline{C}_{i1}$, $\overline{D}_{i1}$ and $\overline{\overline{A}}_{i1}$, $\overline{\overline{B}}_{i1}$, $\overline{\overline{C}}_{i1}$, $\overline{\overline{D}}_{i1}$ $(i = 1, 2, 3)$ are completely identical, we deal with part of the solution:

for $\cos\theta$:

$$\begin{cases} -\overline{A}_{11} r_0^- + \overline{B}_{11} r_0^- + \dfrac{\overline{C}_{11}}{r_0^-} = 0, \\[3mm] -\lambda \overline{A}_{11} + \overline{B}_{11} - \dfrac{\overline{C}_{11}}{r_0^{-2}} = 0, \\[3mm] \overline{B}_{11} r_0^+ + \dfrac{\overline{C}_{11}}{r_0^+} - \dfrac{\overline{D}_{11}}{r_0^+} = 0, \\[3mm] \overline{B}_{11} - \dfrac{\overline{C}_{11}}{r_0^{+2}} + \tilde{\lambda}_0 \dfrac{\overline{D}_{11}}{r_0^{+2}} = -\tilde{\lambda}_1 \dfrac{\overline{D}_{10}}{r_0^{+2}}, \end{cases} \qquad (3.138)$$

for cos 3θ:

$$
\begin{cases}
-\overline{A}_{21}\, r_0^{-3} + \overline{B}_{21}\, r_0^{-3} + \dfrac{\overline{C}_{21}}{r_0^{-3}} = \dfrac{r_0^{-}}{2}\left(\overline{A}_{10} - \overline{B}_{10} + \dfrac{\overline{C}_{10}}{r_0^{-2}}\right), \\[4mm]
-\lambda\overline{A}_{21}\, r_0^{-2} + \overline{B}_{21}\, r_0^{-2} - \dfrac{\overline{C}_{21}}{r_0^{-4}} = \dfrac{1}{3}\left(-2\lambda\overline{A}_{10} + 2\overline{B}_{10} + \dfrac{\overline{C}_{10}}{r_0^{-2}}\right) - \\[4mm]
\qquad -\dfrac{2}{3}(\lambda - 1)\dfrac{\partial u_0}{\partial x}, \\[4mm]
\overline{B}_{21}\, r_0^{+3} + \dfrac{\overline{C}_{21}}{r_0^{+3}} - \dfrac{\overline{D}_{21}}{r_0^{+3}} = \dfrac{r_0^{+}}{2}\left(-\overline{B}_{10} - \dfrac{\overline{C}_{10}}{r_0^{+2}} - \dfrac{\overline{D}_{10}}{r_0^{+2}}\right), \\[4mm]
\overline{B}_{21}\, r_0^{+2} - \dfrac{\overline{C}_{21}}{r_0^{+4}} + \tilde{\lambda}_0\dfrac{\overline{D}_{21}}{r_0^{+4}} = \dfrac{1}{3}\left(2\overline{B}_{10} + \dfrac{\overline{C}_{10}}{r_0^{+2}}\right) - \dfrac{2}{3}(\tilde{\lambda}_0 - 1)\dfrac{\partial u_0}{\partial x},
\end{cases}
\tag{3.139}
$$

for cos 5θ:

$$
\begin{cases}
-\overline{A}_{31}\, r_0^{-5} + \overline{B}_{31}\, r_0^{-5} + \dfrac{\overline{C}_{31}}{r_0^{-5}} = \dfrac{r_0^{-}}{2}\left(\overline{A}_{10} - \overline{B}_{10} + \dfrac{\overline{C}_{10}}{r_0^{-2}}\right), \\[4mm]
-\lambda\overline{A}_{31}\, r_0^{-4} + \overline{B}_{31}\, r_0^{-4} - \dfrac{\overline{C}_{31}}{r_0^{-6}} = \dfrac{1}{5}\left(2\lambda\overline{A}_{10} - 2\overline{B}_{10} - 3\dfrac{\overline{C}_{10}}{r_0^{-2}}\right) + \\[4mm]
\qquad +\dfrac{2}{5}(\lambda - 1)\dfrac{\partial u_0}{\partial x}, \\[4mm]
\overline{B}_{31}\, r_0^{+5} + \dfrac{\overline{C}_{31}}{r_0^{+5}} - \dfrac{\overline{D}_{31}}{r_0^{+5}} = \dfrac{r_0^{+}}{2}\left(-\overline{B}_{10} + \dfrac{\overline{C}_{10}}{r_0^{+2}} - \dfrac{\overline{D}_{10}}{r_0^{+2}}\right), \\[4mm]
\overline{B}_{31}\, r_0^{+4} - \dfrac{\overline{C}_{31}}{r_0^{+6}} + \tilde{\lambda}_0\dfrac{\overline{D}_{31}}{r_0^{+6}} = \dfrac{1}{5}\left(-2\overline{B}_{10} - 3\dfrac{\overline{C}_{10}}{r_0^{+2}} + 3\tilde{\lambda}_0\dfrac{\overline{D}_{10}}{r_0^{+2}}\right) + \\[4mm]
\qquad +\dfrac{2}{5}(\tilde{\lambda}_0 - 1)\dfrac{\partial u_0}{\partial x}.
\end{cases}
\tag{3.140}
$$

From (3.136) and (3.122), we have

$$
\overline{D}_{10} = 0.
\tag{3.141}
$$

Thus, from the system of equations (3.138), we directly find that

$$
\overline{A}_{11} = \overline{B}_{11} = \overline{C}_{11} = \overline{D}_{11} = 0,
\tag{3.142}
$$

and therefore

$$
\overline{\overline{A}}_{11} = \overline{\overline{B}}_{11} = \overline{\overline{C}}_{11} = \overline{\overline{D}}_{11} = 0.
\tag{3.143}
$$

Taking into account relations (3.122), (3.136), after algebraic simplifications, the systems of equations (3.139), (3.140) are transformed to the following form

$$\text{for} \quad \cos 3\theta: \begin{cases} -\overline{A}_{21}\,r_0^{-6} + \overline{B}_{21}\,r_0^{-6} + \overline{C}_{21} = \overline{C}_{10}\,r_0^{-2} \\ -\lambda\overline{A}_{21}\,r_0^{-6} + \overline{B}_{21}\,r_0^{-6} - \overline{C}_{21} = \overline{C}_{10}\,r_0^{-2} \\ \overline{B}_{21}\,r_0^{+6} + \overline{C}_{21} - \overline{D}_{21} = \overline{C}_{10}\,r_0^{+2} \\ \overline{B}_{21}\,r_0^{+6} - \overline{C}_{21} + \tilde{\lambda}_0\overline{D}_{21} = \overline{C}_{10}\,r_0^{+2} \end{cases} \tag{3.144}$$

$$\text{for} \quad \cos 5\theta: \begin{cases} -\overline{A}_{31}\,r_0^{-10} + \overline{B}_{31}\,r_0^{-10} + \overline{C}_{31} = \overline{C}_{10}\,r_0^{-4} \\ -\lambda\overline{A}_{31}\,r_0^{-10} + \overline{B}_{31}\,r_0^{-10} - \overline{C}_{31} = -\overline{C}_{10}\,r_0^{-4} \\ \overline{B}_{31}\,r_0^{+10} + \overline{C}_{31} - \overline{D}_{31} = \overline{C}_{10}\,r_0^{+4} \\ \overline{B}_{31}\,r_0^{+10} - \overline{C}_{31} + \tilde{\lambda}_0\overline{D}_{31} = -\overline{C}_{10}\,r_0^{+4} \end{cases} \tag{3.145}$$

Solution of systems of equations (3.144), (3.145) $\overline{A}_{i1}$, $\overline{B}_{i1}$, $\overline{C}_{i1}$, $\overline{D}_{i1}$ $(i=2,3)$ read

$$\overline{A}_{21} = A_{21}^* \frac{\partial u_0}{\partial x} = \frac{\pi}{2a^2}\left(\lambda^2 - 1\right)\left(1 - a^4\right)\Delta\Delta_1 \frac{\partial u_0}{\partial x},$$

$$\overline{B}_{21} = B_{21}^* \frac{\partial u_0}{\partial x} = -\frac{\pi}{8}\left(\lambda - 1\right)a^2\left[\left(\lambda + 1\right) + \left(\lambda - 1\right)a^2\right]\Delta_1 \frac{\partial u_0}{\partial x}, \tag{3.146}$$

$$\overline{C}_{21} = C_{21}^* \frac{\partial u_0}{\partial x} = -\frac{16}{\pi^2}\left(\lambda - 1\right)^2\left(\lambda + 1\right)a^4\left(1 - a^4\right)\Delta\Delta_1 \frac{\partial u_0}{\partial x},$$

$$\overline{D}_{21} = D_{21}^* \frac{\partial u_0}{\partial x} = -\frac{16}{\pi^2}\left(\lambda - 1\right)^2 a^4\left(1 - a^4\right)\Delta_1 \frac{\partial u_0}{\partial x},$$

where

$$\Delta_1 = \left((\lambda + 1)^2 - (\lambda - 1)^2 a^8\right)^{-1},$$

$$\overline{A}_{31} = A_{31}^* \frac{\partial u_0}{\partial x} = -\frac{\pi^2}{8}\left(\lambda - 1\right)^2 a^4\left(1 - a^4\right)\Delta\Delta_2 \frac{\partial u_0}{\partial x},$$

$$\overline{B}_{31} = B_{31}^* \frac{\partial u_0}{\partial x} = -\frac{\pi^2}{16}\left(\lambda - 1\right)^2\left(\lambda + 1\right)a^4\left(1 - a^4\right)\Delta\Delta_2 \frac{\partial u_0}{\partial x}, \tag{3.147}$$

$$\overline{C}_{31} = C_{31}^* \frac{\partial u_0}{\partial x} = -\frac{64}{\pi^3}\left(\lambda - 1\right)a^6\frac{\Delta\Delta_2}{\Delta_1}\frac{\partial u_0}{\partial x},$$

$$\overline{D}_{31} = D_{31}^* \frac{\partial u_0}{\partial x} = \frac{64}{\pi^3}\left(\lambda^2 - 1\right)a^2\left(1 - a^4\right)\Delta_2 \frac{\partial u_0}{\partial x},$$

$$\Delta_2 = \left((\lambda + 1)^2 - (\lambda - 1)^2 a^{12}\right)^{-1}. \tag{3.148}$$

Obviously, for the constants $\overline{\overline{A}}_{i1}$, $\overline{\overline{B}}_{i1}$, $\overline{\overline{C}}_{i1}$, $\overline{\overline{D}}_{i1}$ $(i=2,3)$, we have

$$\overline{\overline{A}}_{i1} = \overline{A}_{i1}, \quad \overline{\overline{B}}_{i1} = \overline{B}_{i1}, \quad \overline{\overline{C}}_{i1} = \overline{C}_{i1}, \quad \overline{\overline{D}}_{i1} = \overline{D}_{i1} \quad \left(\frac{\partial u_0}{\partial x} \to \frac{\partial u_0}{\partial y}\right). \tag{3.149}$$

Homogenized equation can be written as follows:

$$
\frac{1}{|\Omega_1^*|} \left\{ \iint\limits_{\Omega_{i0}^+} \left(\frac{\partial^2 u_0}{\partial x^2} + \frac{\partial^2 u_0}{\partial y^2} + \frac{\partial^2 u_{10}^+}{\partial x \partial \xi} + \frac{\partial^2 u_{10}^+}{\partial y \partial \eta} \right) d\xi\, d\eta + \right.
$$

$$
+ \lambda \iint\limits_{\Omega_{i0}^-} \left(\frac{\partial^2 u_0}{\partial x^2} + \frac{\partial^2 u_0}{\partial y^2} + \frac{\partial^2 u_{10}^-}{\partial x \partial \xi} + \frac{\partial^2 u_{10}^-}{\partial y \partial \eta} \right) d\xi\, d\eta +
$$

$$
+ \tilde{\lambda}_0 \iint\limits_{\tilde{\Omega}_0} \left(\frac{\partial^2 u_0}{\partial x^2} + \frac{\partial^2 u_0}{\partial y^2} + \frac{\partial^2 \tilde{u}_{10}}{\partial x \partial \xi} + \frac{\partial^2 \tilde{u}_{10}}{\partial y \partial \eta} \right) d\xi\, d\eta +
$$

$$
+ \varepsilon_1 \left[\iint\limits_{\Omega_{i0}^+} \left(\frac{\partial^2 u_{11}^+}{\partial x \partial \xi} + \frac{\partial^2 u_{11}^+}{\partial y \partial \eta} \right) d\xi\, d\eta + \lambda \iint\limits_{\Omega_{i0}^-} \left(\frac{\partial^2 u_{11}^-}{\partial x \partial \xi} + \frac{\partial^2 u_{11}^-}{\partial y \partial \eta} \right) d\xi\, d\eta + \right.
$$

$$
+ \tilde{\lambda}_0 \iint\limits_{\tilde{\Omega}_0} \left(\frac{\partial^2 \tilde{u}_{11}}{\partial x \partial \xi} + \frac{\partial^2 \tilde{u}_{11}}{\partial y \partial \eta} \right) d\xi\, d\eta + \tag{3.150}
$$

$$
+ \iint\limits_{\Omega_{i1}^+ \setminus \Omega_{i0}^+} \left(\frac{\partial^2 u_0}{\partial x^2} + \frac{\partial^2 u_0}{\partial y^2} + \frac{\partial^2 u_{10}^+}{\partial x \partial \xi} + \frac{\partial^2 u_{10}^+}{\partial y \partial \eta} \right) d\xi\, d\eta +
$$

$$
\underline{+ \lambda \iint\limits_{\Omega_{i1}^- \setminus \Omega_{i0}^-} \left(\frac{\partial^2 u_0}{\partial x^2} + \frac{\partial^2 u_0}{\partial y^2} + \frac{\partial^2 u_{10}^-}{\partial x \partial \xi} + \frac{\partial^2 u_{10}^-}{\partial y \partial \eta} \right) d\xi\, d\eta +}
$$

$$
\underline{+ \tilde{\lambda}_0 \iint\limits_{\tilde{\Omega}_1 \setminus \tilde{\Omega}_0} \left(\frac{\partial^2 u_0}{\partial x^2} + \frac{\partial^2 u_0}{\partial y^2} + \frac{\partial^2 \tilde{u}_{10}}{\partial x \partial \xi} + \frac{\partial^2 \tilde{u}_{10}}{\partial y \partial \eta} \right) d\xi\, d\eta +}
$$

$$
\left. \left. + \tilde{\lambda}_1 \iint\limits_{\tilde{\Omega}_0} \left(\frac{\partial^2 u_0}{\partial x^2} + \frac{\partial^2 u_0}{\partial y^2} + \frac{\partial^2 \tilde{u}_{10}}{\partial x \partial \xi} + \frac{\partial^2 \tilde{u}_{10}}{\partial y \partial \eta} \right) d\xi\, d\eta \right] \right\} =
$$

$$
= \left(\frac{\partial^2 u_0}{\partial x^2} + \frac{\partial^2 u_0}{\partial y^2} \right) \left(\tilde{\lambda}_0 + \varepsilon_1 \tilde{\lambda}_1 \right),
$$

where $|\Omega_1^*| = |\Omega_{i1}^+ \cup \Omega_{i1}^- \cup \tilde{\Omega}_1|$ is a measure of three-phase area. In relation (3.150), the underlined terms determine the correction to the effective coefficient due to the perturbation of the contours $r = r_0^-$ and $r = r_0^+$, i.e. the correction determined by the terms of the order ε_1: $r = \varepsilon_1 r_0^- \cos 4\theta$ and $r = \varepsilon_1 r_0^+ \cos 4\theta$. Regions of integration $\Omega_{i1}^- \setminus \Omega_{i0}^-$, $\Omega_{i1}^+ \setminus \Omega_{i0}^+$, $\tilde{\Omega}_1 \setminus \tilde{\Omega}_0$ shown in Fig. 3.35, while the signs "+" and "-" mark

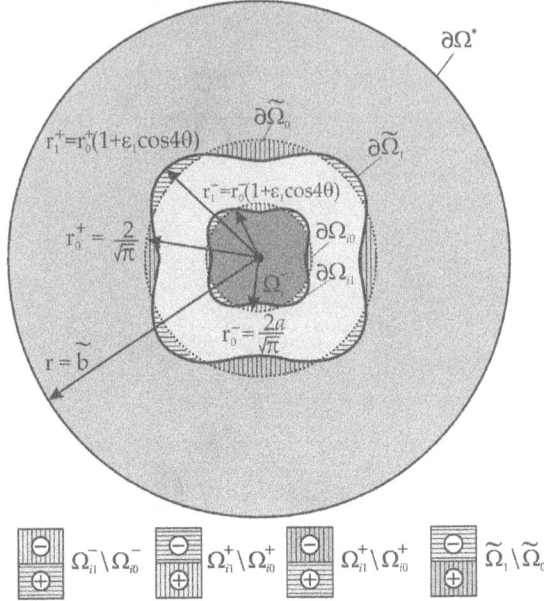

Figure 3.35 Domain with taking into account the first approximation.

positive and negative additions to the corresponding areas. Fig. 3.36 shows the inclusion and cell contours in the zeroth approximation and curves that determine the corrections to them in the first approximation (order of correction is ε_1).

Equating the expressions at ε_1 in (3.150), we obtain the first correction $\tilde{\lambda}_1$ to the homogenized thermal conductivity coefficient:

$$\tilde{\lambda}_1 \Delta u_0 = \iint\limits_{\Omega_{i0}^+} \left(\frac{\partial^2 u_{11}^+}{\partial x \partial \xi} + \frac{\partial^2 u_{11}^+}{\partial y \partial \eta} \right) d\xi \, d\eta +$$

$$+ \lambda \iint\limits_{\Omega_{i0}^-} \left(\frac{\partial^2 u_{11}^-}{\partial x \partial \xi} + \frac{\partial^2 u_{11}^-}{\partial y \partial \eta} \right) d\xi \, d\eta + \tilde{\lambda}_0 \iint\limits_{\tilde{\Omega}_0} \left(\frac{\partial^2 \tilde{u}_{11}}{\partial x \partial \xi} + \frac{\partial^2 \tilde{u}_{11}}{\partial y \partial \eta} \right) d\xi \, d\eta +$$

$$+ \iint\limits_{\Omega_{i1}^+ \setminus \Omega_{i0}^+} \left(\frac{\partial^2 u_0}{\partial x^2} + \frac{\partial^2 u_0}{\partial y^2} + \frac{\partial^2 u_{10}^+}{\partial x \partial \xi} + \frac{\partial^2 u_{10}^+}{\partial y \partial \eta} \right) d\xi \, d\eta +$$

$$+ \lambda \iint\limits_{\Omega_{i1}^- \setminus \Omega_{i0}^-} \left(\frac{\partial^2 u_0}{\partial x^2} + \frac{\partial^2 u_0}{\partial y^2} + \frac{\partial^2 u_{10}^-}{\partial x \partial \xi} + \frac{\partial^2 u_{10}^-}{\partial y \partial \eta} \right) d\xi \, d\eta +$$

$$+ \tilde{\lambda}_0 \iint\limits_{\tilde{\Omega}_1 \setminus \tilde{\Omega}_0} \left(\frac{\partial^2 u_0}{\partial x^2} + \frac{\partial^2 u_0}{\partial y^2} + \frac{\partial^2 \tilde{u}_{10}}{\partial x \partial \xi} + \frac{\partial^2 \tilde{u}_{10}}{\partial y \partial \eta} \right) d\xi \, d\eta +$$

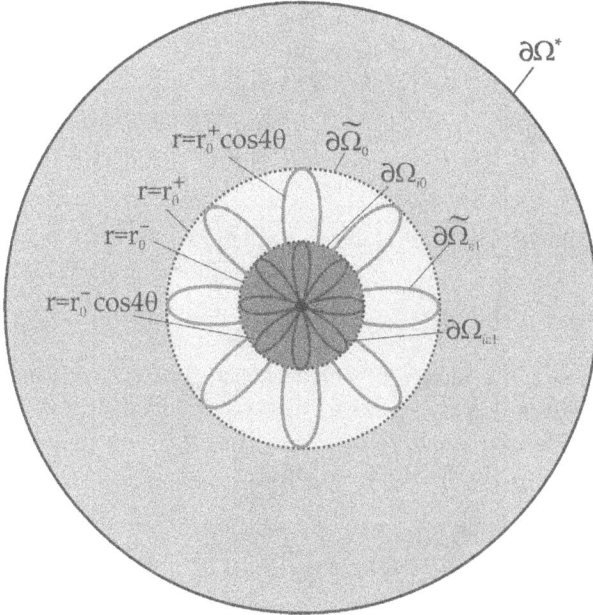

Figure 3.36 Cell and inclusions contour in zero approximation and first corrections to them.

$$\tilde{\lambda}_1 \iint\limits_{\tilde{\Omega}_0} \left(\frac{\partial^2 u_0}{\partial x^2} + \frac{\partial^2 u_0}{\partial y^2} + \frac{\partial^2 \tilde{u}_{10}}{\partial x \partial \xi} + \frac{\partial^2 \tilde{u}_{10}}{\partial y \partial \eta} \right) d\xi \, d\eta,$$

or after obvious transformations

$$\tilde{\lambda}_1 \Delta u_0 = \frac{1}{4b^2} \left[\iint\limits_{\Omega_{i1}^+ \backslash \Omega_{i0}^+} \Delta u_0 \, d\xi \, d\eta + \lambda \iint\limits_{\Omega_{i1}^- \backslash \Omega_{i0}^-} \Delta u_0 \, d\xi \, d\eta + \right.$$

$$+ \tilde{\lambda}_0 \iint\limits_{\tilde{\Omega}_0} \left(\frac{\partial^2 \tilde{u}_{11}}{\partial x \partial \xi} + \frac{\partial^2 \tilde{u}_{11}}{\partial y \partial \eta} \right) d\xi \, d\eta + \qquad (3.151)$$

$$+ \lambda \oint\limits_{\partial \Omega_{i\varepsilon_1}} \frac{\partial u_{10}^-}{\partial \mathbf{n}} \, d\ell + \tilde{\lambda}_0 \oint\limits_{\partial \tilde{\Omega}_{\varepsilon_1}} \frac{\partial \tilde{u}_{10}}{\partial \mathbf{n}} \, d\ell + \oint\limits_{\partial \tilde{\Omega}_0} \frac{\partial u_{11}^+}{\partial \mathbf{n}} \, d\ell +$$

$$\left. + \oint\limits_{\partial \Omega_{i0}} \frac{\partial u_{11}^+}{\partial \mathbf{n}} \, d\ell + \lambda \oint\limits_{\partial \Omega_{i0}} \frac{\partial u_{11}^-}{\partial \mathbf{n}} \, d\ell + \tilde{\lambda}_0 \oint\limits_{\partial \tilde{\Omega}_0} \frac{\partial \tilde{u}_{11}}{\partial \mathbf{n}} \, d\ell \right],$$

where $\mathbf{n}$ is the outer normal to the contour in "slow" variables x, y; $\frac{\partial (\ldots)}{\partial \mathbf{n}} = \frac{\partial (\ldots)}{\partial x} \cos (\mathbf{n}, \, \xi) + \frac{\partial (\ldots)}{\partial y} \cos (\mathbf{n}, \, \eta)$.

Performing integration in (3.151), taking into account the expressions (3.121)–(3.125) and (3.137), (3.146)–(3.149), we have

$$\tilde{\lambda}_1 = \frac{1}{4}\left[2\left(1-a^2\right)\left(1+B_{10}^*\right)+2\lambda a^2\left(1+A_{10}^*\right)-2\tilde{\lambda}_0\right] =$$

$$= \frac{1}{2}\left[\frac{(\lambda+1)\left(1-a^2\right)}{(\lambda+1)-(\lambda-1)a^2}+\frac{2\lambda a^2}{(\lambda+1)-(\lambda-1)a^2}-1\right]=0,$$

i.e.

$$\tilde{\lambda}_1 = 0. \tag{3.152}$$

Thus, in the case of a composite with periodic cylindrical inclusions of a square profile, the first correction to the effective thermal conductivity coefficient $\tilde{\lambda}_0$ determined by the ThPhM is equal to zero.

Therefore, due to (3.131), (3.152), we have

$$\tilde{\lambda} = \tilde{\lambda}_0 + o\left(\varepsilon_1\right) + \ldots = \frac{1-a^2+\lambda\left(1+a^2\right)}{1+a^2+\lambda\left(1-a^2\right)}+o\left(\varepsilon_1\right).$$

Thus, the expression for the effective conductivity q of a composite with cylindrical inclusions of a square profile, obtained on the basis of ThPhM taking into account the first PPB correction, has the following form

$$q = \frac{1-a^2+\lambda\left(1+a^2\right)}{1+a^2+\lambda\left(1-a^2\right)}. \tag{3.153}$$

3.4.5 ASYMPTOTIC RELATIONS AND SOLUTION ANALYSIS

It should be noticed that the expression (3.153) exactly satisfies the Keller's theorem (3.59) for all values of the conductivity of the inclusions, including limit cases $\lambda \to 0$ and $\lambda \to \infty$.

Let us analyze the expression (3.153) for the effective coefficient of thermal conductivity and derive from it, asymptotic representations in the limiting cases of the values of the size of the inclusion a and its conductivity λ.

1) Inclusion sizes are small $(a \to 0)$, in the limit, we have a homogeneous material with matrix conductivity.
   Then, for an arbitrary value of the conductivity of the inclusions λ, we obtain

$$q = 1 + \frac{2(\lambda-1)}{\lambda+1}a^2 + O\left(a^4\right), \tag{3.154}$$

   In particular, from (3.154), we have

   1.1. Inclusions of low conductivity $(\lambda \ll 1)$:

$$q = 1 - 2a^2 + 4a^2\lambda - 4a^2\lambda^2.$$

1.2. Conductivity of inclusions close to the matrix conductivity ($\lambda \sim 1$):

$$q = 1 + a^2 (\lambda - 1) - \frac{a^2}{2} (\lambda - 1)^2.$$

1.3. High conductivity of inclusions ($\lambda \gg 1$):

$$q = 1 + 2a^2 - \frac{4a^2}{\lambda} + \frac{4a^2}{\lambda^2}.$$

2) Large inclusion sizes, $a \to 1$, in the limit – a homogeneous structure, the conductivity of which is equal to the conductivity of the inclusions. Then for any values $0 \leq \lambda < \infty$, we get

$$q = \lambda - (\lambda^2 - 1)(1 - a) + \frac{1}{2}(\lambda^2 - 1)(2\lambda - 1)(1 - a)^2 + O\left((1 - a)^3\right).$$
(3.155)

In special cases of conductivity of inclusions from (3.155), we have
2.1. Inclusions of low conductivity ($\lambda \ll 1$):

$$q = (1 - a)\left(1 + \frac{1 - a}{2}\right) + \left(1 - (1 - a)^2\right)\lambda - (1 - a)\left(1 + \frac{1 - a}{2}\right)\lambda^2.$$
(3.156)

2.2. Conductivities of inclusions and matrices of the same order ($\lambda \sim 1$):

$$q = 1 + \left(1 + (1 - a)^2\right)(\lambda - 1) - (1 - a)\left(1 - \frac{5(1 - a)}{2}\right)(\lambda - 1)^2.$$

2.3. High conductivity inclusions ($1 \ll \lambda \ll \frac{1}{\sqrt{1-a}}$):

$$q = \left(1 - (1 - a)^2\right)\lambda - (1 - a)\left(1 + \frac{1 - a}{2}\right)\lambda^2 +$$
$$+ (1 - a)^2\lambda^3 + (1 - a)\left(1 + \frac{1 - a}{2}\right).$$
(3.157)

The dependence of the effective thermal conductivity coefficient q on the geometric a and physical λ parameters for low conductivity of the inclusions ($\lambda \ll 1$) and their small and large sizes a is shown in Fig. 3.37.

3) Inclusions of low conductivity, $\lambda \to 0$, in the limit with regard to a composite with heat insulators:

$$q = \frac{1 - a^2}{1 + a^2} + \frac{4a^2}{(1 + a^2)^2}\lambda - \frac{4a^2(1 - a^2)}{(1 + a^2)^3}\lambda^2 + O(\lambda^3).$$
(3.158)

In particular, from (3.158), one can obtain asymptotics for inclusions of low conductivity in terms of their size.

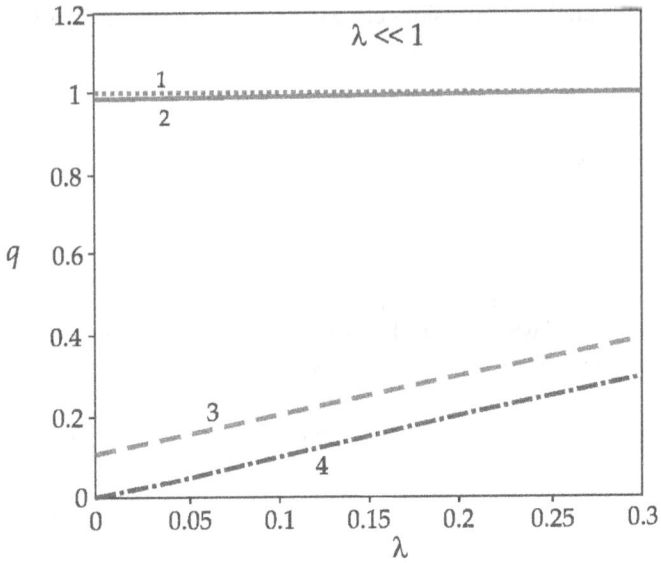

Figure 3.37 Graphs of the effective thermal conductivity parameter q at low conductivity of inclusions $\lambda \ll 1$ and their extremely small and extremely large sizes a ($1 - a = 0$; $2 - a \to 0$; $3 - a \to 1$; $4 - a = 1$).

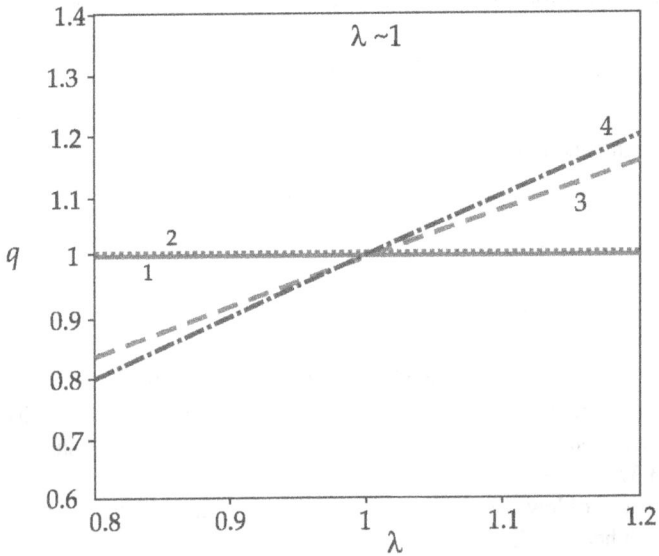

Figure 3.38 Graphs of the effective thermal conductivity parameter q for the conductivity of inclusions $\lambda \sim 1$ and their extremely small and extremely large sizes a ($1 - a = 0$; $2 - a \to 0$; $3 - a \to 1$; $4 - a = 1$).

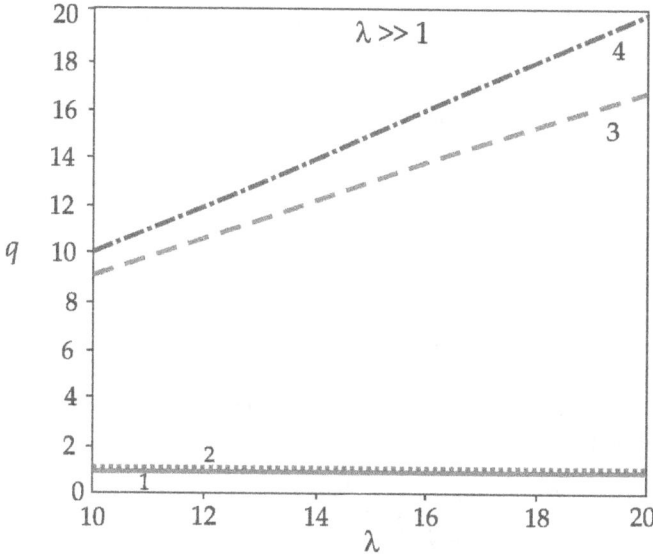

Figure 3.39 Graphs of the effective thermal conductivity parameter q for high conductivity of inclusions and their extremely small and extremely large sizes a ($1 - a = 0$; $2 - a \to 0$; $3 - a \to 1$; $4 - a = 1$).

3.1. Inclusion sizes are small, $a \ll 1$.

Expanding relation (3.158) in a series in powers of the small parameter a, we find

$$q = 1 - 2\left(1 - 2\lambda + 2\lambda^2\right) a^2.$$

For comparison, Voigt averaging in this case is represented by the following relation

$$\bar{q} = 1 - a^2.$$

3.2. Large sizes of inclusions-heat insulators $a \gg 0$.

From (3.158), we have the asymptotic expansion

$$q = \lambda + \left(1 - \lambda^2\right)(1 - a) + \frac{1}{2}\left(1 - 2\lambda - \lambda^2\right)(1 - a)^2, \qquad (3.159)$$

whose leading term at $\lambda = 0$ coincides with the well-known result [3]

$$q = 1 - a. \qquad (3.160)$$

4) Thermal conductivity coefficients of the matrix and inclusions of the same order: $\lambda \to 1$; in the limit – a homogeneous structure (matrix). For arbitrary values of the size of inclusions a, we find

$$q = 1 + a^2\left(\lambda - 1\right) - \frac{a^2}{2}\left(1 - a^2\right)(\lambda - 1)^2 + O\left((\lambda - 1)^3\right). \qquad (3.161)$$

In special cases, for inclusions of small and extremely possible large sizes a, from (3.161), we obtain the following asymptotic relations.

4.1. Small inclusions, $a \ll 1$.
From (3.161), we obtain

$$q = 1 + (\lambda - 1)\left(1 - \frac{\lambda - 1}{2}\right) a^2 + \frac{(\lambda - 1)^2}{2} a^4.$$

4.2. Large inclusion sizes, $a \gg 0$.
In this case, expression (3.161) is transformed to the following form

$$q = \lambda - (\lambda - 1)\left(2 + (\lambda - 1)\right)(1 - a) + (\lambda - 1)\left(1 + \frac{5(\lambda - 1)}{2}\right)(1 - a)^2.$$

5) High conductivity inclusions, $\lambda \to \infty$, in the limit with regard to a composite with absolutely conductive inclusions

$$q = \frac{1 + a^2}{1 - a^2} - \frac{4a^2}{(1 - a^2)^2}\frac{1}{\lambda} - \frac{4a^2\left(1 + a^2\right)}{(1 - a^2)^3}\frac{1}{\lambda^2} + O\left(\frac{1}{\lambda^3}\right). \qquad (3.162)$$

Similarly, we obtain from expression (3.162) various asymptotics with respect to the size of the inclusions.

5.1. Small inclusions, $a \ll 1$.
The asymptotic expansion of the expression (3.162) in a series in powers of the small parameter a leads to the following result

$$q = 1 + 2\left(1 - \frac{2}{\lambda} + \frac{2}{\lambda^2}\right) a^2 + 2\left(1 - \frac{4}{\lambda} + \frac{8}{\lambda^2}\right) a^4.$$

For comparison, at $\lambda \to \infty$, the Reuss averaging in this case is determined by the following relation

$$\underline{q} = 1 + a^2.$$

5.2. Large inclusion sizes, $0 \ll a \ll 1 - 1/\sqrt{\lambda}$.
From (3.161), we obtain the following asymptotic expression for the homogenized coefficient:

$$q = \left(1 + \frac{1}{\lambda} + \frac{1}{2\lambda^2}\right)\frac{1}{1 - a} - \frac{1}{2}\left(1 - \frac{1}{2\lambda}\right) -$$
$$- \frac{1}{\lambda}\left(1 + \frac{3}{2\lambda}\right)\frac{1}{(1 - a)^2} + \frac{1}{\lambda^2}\frac{1}{(1 - a)^3}. \qquad (3.163)$$

The dependence of effective thermal conductivity on the size of inclusions and their conductivity according to the obtained analytical solution (3.153) is shown in Figs. 3.40–3.42.
The results of calculation by formula (3.153) are compared with the results of FEM solution [60] for the concentration of inclusions $c = a^2 = 1/9$ (see Figure 3.43).

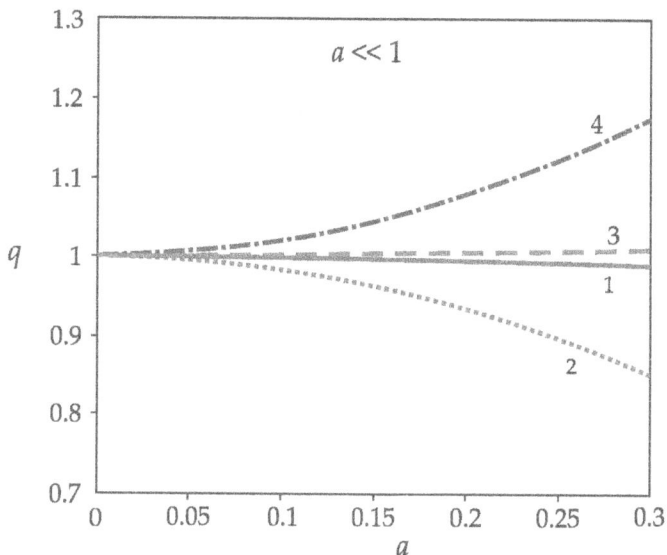

Figure 3.40 Graphs of the effective thermal conductivity parameter q for small sizes of inclusions a of different conductivity λ (1– $\lambda \sim 1(\lambda < 1)$; 2– $\lambda << 1$; 3– $\lambda \sim 1(\lambda > 1)$; 4– $\lambda >> 1$).

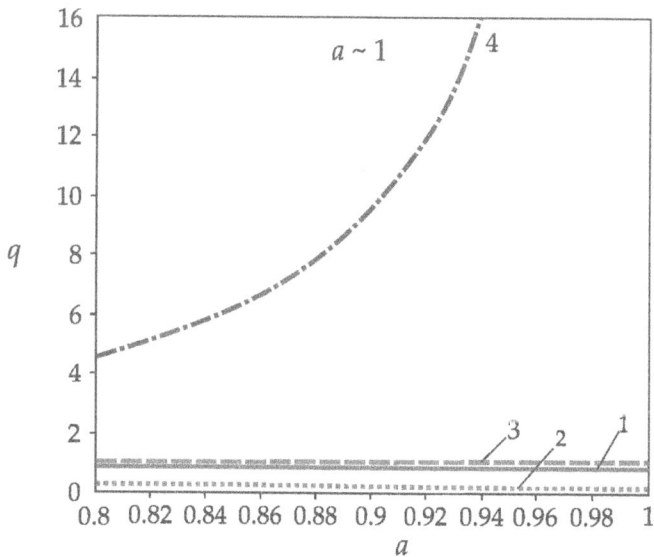

Figure 3.41 Graphs of the effective thermal conductivity parameter q for small sizes of inclusions a of different conductivity λ and $a \sim 1$ (1– $\lambda \sim 1(\lambda < 1)$; 2– $\lambda << 1$; 3– $\lambda \sim 1(\lambda > 1)$; 4– $\lambda >> 1$).

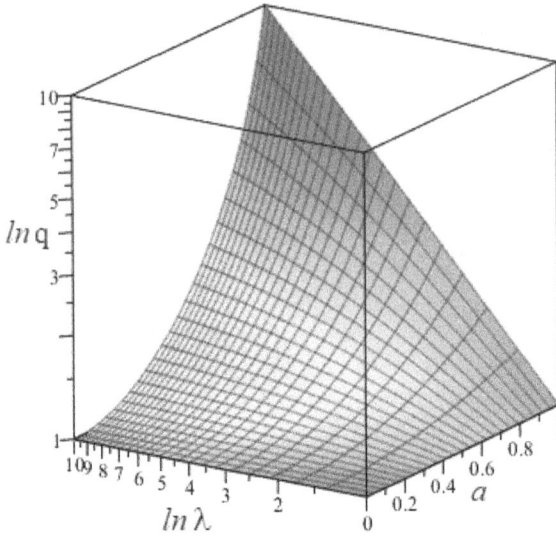

Figure 3.42 Dependence of the effective thermal conductivity q on the size of the inclusions a and their conductivity λ according to the analytical solution (3.153).

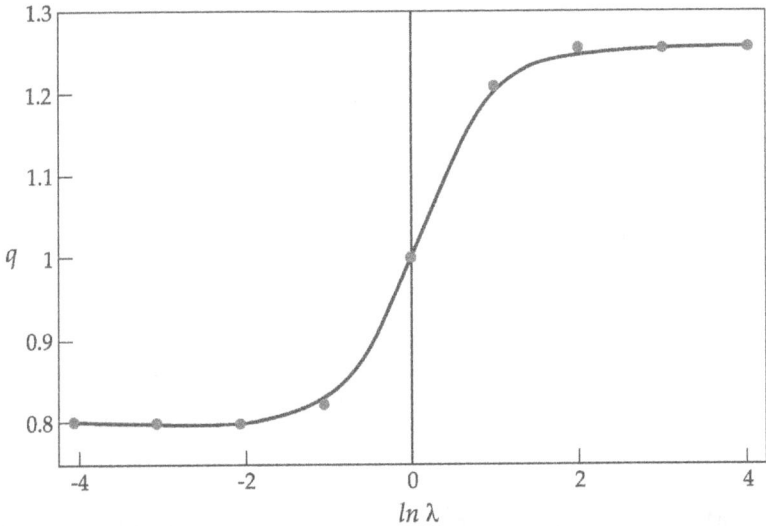

Figure 3.43 Effective thermal conductivity parameter for inclusion concentration $c = 1/9$ (analytical solution – solid line (3.153), numerical solution [29] – dots).

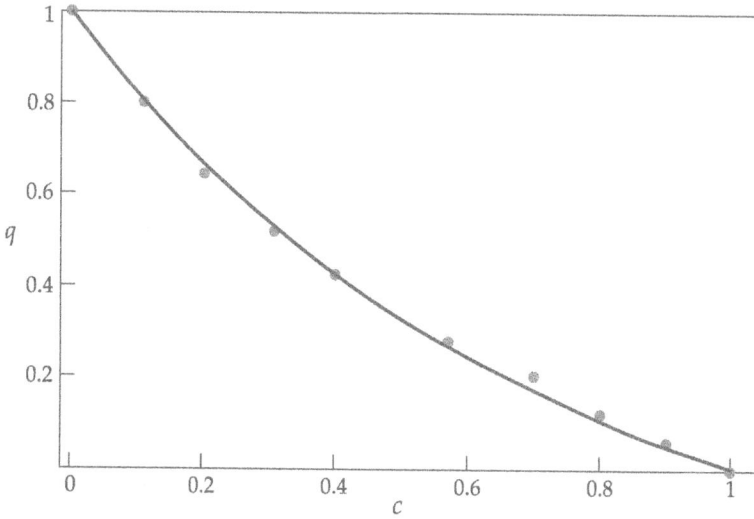

Figure 3.44 Effective thermal conductivity parameter for non-conductive inclusions (analytical solution – solid line (3.153), experimental data [33] – dots).

Table 3.5

Non-conducting inclusions.

The effective thermal conductivity q			
$\lambda = 0$; $a = \frac{1}{3}$			
Bubnov-Galerkin method [25]	R-function method [26]	FEM solution [60]	ThPhM formula (3.153)
0.8232	0.8187	0.81	0.8000

The calculation of the effective parameter for the particular case of non-conductive inclusions ($\lambda = 0$) is shown in Fig. 3.44, where the analytical solution (3.153) is compared with experimental data [73].

We also compare the numerical values of the effective thermal conductivity parameter calculated by formula (3.153) with known data (Tables 3.5, 3.6).

3.5 MODIFIED THREE-PHASE COMPOSITE MODEL

MThPhM does not allow overcoming the limitations of the ThPhM model for large sizes of inclusions of extremely high conductivity.

In this regard, a modernized algorithm has been developed, the essence and main provisions of which are illustrated on the problem of the heat conductivity for a composite with cylindrical inclusions, circular in cross section.

Table 3.6

Inclusion of high conductivity, including the case of absolute conductivity.

The effective thermal conductivity q		
$\lambda = 10;\ a = \frac{1}{2}$		
FEM solution [60]	ThPhM formula (3.153)	ThPhM decision error
1.548	1.5143	−2.18%
$\lambda = 114;\ a = \frac{1}{3}$		
FEM solution [60]	ThPhM formula (3.153)	ThPhM error
1.2500	1.2451	−0.39%
$\lambda \to \infty;\ a = \frac{1}{3}$		
FEM solution [60]	ThPhM formula (3.153)	ThPhM error
1.3	1.2500	−3.85%

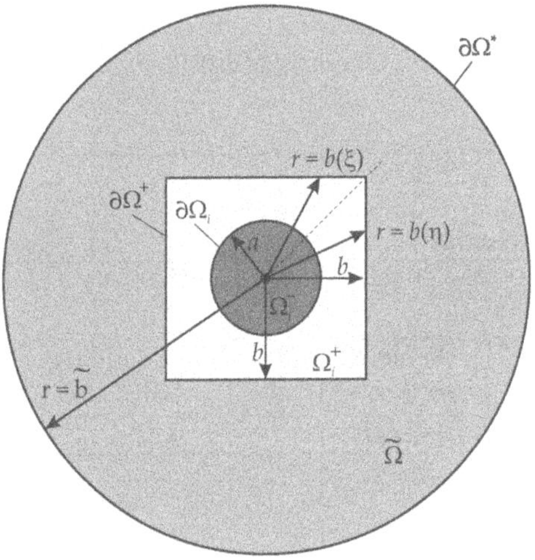

Figure 3.45 MThPhM scheme.

3.5.1 THE PROBLEM OF THERMAL CONDUCTIVITY FOR A 2D COMPOSITE

The physical formulation of the problem and its mathematical description in terms of homogenization theory are given in Section 3.2.1. Consider the solution of the local problem (3.30)–(3.32) based on the MThPhM.

The essence of MThPhM is as follows: all but one of the cells of the composite structure are replaced by a homogeneous medium $\tilde{\Omega}$, having an unknown thermal conductivity coefficient $\tilde{\lambda}$. In this case, we consider the square contour of the cell as a curve of variable radius (Fig. 3.45):

$$r = \begin{cases} b(\eta) = \sqrt{1+\eta^2}, & 0 \le \eta \le 1 \quad \text{at} \quad 0 \le \theta \le \pi/4 \\ b(\xi) = \sqrt{1+\xi^2}, & 0 \le \xi \le 1 \quad \text{at} \quad \pi/4 \le \theta \le \pi/2 \end{cases} \tag{3.164}$$

The cell problem (3.30)–(3.32) in "fast" polar coordinates r, θ is written as follows:

$$\frac{\partial^2 u_1^+}{\partial r^2} + \frac{1}{r}\frac{\partial u_1^+}{\partial r} + \frac{1}{r^2}\frac{\partial^2 u_1^+}{\partial \theta^2} = 0 \quad \text{in} \quad \Omega_i^+, \tag{3.165}$$

$$\frac{\partial^2 u_1^-}{\partial r^2} + \frac{1}{r}\frac{\partial u_1^-}{\partial r} + \frac{1}{r^2}\frac{\partial^2 u_1^-}{\partial \theta^2} = 0 \quad \text{in} \quad \Omega_i^-, \tag{3.166}$$

$$\frac{\partial^2 \tilde{u}_1}{\partial r^2} + \frac{1}{r}\frac{\partial \tilde{u}_1}{\partial r} + \frac{1}{r^2}\frac{\partial^2 \tilde{u}_1}{\partial \theta^2} = 0 \quad \text{in} \quad \tilde{\Omega}, \tag{3.167}$$

at $r = a$:

$$u_1^+ = u_1^-, \quad \frac{\partial u_1^+}{\partial r} - \lambda \frac{\partial u_1^-}{\partial r} = (\lambda - 1)\left(\frac{\partial u_0}{\partial x}\cos\theta + \frac{\partial u_0}{\partial y}\sin\theta\right), \tag{3.168}$$

at $r = b$:

$$u_1^+ = \tilde{u}_1, \quad \frac{\partial u_1^+}{\partial r} - \tilde{\lambda}\frac{\partial \tilde{u}_1}{\partial r} = \left(\tilde{\lambda} - 1\right)\left(\frac{\partial u_0}{\partial x}\cos\theta + \frac{\partial u_0}{\partial y}\sin\theta\right), \tag{3.169}$$

$$\tilde{u}_1 \to 0, \quad \frac{\partial \tilde{u}_1}{\partial r} \to 0 \quad \text{at} \quad r = \tilde{b} \to \infty. \tag{3.170}$$

The solution of problem (3.165)-(3.170) has the following form

$$u_1^- = A_1 r \cos\theta + A_2 r \sin\theta,$$

$$u_1^+ = \left(B_1 r + \frac{C_1}{r}\right)\cos\theta + \left(B_2 r + \frac{C_2}{r}\right)\sin\theta, \tag{3.171}$$

$$\tilde{u}_1 = \frac{D_1}{r}\cos\theta + \frac{D_2}{r}\sin\theta,$$

where

$$A_1 = -\left(1 + \frac{4\tilde{\lambda}b^2}{\left(\tilde{\lambda} - 1\right)(\lambda - 1)a^2 - \left(\tilde{\lambda} + 1\right)(\lambda + 1)b^2}\right)\frac{\partial u_0}{\partial x},$$

$$B_1 = -\left(1 + \frac{2\tilde{\lambda}(\lambda + 1)b^2}{\left(\tilde{\lambda} - 1\right)(\lambda - 1)a^2 - \left(\tilde{\lambda} + 1\right)(\lambda + 1)b^2}\right)\frac{\partial u_0}{\partial x},$$

$$C_1 = \frac{2\tilde{\lambda}(\lambda - 1)a^2 b^2}{\left(\tilde{\lambda} - 1\right)(\lambda - 1)a^2 - \left(\tilde{\lambda} + 1\right)(\lambda + 1)b^2}\frac{\partial u_0}{\partial x}, \tag{3.172}$$

$$D_1 = b^2 \left(1 + 2 \frac{(\lambda+1)\,b^2 + (\lambda-1)\,a^2}{\left(\tilde{\lambda}-1\right)(\lambda-1)\,a^2 - \left(\tilde{\lambda}+1\right)(\lambda+1)\,b^2} \right) \frac{\partial u_0}{\partial x},$$

$$A_2 = A_1, \quad B_2 = B_1, \quad C_2 = C_1, \quad D_2 = D_1 \left(\frac{\partial u_0}{\partial x} \to \frac{\partial u_0}{\partial y} \right).$$

During averaging the equation

$$\frac{\partial^2 u_0}{\partial x^2} + \frac{\partial^2 u_0}{\partial y^2} + 2\frac{\partial^2 u_1^+}{\partial x \partial \xi} + 2\frac{\partial^2 u_1^+}{\partial y \partial \eta} + \frac{\partial^2 u_2^+}{\partial \xi^2} + \frac{\partial^2 u_2^+}{\partial \eta^2} +$$

$$+\lambda \left(\frac{\partial^2 u_0}{\partial x^2} + \frac{\partial^2 u_0}{\partial y^2} + 2\frac{\partial^2 u_1^-}{\partial x \partial \xi} + 2\frac{\partial^2 u_1^-}{\partial y \partial \eta} + \frac{\partial^2 u_2^-}{\partial \xi^2} + \frac{\partial^2 u_2^-}{\partial \eta^2} \right) +$$

$$+\tilde{\lambda} \left(\frac{\partial^2 u_0}{\partial x^2} + \frac{\partial^2 u_0}{\partial y^2} + 2\frac{\partial^2 \tilde{u}_1}{\partial x \partial \xi} + 2\frac{\partial^2 \tilde{u}_1}{\partial y \partial \eta} + \frac{\partial^2 \tilde{u}_2}{\partial \xi^2} + \frac{\partial^2 \tilde{u}_2}{\partial \eta^2} \right) = F,$$

the integration is performed taking into account the relation (3.164), i.e., in expressions (3.172) we set:

$$b = \begin{cases} b(\xi) = \sqrt{1+\xi^2}, & 0 \le \xi \le 1 \quad \text{in} \quad \Omega_{i1}^+, \Omega_{i1}^-, \tilde{\Omega}_1 \\ b(\eta) = \sqrt{1+\eta^2}, & 0 \le \eta \le 1 \quad \text{in} \quad \Omega_{i2}^+, \Omega_{i2}^-, \tilde{\Omega}_2 \end{cases} \tag{3.173}$$

Fig. 3.46 shows one-fourth of domain over which integration is performed. Taking into account, relations (3.172), (3.173), we carry out the integration

$$\frac{1}{|\Omega^*|} \left[\iint\limits_{\Omega_i^+} \left(\frac{\partial^2 u_0}{\partial x^2} + \frac{\partial^2 u_0}{\partial y^2} + \frac{\partial^2 u_1^+}{\partial x \partial \xi} + \frac{\partial^2 u_1^+}{\partial y \partial \eta} \right) d\xi\, d\eta + \right.$$

$$+\lambda \iint\limits_{\Omega_i^-} \left(\frac{\partial^2 u_0}{\partial x^2} + \frac{\partial^2 u_0}{\partial y^2} + \frac{\partial^2 u_1^-}{\partial x \partial \xi} + \frac{\partial^2 u_1^-}{\partial y \partial \eta} \right) d\xi\, d\eta +$$

$$\left. +\tilde{\lambda} \iint\limits_{\tilde{\Omega}} \left(\frac{\partial^2 u_0}{\partial x^2} + \frac{\partial^2 u_0}{\partial y^2} + \frac{\partial^2 \tilde{u}_1}{\partial x \partial \xi} + \frac{\partial^2 \tilde{u}_1}{\partial y \partial \eta} \right) d\xi\, d\eta \right] = F,$$

where

$$\Omega_i^\pm = \Omega_{i1}^\pm \bigcup \Omega_{i2}^\pm, \quad \tilde{\Omega} = \tilde{\Omega}_1 \bigcup \tilde{\Omega}_2, \quad |\Omega^*| = \left| \Omega_i^+ \bigcup \Omega_i^- \bigcup \tilde{\Omega} \right|.$$

Then, after performing the necessary transformations, we come to the following transcendental equation

$$\tilde{\lambda} = \frac{1 + \left(\frac{1}{\Delta} - \Delta \right) \arctan \frac{1}{\Delta} + \frac{\lambda-1}{\lambda+1} \frac{a^2}{\Delta} \arctan \frac{1}{\Delta}}{1 + \left(\frac{1}{\Delta} - \Delta \right) \arctan \frac{1}{\Delta} - \frac{\lambda-1}{\lambda+1} \frac{a^2}{\Delta} \arctan \frac{1}{\Delta}}, \tag{3.174}$$

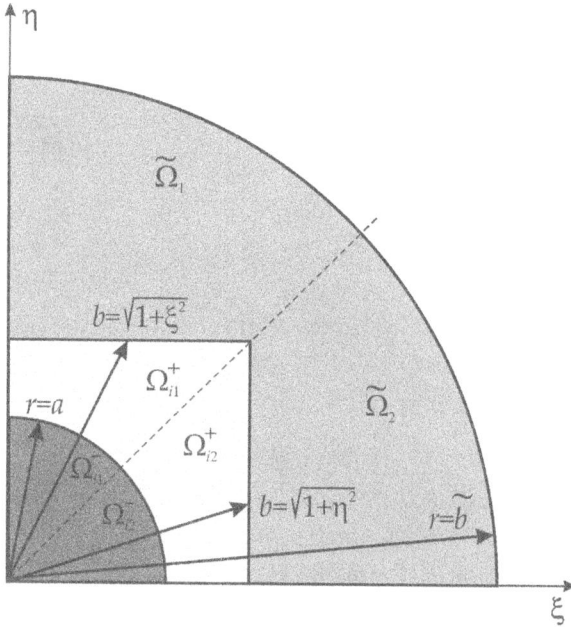

Figure 3.46 Approximation of domain in MThPhM.

where

$$\Delta = \sqrt{1 - \frac{\tilde{\lambda} - 1}{\tilde{\lambda} + 1} \frac{\lambda - 1}{\lambda + 1} a^2}. \tag{3.175}$$

The transcendental equation (3.174), (3.175) can be solved numerically, except for the range of limiting parameter values: $\lambda \to \infty$, $a \to 1$ and $\lambda \to 0$, $a \to 1$. In the latter case, it is worthy to use asymptotic representations, which can further be obtained from expressions (3.174), (3.175).

3.5.2 ASYMPTOTIC RELATIONS OF MTHPHM

1) The expression for the effective thermal conductivity parameter obtained from the transcendental equation (3.174), (3.175) satisfies the Keller's theorem [127]:

$$\tilde{\lambda}(\lambda) = \tilde{\lambda}^{-1}(\lambda^{-1}). \tag{3.176}$$

Indeed,

$$\tilde{\lambda}^{-1}(\lambda^{-1}) = \left(\frac{1 + \left(\frac{1}{\Delta} - \Delta\right) \arctan \frac{1}{\Delta} - \frac{\lambda - 1}{\lambda + 1} \frac{a^2}{\Delta} \arctan \frac{1}{\Delta}}{1 + \left(\frac{1}{\Delta} - \Delta\right) \arctan \frac{1}{\Delta} + \frac{\lambda - 1}{\lambda + 1} \frac{a^2}{\Delta} \arctan \frac{1}{\Delta}} \right)^{-1} =$$

$$= \frac{1 + \left(\frac{1}{\Delta} - \Delta\right) \arctan \frac{1}{\Delta} + \frac{\lambda - 1}{\lambda + 1} \frac{a^2}{\Delta} \arctan \frac{1}{\Delta}}{1 + \left(\frac{1}{\Delta} - \Delta\right) \arctan \frac{1}{\Delta} - \frac{\lambda - 1}{\lambda + 1} \frac{a^2}{\Delta} \arctan \frac{1}{\Delta}},$$

where

$$\Delta\left(\tilde{\lambda}^{-1}, \lambda^{-1}\right) = \Delta\left(\tilde{\lambda}, \lambda\right),$$

i.e., the relation obviously follows (3.176).

Further, in the (3.174), (3.175) expression $\Delta\left(\tilde{\lambda}, \lambda\right)$ can be consided as a small parameter:

$$0 \le \Delta \le 1$$

for any values $0 \le \lambda < \infty$ and $0 \le a \le 1$.

2) Consider a composite in which the conductivity of the inclusions and the conductivity of the matrix are of the same order, i.e. $\lambda \to 1$.
In this case, for any values of the size of inclusions a, we obtain

$$\tilde{\lambda} \sim 1 \quad \Rightarrow \quad \Delta \to 1.$$

Therefore, we obtain that the transcendental equation (3.174), (3.175) reduces to the form:

2.1.   for inclusions of small sizes, $a \to 0$:

$$\tilde{\lambda} = 1 + (\lambda - 1)\,\frac{\pi a^2}{4}.$$

2.2.   for large inclusions, $a \to 1$:

$$\tilde{\lambda} = 1 + \frac{\pi\,(\lambda - 1)}{4} - \frac{\pi\,(\lambda - 1)}{2}\,(1 - a).$$

3) Consider the case of absolutely conducting inclusions, $\lambda \to \infty$.

3.1.   If the sizes of inclusions are small, i.e. $a \to 0$, then we have:

$$\tilde{\lambda} \sim 1 \quad \Rightarrow \quad \Delta \to 1,$$

hence,

$$\tilde{\lambda} = 1 + \left(1 - \frac{2}{\lambda}\right)\frac{\pi a^2}{2}.$$

3.2.   Inclusions of large sizes, $0 << a < 1$.
From the physical essence of the problem, it is obvious that in the case of inclusions of infinitely high conductivity ($\lambda \to \infty$) with sizes close to extremely large ($a \sim 1$), effective conductivity $\tilde{\lambda}$ will also be infinitely large:

$$\tilde{\lambda} \to \infty \quad \Rightarrow \quad \Delta \to 0.$$

Therefore, the transcendental equation (3.174), (3.175) is transformed to the following form

$$\tilde{\lambda} = 1 + \frac{2a^2}{\sqrt{1 - a^2}}\,\arctan\frac{1}{\sqrt{1 - a^2}}, \qquad (3.177)$$

i.e. for $a \to 1$, we have

$$\tilde{\lambda} = \frac{\pi}{\sqrt{1-a^2}} + 1. \tag{3.178}$$

The main term of the expression (3.178) coincides with the asymptotic solution $q_{as}^{(\infty)}$ (3.68), obtained in [158] (up to the normalization adopted in this article) for the effective thermal conductivity of a composite with cylindrical absolutely conducting inclusions of large size, circular in cross section.

4) Consider the case of non-conducting inclusions, $\lambda \to 0$.

4.1. For small inclusions ($a \to 0$), we get

$$\tilde{\lambda} \sim 1 \quad \Rightarrow \quad \Delta \to 1,$$

hence

$$\tilde{\lambda} = 1 - (1 - 2\lambda)\frac{\pi a^2}{2}. \tag{3.179}$$

Relation (3.179) for $\lambda = 0$ coincides up to terms of the order a^2 inclusive with expression

$$q = 1 - \frac{\pi a^2}{2},$$

obtained in [2] for the effective conductivity of a composite with absolutely non-conductive inclusions.

4.2. Large inclusions, $0 << a < 1$.

Similarly to how it was done in Section 3.3.2, the passage to the limit is carried out in the relations (3.174), (3.175) for large non-conductive inclusions: $\lambda \to 0$; $a \sim 1$. In this case, the effective conductivity $\tilde{\lambda}$ is infinitely small:

$$\tilde{\lambda} \to 0 \quad \Rightarrow \quad \Delta \to 0.$$

Then we have:

$$\tilde{\lambda} = \frac{1}{1 + \frac{2a^2}{\sqrt{1-a^2}} \arctan \frac{1}{\sqrt{1-a^2}}}, \tag{3.180}$$

i.e. for $a \to 1$:

$$\tilde{\lambda} = \frac{\sqrt{1-a^2}}{\pi + \sqrt{1-a^2}}. \tag{3.181}$$

The main term of the expression (3.181) coincides with the asymptotics of a large non-conducting inclusion (3.99), obtained by transforming relation (3.68) using the Keller's theorem [127].

5) Let us analyse now composite with small inclusiond, $a \to 0$.

From physical considerations, we conclude that in this case for inclusions of any conductivity, $0 < \lambda < \infty$, including infinitely large ($\lambda \to \infty$) and small ($\lambda \to 0$), the effective conductivity is described by the relation:

$$\tilde{\lambda} \sim 1 \quad (\Delta \to 1). \tag{3.182}$$

Then, taking into account (3.182), we obtain:

5.1. for inclusions of infinite conductivity, $\lambda \to \infty$:

$$\tilde{\lambda} = 1 + \frac{\pi a^2}{2} - \frac{\pi a^2}{\lambda};$$

5.2. for inclusions of small conductivity, $\lambda \to 0$:

$$\tilde{\lambda} = 1 - \frac{\pi a^2}{2} + \pi a^2 \lambda;$$

5.3. for inclusions whose conductivity is of the order of the matrix conductivity, $\lambda \sim 1$:

$$\tilde{\lambda} = 1 + \frac{\pi a^2}{4} (\lambda - 1).$$

6) For a composite with large inclusions, we obtain that for any conductivity of the inclusions ($0 \leq \lambda < \infty$), effective conductivity $\tilde{\lambda}$ can be estimated as follows:

$$\frac{\tilde{\lambda} - 1}{\tilde{\lambda} + 1} \sim \frac{\lambda - 1}{\lambda + 1}$$

because

$$\frac{\tilde{\lambda} - 1}{\tilde{\lambda} + 1} \sim \frac{\lambda - 1}{\lambda + 1} \approx -1 \quad \text{at} \quad \lambda \to 0,$$

$$\frac{\tilde{\lambda} - 1}{\tilde{\lambda} + 1} \sim \frac{\lambda - 1}{\lambda + 1} \approx 0 \quad \text{at} \quad \lambda \sim 1,$$

$$\frac{\tilde{\lambda} - 1}{\tilde{\lambda} + 1} \sim \frac{\lambda - 1}{\lambda + 1} \approx 1 \quad \text{at} \quad \lambda \to \infty,$$

Then the transcendental equation (3.174) reduces to a relation, in which the expression Δ (3.175) takes the form:

$$\Delta = \sqrt{1 - \left(\frac{\lambda - 1}{\lambda + 1}\right)^2 a^2}. \tag{3.183}$$

In particular, we have

6.1. for inclusions of high conductivity, $\lambda \gg 1$:

$$\tilde{\lambda} = \frac{\pi}{\sqrt{2}} \sqrt{\lambda} - 1;$$

6.2. for inclusions of low conductivity, $\lambda \ll 1$:

$$\tilde{\lambda} = \frac{\sqrt{2}}{\pi} \sqrt{\lambda} + \frac{2}{\pi^2} \lambda;$$

6.3. for inclusions with conductivity of the order of matrix conductivity, $\lambda \sim 1$:

$$\tilde{\lambda} = 1 + \frac{\pi}{4} (\lambda - 1).$$

3.5.3 ANALYSIS OF THE RESULTS

1) Figs. 3.47a–j show graphs of the effective thermal conductivity calculated by MThPhM for different values of the conductivity of the inclusions λ:
   (i) $\lambda << 1$ $(\lambda = 10^{-2};\ \lambda = 10^{-1})$ (Figs. 3.47a,b);
   (ii) $\lambda < 1$ $(\lambda = 0.2;\ \lambda = 0.5)$ (Figs. 3.47c,d);
   (iii) $\lambda \sim 1$ $(\lambda = 0.8;\ \lambda = 1.25)$ (Figs. 3.47e,f);
   (iv) $\lambda > 1$ $(\lambda = 2;\ \lambda = 5)$ (Figs. 3.47g,h);
   (v) $\lambda >> 1$ $(\lambda = 10;\ \lambda = 10^2)$ (Figs. 3.47i,j).
   For comparison, the Hashin–Shtrikman bounds are also given, defined by the relations (3.92), (3.93).
   The solution of the transcendental equation (3.174), (3.175) was determined numerically.
2) For cases of inclusions of extremely large sizes $(a \to 1)$ and conductivity close to the maximum value $(\lambda \to \infty)$ or extremely small $(\lambda \to 0)$, Figs. 3.48 and 3.49 show graphs of the effective thermal conductivity, constructed in accordance with MThPhM, as well as the corresponding asymptotic solutions $q_{as}^{(\infty)}$ (3.68) and $q_{as}^{(0)}$ (3.99).
   Graphs for $\lambda \to \infty$, $a \to 1$ and $\lambda \to 0$, $a \to 1$ are constructed using the asymptotic relations (3.177) and (3.180), respectively.
3) For small inclusions $(a << 1)$ and various conductivities λ in Figs. 3.50a–h graphs of the effective thermal conductivity parameter, obtained according to MThPhM, are presented in comparison with the solution found by the Schwarz alternating method.
4) Figs. 3.51a–d show graphs of the effective thermal conductivity calculated by MThPhM for various inclusion sizes a:
   (i) small inclusion: $a = 0.2$ (Fig. 3.51a);
   (ii) inclusions of medium sizes: $a = 0.4$, $a = 0.6$ (Fig. 3.51b,c);
   (iii) large inclusion: $a = 0.8$ (Fig. 3.51d).
   The same figures show the upper $\bar{q}_{HS}$ and lower $\underline{q}_{HS}$ Hashin-Shtrikman bounds, as well as the ThPhM solution, which is the same as $\underline{q}_{HS}$ for $1 \leq \lambda < \infty$ and $\bar{q}_{HS}$ for $0 \leq \lambda \leq 1$.

5) Graphs in Figs. 3.52a–f illustrate the effective thermal conductivity coefficient depending on the conductivity of λ for extremely large inclusions $(a \to 1)$. When constructing the graphs, the asymptotic relation (3.174), (3.183) was used. Graphs are also shown for comparison $q^{(\infty)}$ and $q^{(0)}$ constructed from the asymptotic relations (3.68) and (3.99) [127, 158], and the Hashin-Shtrikman bounds (3.92) for $0 \leq \lambda \leq 1$ and (3.93) for $1 \leq \lambda < \infty$.
6) For the case of a composite with absolutely conducting inclusions $(\lambda \to \infty)$ of large sizes $(a \to 1)$, comparison of the results, obtained by MThPhM, with the data of other authors is shown in Table 3.7 and Table 3.9.
7) Table. 3.8 shows the effective coefficient calculated by MThPhM in comparison with the asymptotic solution [158], for cases of large inclusions $(0 << a < 1)$ and high conductivity $(0 << \lambda < \infty)$.

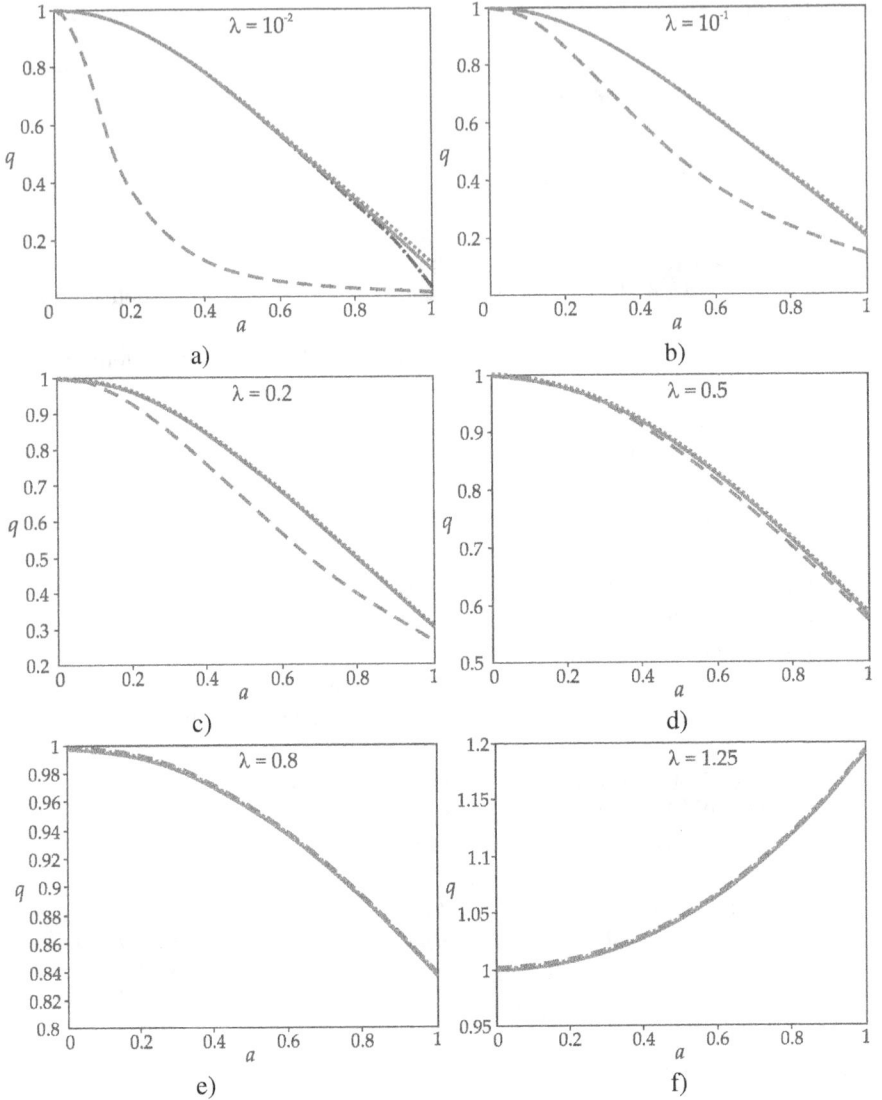

Figure 3.47 Graphs of the effective thermal conductivity at: (a) $\lambda = 10^{-2}$, (b) $\lambda = 10^{-1}$, (c) $\lambda = 0.2$, (d) $\lambda = 0.5$, (e) $\lambda = 0.8$, (f) $\lambda = 1.25$ (solid line – q_{MThPM} (3.174), (3.175), dotted line – $\bar{q}_{HS}(q_{ThPhM})$, dashed line – $\underline{q}_{HS}$, pointed line – $q_{as}^{(0)}$ [127, 158], two-pointed line – $q_{as}^{(\infty)}$ [158]).

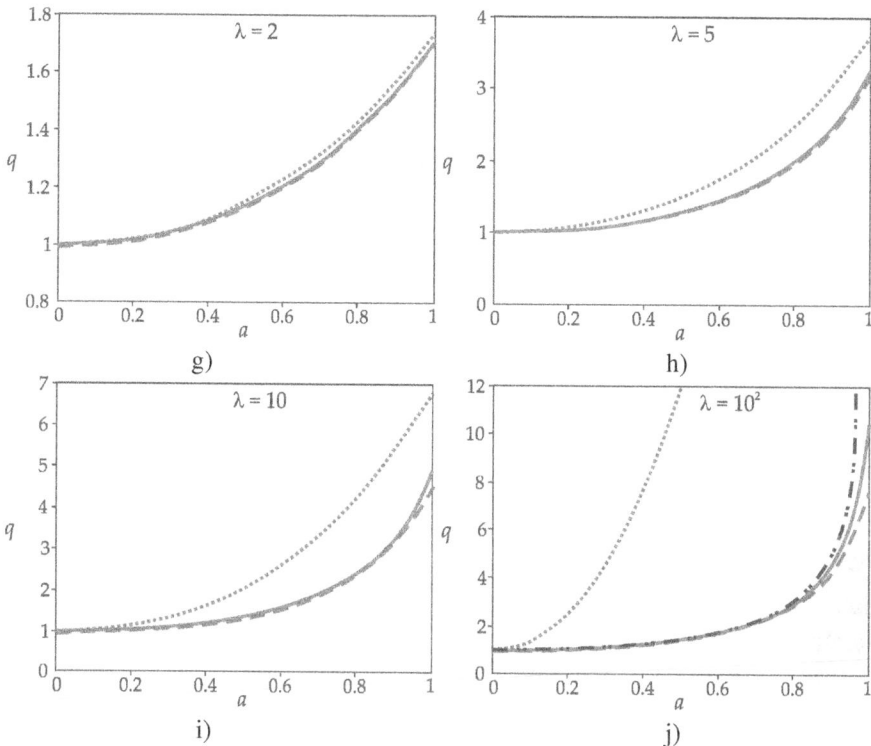

Fig. 3.47 (continued): (g) $\lambda = 2$, (h) $\lambda = 5$, (i) $\lambda = 10$, (j) $\lambda = 10^2$.

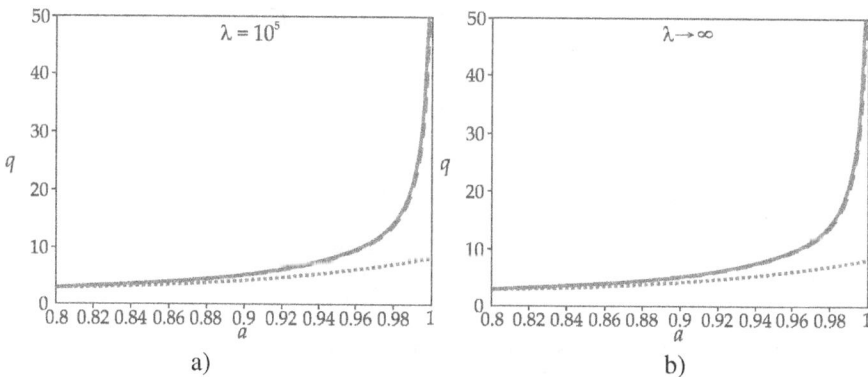

Figure 3.48 Graphs of the effective coefficient of thermal conductivity at: (a) $\lambda = 10^5$, $a \to 1$; (b) $\lambda \to \infty, a \to 1$ (solid line – q_{MThPM} (3.177), dotted line – $\underline{q}_{HS}(q_{ThPhM})$, dashed line – $q_{as}^{(\infty)}$ [158]).

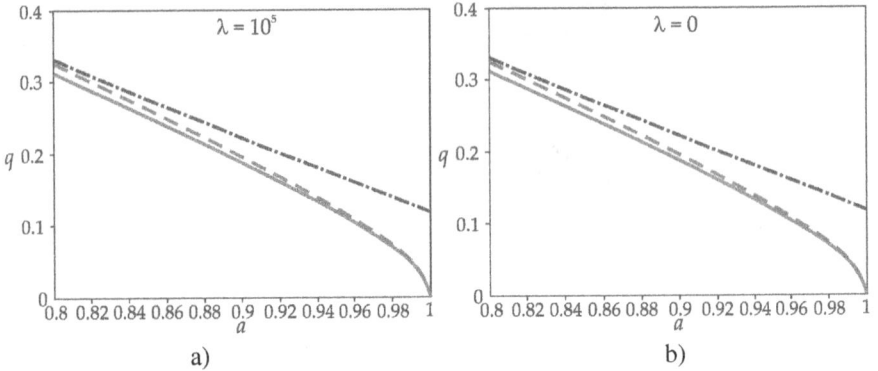

Figure 3.49   Graphs of the effective coefficient of thermal conductivity at: (a) $\lambda = 10^{-5}, a \rightarrow 1$; (b) $\lambda \rightarrow 0, a \rightarrow 1$ (solid line – q_{MThPM} (3.180), dashed line – $q_{as}^{(0)}$ [127, 158]), pointed line – $\bar{q}_{HS}(q_{ThPhM})$).

Table 3.7

Results of calculations by various methods of the effective thermal conductivity coefficient for absolutely conductive inclusions.

Inclusion concentration c	Inclusion size a	Formula (3.9) [119]	Formula (3.12) [119]	Formula (3.13) [119]	Numerical solution [194]	Asymptotics [158]	MThPhM
0.1	0.3568	1.210	1.247	1.223	1.222	1.2214	1.2234
0.2	0.5046	1.470	1.544	1.506	1.500	1.4973	1.5065
0.3	0.6180	1.811	1.918	1.879	1.860	1.8546	1.8790
0.4	0.7136	2.306	2.417	2.395	2.351	2.3432	2.3955
0.5	0.7979	3.270	3.145	3.172	3.080	3.0700	3.1720
0.6	0.8740	7.106	4.386	4.517	4.342	4.3245	4.5175
0.7	0.9441	73.92	7.409	7.769	7.433	7.3857	7.7695
0.74	0.9707	–	10.91	11.46	11.01	10.9254	11.4624
0.76	0.9837	–	15.29	15.99	15.44	15.3284	15.9902
0.77	0.9901	–	20.18	21.04	20.43	20.2952	21.0488
0.78	0.9966	–	35.01	36.60	35.93	35.7525	36.6519

8) For small inclusion sizes ($a \ll 1$), an expression for the effective thermal conductivity was obtained using the Schwarz alternating method.
   Table 3.10 shows the effective thermal conductivity coefficient calculated by ThPhM, by the Schwarz alternating method [26], and by the MThPhM for various cases of conductivity of inclusions λ and small sizes of inclusions a.

9) In Table 3.11 for various parameters of composite, the effective thermal conductivity coefficient obtained by MThPhM is given in comparison with the results of other studies.

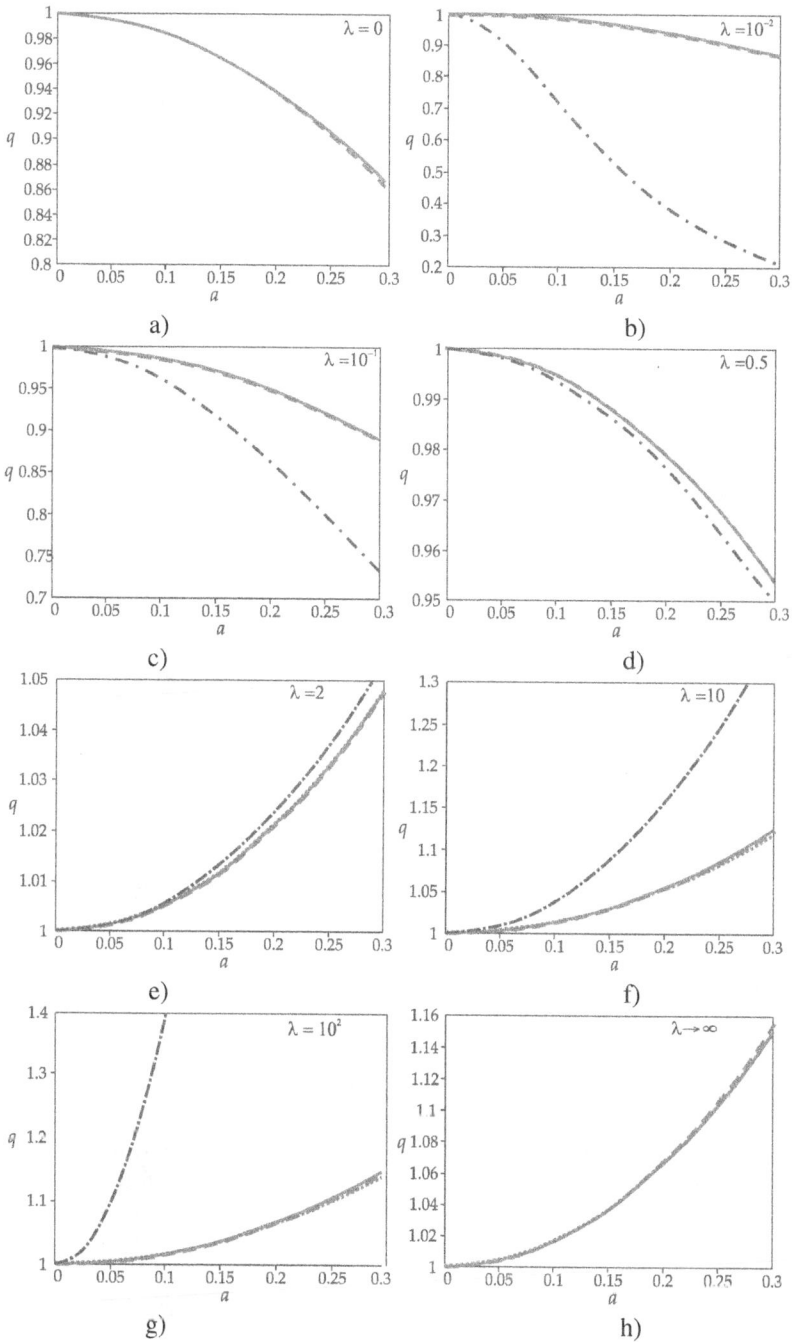

Figure 3.50 Graphs of the effective coefficient of thermal conductivity at: (a) $\lambda = 0$, $a <<$ 1, (b) $\lambda = 10^{-2}$, $a << 1$, (c) $\lambda = 10^{-1}$, $a << 1$, (d) $\lambda = 0.5$, $a << 1$, (e) $\lambda = 2$, $a << 1$, (f) $\lambda = 10$, $a << 1$, (g) $\lambda = 10^2$, $a << 1$, (h) $\lambda = \infty$, $a << 1$ (q_{MThPhM} (3.174), (3.175) – solid line; Schwarz alternating method [26] – dashed line; $\underline{q}_{HS}(q_{ThPhM})$ – dotted line; $\bar{q}_{HS}$ – pointed line).

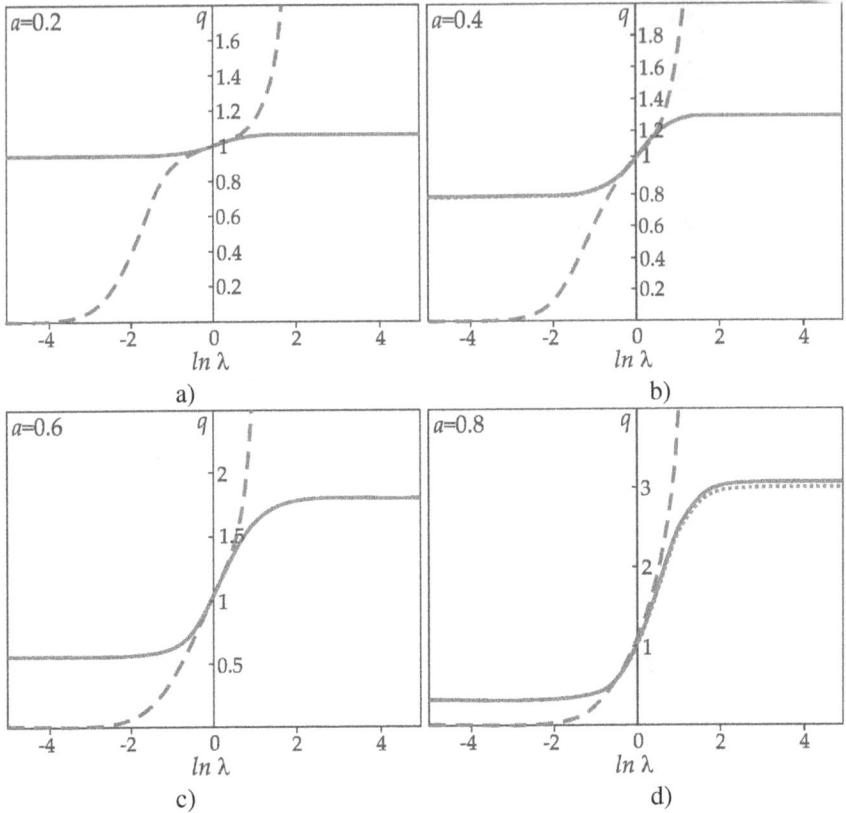

Figure 3.51 Graphs of the effective coefficient of thermal conductivity at: (a) $a = 0.2$, (b) $a = 0.4$, (c) $a = 0.6$, (d) $a = 0.8$ (q_{MThPhM} (3.174), (3.175) – solid line; $q_{HS}(q_{ThPhM})$ – dotted line; q_{HS} – dashed line).

3.6   THE EFFECTIVE CHARACTERISTICS OF 3D COMPOSITE STRUCTURES

3.6.1   USING OF THPHM FOR COMPOSITE WITH CUBIC INCLUSIONS

We consider the problem of determining the effective thermal conductivity of a three-dimensional composite array with periodically arranged cubic inclusions with a side $2a$. The period of the structure in the direction of all coordinate axes is equal to $2b$ and is assumed to be small compared to the characteristic size of the domanin Ω: $\varepsilon = \frac{2b}{l} \ll 1$. The thermal conductivities of the composite phases are λ^+ and λ^- in the matrix and inclusions, respectively, and $\frac{\lambda^-}{\lambda^+} = \lambda$ (see Fig. 3.53).

The original governing equations are [71]

$$\frac{\partial^2 u^+}{\partial \xi^2} + \frac{\partial^2 u^+}{\partial \eta^2} + \frac{\partial^2 u^+}{\partial \zeta^2} = F \quad \text{in} \quad \Omega^+, \tag{3.184}$$

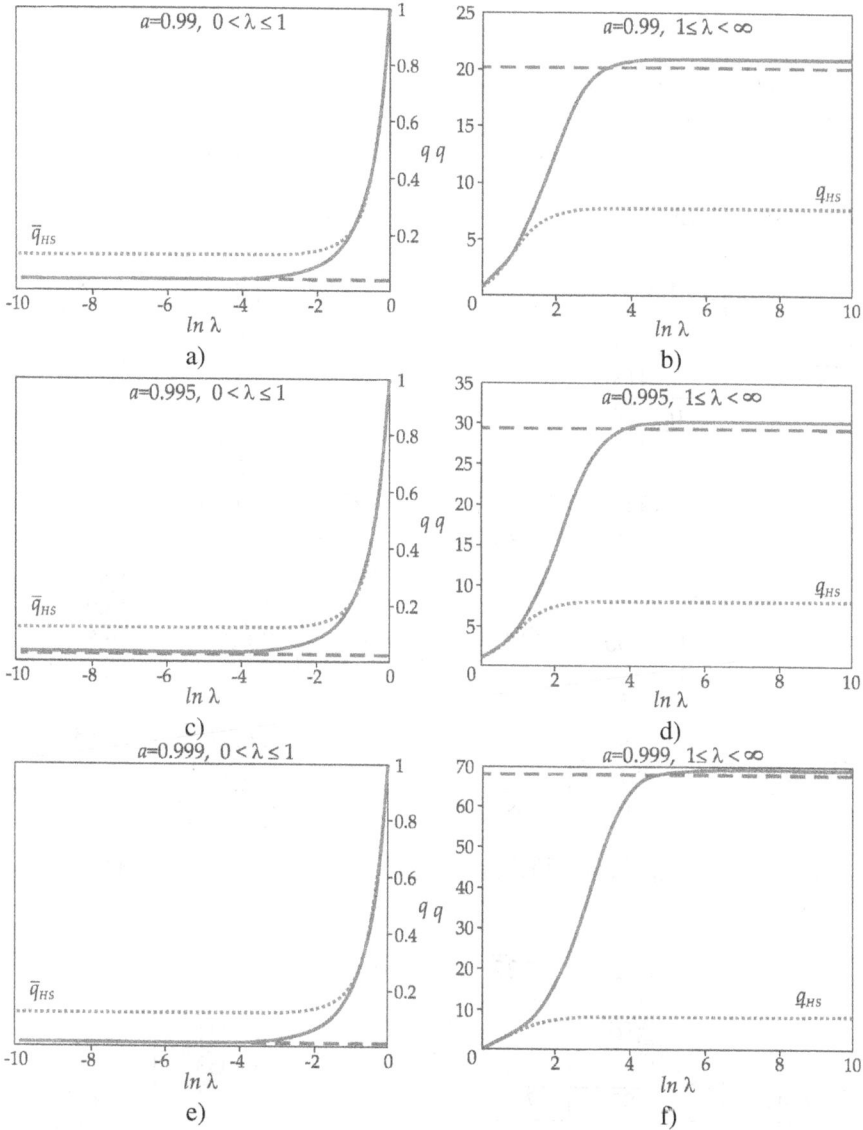

Figure 3.52 Graphs of the effective coefficient of thermal conductivity at: (a) $a = 0.99$, $0 < \lambda \leq 1$; (b) $a = 0.99$, $1 \leq \lambda < \infty$; (c) $a = 0.995$, $0 < \lambda \leq 1$; (d) $a = 0.995$, $1 \leq \lambda < \infty$; (e) $a = 0.999$, $0 < \lambda \leq 1$; (f) $a = 0.999$, $1 \leq \lambda < \infty$ (q_{MThPhM} (3.174), (3.183) – solid line, $q_{as}^{(0)}$ [127, 158], $q_{as}^{(\infty)}$ [158] – dashed line, $\bar{q}_{HS}(q_{ThPhM})$, and $\underline{q}_{HS}(q_{ThPhM})$ – dotted line).

$$\lambda \left(\frac{\partial^2 u^-}{\partial \xi^2} + \frac{\partial^2 u^-}{\partial \eta^2} + \frac{\partial^2 u^-}{\partial \zeta^2} \right) = F \quad \text{in} \quad \Omega^-, \tag{3.185}$$

$$u^+ = u^-, \quad \frac{\partial u^+}{\partial \mathbf{n}} = \lambda \frac{\partial u^-}{\partial \mathbf{n}} \quad \text{at} \quad \partial \Omega_i, \tag{3.186}$$

Table 3.8

Effective thermal conductivity coefficient for inclusions of large sizes with high conductivity.

Conductivity of inclusions $\lambda = 10^2$			Conductivity of inclusions $\lambda = 5 \cdot 10^2$		
Inclusion size a	Asymptotic solution [158]	MThPhM	Inclusion size a	Asymptotic solution [158]	MThPhM
0.9	4.9108	5.0044	0.9	5.0347	5.2456
0.91	5.2473	5.3449	0.91	5.3980	5.6282
0.92	5.6413	5.7411	0.92	5.8277	6.0792
0.93	6.1112	6.2097	0.93	6.3467	6.6213
0.94	6.6841	6.7753	0.94	6.9901	7.2893
0.95	7.4037	7.4765	0.95	7.8164	8.1407
0.96	8.3438	8.3770	0.96	8.9315	9.2776
0.97	9.6432	9.5934	0.97	10.5536	10.9067
0.98	11.5936	11.3680	0.98	13.2351	13.5334
0.99	14.8171	14.3260	0.99	19.0663	18.9367
0.991	15.2200	14.7431	0.991	20.1058	19.8518
0.992	15.6149	15.1940	0.992	21.3187	20.8999
0.993	15.9721	15.6836	0.993	22.7595	22.1169
0.994	16.2297	16.2179	0.994	24.5101	23.5539
0.995	16.2513	16.8041	0.995	26.7012	25.2868
Conductivity of inclusions $\lambda = 10^3$			Conductivity of inclusions $\lambda = 10^5$		
Inclusion size a	Asymptotic solution [158]	MThPhM	Inclusion size a	Asymptotic solution [158]	MThPhM
0.9	5.0502	5.2776	0.9	5.0656	5.3099
0.91	5.4168	5.6661	0.91	5.4355	5.7043
0.92	5.8510	6.1249	0.92	5.8741	6.1709
0.93	6.3761	6.6774	0.93	6.4053	6.7342
0.94	7.0283	7.3605	0.94	7.0662	7.4325
0.95	7.8680	8.2343	0.95	7.9190	8.3297
0.96	9.0049	9.4080	0.96	9.0776	9.5418
0.97	10.6674	11.1050	0.97	10.7801	11.3109
0.98	13.4403	13.8868	0.98	13.6435	14.2620
0.99	19.5974	19.8425	0.99	20.1233	20.8685
0.991	20.7166	20.8903	0.991	21.3212	22.0828
0.992	22.0317	22.1070	0.992	22.7376	23.5164
0.993	23.6079	23.5437	0.993	24.4478	25.2448
0.994	25.5452	25.2763	0.994	26.5699	27.3860
0.995	28.0074	27.4232	0.995	29.3006	30.1366

Table 3.9

Results of numerical and analytical calculations of the effective thermal conductivity for absolutely conductive inclusions.

Parameter $1/\sqrt{1-a^2}$	Inclusion size a	Numerical solution [158]	Asymptotics [158]	MThPhM
10	0.9949874370	29.4440	29.2743	30.1283
20	0.9987492180	60.8976	60.6903	61.6814
30	0.9994442899	92.3575	92.1062	93.1460
40	0.9996874512	123.7734	123.5221	124.5868
50	0.9997999801	155.1894	154.9380	156.0179
100	0.9999499986	312.6460	312.0177	313.1276
1000	0.9999995001	3142.5927	3142.5927	3140.9037

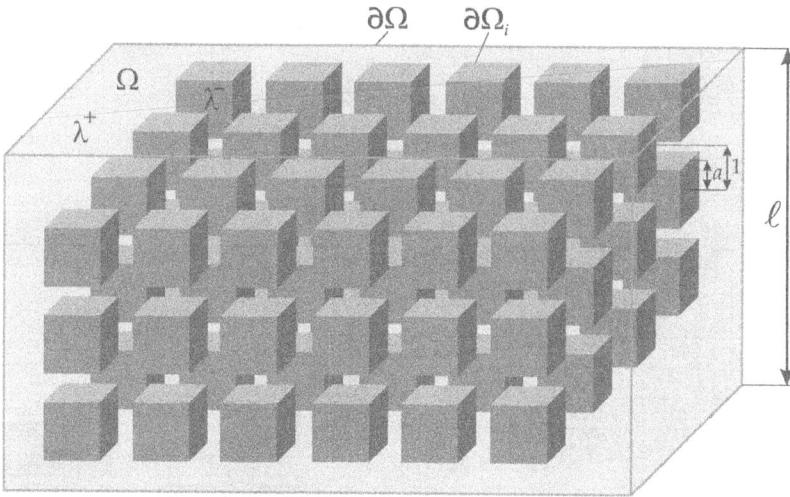

Figure 3.53 Composite with cubic inclusions.

$$u^+ = 0 \quad \text{at} \quad \partial\Omega, \qquad (3.187)$$

where indexes «+» and «-» refer, respectively, to the matrix Ω^+ and inclusions Ω^-; $\mathbf{n}$ is the outer normal to the inclusion contour $\partial\Omega_i$.

We introduce "slow" x, y, z and "fast" ξ, η, ζ variables:

$$\xi = \frac{x}{\varepsilon}, \quad \eta = \frac{y}{\varepsilon}, \quad \zeta = \frac{z}{\varepsilon},$$

Table 3.10

Effective thermal conductivity coefficient for small sizes of inclusions.

Conductivity of inclusions $\lambda = 0$				Conductivity of inclusions $\lambda = 10^{-2}$			
Inclusion size a	ThPhM	Schwarz Altern. Method [26]	MThPhM	Inclusion size a	ThPhM	Schwarz Altern. Method [26]	MThPhM
0.05	0.9961	0.9961	0.9961	0.05	0.9962	0.9962	0.9962
0.1	0.9844	0.9844	0.9844	0.1	0.9847	0.9847	0.9847
0.15	0.9653	0.9650	0.9653	0.15	0.9659	0.9657	0.9659
0.2	0.9391	0.9382	0.9391	0.2	0.9403	0.9394	0.9402
0.25	0.9064	0.9044	0.9064	0.25	0.9082	0.9063	0.9082
0.3	0.8680	0.8639	0.8679	0.3	0.8704	0.8666	0.8704

Conductivity of inclusions $\lambda = 10^{-1}$				Conductivity of inclusions $\lambda = 0.5$			
Inclusion size a	ThPhM	Schwarz Altern. Method [26]	MThPhM	Inclusion size a	ThPhM	Schwarz Altern. Method [26]	MThPhM
0.05	0.9968	0.9968	0.9968	0.05	0.9987	0.9987	0.9987
0.1	0.9872	0.9872	0.9872	0.1	0.9948	0.9948	0.9948
0.15	0.9715	0.9714	0.9715	0.15	0.9883	0.9883	0.9883
0.2	0.9499	0.9494	0.9499	0.2	0.9793	0.9794	0.9793
0.25	0.9228	0.9218	0.9228	0.25	0.9678	0.9681	0.9678
0.3	0.8907	0.8887	0.8906	0.3	0.9540	0.9546	0.9540

Conductivity of inclusions $\lambda = 2$				Conductivity of inclusions $\lambda = 10$			
Inclusion size a	ThPhM	Schwarz Altern. Method [26]	MThPhM	Inclusion size a	ThPhM	Schwarz Altern. Method [26]	MThPhM
0.05	1.0013	1.0013	1.0013	0.05	1.0032	1.0032	1.0032
0.1	1.0052	1.0052	1.0052	0.1	1.0129	1.0130	1.0129
0.15	1.0119	1.0118	1.0119	0.15	1.0293	1.0295	1.0293
0.2	1.0212	1.0210	1.0212	0.2	1.0528	1.0532	1.0528
0.25	1.0333	1.0329	1.0333	0.25	1.0839	1.0849	1.0837
0.3	1.0483	1.0475	1.0483	0.3	1.1228	1.1253	1.1228

Conductivity of inclusions $\lambda = 10^2$				Conductivity of inclusions $\lambda \to \infty$			
Inclusion size a	ThPhM	Schwarz Altern. Method [26]	MThPhM	Inclusion size a	ThPhM	Schwarz Altern. Method [26]	MThPhM
0.05	1.0039	1.0039	1.0039	0.05	1.0039	1.0039	1.0039
0.1	1.0155	1.0156	1.0155	0.1	1.0158	1.0159	1.0158
0.15	1.0353	1.0355	1.0353	0.15	1.0360	1.0363	1.0360
0.2	1.0635	1.0645	1.0635	0.2	1.0649	1.0659	1.0649
0.25	1.1011	1.1034	1.1011	0.25	1.1032	1.1057	1.1033
0.3	1.1489	1.1539	1.1489	0.3	1.1521	1.1575	1.1522

Table 3.11

Comparison of the effective thermal conductivity parameter, obtained by the MThPhM, with known data.

		medium size : $0.3568 \leq a \leq 0.8740$		
	1. Inclusions :			
		absolute conductivity : $\lambda \to \infty$		
Formula (3.9) [119]	Formula (3.12) [119]	Formula (3.13) [119]	Numerical solutions [194]	Asymptotic [158]
3.2790	2.3457	0.0119	2.4839	2.8573

		mediun size : $0.9441 \leq a \leq 0.9966$	
	2. Inclusions :		
		absolute conductivity : $\lambda \to \infty$	
Formula (3.12) [119]	Formula (3.13) [119]	Numerical solutions [194]	Asymptotics [158]
4.6594	0.0515	3.1776	3.8654

extremely large sizes:

3. Inclusions : $0.99498744 \leq a \leq 0.99999950$

absolute conductivity : $\lambda \to \infty$

Numerical calculation [158]	Asymptotics [158]
0.5899	0.7884

large sizes : $0.9 \leq a \leq 0.995$

4. Inclusions :

high conductivity : $10^2 \leq \lambda \leq 10^5$

Asymptotics [158]
3.1354

small sizes : $0 \leq a \leq 0.3$

5. Inclusions :

non-conductivity : $\lambda = 0$

ThPhM	Schwarz Alternating Method [26]
0.0016	0.1159

small sizes : $0 \leq a \leq 0.3$

6. Inclusions :

low conductivity : $10^{-2} \leq \lambda \leq 10^{-1}$

ThPhM	Schwarz Alternating Method [26]
0.0016	0.0814

small sizes : $0 \leq a \leq 0.3$

7. Inclusions :

conductivity order of matrix conductivity:

$0.5 \leq \lambda \leq 2$

ThPhM	Schwarz Alternating Method [26]
0.0000	0.0178

small sizes : $0 \leq a \leq 0.3$

8. Inclusions :

high conductivity: $10 \leq \lambda \leq 100$

ThPhM	Schwarz Alternating Method [26]
0.0561	0.2585

small sizes : $0 \leq a \leq 0.3$

9. Inclusions :

absolute conductivity: $\lambda \to \infty$

ThPhM	Schwarz Alternating Method [26]
0.0030	0.1346

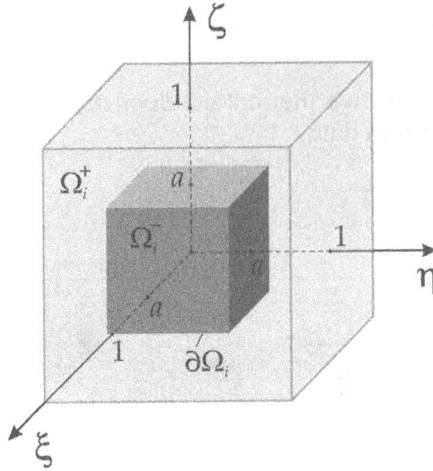

Figure 3.54 Periodically repeating cell of a composite with cubic inclusions.

and represent the solution of the original boundary value problem (3.184)–(3.187) in the form of expansion:

$$u^{\pm} = u_0(x,y,z) + \varepsilon u_1^{\pm}(x,y,z,\xi,\eta,\zeta) + \varepsilon^2 u_2^{\pm}(x,y,z,\xi,\eta,\zeta) + \dots \qquad (3.188)$$

Substituting the expansions (3.184) into the boundary value problem (3.184)–(3.187) and taking into account the following relations for the derivatives:

$$\frac{\partial}{\partial x} = \frac{\partial}{\partial x} + \frac{1}{\varepsilon}\frac{\partial}{\partial \xi}, \quad \frac{\partial}{\partial y} = \frac{\partial}{\partial y} + \frac{1}{\varepsilon}\frac{\partial}{\partial \eta}, \quad \frac{\partial}{\partial z} = \frac{\partial}{\partial z} + \frac{1}{\varepsilon}\frac{\partial}{\partial \zeta},$$

after splitting in powers of the small parameter ε, we arrive at a recurrent sequence of boundary value problems.

The local problem, i.e., the problem on a periodically repeating cell (Fig. 3.54) is formulated as follows:

$$\frac{\partial^2 u_1^+}{\partial \xi^2} + \frac{\partial^2 u_1^+}{\partial \eta^2} + \frac{\partial^2 u_1^+}{\partial \zeta^2} = 0 \quad \text{in} \quad \Omega_i^+,$$

$$\frac{\partial^2 u_1^-}{\partial \xi^2} + \frac{\partial^2 u_1^-}{\partial \eta^2} + \frac{\partial^2 u_1^-}{\partial \zeta^2} = 0 \quad \text{in} \quad \Omega_i^-,$$

$$u_1^+ = u_1^-, \quad \frac{\partial u_1^+}{\partial \bar{n}} + \frac{\partial u_0^+}{\partial n} = \lambda \left(\frac{\partial u_1^-}{\partial \bar{n}} + \frac{\partial u_0^-}{\partial n} \right) \quad \text{at} \quad \partial \Omega_i, \qquad (3.189)$$

$$u_1^+ \Big|_{\xi=1} = u_1^+ \Big|_{\xi=-1}, \quad u_1^+ \Big|_{\eta=1} = u_1^+ \Big|_{\eta=-1}, \quad u_1^+ \Big|_{\zeta=1} = u_1^+ \Big|_{\zeta=-1},$$

$$\frac{\partial u_1^+}{\partial \xi} \Big|_{\xi=1} = \frac{\partial u_1^+}{\partial \xi} \Big|_{\xi=-1}, \quad \frac{\partial u_1^+}{\partial \eta} \Big|_{\eta=1} = \frac{\partial u_1^+}{\partial \eta} \Big|_{\eta=-1}, \rightarrow \frac{\partial u_1^+}{\partial \zeta} \Big|_{\zeta=1} = \frac{\partial u_1^+}{\partial \zeta} \Big|_{\zeta=-1},$$

where $\bar{\mathbf{n}}$ is the outer normal in "fast" variables,

$$\frac{\partial}{\partial \bar{\mathbf{n}}} = \frac{\partial}{\partial \xi} \cos \alpha + \frac{\partial}{\partial \eta} \cos \beta + \frac{\partial}{\partial \zeta} \cos \gamma.$$

To solve the local problem (3.189), we use the ThPhM and perturbation position of the boundary approach. In order to get the first approximation, we replace the contour of the cube with a sphere and arrive at "fast" spherical coordinates r, θ, φ to the following boundary value problem:

$$\frac{\partial^2 u_1^+}{\partial r^2} + \frac{2}{r}\frac{\partial u_1^+}{\partial r} + \frac{\cos \theta}{r^2 \sin \theta}\frac{\partial u_1^+}{\partial \theta} + \frac{1}{r^2}\frac{\partial^2 u_1^+}{\partial \theta^2} + \frac{1}{r^2 \sin^2 \theta}\frac{\partial u_1^+}{\partial \varphi^2} = 0 \quad \text{in} \quad \Omega_{i0}^+, \quad (3.190)$$

$$\frac{\partial^2 u_1^-}{\partial r^2} + \frac{2}{r}\frac{\partial u_1^-}{\partial r} + \frac{\cos \theta}{r^2 \sin \theta}\frac{\partial u_1^-}{\partial \theta} + \frac{1}{r^2}\frac{\partial^2 u_1^-}{\partial \theta^2} + \frac{1}{r^2 \sin^2 \theta}\frac{\partial u_1^-}{\partial \varphi^2} = 0 \quad \text{in} \quad \Omega_i^-, \quad (3.191)$$

$$\frac{\partial^2 \tilde{u}_1}{\partial r^2} + \frac{2}{r}\frac{\partial \tilde{u}_1}{\partial r} + \frac{\cos \theta}{r^2 \sin \theta}\frac{\partial \tilde{u}_1}{\partial \theta} + \frac{1}{r^2}\frac{\partial^2 \tilde{u}_1}{\partial \theta^2} + \frac{1}{r^2 \sin^2 \theta}\frac{\partial \tilde{u}_1}{\partial \varphi^2} = 0 \quad \text{in} \quad \tilde{\Omega}_0, \quad (3.192)$$

$$\text{at} \quad \partial \Omega_i: \quad r = a\sqrt[3]{\frac{6}{\pi}}: \quad (3.193)$$

$$u_{10}^+ = u_{10}^-,$$

$$\frac{\partial u_1^+}{\partial r} - \lambda \frac{\partial u_1^-}{\partial r} = (\lambda - 1)\left(\frac{\partial u_0}{\partial x}\cos \theta \sin \varphi + \frac{\partial u_0}{\partial y}\sin \theta \sin \varphi + \frac{\partial u_0}{\partial z}\cos \varphi\right);$$

$$\text{at} \quad \partial \Omega_i: \quad r = \sqrt[3]{\frac{6}{\pi}}: \quad (3.194)$$

$$u_1^+ = \tilde{u}_1,$$

$$\frac{\partial u_1^+}{\partial r} - \tilde{\lambda}\frac{\partial \tilde{u}_1}{\partial r} = (\tilde{\lambda} - 1)\left(\frac{\partial u_0}{\partial x}\cos \theta \sin \varphi + \frac{\partial u_0}{\partial y}\sin \theta \sin \varphi + \frac{\partial u_0}{\partial z}\cos \varphi\right);$$

$$\tilde{u}_1 \to 0, \quad \frac{\partial \tilde{u}_1}{\partial r} \to 0 \quad \text{at} \quad r \to \infty. \quad (3.195)$$

The solution of the boundary value problem (3.190)–(3.195) can be represented as:

$$u_1^- = A_1 r \cos \theta \sin \varphi + A_2 r \sin \theta \sin \varphi + A_3 r \cos \varphi,$$

$$\tilde{u}_1 = \frac{B_1}{r}\cos \theta \sin \varphi + \frac{B_2}{r}\sin \theta \sin \varphi + \frac{B_3}{r}\cos \varphi,$$

$$u_1^+ = \left(\frac{C_1}{r} + D_1 r\right)\cos \theta \sin \varphi + \left(\frac{C_2}{r} + D_2 r\right)\sin \theta \sin \varphi + \left(\frac{C_3}{r} + D_3 r\right)\cos \varphi,$$

where

$$A_1 = -\left(1 - 9\tilde{\lambda}\Delta\right)\frac{\partial u_0}{\partial x},$$

$$B_1 = \frac{3}{\pi}\left(1 + 3\left(2a^3\left(\lambda - 1\right) + \lambda + 2\right)\Delta\right)\frac{\partial u_0}{\partial x},$$

$$C_1 = \frac{18a^3}{\pi}\left(\lambda - 1\right)\tilde{\lambda}\Delta\frac{\partial u_0}{\partial x},$$

$$D_1 = -\left(1 + 3\left(\lambda + 2\right)\tilde{\lambda}\Delta\right)\frac{\partial u_0}{\partial x},$$

$$\Delta = \left(2\left(\tilde{\lambda} - 1\right)\left(\lambda - 1\right)a^3 - \left(\tilde{\lambda} + 1\right)\left(\lambda + 2\right)\right)^{-1},$$

$$A_1 = A_2 = A_3, \quad B_1 = B_2 = B_3, \quad C_1 = C_2 = C_3, \quad D_1 = D_2 = D_3$$

using replacement $\frac{\partial u_0}{\partial x} \to \frac{\partial u_0}{\partial y} \to \frac{\partial u_0}{\partial z}$.

The effective parameter of a homogeneous medium is yielded by the following equation

$$\Phi\left(u^+\right) + \lambda\,\Phi\left(u^-\right) + \tilde{\lambda}\,\Phi\left(\tilde{u}\right) = F, \tag{3.196}$$

where $\Phi\left(u\right) = \Delta_{xyz}u_0 + 2\tilde{\Delta}u_1 + \Delta_{\xi\eta\zeta}u_2$; Δ_{xyz}, $\Delta_{\xi\eta\zeta}$ are the Laplace operators in "slow" and "fast" variables, respectively; $\tilde{\Delta} = \frac{\partial^2}{\partial x\partial\xi} + \frac{\partial^2}{\partial y\partial\eta} + \frac{\partial^2}{\partial z\partial\zeta}$.

We apply to relation (3.196) the averaging operator

$$\tilde{\Phi}\left(u\right) = \frac{1}{\left|\Omega_i^*\right|}\left(\int_{\Omega_i^+}\Phi\left(u^+\right)dV + \lambda\int_{\Omega_i^-}\Phi\left(u^-\right)dV + \tilde{\lambda}\int_{\tilde{\Omega}}\Phi\left(\tilde{u}\right)dV\right),$$

where $\Omega_i^* = \Omega_i^+ \bigcup \Omega_i^- \bigcup \tilde{\Omega}$; $dV = d\xi\,d\eta\,d\zeta$.

Then, homogenized equation can be written as follows:

$$\Delta u_0 + \frac{1}{\left|\Omega_i^*\right|}\left(\int_{\Omega_i^+}\tilde{\Delta}u_1^+\,dV + \lambda\int_{\Omega_i^-}\tilde{\Delta}u_1^-\,dV + \tilde{\lambda}\int_{\tilde{\Omega}}\tilde{\Delta}\tilde{u}_1\,dV\right) = F.$$

After performing the necessary transformations, the effective thermal conductivity coefficient is

$$\tilde{\lambda} = q = \frac{\lambda\left(1 + 2a^3\right) + 2\left(1 - a^3\right)}{\lambda\left(1 - a^3\right) + 2 + a^3}. \tag{3.197}$$

This expression coincides with the Maxwell formula for the effective parameter of the 3D composite structure and with the Hashin-Shtrikman upper bound at $0 \leq \lambda \leq 1$ and with lower bound at $1 \leq \lambda < \infty$.

Let us analyze the expression (3.197) and obtain from it the asymptotic relations in the limiting cases of the values of the geometric and physical parameters of the composite.

1. Inclusion sizes are small, $a \to 0$, in the limit one gets a homogeneous material with conductivity:

$$q = 1 + \frac{3(\lambda - 1)}{\lambda + 2} a^3 + O\left(a^6\right). \tag{3.198}$$

From (3.198), in particular, we obtain:

1.1. Inclusions of low conductivity, $\lambda << 1$:

$$q = 1 - \frac{3}{2}a^3 + \frac{9}{4}a^3\lambda + O(\lambda^2).$$

1.2. Conductivities of inclusions and matrices of the same order, $\lambda \sim 1$:

$$q = 1 + a^3(\lambda - 1) - \frac{1}{2}a^3(\lambda - 1)^2 + O\left((\lambda - 1)^3\right).$$

1.3. The conductivity of the inclusions is much larger than the conductivity of the matrix, $\lambda >> 1$:

$$q = 1 + 3a^3 - 9a^3\left(\frac{1}{\lambda} - \frac{2}{\lambda^2}\right) + O\left(\frac{1}{\lambda^3}\right).$$

2. Large inclusion sizes, $a \to 1$, in the limit one obtain a homogeneous material with conductivity of inclusions:

$$q = \lambda - (\lambda + 2)(\lambda - 1)(1 - a) + $$
$$+ \lambda(\lambda + 2)(\lambda - 1)(1 - a)^2 + O\left((1 - a)^3\right). \tag{3.199}$$

From (3.199) in special cases of conductivity of inclusions, we have

2.1. Inclusions of low conductivity, $\lambda << 1$:

$$q = 2(1 - a) + \left(a - 2(1 - a)^2\right)\lambda - a(1 - a)\lambda^2 + O(\lambda^3). \tag{3.200}$$

2.2. Conductivities of inclusions and matrices of the same order, $\lambda \sim 1$:

$$q = 1 + (1 - 3a(1 - a))(\lambda - 1) - $$
$$- (1 - a)(a - (1 - a))(\lambda - 1)^2 + O\left((\lambda - 1)^3\right) \tag{3.201}$$

2.3. Inclusions of large but finite conductivity, $1 << \lambda << \frac{1}{\sqrt{1-a}}$:

$$q = a\lambda - (1 - a)(\lambda^2 - 2) + (1 - a)^2\lambda(\lambda + 2)(\lambda - 1). \tag{3.202}$$

3. Inclusions of low conductivity, $\lambda \to 0$, in the limit one gets non-conductive inclusions:

$$q = \frac{2(1 - a^3)}{2 + a^3} + \frac{9a^3}{(2 + a^3)^2}\lambda - \frac{9a^3(1 - a^3)}{(2 + a^2)^3}\lambda^2 + O(\lambda^3). \tag{3.203}$$

In particular cases, from (3.203), we have

3.1. Inclusion sizes are small, $a \ll 1$:

$$q = 1 - \frac{3}{2}a^3 + \frac{9}{4}\lambda\left(1 - \frac{\lambda}{2}\right)a^3 + O\left(a^4\right).$$

3.2. Large inclusion sizes, $a \gg 0$:

$$q = \lambda - (\lambda - 1)(\lambda + 2)(1 - a) + \lambda(\lambda - 2)(1 - a)^2 + O\left((1 - a)^3\right).$$

(3.204)

Note that from the asymptotic expansions (3.200) and (3.204) for $\lambda = 0$ we get the ratio:

$$q = 2(1 - a) \quad \text{at} \quad \lambda = 0, \quad a \to 1,$$

coinciding with obtained in [38].

4. Thermal conductivity of the matrix and inclusions are of the same order, $\lambda \to 1$, in the limit – homogeneous material:

$$q = 1 + a^3(\lambda - 1) - \frac{1}{3}a^3(1 - a^3)(\lambda - 1)^2 + O\left((\lambda - 1)^3\right). \quad (3.205)$$

From (3.205), we obtain in special cases:

4.1. Small inclusions, $a \ll 1$:

$$q = 1 + (\lambda - 1)\left(1 - \frac{1}{3}(\lambda - 1)\right)a^3 + O\left(a^6\right).$$

4.2. Large inclusion sizes, $a \gg 0$:

$$q = \lambda - (\lambda - 1)(\lambda + 2)(1 - a) +$$
$$+ (\lambda - 1)(4\lambda - 1)(1 - a)^2 + O\left((1 - a)^3\right).$$

(3.206)

5. Inclusions of extremely high conductivity, $\lambda \gg \frac{1}{(1-a)^2}$, in the limit we get absolutely conductive inclusions:

$$q = \frac{1 + 2a^3}{1 - a^3} - \frac{9a^3}{(1 - a^3)^2}\frac{1}{\lambda} + \frac{9a^3(2 + a^3)}{(1 - a^3)^3}\frac{1}{\lambda^2} + O\left(\frac{1}{\lambda^3}\right). \quad (3.207)$$

Similarly, we obtain from expression (3.207) various asymptotics with respect to the size of inclusions:

5.1. Small inclusions, $a \ll 1$:

$$q = 1 + 3a^3 - \frac{9}{\lambda}\left(1 - \frac{2}{\lambda}\right)a^2 + O\left(a^6\right).$$

5.2. Large inclusion, $0 \ll a \ll 1 - \frac{1}{\sqrt{\lambda}}$:

$$q = \left(1 + \frac{1}{\lambda}\right)\frac{1}{1 - a} - \frac{1}{\lambda}\left(1 + \frac{1}{\lambda}\right)\frac{1}{(1 - a)^2} + \frac{1}{\lambda^2}\frac{1}{(1 - a)^3} -$$
$$- \left(1 - \frac{2}{3\lambda} + \frac{1}{3\lambda^2}\right) + O((1 - a)).$$

(3.208)

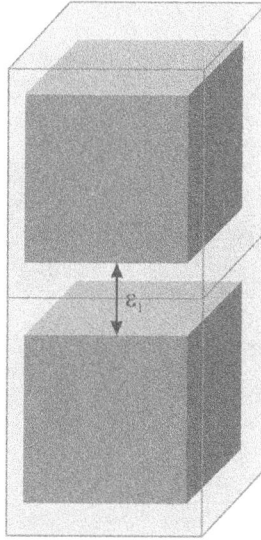

Figure 3.55 Small parameter characterizes the distance between two cubes.

3.6.2 ASYMPTOTICS FOR LARGE INCLUSION SIZES

Consider the case of large inclusions. In this case, a new small parameter appears, ε_1 which characterizes the distance between two cubes (Fig. 3.55). Then, to construct an asymptotic solution, we use the lubrication approach (see Chapter 4).

Due to the symmetry, it is possible to consider only one of the layers, for example, Ω_{i3}^+ (Fig. 3.56); solutions in the direction of the axes ξ, η can be written by taking into account the replacement $\zeta \to \xi \to \eta, z \to x \to y$.

For the function u_{13}^+, defined in the area under consideration Ω_{i3}^+, the following estimations hold

$$\frac{\partial^2 u_{13}^+}{\partial \xi^2} << \frac{\partial^2 u_{13}^+}{\partial \zeta^2}, \qquad \frac{\partial^2 u_{13}^+}{\partial \eta^2} << \frac{\partial^2 u_{13}^+}{\partial \zeta^2}.$$

Therefore, the cell problem is defined by the following equations

$$\frac{\partial^2 u_{13}^+}{\partial \zeta^2} = 0 \quad \text{in} \quad \Omega_{i3}^+, \quad \frac{\partial^2 u_1^-}{\partial \xi^2} + \frac{\partial^2 u_1^-}{\partial \eta^2} + \frac{\partial^2 u_1^-}{\partial \zeta^2} = 0 \quad \text{in} \quad \Omega_i^-,$$

$$u_{13}^+ = u_1^-, \quad \frac{\partial u_{13}^+}{\partial \zeta} - \lambda \frac{\partial u_1^-}{\partial \zeta} = (\lambda - 1) \frac{\partial u_0}{\partial z} \quad \text{at} \quad \zeta = a, \qquad (3.209)$$

$$u_{13}^+ = 0 \quad \text{at} \quad \zeta = 0.$$

The solution of the boundary value problem (3.209) has the following form

$$u_{13}^+ = A + B\zeta, \qquad u_1^- = C\zeta,$$

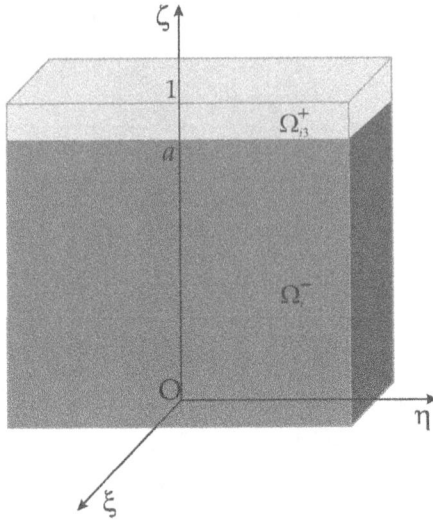

Figure 3.56 Model of the composite in the case of large sizes of inclusions.

where

$$A = (1 - \lambda \Delta) \frac{\partial u_0}{\partial z}, \quad B = -(1 - \lambda \Delta) \frac{\partial u_0}{\partial z};$$

$$C = -(1 - \Delta) \frac{\partial u_0}{\partial z}, \quad \Delta = (a + \lambda (1 - a))^{-1}.$$

The homogenized equation is as follows:

$$\Delta u_0 + \frac{1}{|\Omega_i^*|} \left(\int_{\Omega_{i3}^+} \tilde{\Delta} u_{13}^+ \, dV + \lambda \int_{\Omega_{i3}^-} \tilde{\Delta} u_1^- \, dV \right) = F,$$

where $\Omega_i^* = \Omega_{i3}^+ \cup \Omega_i^-$.

The effective characteristic

$$q = \frac{\lambda (1 - a^2 + a^4) + a^2 (1 - a^2)}{\lambda (1 - a) + a}. \tag{3.210}$$

In limiting cases, based on formula (3.210), one gets:

1. Inclusion sizes are close to extremely large, $a \to 1$:

$$q = \lambda - (\lambda + 2) (\lambda - 1) (1 - a) +$$
$$+ (\lambda^2 + \lambda + 3) (\lambda - 1) (1 - a)^2 + O\left((1 - a)^3\right). \tag{3.211}$$

From (3.211) for different values of the conductivities of the inclusions, we find

1.1. Inclusions of low conductivity, $\lambda \ll 1$:

$$q = 2(1-a) - 3(1-a)^2 + \left(a + 2(1-a)^2\right)\lambda - (1-a)\lambda^2 + O\left(\lambda^3\right).$$
(3.212)

1.2. Inclusions with conductivity of the order of matrix conductivity, $\lambda \sim 1$:

$$q = 1 + \left(1 - 3(1-a) + 5(1-a)^2\right)(\lambda - 1) -$$
$$- (1-a)(1 - 3(1-a))(\lambda - 1)^2 + O\left((\lambda - 1)^3\right).$$
(3.213)

1.3. Inclusions of large but finite conductivity, $1 \ll \lambda \ll \frac{1}{\sqrt{1-a}}$:

$$q = a\lambda - (1-a)\left(\lambda^2 - 2\right) + (1-a)^2(\lambda - 1)\left(\lambda^2 + \lambda + 3\right). \quad (3.214)$$

2. Inclusions of low conductivity, $\lambda \to 0$:

$$q = a\left(1 - a^2\right) + \frac{a^3 - a + 1}{a}\lambda - \frac{\left(a^3 - a + 1\right)(1-a)}{a^2}\lambda^2 + O\left(\lambda^3\right).$$
(3.215)

2.1. From (3.215) with $a \to 1$, we have

$$q = \lambda - (\lambda - 1)(\lambda + 2)(1-a) + (2\lambda - 3)(1-a)^2 + O\left((1-a)^3\right).$$
(3.216)

3. Conductivities of inclusions and matrices of the same order, $\lambda \sim 1$:

$$q = 1 + a\left(1 - a + a^3\right)(\lambda - 1) -$$
$$- a(1-a)\left(1 - a + a^3\right)(\lambda - 1)^2 + O\left((\lambda - 1)^3\right).$$
(3.217)

3.1. In particular, for $a \to 1$ from (3.217), we obtain

$$q = \lambda - (\lambda - 1)(\lambda + 2)(1-a) +$$
$$+ (\lambda - 1)(3\lambda + 2)(1-a)^2 + O\left((1-a)^3\right).$$
(3.218)

4. Inclusions of extremely high conductivity, $\lambda \gg \frac{1}{(1-a)^2}$; in the limit we obtain absolutely conductive inclusions, $\lambda \to \infty$:

$$q = \frac{1 - a^2 + a^4}{1-a} - \frac{a\left(1 - a + a^3\right)}{(1-a)^2}\frac{1}{\lambda} + \frac{a^2\left(1 - a + a^3\right)}{(1-a)^3}\frac{1}{\lambda^2} + O\left(\frac{1}{\lambda^3}\right).$$
(3.219)

4.1. For large inclusions, $0 \ll a \ll 1 - \frac{1}{\sqrt{\lambda}}$, from relation (3.219) follows:

$$q = \left(1 + \frac{3}{\lambda} + \frac{8}{\lambda^2}\right)\frac{1}{1-a} - \frac{1}{\lambda}\left(1 + \frac{4}{\lambda}\right)\frac{1}{(1-a)^2} + \frac{1}{\lambda^2}\frac{1}{(1-a)^3}$$
$$+ O\left((1-a)^0\right).$$
(3.220)

Comparison of asymptotic expressions (3.199)–(3.202), (3.204), (3.206), (3.208) and (3.211)–(3.214), (3.216), (3.218), (3.220) shows that in the case of cubic inclusions, the ThPhM solution correctly describes the effective characteristics of the composite for large inclusion sizes, $a \to 1$, and for any values of their conductivity. The obtained results can be summarized in the following way:

1. In formulas (3.199) (ThPhM) and (3.211) (asymptotic solution) for inclusion sizes close to the maximally possible large $(a \to 1)$ and arbitrary finite conductivity $(0 \le \lambda < \infty)$ the leading term of the expansion and the first correction to it coincide

$$q \sim \lambda - (\lambda + 2)(\lambda - 1)(1 - a) \quad \text{at} \quad a \to 1.$$

2. For extremely large non-conductive inclusions $(a \to 1)$ from (3.200), (3.204) and (3.212), (3.216), the following relation is obtained:

$$q = 2(1 - a) \quad \text{at} \quad \lambda = 0, \quad a \to 1,$$

coinciding with the known result [38].

3. In the case of extremely large inclusions $(a \to 1)$ and their conductivity close to that of the matrix $(\lambda \sim 1)$, from (3.201), (3.206) and (3.213), (3.218), we find the asymptotics of the form:

$$q \sim 1 - 3(\lambda - 1)(1 - a) \quad \text{at} \quad \lambda \sim 1, \quad a \to 1.$$

4. For extremely large inclusions of large but finite conductivity $(1 << \lambda << \frac{1}{\sqrt{1-a}})$, relations (3.202) and (3.214) yield

$$q \sim a\lambda - (1 - a)(\lambda^2 - 2) \quad \text{at} \quad 1 << \lambda << \frac{1}{\sqrt{1-a}}, \quad a \to 1.$$

5. For absolutely conductive inclusions of large sizes $\left(0 << a << 1 - \frac{1}{\sqrt{\lambda}}\right)$ from relations (3.208) and (3.220), we have

$$q \sim \frac{1}{1 - a} \quad \text{at} \quad \lambda \to \infty, \quad 0 << a << 1 - \frac{1}{\sqrt{\lambda}}.$$

3.6.3 USING MULTI-POINT PADÉ APPROXIMANTS

In order to estimate the accuracy of the ThPhM at intermediate values of the parameters, we use the multipoint APs. The following cases are considered:

(i) Voigt and Reuss bounds:

$$q = 1 - a^3 \quad \text{at} \quad \lambda \to 0,$$
$$q = 1 + a^3 \quad \text{at} \quad \lambda \to \infty \tag{3.221}$$

(ii) obvious ratio

$$q = 1 \quad \text{at} \quad \lambda = 1. \tag{3.222}$$

Matching expressions (3.221), (3.222) with respect to variable λ and using three-point AP, we obtain

$$q = \frac{1 - a^3 + \lambda + \lambda^2}{1 + \lambda + \lambda^2 (1 - a^3)}. \tag{3.223}$$

Next, matching over the variable a and using the two-point AP expression, we get: AP (3.223) at $a \to 0$; asymptotic solution (3.210) at $a \to 1$.

Then, after performing the necessary transformations, we have

$$q = \left[(\lambda^2 + \lambda + 1)(\lambda + 2) - 2\lambda^2(\lambda + 2)a + (2\lambda^2 - 1)(\lambda + 2)a^2 + \right.$$
$$+ (3\lambda^2 - \lambda + 1)(\lambda - 1)a^3 - 10\lambda^2(\lambda - 1)a^4 + 5(2\lambda^2 - 1)(\lambda - 1)a^5 -$$
$$\left. -4(\lambda - 1)^2(\lambda + 1)a^6 \right] / \left[(\lambda^2 + \lambda + 1)(\lambda + 2) - 2\lambda^2(\lambda + 2)a + \right. \tag{3.224}$$
$$+ (2\lambda^2 - 1)(\lambda + 2)a^2 - (2\lambda^2 + 4\lambda + 3)(\lambda - 1)a^3 + 2\lambda^2(\lambda - 1)a^4 -$$
$$\left. - (2\lambda^2 - 1)(\lambda - 1)a^5 + (\lambda - 1)^2(\lambda + 1)a^6 \right].$$

Figures 3.57–3.58 present graphs of the effective thermal conductivity parameter for a composite with cubic inclusions, calculated by ThPhM (formula (3.197)), and its estimate (3.224). For comparison, the Hashin–Shtrikman bounds are reported below:

(i) for $1 \leq \lambda < \infty$:

$$\frac{\lambda(1 + 2a^3) + 2(1 - a^3)}{\lambda(1 - a^3) + 2 + a^3} = \underline{q}_{HS} \leq q \leq \bar{q}_{HS} = \lambda \frac{3 - 2a^3 + 2\lambda a^3}{a^3 + \lambda(3 - a^3)}; \tag{3.225}$$

(ii) for $0 \leq \lambda \leq 1$:

$$\lambda \frac{3 - 2a^3 + 2\lambda a^3}{a^3 + \lambda(3 - a^3)} = \underline{q}_{HS} \leq q \leq \bar{q}_{HS} = \frac{\lambda(1 + 2a^3) + 2(1 - a^3)}{\lambda(1 - a^3) + 2 + a^3}. \tag{3.226}$$

As follows from the calculated data, the values of the effective coefficient q determined by formula (3.224) coincide well with the results obtained on the basis of ThPhM: for any inclusion of size a, expression (3.224) fits well with the Hashin-Shtrikman bounds (3.225), (3.226) and gives an upper bound for ThPhM at $\lambda > 1$ and the lower bound $\lambda < 1$.

Thus, in the case of a composite with cubic inclusions, the following conclusions hold:

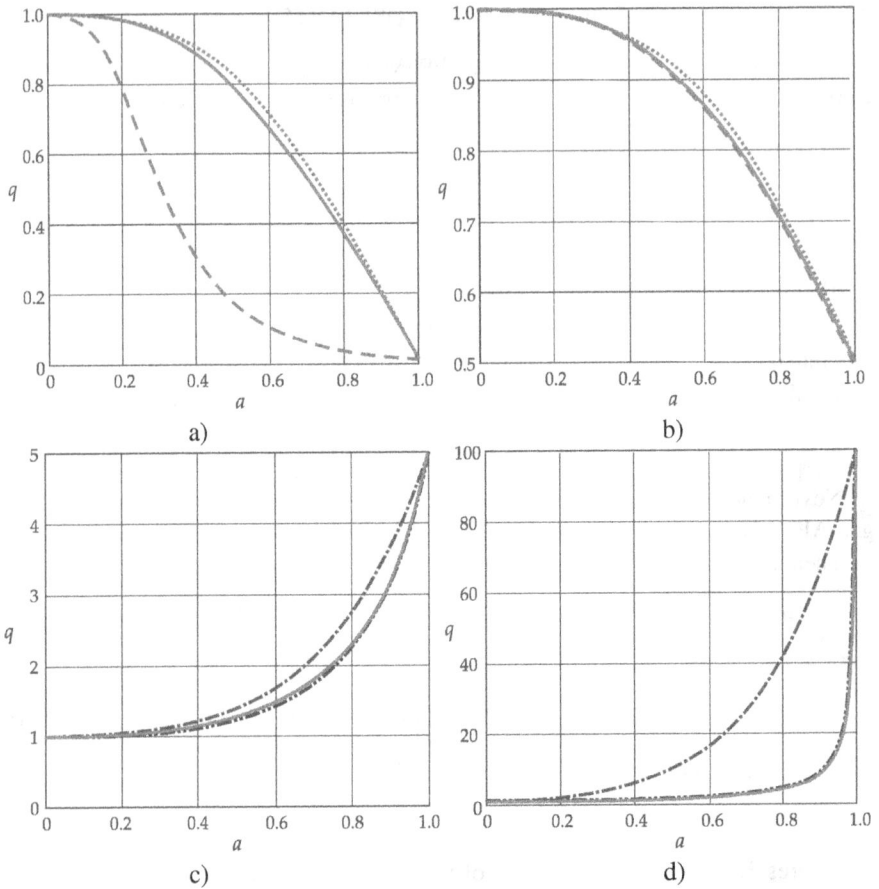

Figure 3.57 Effective thermal conductivity parameter calculated using ThPhM and its estimate (3.224) for $\lambda = 0.01$ (a), $\lambda = 0.5$ (b) ($q(3.224)$ – solid line, $q_{ThPhM} \equiv \bar{q}_{HS}$ – dotted line, $\underline{q}_{HS}$ – dashed line), $\lambda = 5$ (c), $\lambda = 100$ (d) ($q(3.224)$ – solid line, $q_{ThPhM} \equiv \underline{q}_{HS}$ – two-pointed line, $\bar{q}_{HS}$ – pointed line).

(i) ThPhM allows to correctly determine the effective characteristics of the composite for any values of its physical and geometric characteristics, including limiting cases;

(ii) the multi-point AP makes it possible to determine the upper (at $\lambda > 1$) and lower (at $\lambda < 1$) bounds of effective thermal conductivity, and the "fork" between the ThPhM solution and the estimate (3.224) are much narrower than the Hashin–Shtrikman bounds (3.225), (3.226).

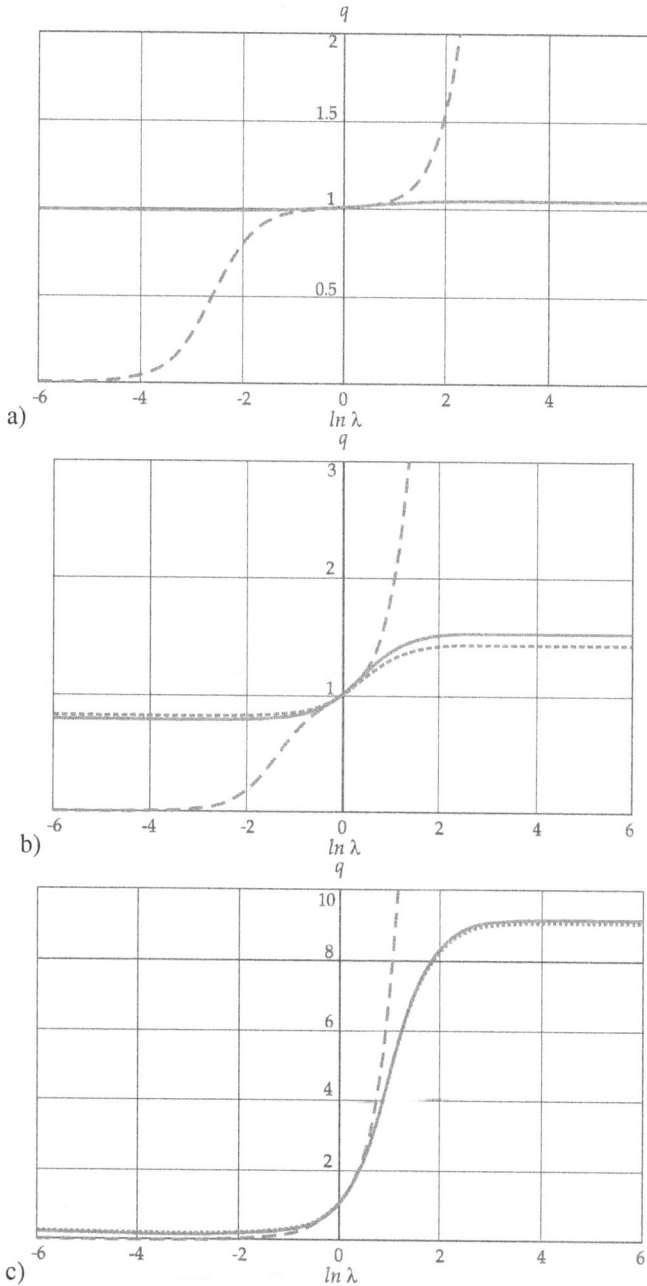

Figure 3.58 Effective thermal conductivity parameter calculated using ThPhM and its estimate (3.224) for $a = 0.2$ (a), $a = 0.5$ (b), $a = 0.9$ (c) (dotted line – $q_{ThPhM} \equiv \bar{q}_{HS}$ for $\ln \lambda < 0$ and $q_{ThPhM} \equiv \underline{q}_{HS}$ for $\ln \lambda > 0$, dashed line – $\underline{q}_{HS}$ for $\ln \lambda < 0$ and $\bar{q}_{HS}$ for $\ln \lambda > 0$, solid line – $q(3.224)$).

4 Lubrication Approach

This chapter is devoted to lubrication approach (LA). First, a brief history of theoretical aspects of composite with cylindrical inclusion of a square profile is presented; and then application of the LA to find a solution to the cell problem of heat conduction for a two phase composite composed of a continuous matrix with cylindrical inclusions and square profiles is described. The considerations include asymptotic relations of the LA, two- and three-point Padé approximations bilateral estimates of the effective thermal conductivity coefficient, as well as the numerical calculations and results evaluation. Next, 2D composites of hexagonal structure are investigated. The study contains statement of the problem, solving the cell problem by LA, links of effective thermal conductivity parameters in various directions, generalization of the LA to the case of finite values of the conductivity of inclusions, modified relations of the LA for medium-sized inclusions, asymptotic formulas in the case of extremaly small conductivity of inclusions, analytical expressions of effective parameters for low conductivities of inclusions, applications of ThPhM for solving local problem for small sizes of inclusions, asymptotic analysis of solutions with regard to small inclusions, numerical results, as well as two-sided estimates of the asymptotics of the effective thermal conductivity parameter for a hexagonal structure. Finally, natural vibrations of composite membranes are analyzed. The consideration includes membranes with a hexagonal array of circle inclusions, construction of analytical relations for eigenmodes and oscillation frequencies, composite membrane with a square lattice of circle inclusions, as well as composite membrane with square holes.

4.1 COMPOSITES WITH CYLINDRICAL INCLUSIONS OF A SQUARE PROFILE

In the seventieth of the previous century, composite materials with cylindrical inclusions [158, 194] became a subject of increased interest of scientists due to their new applications, such as solar energy absorbers [185].

The trend of modern technologies is focused on the creation of composite materials with a given combination of their physical properties (deformation, strength, electrical, thermal conductivity, etc.) and structural and geometric characteristics which is realized by varying the internal microstructure of the composite, using various constituent materials, and employing various technologies and production conditions.

The importance, from a point of view of theory and application, relies on a study of composites with parallelepiped inclusions in relation to solving a wide range of problems in the theory of elasticity, thermal and electrical conductivity, etc., and it is reflected in a large number of diverse and multidisciplinary scientific publications and monographs.

In particular, Torquato [226] considered composites with various microstructures, including arrays of oriented rectangles or cylinders of arbitrary, but constant cross

DOI: 10.1201/9781003391029-4

section. Chaikin and Lubensky [65] studied idealized nematic liquid crystals using the arrangement of oriented rectangles and oriented parallelepipeds for the 2D and 3D cases. Milton [161] analyzed heat conductivity problems for periodic 2D microstructures, which have square inclusions. Bakhvalov and Panasenko [38] analyzed frame structures, which are widely used in engineering. Such systems can be considered as composites with parallelepiped inclusions of square or rectangular cross section. Using an asymptotic procedure for large (close to the maximum possible size) cylindrical cavities of a rectangular cross section, the limiting case was investigated and simple analytical expressions for the effective parameters were obtained.

The book [53] considers models for filled polymers. In the absence of adhesion between the filler and the polymer, a composite model with a square inclusion is used. A formula is given that describes the elongation of the composite depending on the volume fraction of the filler, and it is noted that such dependences are experimentally observed when filling elastomers with dispersed fillers. Composite materials with square or rectangular fibers were studied by Guz et al. [106] using the mesh method and Bourgat [60] using the finite element method. In the reference book on composite materials [151], it is noted that glass fibers are widely used in the creation of non-metallic structural composites like fiberglass exhibiting a relatively low density, high strength, low thermal conductivity, heat resistance, and resistance to chemical and biological action. The cross-sectional shapes of glass fibers can be, in particular, a square or a rectangle. Also the handbook on composite materials [151] describes the form of composite materials called structural sandwich structures. As a filler in such structures, metals, plastics, reinforced plastics, etc., are used. The cell shape can have a different configuration depending on the fillers of the composites, including a square profile. It should be also noticed that material scientists at Harvard have developed honeycomb composite materials of unprecedented light weight and stiffness, whose cells can be square. Harvard engineers use new resin paints and 3D printing to create lightweight cellular composites.

4.1.1 THEORETICAL FOUNDATIONS

The solution of the homogenization problem is simpler for small inclusion sizes. In this case, the shape of the inclusions does not significantly affect the averaged parameter, since its value is determined to a greater extent by the concentration of inclusions than by their shape. This observation allows one to obtain fairly good approximations for inclusions of various shapes and conductivity using asymptotic [26, 27, 38, 119], variational [49], or numerical algorithms [60, 194].

A nontrivial situation is the question of studying the composite structures for extremely large size of inclusions or their concentration, especially in cases of extremely large or extremely small physical characteristics of inclusions.

In this chapter, a method for solving a local problem for the heat conduction equation for a composite array with cylindrical square-section inclusions is proposed. The considered approach is based on the LA [71] and is applicable, due to its physical meaning, for the asymptotic study of composites with large inclusions of high conductivity.

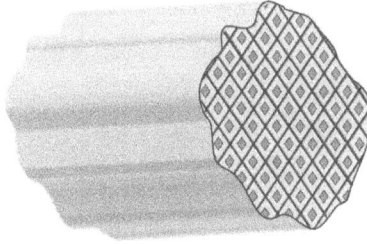

Figure 4.1 Composite structure with cylindrical inclusions of a square profile.

The essence of the LA is to replace the boundary value problem in the original characteristic cell of the composite with a problem in a region with a simpler geometry. Further, the value of the "transformed" geometric parameter is considered to be a function of the coordinates, and all the analytical relationships are built taking into account this dependence in the original cell of the structure.

As example, let us mention that the solution of the problem of heat transfer between two highly conductive particles in the vicinity of their contact point was considered in [42]. The temperature of the particles is uniform. The particles are separated from each other by a matrix layer, the minimum width of which is small compared to the radii of curvature of the particle surfaces. It is assumed that the width of the matrix layer is approximately described by the quadratic function of coordinates. The use of LA allows one to obtain the main asymptotic term in solving the problem of heat transfer between two particles.

4.1.2 APPLICATION OF THE LA TO THE SOLUTION OF THE HOMOGENIZATION PROBLEM

Let us consider the problem of heat conduction for a two-phase composite consisting of a continuous matrix with cylindrical inclusions of a square profile periodically located in it. The structure is doubly periodic, with the period the same in both directions, and inclusions are arranged in a square lattice (Fig. 4.1).

We obtain a solution to the cell problem using the LA approximation for the case of inclusions of large sizes $(a >> 0)$ and high conductivity $(\lambda >> 1)$. The structural cell of the composite is shown in Fig. 4.2, and the cell problem is governed by relations (3.105)–(3.107).

Let us separate the heat fluxes in the direction of the coordinate axes ξ, η, and consider one of them, for example, in the direction of the axis $O\eta$.

The legitimacy and expediency of using LA in this case follows from physical considerations: the heat flux in the region of inclusions (large size and high conductivity) will make a contribution to the effective parameter much greater than the

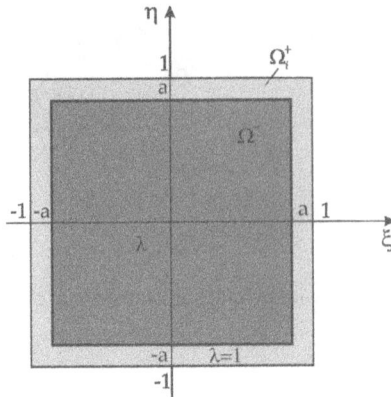

Figure 4.2 Structural cell of the composite.

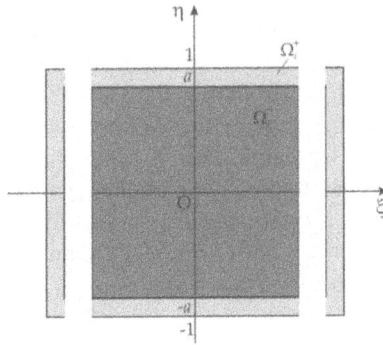

Figure 4.3 Scheme of heat flux distribution in the cell in the direction of axis $O\eta$.

contribution from the heat flux in the region of the matrix, being much smaller in geometric size and much lower with regard to conductivity (Fig. 4.3)

Due to the symmetry of the structure, we consider $1/2$ of the transformed cell, for which $\eta \geq 0$, denoting the area of the matrix Ω_{i1}^+ (Fig. 4.4). Within the accepted model, the local problem (3.105)–(3.107) can be symplified taking into account the following considerations:

1) Obviously, in the matrix area Ω_{i1}^+ of the transformed cell $\Omega_{i1} = \Omega_{i1}^+ \bigcup \Omega_i^-$ variables ξ, η have a different scale:

$$a \leq \eta \leq 1, \qquad -a \leq \xi \leq a.$$

This means that ξ and η can be interpreted as "slow" and "fast" variables, respectively. Then, for the temperature distribution function in matrix u_{11}^+,

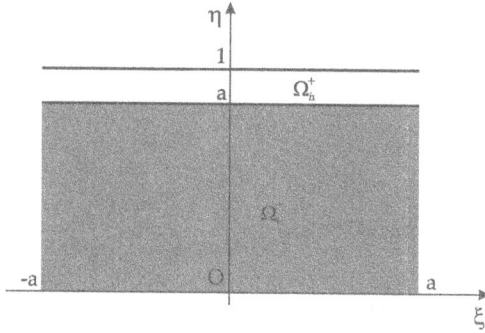

Figure 4.4 Model of LA for a cell with a square inclusion.

we have

$$\frac{\partial^2 u_{11}^+}{\partial \xi^2} << \frac{\partial^2 u_{11}^+}{\partial \eta^2} \quad \text{in} \quad \Omega_{i1}^+.$$

2) Due to the symmetry of the original cell, the temperature distribution function in the inclusions u_{11}^- must satisfy the symmetry condition:

$$u_{11}^- = 0 \quad \text{at} \quad \eta = 0.$$

3) The periodicity conditions (3.107) for the function u_{11}^+ can be presented in the following form

$$u_{11}^+ = 0 \quad \text{at} \quad \eta = 1.$$

Thus, the local problem (3.105)–(3.107) takes the following form

$$\frac{\partial^2 u_{11}^+}{\partial \eta^2} = 0 \quad \text{in} \quad \Omega_{i1}^+,$$

$$\frac{\partial^2 u_{11}^-}{\partial \xi^2} + \frac{\partial^2 u_{11}^-}{\partial \eta^2} = 0 \quad \text{in} \quad \Omega_i^-,$$

(4.1)

$$u_{11}^+ = u_{11}^-, \quad \frac{\partial u_1^+}{\partial \eta} - \lambda \frac{\partial u_1^-}{\partial \eta} = (\lambda \ 1) \frac{\partial u_0}{\partial \eta} \quad \text{at} \quad \eta = a, \qquad (4.2)$$

$$u_{11}^+ = 0 \quad \text{at} \quad \eta = 1, \qquad (4.3)$$

$$u_{11}^- = 0 \quad \text{at} \quad \eta = 0. \qquad (4.4)$$

The general solutions of equations (4.1) are

$$u_{11}^+ = A_0 + B_0 \eta, \qquad (4.5)$$

$$u_{\bar{1}1} = C_0 + D_0\eta + \sum_{n=1}^{\infty} \left[\left(C_n \cosh \pi n\eta + D_n \sinh \pi n\eta \right) \cos \pi n\xi + \right.$$

$$\left. + \left(\bar{C}_n \cosh \pi n\eta + \bar{D}_n \sinh \pi n\eta \right) \sin \pi n\xi \right], \tag{4.6}$$

where A_0, B_0, C_n, D_n, $\bar{C}_n$, $\bar{D}_n$ $(n = 0, 1, 2, ...)$ are the constants determined from conditions (4.2) and (4.3), (4.4). In particular, the conditions (4.2) directly imply

$$C_n = D_n = \bar{C}_n = \bar{D}_n = 0, \quad n = 1, 2, ..., \tag{4.7}$$

since the expression (4.5) contains only zero terms in the expansion of the function $u_{\bar{1}1}^+$ to Fourier series by variable ξ.

Further, by virtue of the symmetry condition (4.4), we have

$$C_0 = 0. \tag{4.8}$$

The remaining three arbitrary constants A_0, B_0, D_0 are determined from the three conditions (4.2), (4.3). Solving the resulting system of equations

$$\begin{cases} A_0 + B_0 = 0, \\[2mm] A_0 + B_0 a - D_0 a = 0, \\[2mm] B_0 - \lambda D_0 = \dfrac{\partial u_0}{\partial y}(\lambda - 1), \end{cases}$$

we find the constants in the following form

$$A_0 = (1 - \lambda\Delta_1)\frac{\partial u_0}{\partial y},$$

$$B_0 = -(1 - \lambda\Delta_1)\frac{\partial u_0}{\partial y}, \tag{4.9}$$

$$D_0 = -(1 - \Delta_1)\frac{\partial u_0}{\partial y},$$

where

$$\Delta_1 = (a + \lambda(1 - a))^{-1}. \tag{4.10}$$

The solution for the functions u_{12}^+, u_{12}^-, determining the heat flux in the direction of axis $O\xi$ can be obtained in a similar way and has the following form

$$u_{12}^+ = u_{11}^+, \quad u_{12}^- = u_{11}^- \left(\frac{\partial u_0}{\partial y} \rightarrow \frac{\partial u_0}{\partial x} \right). \tag{4.11}$$

Further, in accordance with the LA, we consider $a = a(\xi)$. From a mathematical point of view, this means that when applying the averaging operator

$$\widetilde{(...)} = \frac{1}{|\Omega_i^*|}\left(\iint\limits_{\Omega_i^+} (...) \, d\xi \, d\eta + \lambda \iint\limits_{\Omega_i^-} (...) \, d\xi \, d\eta \right),$$

integration is performed over the original cell area $\Omega_i^* = \Omega_i^+ \cup \Omega_i^-$, where $|\Omega_i^*| = 4$.

Thus, the expression for the effective parameter q has the following form

$$q = \frac{1}{|\Omega_i^*|} \left[\iint_{\Omega_i^+} \left(1 + \frac{\partial u_1^{+*}}{\partial \eta} \right) ds + \lambda \iint_{\Omega_i^-} \left(1 + \frac{\partial u_1^{-*}}{\partial \eta} \right) ds \right], \quad (4.12)$$

where

$$u_1^+ = u_{11}^+ + u_{12}^+, \quad u_1^- = u_{11}^- + u_{12}^-, \quad u_1^\pm = \frac{\partial u_0}{\partial y} u_1^{\pm*}. \quad (4.13)$$

Taking into account the solutions found for u_{11}^+, u_{11}^- (4.5)–(4.11), (4.13), the relation for the parameter (4.12) is transformed as follows:

$$q = 1 - a^2 + \lambda a^2 + \frac{1}{4} \left(\iint_{\Omega_i^+} B d\xi d\eta + \lambda \iint_{\Omega_i^-} D d\xi d\eta \right), \quad (4.14)$$

where

$$B = \left(\frac{\partial u_0}{\partial y} \right)^{-1} B_0, \quad D = \left(\frac{\partial u_0}{\partial y} \right)^{-1} D_0. \quad (4.15)$$

The final expression for the effective thermal conductivity is obtained by integrating in (4.14), and by taking into account the relations for the integration constants (4.9), (4.10), (4.15):

$$q = \frac{\lambda \left(1 - a^2 + a^3 \right) + a^2 \left(1 - a \right)}{\lambda \left(1 - a \right) + a}. \quad (4.16)$$

4.1.3 ASYMPTOTIC RELATIONS OF LA

From the solution (4.16) for inclusions of large sizes of high conductivity, we derive asymptotic representations in the limiting cases for the values of the parameter of the composite.

1. $a \to 1$ – large sizes of inclusions, in the limit one gets a homogeneous structure with the conductivity of the inclusions. Then for any values λ, we have

$$q = \lambda - \left(\lambda^2 - 1 \right) \left(1 - a \right) + \left(\lambda^2 + 1 \right) \left(\lambda - 1 \right) \left(1 - a \right)^2 + O \left(\left(1 - a \right)^3 \right). \quad (4.17)$$

1.1. For inclusions of high conductivity $\left(1 << \lambda << \frac{1}{\sqrt{1-a}} \right)$ from (4.17), we find

$$q = \left(1 + \left(1 - a \right)^2 \right) \lambda - \left(1 - a \right) \left(1 + \left(1 - a \right) \right) \lambda^2 + \\ + \left(1 - a \right)^2 \lambda^3 + \left(1 - a \right) \left(1 - \left(1 - a \right) \right). \quad (4.18)$$

For comparison, a similar asymptotic relation (3.157), obtained on the basis of ThPhM, coincides with expression (4.18) up to terms of order $\lambda^2 (1-a)$ inclusive

$$q_{ThPhM} \sim q_{LA} = \lambda - \lambda^2 (1-a) + (1-a) \quad \text{at} \quad a \to 1,$$

$$1 << \lambda << \frac{1}{\sqrt{1-a}}.$$

2. $\lambda \to \infty$ – inclusions of high conductivity, in the limit it corresponds to a composite with ideally conducting inclusions:

$$q = \frac{1-a^2+a^3}{1-a} - \frac{a\left(1-a^2+a^3\right)}{(1-a)^2} \frac{1}{\lambda} + \frac{a^2\left(1-a^2+a^3\right)}{(1-a)^3} \frac{1}{\lambda^2} + O\left(\frac{1}{\lambda^3}\right).$$
(4.19)

2.1. From relation (4.19), we obtain the asymptotics in the case of ideal conductivity of inclusions and their large geometric sizes, $0 << a << 1 - \frac{1}{\sqrt{\lambda}}$:

$$q = \left(1+\frac{2}{\lambda}+\frac{4}{\lambda^2}\right)\frac{1}{1-a} - \left(1+\frac{2}{\lambda}+\frac{3}{\lambda^2}\right) -$$
$$- \frac{1}{\lambda}\left(1+\frac{3}{\lambda}\right)\frac{1}{(1-a)^2} + \frac{1}{\lambda^2}\frac{1}{(1-a)^3}.$$
(4.20)

For comparison, the asymptotic relation (3.163), obtained from ThPhM, coincides with expression (4.20) up to terms of order $\frac{1}{1-a}$ inclusive:

$$q_{ThPhM} \sim q_{LA} = \frac{1}{1-a} \quad \text{for} \quad \lambda \to \infty, \quad 0 << a << 1 - \frac{1}{\sqrt{\lambda}}.$$

Thus, the leading terms of the asymptotics in the expressions of the effective parameter (4.16), obtained by LA, and in the solution (3.153) found using ThPhM, coincide.

We note one more fact confirming the adequacy of the LA approximations. The model of the structure in this case is built on the assumption of a high conductivity of the inclusions. At low conductivity of the inclusions, the physical processes occurring in the composite changes dramatically, and the model does not work in this case.

But, using solution (4.16) for large inclusions of high conductivity ($\lambda >> 1$), based on the Keller's theorem, we can write the expression for the effective thermal conductivity for large inclusions of low conductivity ($\lambda << 1$). This relation, as follows from (4.16) and (3.59), has the following form

$$q = \frac{1-a+a\lambda}{1-a^2+a^3+a^2(1-a)\lambda}.$$
(4.21)

The asymptotic expansions, obtained from (4.21):

(i) at $a \to 1$, $\lambda \ll 1$:

$$q = (1-a)(1+(1-a)) + \left(1 - 3(1-a)^2\right)\lambda - (1-a)(1-(1-a))\lambda^2;$$

(ii) at $\lambda \to 0$, $a \gg 0$:

$$q = \lambda + \left(1-\lambda^2\right)(1-a) + \left(1 - 3\lambda + \lambda^2\right)(1-a)^2, \qquad (4.22)$$

coincide, respectively, with the ThPhM relations (3.156) and (3.159) up to terms of order $\lambda^2(1-a)$ inclusive:

$$q_{ThPhM} \sim q_{LA} = 1 - a + \lambda - (1-a)\lambda^2 \quad \text{at} \quad a \to 1, \quad \lambda \ll 1,$$

$$q_{ThPhM} \sim q_{LA} = \lambda + \left(1-\lambda^2\right)(1-a) \quad \text{at} \quad \lambda \to 0, \quad a \gg 0.$$

Moreover, the main term of the asymptotics (4.22) coincides for $\lambda = 0$ with the known solution (3.160) [38].

4.1.4 TWO- AND THREE-POINT APs

The notion of a three-point AP is defined in [30, 31].
    Let

$$F(\varepsilon) = \sum_{i=0}^{\infty} a_i \varepsilon^i \quad \text{at} \quad \varepsilon \to 0, \qquad (4.23)$$

$$F(\varepsilon) = \sum_{i=0}^{\infty} b_i(-1+\varepsilon)^i \quad \text{at} \quad \varepsilon \to 1, \qquad (4.24)$$

$$F(\varepsilon) = \sum_{i=0}^{\infty} c_i \varepsilon^{-i} \quad \text{at} \quad \varepsilon \to \infty. \qquad (4.25)$$

The three-point AP is represented by the function

$$F(\varepsilon) = \frac{\sum\limits_{i=0}^{m} \alpha_i \varepsilon^i}{\sum\limits_{j=0}^{n} \beta_j \varepsilon^j},$$

where k_1, k_2 and k_3 $(k_1 + k_2 + k_3 = m + n + 1)$ expansion coefficients in the Taylor series for $\varepsilon \to 0$, $\varepsilon \to 1$ and $\varepsilon \to \infty$ coincide with the corresponding coefficients of the series (4.23), (4.24), and (4.25), respectively.

Let us use for the composite with cylindrical square inclusions, under assumption $a \to 0$, the well-known Voigt-Reuss estimates [71]:

$$\bar{q} = \frac{1}{|\Omega_i^*|}\left(\iint\limits_{\Omega_i^+} q\, d\xi\, d\eta + \iint\limits_{\Omega_i^-} \lambda q\, d\xi\, d\eta\right),$$

$$\underline{q} = \left\{ \frac{1}{|\Omega_i^*|} \left(\iint\limits_{\Omega_i^+} q^{-1} d\xi\, d\eta + \iint\limits_{\Omega_i^-} (\lambda q)^{-1} d\xi\, d\eta \right) \right\}^{1}.$$

Then we get

$$\bar{q} = 1 - a^2 \quad \text{at} \quad \lambda \to 0, \tag{4.26}$$

$$\underline{q} = 1 + a^2 \quad \text{at} \quad \lambda \to \infty. \tag{4.27}$$

We supplement relations (4.26), (4.27) with the obvious condition

$$q = 1 \quad for \quad \lambda = 1 \tag{4.28}$$

and matching expressions (4.26)–(4.28) over variable λ using a three-point AP yields

$$q = \frac{1 - a^2 + \lambda + \lambda^2}{1 + \lambda + \lambda^2 (1 - a^2)} \quad \text{for} \quad a \to 0. \tag{4.29}$$

Next, using the asymptotic formula (4.16) for $a \to 1$, we match expressions (4.16) and (4.29) with respect to the variable a using the two-point AP, and we get

$$q = \frac{\left(1 + \lambda + \lambda^2\right) - a\left(1 + \lambda^2\right) + a^2 \lambda^2}{\left(1 + \lambda + \lambda^2\right) - a\left(1 + \lambda^2\right) + a^2}. \tag{4.30}$$

From formula (4.30), all the above asymptotics can be obtained in the limiting cases of inclusion sizes a and their conductivity λ.

The values of the effective thermal conductivity calculated using formula (4.30) are in good agreement with the results obtained using ThPhM (3.153):

1) Comparison of the effective coefficients at a fixed conductivity value shows the maximum discrepancy (10%) for $a \approx 0.5$. This discrepancy decreases with decreasing contrast of the composite phases, i.e., the difference between the conductivities of the matrix and inclusions.
2) For a fixed size of inclusions a, the maximum discrepancies in the results (10%) occur in the limiting cases, i.e. for non-conductive ($\lambda = 0$) and absolutely conductive ($\lambda \to \infty$) inclusions.

4.1.5 BILATERAL ESTIMATES OF THE EFFECTIVE THERMAL CONDUCTIVITY COEFFICIENT

Two-sided estimates obtained by Hashin and Shtrikman [110, 111] are widely used in the theory of composite materials [71]. However, these estimates do not take into account the shape of the inclusions and, in the limiting cases of the conductivity of the inclusions ($\lambda \to 0$ and $\lambda \to \infty$) give trivial results. As will be shown below, the use of the AP technique allows to find estimates of effective parameters for given shapes of inclusions for any values of the size of inclusions and their conductivity.

The ThPhM solution (3.153) is taken as a constitutive relation for the effective thermal conductivity parameter. For any inclusion size a, formula (4.30) gives the upper bound ThPhM for $\lambda < 1$ and the lower bound at $\lambda > 1$. Next, we obtain a lower bound for the ThPhM solution for $\lambda < 1$ and upper bound for $\lambda > 1$.

We deal with small inclusions $(a << 1)$, and use the solution for a small circle inclusion obtained in Chapter 6 by the Schwarz alternating method [160] (taking into account two approximations). This solution gives a lower bound for the conductivity of inclusions $\lambda < 1$ and upper bound for $\lambda > 1$.

Let us assume that the radius of inclusions a is a function of the coordinates, i.e. $a = a(\xi, \eta)$. The solution of the local problem is written as follows:

$$u_1^- = \frac{\lambda - 1}{\lambda + 1} \left(\frac{\partial u_0}{\partial x} \xi + \frac{\partial u_0}{\partial y} \eta \right),$$

$$u_1^+ = u_{11}^+ + u_{12}^+, \tag{4.31}$$

$$u_{11}^+ = \frac{\lambda - 1}{\lambda + 1} \left(\frac{\xi}{\xi^2 + \eta^2} - \frac{\pi}{4} \xi \right) (a^2 + \eta^2) \frac{\partial u_0}{\partial x},$$

$$u_{11}^+ = u_{12}^+ \ (\xi \leftrightarrows \eta; \ x \leftrightarrows y).$$

After performing the standard homogenization procedure (3.25), we obtain an expression for the effective conductivity coefficient

$$q = 1 + \frac{\lambda - 1}{\lambda + 1} a^2 \left(1 + \frac{\pi}{4} + \frac{\pi a}{6} - \frac{5\pi a^2}{12} \right). \tag{4.32}$$

Matching by two-point AP, the asymptotic expression for the large (4.16) and small (4.32) inclusions becomes

$$q = \frac{4(1 + \lambda) - 4(1 + \lambda) a + \lambda(4 + \pi) a^2 - \lambda \pi a^3}{4(1 + \lambda) - 4(1 + \lambda) a + (4 + \pi) a^2 - \pi a^3}. \tag{4.33}$$

Formula (4.33) gives a lower bound for the solution ThPhM for $\lambda < 1$ and upper bound for $\lambda > 1$. The joint use of expressions (4.30) and (4.33) gives for any values of the size of inclusions and their conductivity, two-sided estimates of the effective thermal conductivity parameter found from ThPhM, and the "fork" between them does not exceed 10% even in limiting cases $\lambda \to 0$, $\lambda \to \infty$ and for $a \approx 0.5$.

4.1.6 NUMERICAL CALCULATIONS AND EVALUATION OF RESULTS

Figures 4.5–4.10 present graphs of the effective thermal conductivity q, calculated by ThPhM (3.153), and its two-sided bounds, determined by formulas (4.30) and (4.33).

Figures 4.5–4.7 show the dependence of the effective coefficient on the size of inclusions $q = q(a)$ for fixed values of its conductivity, corresponding to:

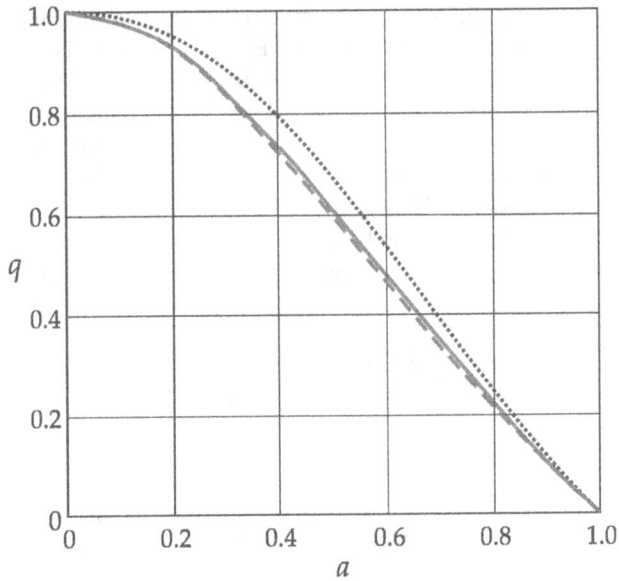

Figure 4.5 Effective parameter of a composite with low conductivity of inclusions ($\lambda = 10^{-4}$) and its two-sided estimates (ThPhM (3.153) – solid line, (4.30) – dotted line, (4.33) – dashed line).

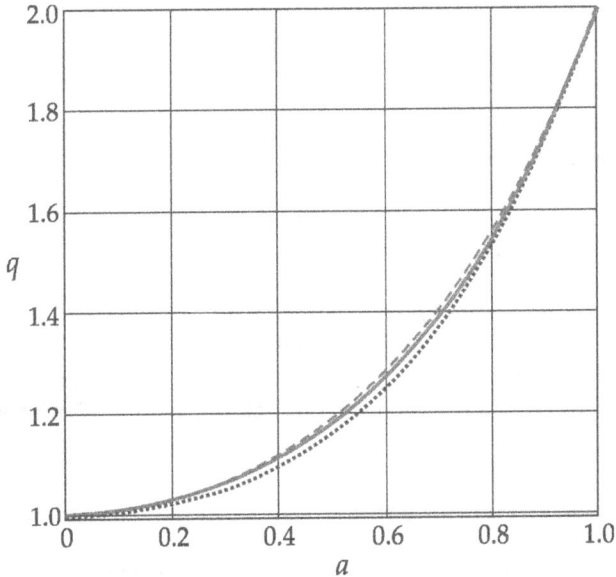

Figure 4.6 Effective parameter of a composite with conductivity of inclusions of the order of matrix conductivity ($\lambda = 2$) and its two-sided estimates (ThPhM (3.153) – solid line, (4.30) – dotted line, (4.33) – dashed line).

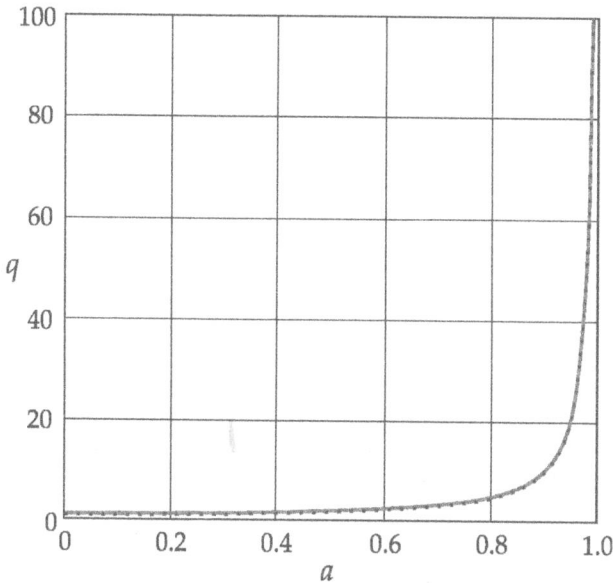

Figure 4.7 The effective parameter of a composite with high conductivity of inclusions $(\lambda = 10^3)$ and its two-sided estimates (ThPhM (3.153) – solid line, (4.30) – dotted line, (4.33) – dashed line).

    (i) inclusions of low conductivity, $\lambda \ll 1$;
   (ii) conductivity of inclusions of the order of conductivity of the matrix, $\lambda \sim 1$;
  (iii) inclusions of high conductivity, $\lambda \gg 1$.

On the other hand, Figures 4.8–4.10 present the dependence of the effective conductivity on conductivity of inclusions $q = q\,(\lambda)$ for fixed values of its size, corresponding to:

    (i) small inclusions, $a \ll 1$;
   (ii) inclusions of medium size, $a \sim 0.5$;
  (iii) large inclusions, $a \gg 0$.

    An analysis of the obtained analytical relations and numerical data showed that LA correctly describes the behavior of an effective homogeneous medium at large values of the geometric and physical parameters of the structure, including limiting cases. This conclusion is confirmed by the coincidence of the asymptotic expressions for the effective thermal conductivity parameter in the limiting cases of the conductivity of the inclusions and their sizes with the approximations constructed using ThPhM.

    The developed method can be generalized to calculate other effective coefficients of composite materials with a periodic structure.

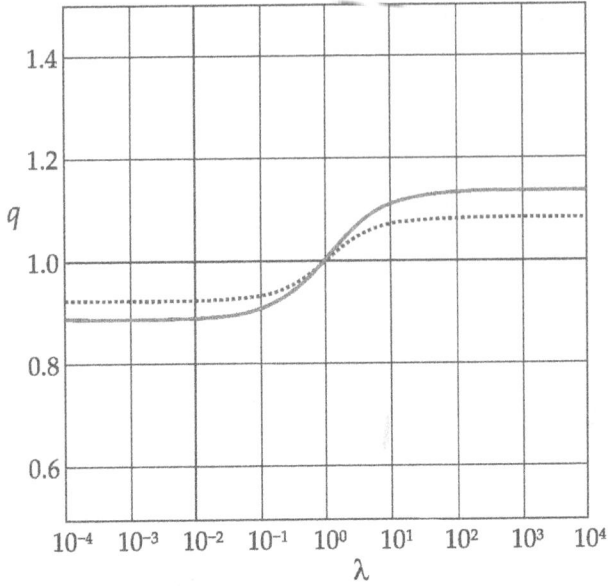

Figure 4.8 The effective parameter of a composite with small inclusions ($a = 0.25$) and its two-sided estimates (ThPhM (3.153) – solid line, (4.30) – dotted line, (4.33) – dashed line).

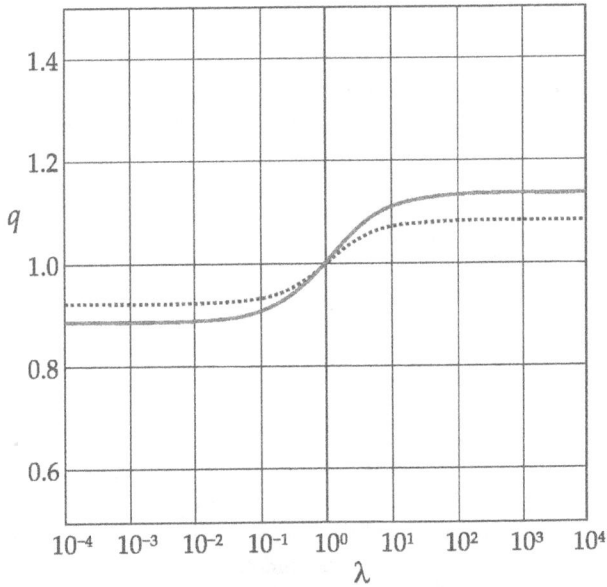

Figure 4.9 Effective parameter of a composite with inclusions of average size ($a = 0.5$) and its two-sided estimates (ThPhM (3.153) – solid line, (4.30) – dotted line, (4.33) – dashed line).

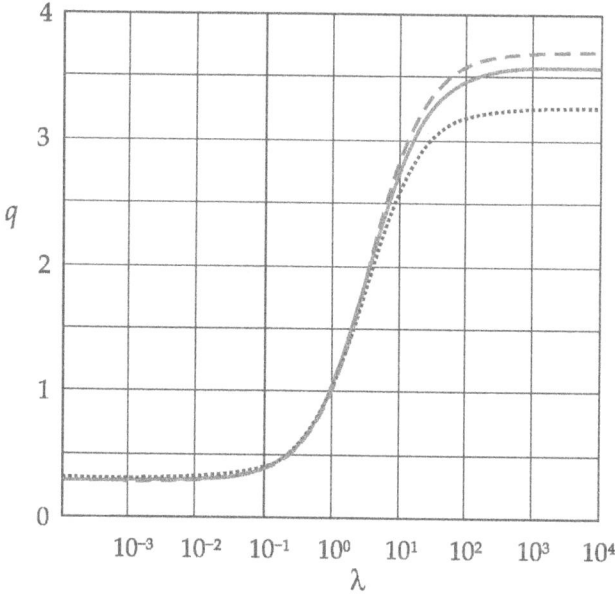

Figure 4.10   Effective parameter of a composite with large inclusions ($a = 0.75$) and its two-sided estimates (ThPhM (3.153) – solid line, (4.30) – dotted line, (4.33) – dashed line).

4.2 2D COMPOSITES OF HEXAGONAL STRUCTURE

The study of composite with a hexagonal structure is presented in a number of publications from different points of view and in different formulations of the problem. For example, in [194], the effective conductivity of a composite structure with highly conductive cylindrical inclusions of a circle profile is studied in the two-dimensional case for the case of inclusion concentration close to the limitingly high one. The solution of the problem is based on Rayleigh method [181] in combination with numerical methods.

In [104], using the method of a complex variable and the properties of Weierstrass elliptic functions, expressions were obtained for the effective elastic coefficients of a transversally isotropic structure with circle cylindrical inclusions. The constitutive relations for the hexagonal structure were obtained in [67] on the basis of the homogenization theory.

In [161], a formula was given for the effective conductivity of an array with cylindrical inclusions, which can be used in a wide range of the geometric sizes of inclusions and their conductivity, except for the cases of contacting inclusions of high conductivity.

In [54, 55, 58], the effective properties of high contrast densely packed composite arrays with a stochastic arrangement of inhomogeneities are studied; in this case, the distance between inclusions is taken as a small parameter. In particular, in [58] an asymptotic formula was obtained for the effective conductivity of a composite of a

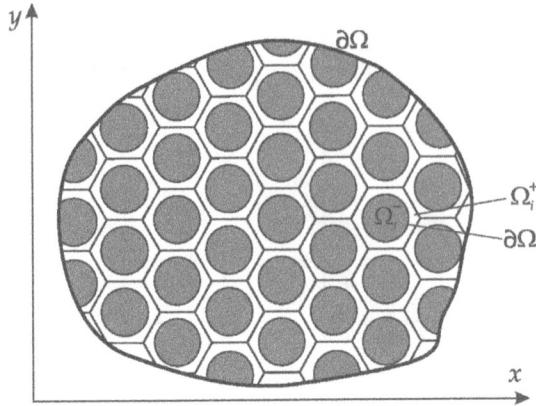

Figure 4.11 Composite material with a hexagonal structure of circular inclusions.

hexagonal structure with circle superconducting inclusions as the distance between them tends to zero.

A number of works are devoted to studying the effective properties of composites of various structures using the AP [12, 225], quasifractional approximants [19, 20], and elliptic functions [175].

4.2.1 STATEMENT OF THE PROBLEM

The problem of thermal conductivity of a composite with periodically arranged circular cylindrical inclusions constituting a hexagonal lattice is considered (Fig. 4.11).

The problem of heat conductivity in such a structure is defined as follows:

$$\lambda^+ \left(\frac{\partial^2 u^+}{\partial x^2} + \frac{\partial^2 u^+}{\partial y^2} \right) = F \quad \text{in} \quad \Omega_i^+, \tag{4.34}$$

$$\lambda^- \left(\frac{\partial^2 u^-}{\partial x^2} + \frac{\partial^2 u^-}{\partial y^2} \right) = F \quad \text{in} \quad \Omega_i^-, \tag{4.35}$$

$$u^+ = u^-, \quad \lambda^+ \frac{\partial u^+}{\partial \mathbf{n}} = \lambda^- \frac{\partial u^-}{\partial \mathbf{n}} \quad \text{at} \quad \partial \Omega_i, \tag{4.36}$$

where u^+, u^- are the temperature distribution functions in the matrix and inclusions; λ_i^+, λ_i^- are the thermal conductivities of the phases of the composite, $\frac{\lambda^-}{\lambda^+} = \lambda$; F is the heat source density; $\mathbf{n}$ is the outer normal to the inclusion contour.

The solution of the boundary value problem (4.34)–(4.36) is represented as asymptotic series in powers of a dimensionless small parameter ε ($\varepsilon \ll 1$), characterizing the period of the structure

$$u^\pm = u_0(x, y) + \varepsilon u_1^\pm(x, y, \xi, \eta) + \varepsilon^2 u_2^\pm(x, y, \xi, \eta) + \dots, \tag{4.37}$$

where ξ, η is the "fast" variables, $\xi = \frac{x}{\varepsilon}, \eta = \frac{y}{\varepsilon}$.

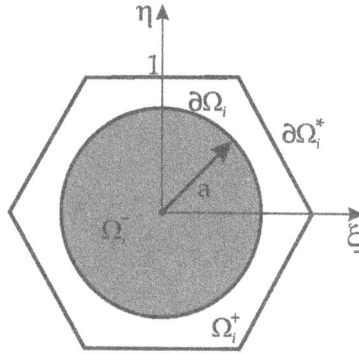

Figure 4.12   A characteristic cell of a composite of a hexagonal structure: Ω_i^+ – matrix area; Ω_i^- – inclusion area.

Taking into account the representation (4.37), after splitting into ε relations (4.34)–(4.36), the solution of the problem is divided into two main stages.

At the first stage, the solution of the local problem is determined, i.e., the problem in the region of a periodically repeating cell of the composite (Fig. 4.12):

$$\frac{\partial^2 u_1^\pm}{\partial \xi^2} + \frac{\partial^2 u_1^\pm}{\partial \eta^2} = 0 \quad \text{in} \quad \Omega_i^\pm, \tag{4.38}$$

$$u_1^+ = u_1^-, \quad \frac{\partial u_1^+}{\partial \bar{n}} - \lambda \frac{\partial u_1^-}{\partial \bar{n}} = (\lambda - 1) \frac{\partial u_0}{\partial n} \quad \text{at} \quad \partial \Omega_i, \tag{4.39}$$

$$u_1^+ = 0 \quad \text{at} \quad \partial \Omega_i^*. \tag{4.40}$$

The second stage is focused on finding the main ("slow") part of solution $u_0(x, y)$ from the following homogenized equation

$$\bar{q} \left(\frac{\partial^2 u_0}{\partial x^2} + \frac{\partial^2 u_0}{\partial y^2} \right) + \tag{4.41}$$

$$+ \frac{1}{|\Omega_i^*|} \left[\iint_{\Omega_i^+} \left(\frac{\partial^2 u_1^+}{\partial x \partial \xi} + \frac{\partial^2 u_1^+}{\partial y \partial \eta} \right) d\xi d\eta + \lambda \iint_{\Omega_i^-} \left(\frac{\partial^2 u_1^-}{\partial x \partial \xi} + \frac{\partial^2 u_1^-}{\partial y \partial \eta} \right) d\xi d\eta \right] = F,$$

where $\Omega_i^* = \Omega_i^+ \bigcup \Omega_i^-$; $\bar{q} = \dfrac{|\Omega_i^+| + \lambda |\Omega_i^-|}{|\Omega_i^*|}$ – Voigt-averaged parameter.

Taking into account the expressions u_1^+, u_1^-, determined by the solution of the cell problem (4.38)–(4.40), the homogenized equation in general form is transformed to the following one

$$q_x \frac{\partial^2 u_0}{\partial x^2} + q_y \frac{\partial^2 u_0}{\partial y^2} = F \quad \text{in} \quad \Omega^*, \tag{4.42}$$

where q_x, q_y are the homogenized parameters

$$q_x = \bar{q} + \frac{1}{|\Omega_i^*|} \left(\iint_{\Omega_i^+} \frac{\partial u_{1(1)}^+}{\partial \xi} d\xi d\eta + \lambda \iint_{\Omega_i^-} \frac{\partial u_{1(1)}^-}{\partial \xi} d\xi d\eta \right), \tag{4.43}$$

$$q_y = \bar{q} + \frac{1}{|\Omega_i^*|} \left(\iint_{\Omega_i^+} \frac{\partial u_{1(2)}^+}{\partial \eta} d\xi d\eta + \lambda \iint_{\Omega_i^-} \frac{\partial u_{1(2)}^-}{\partial \eta} d\xi d\eta \right), \tag{4.44}$$

$u_{1(i)}^\pm$ ($i = 1, 2$) are the solutions of the local problem (4.38)–(4.40), written up to constant factors corresponding to the slow component of the solution:

$$u_1^\pm = u_{1(1)}^\pm (\xi, \eta) \frac{\partial u_0}{\partial x} + u_{1(2)}^\pm (\xi, \eta) \frac{\partial u_0}{\partial y}.$$

Thus, when constructing a solution to the problem of the thermal conductivity of a composite with periodically arranged circular cylindrical inclusions constituting a hexagonal lattice, the main problem consists in solving the cell problem.

4.2.2 SOLVING THE CELL PROBLEM BY LA

The proposed approach is based on the homogenization theory and LA and is applicable, due to its physical essence, for the asymptotic study of composite structures with inclusions of large sizes close to the limit value ($a \to 1$) and extremely high conductivity ($\lambda \to \infty$).

Let us determine the effective parameter of the composite q_y in the direction of the axis y (η in "fast" coordinates). To solve the problems, the following steps are utilized:

1) we replace the outer hexagonal contour of the cell with a circle of radius b (Fig. 4.13).
   The cell problem (4.38)–(4.40) written in fast polar coordinates r, θ in the following way

$$\frac{\partial^2 u_1^+}{\partial r^2} + \frac{1}{r} \frac{\partial u_1^+}{\partial r} + \frac{1}{r^2} \frac{\partial^2 u_1^+}{\partial \theta^2} = 0 \quad \text{in} \quad \Omega_i^+, \tag{4.45}$$

$$\frac{\partial^2 u_1^-}{\partial r^2} + \frac{1}{r} \frac{\partial u_1^-}{\partial r} + \frac{1}{r^2} \frac{\partial^2 u_1^-}{\partial \theta^2} = 0 \quad \text{in} \quad \Omega_i^-, \tag{4.46}$$

$$u_1^+ = u_1^-,$$
$$\frac{\partial u_1^+}{\partial r} - \lambda \frac{\partial u_1^-}{\partial r} = (\lambda - 1) \left(\frac{\partial u_0}{\partial x} \cos \theta + \frac{\partial u_0}{\partial y} \sin \theta \right) \quad \text{at} \quad r = a, \tag{4.47}$$

$$u_1^+ = 0 \quad \text{at} \quad r = b. \tag{4.48}$$

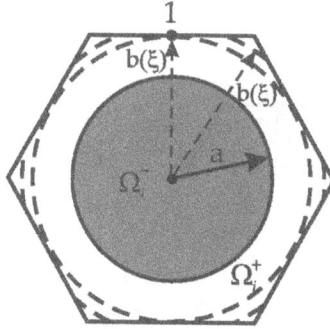

Figure 4.13 Computational model of the LA for a hexagonal lattice.

The solution of problem (4.45)–(4.48) has the form

$$u_1^- = A_1 r \cos\theta + A_2 r \sin\theta, \qquad (4.49)$$

$$u_1^+ = \left(B_1 r + \frac{C_1}{r}\right)\cos\theta + \left(B_2 r + \frac{C_2}{r}\right)\sin\theta, \qquad (4.50)$$

where A_i, B_i, C_i $(i = 1, 2)$ are the constants determined from the conditions (4.47) and (4.48) in the form

$$A_1 = -\frac{(\lambda - 1)\,(b^2 - a^2)}{(b^2 + a^2) + \lambda\,(b^2 - a^2)}\,\frac{\partial u_0}{\partial x},$$

$$B_1 = \frac{(\lambda - 1)\,a^2}{(b^2 + a^2) + \lambda\,(b^2 - a^2)}\,\frac{\partial u_0}{\partial x},$$

$$C_1 = -\frac{(\lambda - 1)\,a^2 b^2}{(b^2 + a^2) + \lambda\,(b^2 - a^2)}\,\frac{\partial u_0}{\partial x}, \qquad (4.51)$$

$$A_2 = A_1, \quad B_2 = B_1, \quad C_2 = C_1 \left(\frac{\partial u_0}{\partial x} \to \frac{\partial u_0}{\partial y}\right).$$

2) In accordance with LA, we consider the outer contour of the cell b as a contour of variable radius:

$$b = \begin{cases} b(\xi) = \sqrt{1 + \xi^2} & \text{for} \quad 0 \le \xi \le \frac{1}{\sqrt{3}} \\ b(\xi) = 2\sqrt{\xi^2 - \sqrt{3}\xi + 1} & \text{for} \quad \frac{1}{\sqrt{3}} \le \xi \le \frac{2}{\sqrt{3}} \end{cases} \qquad (4.52)$$

Fig. 4.14 shows 1/4 of the cell area, over which integration is performed.
3) Integration in expression (4.41) is performed over the original area of the hexagonal cell, taking into account relation (4.52), i.e., assuming

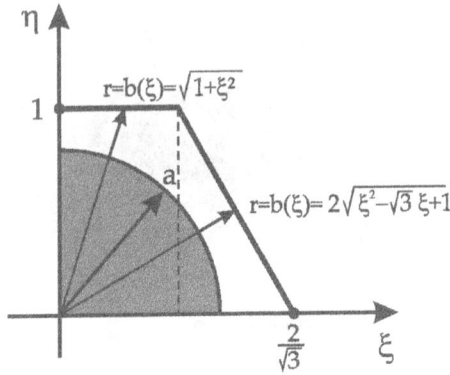

Figure 4.14 Hexagonal cell contour approximation.

A_2, B_2, C_2 as variable functions ξ. After performing the necessary transformations and passing to the limit at $\lambda \to \infty$, we find the asymptotic expression of the parameter $q_y^{(\infty)}$ as a function $q_y^{(\infty)} = q(a)$ of inclusion size in the following form

$$q_y^{(\infty)} = \frac{2\sqrt{3}\,a^2}{\sqrt{1-a^2}} \arctan \frac{\sqrt{3}}{3\sqrt{1-a^2}} + 1 + \frac{\sqrt{3}\,a^2}{3}\left(\frac{\pi}{4} - \frac{3}{2}\arcsin\frac{\sqrt{3}}{3a}\right) +$$

$$+ \frac{4\sqrt{3}\,a^2}{3\sqrt{1-a^2}}\left[\arctan\left(\left(\sqrt{3}\,a - \sqrt{3a^2-1}\right)\sqrt{\frac{1-a}{1+a}}\right) - \right. \tag{4.53}$$

$$-\frac{1}{4}\arctan\frac{2\left(\sqrt{3}\,a - 1 - \sqrt{3a^2-1}\right)\sqrt{1-a^2}}{(1+\sqrt{3})(1-\sqrt{3}\,a)\,a + \sqrt{3a^2-1}\,(a-2+\sqrt{3}\,a)+2} -$$

$$-\frac{1}{8}\arctan\frac{2\sqrt{1-a^2}}{a}\left.\right] - \frac{a^2}{4}\ln\frac{\left(2+3a^2+2\sqrt{3(3a^2-1)}\right)}{(4-3a^2)}.$$

In the limiting case $(a \to 1)$ relation (4.53), taking into account the main term of the asymptotics and the first correction to it, takes the form

$$q_{y\,as}^{(\infty)} = \frac{\sqrt{3}\,\pi}{\sqrt{1-a^2}} + 1 + \frac{\sqrt{3}\,\pi}{12} - \frac{\sqrt{3}}{2}\arcsin\frac{\sqrt{3}}{3} -$$

$$- \frac{1}{4}\ln\left(5 + 2\sqrt{6}\right) - \sqrt{3}\left(\sqrt{3} + \sqrt{2}\right). \tag{4.54}$$

The leading term of the asymptotic representation (4.54) coincides with the asymptotic formula obtained in [96] by generalizing the method applied by Keller [126, 127] for a square array to the case of a hexagonal lattice, and we get

$$q_{as} = \frac{\sqrt{3}\,\pi}{\sqrt{1-a^2}}. \tag{4.55}$$

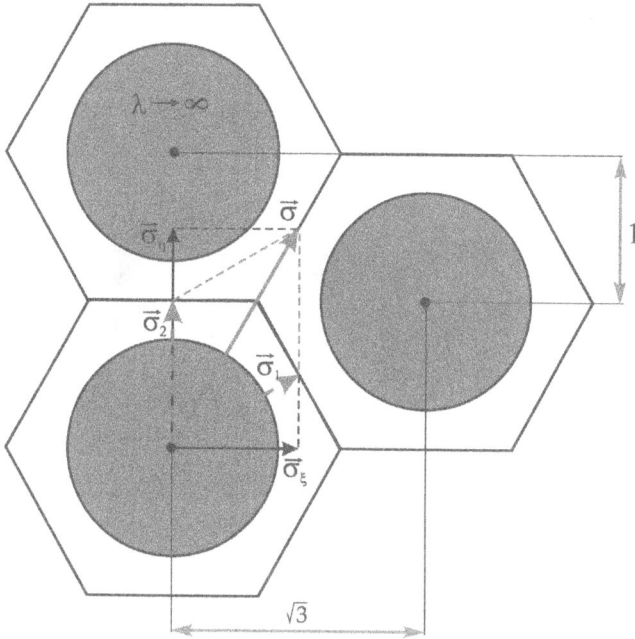

Figure 4.15 Flux distribution at 1/4 cell at $\lambda \to \infty$, $a \gg 0$.

4.2.3 LINKS OF EFFECTIVE THERMAL CONDUCTIVITY PARAMETERS IN VARIOUS DIRECTIONS

The so far obtained results can be summarized as follows:

1) Asymptotic representation of the effective parameter q_x in axis direction x for the case of large sizes of inclusions close to the limit value ($a \to 1$) with extremely high conductivity ($\lambda \to \infty$), can be obtained by considering the heat flux $I(\lambda, a)$ in the cell. The flux distribution at 1/4 cell is illustrated in Fig. 4.15.

Since the intensity of flux $I(\lambda, a)$ is the same in the directions $\vec{\sigma}_1 \left(\frac{\sqrt{3}}{2}; \frac{1}{2} \right)$ and $\vec{\sigma}_2 (0; 1)$, then the resulting stream $\vec{\sigma}$ is defined as follows

$$I(\lambda, a)\, \vec{\sigma}_1 \left(\frac{\sqrt{3}}{2}; \frac{1}{2} \right) + I(\lambda, a)\, \vec{\sigma}_2 (0; 1) = \vec{\sigma} \left(\frac{\sqrt{3}}{2} I(\lambda, a);\ \frac{3}{2} I(\lambda, a) \right).$$

Then the projections $\vec{\sigma}$ on the axis ξ, η are

$$\vec{\sigma}_\xi \left(\frac{\sqrt{3}}{2} I(\lambda, a);\ 0 \right), \qquad \vec{\sigma}_\eta \left(0;\ \frac{3}{2} I(\lambda, a) \right),$$

and, therefore, due to the geometric structure of the hexagonal lattice (Fig. 4.15), the effective parameters in the direction of the x and y axes will be

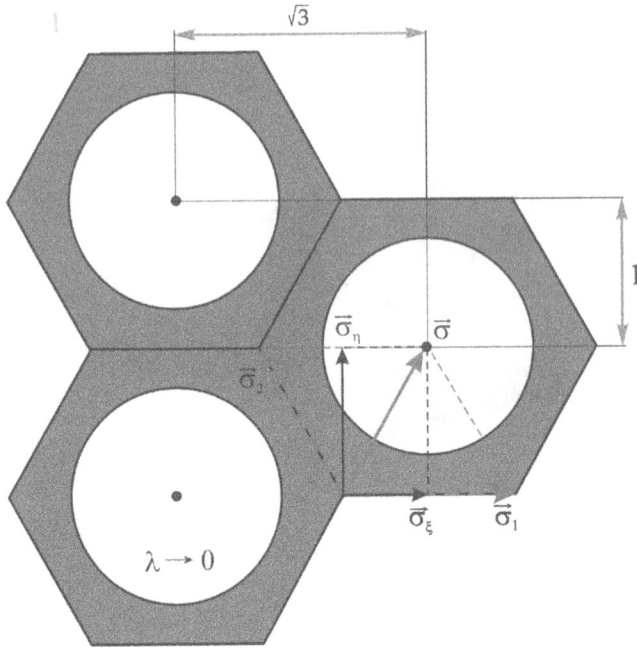

Figure 4.16 Flux distribution for 1/4 cell at $\lambda \to 0$, $a \gg 0$.

the same, since

$$q_x^{(\infty)} = q_y^{(\infty)} = q^{(\infty)},\qquad (4.56)$$

and the homogenized structure is isotropic.

2) Keller [127] provides a theorem that determines, for the case of a composite with cylindrical inclusions in a rectangular lattice, the relationship between the effective thermal conductivity parameters $q_x(\lambda)$ and $q_y(\lambda^{-1})$, which is determined by the ratio

$$q_y(\lambda^{-1}) = q_x^{-1}(\lambda).\qquad (4.57)$$

This result was generalized in [194], where the validity of expression (4.57) is analytically proved also for the hexagonal structure of inclusions.
By virtue of relations (4.56), (4.57), we obtain the asymptotic representations of the effective parameters in the case of extremely small conductivity of the inclusions in the following form

$$q_x^{(0)} = q_y^{(0)} = q^{(0)}.\qquad (4.58)$$

The geometric interpretation of the flux that determine the homogenized parameters that satisfy relation (4.58) is shown in Fig. 4.16.

As follows from Fig. 4.16,

$$I\left(\lambda, a\right) \vec{\sigma}_1 \left(\frac{2\sqrt{3}}{3}; 0\right) + I\left(\lambda, a\right) \vec{\sigma}_2 \left(-\frac{\sqrt{3}}{3}; 1\right)$$

$$= \vec{\sigma}\left(\frac{\sqrt{3}}{3} I\left(\lambda, a\right); I\left(\lambda, a\right)\right)$$

and

$$\vec{\sigma}_\xi \left(\frac{\sqrt{3}}{3} I\left(\lambda, a\right); 0\right), \qquad \vec{\sigma}_\eta \left(0; I\left(\lambda, a\right)\right).$$

4.2.4 GENERALIZATION OF THE LA TO THE CASE OF FINITE VALUES OF THE CONDUCTIVITY OF INCLUSIONS

Formula (4.53) was obtained under the assumption of its extremely large conductivity of inclusions, $\lambda \to \infty$. In this case, the approach described in the framework of the LA allows generalization to the case of finite values, $\lambda \gg 1$.

The application of LA similar to the scheme described above leads in this case to the following expressions for the effective parameter:

(i) at $\lambda \gg 1$, $a \gg 0$:

$$q = \frac{2\sqrt{3}\,\gamma a^2}{\sqrt{1-\gamma a^2}} \arctan \frac{\sqrt{3}}{3\sqrt{1-\gamma a^2}} + 1 + \frac{\gamma a^2 \sqrt{3}}{3} \left\{ \frac{\pi}{4} - \frac{3}{2} \arcsin \frac{\sqrt{3}}{3\sqrt{\gamma}a} + \right.$$

$$+ \frac{4}{\sqrt{1-\gamma a^2}} \left[\arctan \left(\left(\sqrt{3}\gamma a - \sqrt{3\,\gamma a^2 - 1}\right) \frac{\sqrt{1-\sqrt{\gamma}a}}{\sqrt{1+\sqrt{\gamma}a}} \right) - \right. \tag{4.59}$$

$$-\frac{1}{4} \arctan \frac{2\left(\sqrt{3}\gamma a - 1 - \sqrt{3\,\gamma a^2 - 1}\right)\sqrt{1-\gamma a^2}}{(1+\sqrt{3})\,(1-\sqrt{3}\gamma a)\,\sqrt{\gamma}a + \sqrt{3\,\gamma a^2 - 1}\,\left(\sqrt{\gamma}a - 2 + \sqrt{3}\gamma a\right) + 2} -$$

$$\left. - \frac{1}{8} \arctan \left(2\sqrt{\frac{1}{\gamma a^2} - 1} \right) \right] - \frac{\sqrt{3}}{4} \ln \frac{\left(2 + 3\gamma a^2 + 2\sqrt{3\,(3\,\gamma a^2 - 1)}\right)}{(4 - 3\gamma a^2)} \right\},$$

where

$$\gamma = \frac{\lambda - 1}{\lambda + 1} \tag{4.60}$$

and the area of applicability is determined by the relation $\sqrt{\gamma}a \geq \frac{1}{\sqrt{3}}$;

(ii) at $\lambda \gg 1$, $a \to 1$:

$$q = \frac{\sqrt{6\lambda}\,\pi}{2} + 1 + \frac{\sqrt{3}\,\pi}{12} - \frac{\sqrt{3}}{2} \arcsin \frac{\sqrt{3}}{3} - $$

$$- \frac{1}{4} \ln \left(5 + 2\sqrt{6}\right) - \sqrt{3}\left(\sqrt{3} + \sqrt{2}\right), \tag{4.61}$$

the scope of which is determined by the values $\lambda \geq 2$.

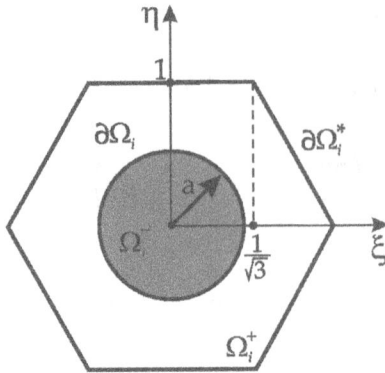

Figure 4.17 Characteristic cell of a composite with an inclusion of medium size.

Obviously, when passing to the limit $\lambda \to \infty$ in expressions (4.59), (4.60), relation (4.59) will coincide with that obtained earlier (4.53).

4.2.5 GENERALIZING RELATIONS OF THE LA FOR MEDIUM-SIZED INCLUSIONS

Note that the asymptotic expression for the parameter (4.53) was obtained for large inclusion sizes a close to the limit value $a = 1$, and it can be used to the value $a \geq \frac{1}{\sqrt{3}} \approx 0.5774$. In this regard, it is of interest to generalize the proposed approach of the LA to the case of medium-sized inclusions $a \leq \frac{1}{\sqrt{3}}$ (Fig. 4.17).

The use of relations (4.49)–(4.51), taking into account the conditions (4.52), leads to the following expression describing the effective parameter $q_{m.\,incl.}$ of composite with extremely large conductivity ($\lambda \to \infty$) for medium sized inclusions ($a \leq \frac{1}{\sqrt{3}}$):

$$q_{m.incl.} = 1 + \frac{\sqrt{3}\,a^2}{\sqrt{1-a^2}}\arctan\frac{\sqrt{3}}{3\sqrt{1-a^2}} +$$

$$+ \frac{\sqrt{3}\,a^2}{3}\left[\frac{2}{\sqrt{1-a^2}}\left(2\arctan\sqrt{\frac{1-a}{1+a}} + \arctan\frac{a}{\sqrt{1-a^2}}\right) - \frac{\pi}{2}\right]. \tag{4.62}$$

Further development of the described approach includes the conductivity of inclusion with large, but finite values ($\lambda \gg 1$). In this case, for medium sizes of inclusions, the effective parameter is described by the following expression

$$q_{m.incl.} = 1 + \frac{\sqrt{3}\,\gamma a^2}{\sqrt{1-\gamma a^2}}\arctan\frac{\sqrt{3}}{3\sqrt{1-\gamma a^2}} + \tag{4.63}$$

$$+ \frac{\sqrt{3}\,\gamma a^2}{3}\left[\frac{2}{\sqrt{1-\gamma a^2}}\left(2\arctan\frac{\sqrt{1-\sqrt{\gamma}a}}{\sqrt{1+\sqrt{\gamma}a}} + \arctan\frac{\sqrt{\gamma}a}{\sqrt{1-\gamma a^2}}\right) - \frac{\pi}{2}\right],$$

which is correct for $\sqrt{\gamma}a = \sqrt{\frac{\lambda-1}{\lambda+1}}\,a \leq \frac{1}{\sqrt{3}}$.

4.2.6 ASYMPTOTIC EXPRESSIONS IN THE CASE OF EXTREMELY SMALL CONDUCTIVITY OF INCLUSIONS

Using the solutions (4.53), (4.54), (4.62), and relations (4.57), (4.58), we can formulate the problem for obtaining asymptotic expressions for the effective parameters at $\lambda \to 0$ in the following form

(i) at $\lambda \to 0$, $a \geq \frac{1}{\sqrt{3}}$:

$$q^{(0)} = \left\{ \frac{2\sqrt{3}a^2}{\sqrt{1-a^2}} \arctan \frac{\sqrt{3}}{3\sqrt{1-a^2}} + 1 + \frac{\sqrt{3}a^2}{3} \left(\frac{\pi}{4} - \frac{3}{2} \arcsin \frac{\sqrt{3}}{3a} \right) + \right.$$

$$+ \frac{4\sqrt{3}a^2}{3\sqrt{1-a^2}} \left[\arctan \left(\left(\sqrt{3}a - \sqrt{3a^2-1} \right) \sqrt{\frac{1-a}{1+a}} \right) - \right. \tag{4.64}$$

$$-\frac{1}{4} \arctan \frac{2 \left(\sqrt{3}a - 1 - \sqrt{3a^2-1} \right) \sqrt{1-a^2}}{\left(1+\sqrt{3}\right) \left(1-\sqrt{3}a\right) a + \sqrt{3a^2-1} \left(a-2+\sqrt{3}a\right) + 2} -$$

$$\left. -\frac{1}{8} \arctan \frac{2\sqrt{1-a^2}}{a} \right] - \frac{a^2}{4} \ln \frac{\left(2+3a^2+2\sqrt{3 \left(3a^2-1\right)}\right)}{(4-3a^2)} \right\}^{-1} ;$$

(ii) at $\lambda \to 0$, $a \to 1$:

$$q_{as}^{(0)} = \sqrt{1-a^2} \left[\sqrt{3}\pi + \sqrt{1-a^2} \left(1 + \frac{\sqrt{3}\pi}{12} - \frac{\sqrt{3}}{2} \arcsin \frac{\sqrt{3}}{3} - \right. \right.$$

$$\left. \left. -\frac{1}{4} \ln \left(5+2\sqrt{6}\right) - \sqrt{3} \left(\sqrt{3}+\sqrt{2}\right) \right) \right]^{-1} ; \tag{4.65}$$

(iii) at $\lambda \to 0$, $a \leq \frac{1}{\sqrt{3}}$:

$$q_{m.incl.} = \left\{ 1 + \frac{\sqrt{3}a^2}{\sqrt{1-a^2}} \arctan \frac{\sqrt{3}}{3\sqrt{1-a^2}} + \right. \tag{4.66}$$

$$\left. + \frac{\sqrt{3}a^2}{3} \left[\frac{2}{\sqrt{1-a^2}} \left(2\arctan \sqrt{\frac{1-a}{1+a}} + \arctan \frac{a}{\sqrt{1-a^2}} \right) - \frac{\pi}{2} \right] \right\}^{-1} .$$

4.2.7 ANALYTICAL EXPRESSIONS OF EFFECTIVE PARAMETERS IN THE CASE OF LOW CONDUCTIVITIES OF INCLUSIONS

The considerations given in Sec. 4.2.3 allow us to generalize the results, using the Keller's theorem (3.59) and to obtain analytical expressions for the effective parameters of a composite in the case of small physical characteristics of the inclusions. Taking into account the solutions (4.59)–(4.61), (4.63), one gets:

(i) expression for large inclusion sizes ($a \gg 0$) and their small conductivity ($\lambda \ll 1$):

$$q = \left\{ -\frac{2\sqrt{3}\,\gamma a^2}{\sqrt{1+\gamma a^2}} \arctan \frac{\sqrt{3}}{3\sqrt{1+\gamma a^2}} + 1 - \frac{\gamma a^2 \sqrt{3}}{3} \times \right.$$

$$\times \left[\frac{\pi}{4} - \frac{3}{2} \arcsin \frac{\sqrt{3}}{3\sqrt{-\gamma a}} + \frac{4}{\sqrt{1+\gamma a^2}} \times \right.$$

$$\times \left(\arctan \left(\left(\sqrt{-3\gamma a} - \sqrt{-3\gamma a^2 - 1} \right) \frac{\sqrt{1-\sqrt{-\gamma a}}}{\sqrt{1+\sqrt{-\gamma a}}} \right) - \right. \tag{4.67}$$

$$-\frac{1}{4} \arctan \left(2 \left(\sqrt{-3\gamma a} - 1 - \sqrt{-3\gamma a^2 - 1} \right) \sqrt{1+\gamma a^2} \left((1+\sqrt{3}) \times \right. \right.$$

$$\times \left(1 + \sqrt{-3\gamma a} \right) \sqrt{-\gamma a} + \sqrt{-3\gamma a^2 - 1} \left(\sqrt{-\gamma a} - 2 + \sqrt{-3\gamma a} \right) + 2 \right)^{-1} \Bigg) -$$

$$-\frac{1}{8} \arctan \left(2\sqrt{-\frac{1}{\gamma a^2} - 1} \right) \Bigg) - \frac{\sqrt{3}}{4} \ln \frac{\left(2 - 3\gamma a^2 + 2\sqrt{-3\,(3\gamma a^2 + 1)} \right)}{(4+3\gamma a^2)} \Bigg]^{-1} \right\}^{-1},$$

which is applicable for $\sqrt{-\gamma a} = \sqrt{\frac{1-\lambda}{\lambda+1}}\; a \geq \frac{1}{\sqrt{3}}$;

(ii) result for extremely large sizes of inclusions ($a \to 1$) with small conductivities ($\lambda \ll 1$), we have

$$q = 12\sqrt{\lambda}\left[6\sqrt{6}\,\pi + \left(12 + \sqrt{3}\,\pi - \right. \right. \tag{4.68}$$

$$\left. -6\sqrt{3}\arcsin\frac{\sqrt{3}}{3} - 3\ln\left(5+2\sqrt{6}\right) - 12\sqrt{3}\left(\sqrt{3}+\sqrt{2}\right)\right)\sqrt{\lambda} \right]^{-1},$$

for the value $\lambda \leq \frac{1}{2}$;

(iii) in the case of inclusions of medium size and conductivity of small but finite values ($0 < \lambda \ll 1$), for $\sqrt{-\gamma a} = \sqrt{\frac{1-\lambda}{1+\lambda}}\; a \leq \frac{1}{\sqrt{3}}$ we get

$$q_{m.incl.} = \left\{ 1 - \frac{\sqrt{3}\,\gamma a^2}{\sqrt{1+\gamma a^2}} \arctan \frac{\sqrt{3}}{3\sqrt{1+\gamma a^2}} - \frac{\sqrt{3}\,\gamma a^2}{3} \times \right. \tag{4.69}$$

$$\times \left[\frac{2}{\sqrt{1+\gamma a^2}} \left(2\arctan \frac{\sqrt{1-\sqrt{-\gamma a}}}{\sqrt{1+\sqrt{-\gamma a}}} + \arctan \frac{\sqrt{-\gamma a}}{\sqrt{1+\gamma a^2}} \right) - \frac{\pi}{2} \right]^{-1} \right\}.$$

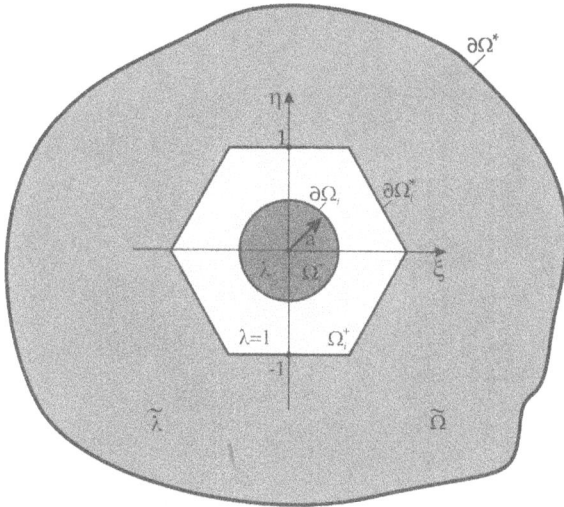

Figure 4.18 Geometric interpretation of a ThPhM for composite with a hexagonal structure.

4.2.8 APPLICATION OF THE THPHM FOR SOLVING A LOCAL PROBLEM IN THE CASE OF SMALL SIZES OF INCLUSIONS

For the considered composite material, we will replace, in accordance with ThPhM, the entire periodic array, with the exception of one cell, by a homogeneous medium with an unknown effective coefficient $\tilde{\lambda}$ (Fig. 4.18).

Mathematically, the ThPhM leads to:

1) the need to expand in an asymptotic series in powers of ε function $\tilde{u}(x, y, \xi, \eta)$

$$\tilde{u} = u_0\,(x,\,y) + \varepsilon\,\tilde{u}_1\,(x,\,y,\,\xi,\,\eta) + \varepsilon^2 \tilde{u}_2\,(x,\,y,\,\xi,\,\eta) + \dots \qquad (4.70)$$

2) the replacement of the condition on the outer contour of the cell (4.40) by the conditions of conjugation of the matrix with an equivalent homogeneous medium;

3) decaying conditions for the function and its derivative at infinity.

Therefore, when using ThPhM, the local problem (4.38)–(4.40) is transformed to the following one

$$\frac{\partial^2 u_1^\pm}{\partial \xi^2} + \frac{\partial^2 u_1^\pm}{\partial \eta^2} = 0 \quad \text{in} \quad \Omega_i^\pm, \qquad (4.71)$$

$$\frac{\partial^2 \tilde{u}_1}{\partial \xi^2} + \frac{\partial^2 \tilde{u}_1}{\partial \eta^2} = 0 \quad \text{in} \quad \tilde{\Omega}, \qquad (4.72)$$

$$u_1^+ = u_1^-, \quad \frac{\partial u_1^+}{\partial \bar{\mathbf{n}}} - \lambda \frac{\partial u_1^-}{\partial \bar{\mathbf{n}}} = (\lambda - 1) \frac{\partial u_0}{\partial \mathbf{n}} \quad \text{at} \quad \partial\Omega_i, \qquad (4.73)$$

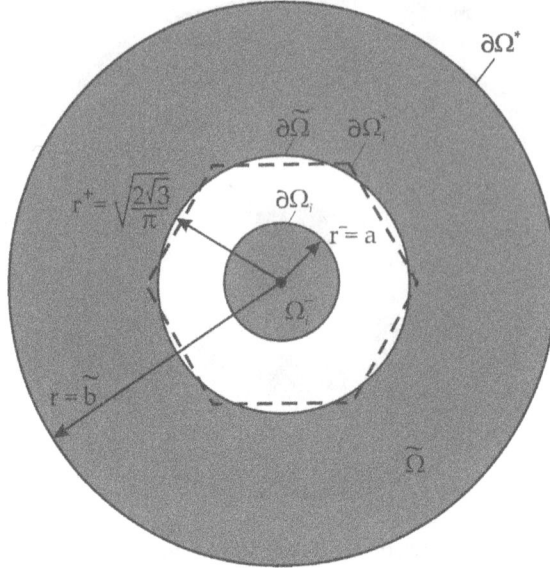

Figure 4.19 ThPhM model for the composite with a hexagonal cell.

$$u_1^+ = \tilde{u}_1, \quad \frac{\partial u_1^+}{\partial \bar{\mathbf{n}}} - \tilde{\lambda} \frac{\partial \tilde{u}_1}{\partial \bar{\mathbf{n}}} = \left(\tilde{\lambda} - 1\right) \frac{\partial u_0}{\partial \mathbf{n}} \quad \text{at} \quad \partial \tilde{\Omega}, \tag{4.74}$$

$$\tilde{u}_1 \to 0, \quad \frac{\partial \tilde{u}_1}{\partial \xi} \to 0 \quad \text{at} \quad \xi = \to \pm\infty,$$

$$\tilde{u}_1 \to 0, \quad \frac{\partial \tilde{u}_1}{\partial \eta} \to 0 \quad \text{at} \quad \eta = \to \pm\infty. \tag{4.75}$$

The outer contour of the cell $\partial \tilde{\Omega}$ in polar coordinates r, θ is governed by an equation of the form [105, 184]:

$$r = r_0 + \varepsilon_1 f(\theta), \tag{4.76}$$

where $r_0 = const > 0$; $f(\theta)$ is the differentiable function characterizing the geometric shape of the considered contour; ε_1 stands for the small parameter ($|\varepsilon_1| \ll 1$).

We restrict ourselves to solving the problem in the zeroth approximation. In this case, the hexagonal contour of the cell is replaced by a circle. The radius of the circle r^+ is chosen in such a way that the equality of the areas of the original and transformed areas of the matrix is preserved:

$$r^+ = \sqrt{\frac{2\sqrt{3}}{\pi}}. \tag{4.77}$$

The model of such a structure is shown in Fig. 4.19. The local problem (4.71)–(4.75) in polar coordinates r, θ can be written as follows

$$\frac{\partial^2 u_1^{\pm}}{\partial r^2} + \frac{1}{r}\frac{\partial u_1^{\pm}}{\partial r} + \frac{1}{r^2}\frac{\partial^2 u_1^{\pm}}{\partial \theta^2} = 0 \quad \text{in} \quad \Omega_i^{\pm}, \tag{4.78}$$

$$\frac{\partial^2 \tilde{u}_1}{\partial r^2} + \frac{1}{r}\frac{\partial \tilde{u}_1}{\partial r} + \frac{1}{r^2}\frac{\partial^2 \tilde{u}_1}{\partial \theta^2} = 0 \quad \text{in} \quad \tilde{\Omega}_0, \tag{4.79}$$

at $\partial \Omega_i$: $r = r^- = a$:

$$u_1^+ = u_1^-,$$

$$\frac{\partial u_1^+}{\partial r} - \lambda \frac{\partial u_1^-}{\partial r} = (\lambda - 1)\left(\frac{\partial u_0}{\partial x}\cos\theta + \frac{\partial u_0}{\partial y}\sin\theta\right), \tag{4.80}$$

at $\partial \tilde{\Omega}$: $r = r^+ = \sqrt{\frac{2\sqrt{3}}{\pi}}$:

$$u_1^+ = \tilde{u}_1,$$

$$\frac{\partial u_1^+}{\partial r} - \tilde{\lambda}\frac{\partial \tilde{u}_1}{\partial r} = \left(\tilde{\lambda}_0 - 1\right)\left(\frac{\partial u_0}{\partial x}\cos\theta + \frac{\partial u_0}{\partial y}\sin\theta\right), \tag{4.81}$$

$$\tilde{u}_1 \to 0, \quad \frac{\partial \tilde{u}_1}{\partial r} \to 0 \quad \text{at} \quad r \to \infty. \tag{4.82}$$

The solution of the boundary value problem (4.78)–(4.82) can be represented as

$$u_1^- = A_1 r\cos\theta + A_2 r\sin\theta,$$

$$u_1^+ = \left(B_1 r + \frac{C_1}{r}\right)\cos\theta + \left(B_2 r + \frac{C_2}{r}\right)\sin\theta, \tag{4.83}$$

$$\tilde{u}_1 = \frac{D_1}{r}\cos\theta + \frac{D_2}{r}\sin\theta,$$

where A_1, A_2, B_1, B_2, C_1, C_2, D_1, D_2 are the arbitrary constants.

Note that in expressions (4.83), u_1^- is written taking into account the boundedness of the function and its derivative $\frac{\partial u_1^-}{\partial r}$ at $r = 0$, while function $\tilde{u}_1$ satisfies the decaying conditions at $r \to \infty$ (4.82).

Relation (4.83) includes eight arbitrary constants, which are determined from the conjugation conditions (4.80), (4.81). Systems of equations for constants A_1, B_1, C_1, D_1 and A_2, B_2, C_2, D_2 are completely identical, so we give the solution of only the first of them:

$$A_1 = A_1^*\frac{\partial u_0}{\partial x}, \quad B_1 = B_1^*\frac{\partial u_0}{\partial x}, \quad C_1 = C_1^*\frac{\partial u_0}{\partial x}, \quad D_1 = D_1^*\frac{\partial u_0}{\partial x}, \tag{4.84}$$

$$A_1^* = -1 + 4\tilde{\lambda}\,\Delta,$$

$$B_1^* = -1 + 2(\lambda + 1)\,\tilde{\lambda}\,\Delta, \tag{4.85}$$

$$C_1^* = -2(\lambda - 1)\,\tilde{\lambda}\,a^2\Delta,$$

$$D_1^* = \frac{4}{\pi}\left(1 - 2\left((\lambda - 1)\frac{\pi a^2}{2\sqrt{3}} + (\lambda + 1)\right)\Delta\right),$$

where

$$\Delta = \left(\left(\tilde{\lambda} + 1\right)(\lambda + 1) - \left(\tilde{\lambda} - 1\right)(\lambda - 1)\frac{\pi a^2}{2\sqrt{3}}\right)^{-1}.$$

For arbitrary constants A_2, B_2, C_2, D_2, we have

$$A_1 = A_2, \quad B_1 = B_2, \quad C_1 = C_2, \quad D_1 = D_2 \left(\frac{\partial u_0}{\partial x} \to \frac{\partial u_0}{\partial y}\right). \tag{4.86}$$

The further solution of the problem is reduced to the construction of homogenized relations; it should be taken into account that the averaging should be carried out over the entire region $\Omega^* = \Omega_i^+ \bigcup \Omega_i^- \bigcup \tilde{\Omega}$. Therefore, the averaging operator in this case is

$$\overline{(...)} = \frac{1}{|\Omega^*|}\left(\iint\limits_{\Omega_i^+}(...)\,d\xi\,d\eta + \lambda\iint\limits_{\Omega_i^-}(...)\,d\xi\,d\eta + \tilde{\lambda}\iint\limits_{\tilde{\Omega}}(...)\,d\xi\,d\eta\right), \tag{4.87}$$

where $|\Omega^*| = |\Omega_i^+ \bigcup \Omega_i^- \bigcup \tilde{\Omega}|$.

Thus, the equation to be averaged is

$$\frac{\partial^2 u_0}{\partial x^2} + \frac{\partial^2 u_0}{\partial y^2} + 2\frac{\partial^2 u_1^+}{\partial x \partial \xi} + 2\frac{\partial^2 u_1^+}{\partial y \partial \eta} + \frac{\partial^2 u_2^+}{\partial \xi^2} + \frac{\partial^2 u_2^+}{\partial \eta^2} +$$

$$+\lambda\left(\frac{\partial^2 u_0}{\partial x^2} + \frac{\partial^2 u_0}{\partial y^2} + 2\frac{\partial^2 u_1^-}{\partial x \partial \xi} + 2\frac{\partial^2 u_1^-}{\partial y \partial \eta} + \frac{\partial^2 u_2^-}{\partial \xi^2} + \frac{\partial^2 u_2^-}{\partial \eta^2}\right) + \tag{4.88}$$

$$+\tilde{\lambda}\left(\frac{\partial^2 u_0}{\partial x^2} + \frac{\partial^2 u_0}{\partial y^2} + 2\frac{\partial^2 \tilde{u}_1}{\partial x \partial \xi} + 2\frac{\partial^2 \tilde{u}_1}{\partial y \partial \eta} + \frac{\partial^2 \tilde{u}_2}{\partial \xi^2} + \frac{\partial^2 \tilde{u}_2}{\partial \eta^2}\right) = F.$$

Applying the averaging operator (4.87) to relation (4.88), we obtain the homogenized equation in the following form

$$\frac{1}{|\Omega^*|}\left[\iint\limits_{\Omega_i^+}\left(\frac{\partial^2 u_0}{\partial x^2} + \frac{\partial^2 u_0}{\partial y^2} + \frac{\partial^2 u_1^+}{\partial x \partial \xi} + \frac{\partial^2 u_1^+}{\partial y \partial \eta}\right)d\xi\,d\eta + \right.$$

$$+ \int\limits_{\partial\Omega_i}\left(\frac{\partial u_2^+}{\partial \bar{\mathbf{n}}} + \frac{\partial u_1^+}{\partial \mathbf{n}}\right)d\ell + \int\limits_{\partial\tilde{\Omega}}\left(\frac{\partial u_2^+}{\partial \bar{\mathbf{n}}} + \frac{\partial u_1^+}{\partial \mathbf{n}}\right)d\ell +$$

$$+\lambda \iint\limits_{\Omega_i^-} \left(\frac{\partial^2 u_0}{\partial x^2} + \frac{\partial^2 u_0}{\partial y^2} + \frac{\partial^2 u_1^-}{\partial x \partial \xi} + \frac{\partial^2 u_1^-}{\partial y \partial \eta} \right) d\xi \, d\eta + \tag{4.89}$$

$$+\lambda \int\limits_{\partial\Omega_i} \left(\frac{\partial u_2^-}{\partial \bar{n}} + \frac{\partial u_1^-}{\partial n} \right) d\ell + \tilde{\lambda} \iint\limits_{\tilde{\Omega}} \left(\frac{\partial^2 u_0}{\partial x^2} + \frac{\partial^2 u_0}{\partial y^2} + \frac{\partial^2 \tilde{u}_1}{\partial x \partial \xi} + \frac{\partial^2 \tilde{u}_1}{\partial y \partial \eta} \right) d\xi \, d\eta +$$

$$\left. + \tilde{\lambda} \int\limits_{\partial\tilde{\Omega}} \left(\frac{\partial \tilde{u}_2}{\partial \bar{n}} + \frac{\partial \tilde{u}_1}{\partial n} \right) d\ell + \tilde{\lambda} \int\limits_{\partial\Omega^*} \left(\frac{\partial \tilde{u}_2}{\partial \bar{n}} + \frac{\partial \tilde{u}_1}{\partial n} \right) d\ell \right] = F.$$

Taking into account the expressions for u_1^+, u_1^-, $\tilde{u}_1$ (4.83)–(4.86), the homogenized equation (4.89) is transformed to the form

$$\frac{1}{|\Omega^*|} \left[|\Omega_i^+| \, (B_1^* + 1) + \lambda \, |\Omega_i^-| \, (A_1^* + 1) + \tilde{\lambda}_0 \, |\tilde{\Omega}| \right] \Delta u_0 = F, \tag{4.90}$$

$$q = \frac{1}{|\Omega^*|} \left[|\Omega_i^+| \, (B_1^* + 1) + \lambda \, |\Omega_i^-| \, (A_1^* + 1) + \tilde{\lambda} \, |\tilde{\Omega}| \right]. \tag{4.91}$$

Coefficient q, defined by expression (4.91), is the effective parameter $\tilde{\lambda}$, which was used in ThPhM to describe an equivalent homogeneous environment $\tilde{\Omega}$. Therefore, equating $\tilde{\lambda}$ to expression (4.91), we arrive at a linear algebraic equation with respect to the coefficient $q = \tilde{\lambda}$. The solution of equation gives the following analytical expression for the effective coefficient:

$$q_{ThPhM}^{hex} = \tilde{\lambda} = \frac{1 - \frac{\pi a^2}{2\sqrt{3}} + \lambda \left(1 + \frac{\pi a^2}{2\sqrt{3}} \right)}{1 + \frac{\pi a^2}{2\sqrt{3}} + \lambda \left(1 - \frac{\pi a^2}{2\sqrt{3}} \right)}. \tag{4.92}$$

We note that relation (4.92) obtained on the basis of ThPhM coincides with the solution given in [158] and represents the Hashin-Shtrikman upper bound at $0 \leq \lambda \leq 1$ and lower band at $1 \leq \lambda < \infty$.

4.2.9 ASYMPTOTIC ANALYSIS OF SOLUTIONS FOR SMALL INCLUSIONS

The solution based on ThPhM (4.92) was obtained under the assumption of small sizes of inclusions. On the other hand, relations (4.62), (4.63), (4.66), (4.69) found using LA, describe the homogenized characteristics for the inclusions of medium size. In this regard, it is of interest to compare the results of the two approaches and estimate the area of their applicability:

1) For a structure with extremely high conductivity ($\lambda \to \infty$), the ThPhM solution (4.92) yields

$$q_{ThPhM}^{hex} = \frac{1 + \frac{\pi a^2}{2\sqrt{3}}}{1 - \frac{\pi a^2}{2\sqrt{3}}}. \tag{4.93}$$

The expansion of expression (4.93) in powers of a for small inclusion sizes $(a \to 0)$ has the form

$$q_{ThPhM}^{hex} = \frac{1 + \frac{\pi a^2}{2\sqrt{3}}}{1 - \frac{\pi a^2}{2\sqrt{3}}} = 1 + \frac{\pi a^2}{\sqrt{3}} + \frac{\pi^2 a^4}{6} + O\left(a^6\right),$$

i.e., it coincides with the expansion of relation (4.62) up to terms of the order a^2 inclusive:

$$q_{m.\,incl.} = 1 + \frac{\sqrt{3}\,a^2}{\sqrt{1-a^2}} \arctan \frac{1}{\sqrt{3\,(1-a^2)}} +$$

$$+ \frac{a^2}{\sqrt{3}} \left[\frac{2}{\sqrt{1-a^2}} \left(2\arctan\left(\sqrt{\frac{1-a}{1+a}}\right) + \arctan\left(\frac{a}{\sqrt{1-a^2}}\right) \right) - \frac{\pi}{2} \right] =$$

$$= 1 + \frac{\pi a^2}{\sqrt{3}} + \frac{3 + 2\pi\sqrt{3}}{8} a^4 + O\left(a^6\right).$$

2) A similar correspondence of solutions also takes place in the case of extremely small conductivity of the inclusions $(\lambda \to 0)$ and their small size $(a \to 0)$. Indeed, from (4.66), (4.92), we have

$$q_{ThPhM}^{hex} = \frac{1 - \frac{\pi a^2}{2\sqrt{3}}}{1 + \frac{\pi a^2}{2\sqrt{3}}} = 1 - \frac{\pi a^2}{\sqrt{3}} + \frac{\pi^2 a^4}{6} + O\left(a^6\right), \qquad (4.94)$$

$$q_{m.\,incl.} = \left\{ 1 + \frac{\sqrt{3}\,a^2}{\sqrt{1-a^2}} \arctan \frac{1}{\sqrt{3\,(1-a^2)}} + \right.$$

$$+ \frac{a^2}{\sqrt{3}} \left[\frac{2}{\sqrt{1-a^2}} \left(2\arctan\left(\sqrt{\frac{1-a}{1+a}}\right) + \arctan\left(\frac{a}{\sqrt{1-a^2}}\right) \right) - \frac{\pi}{2} \right] \right\}^{-1} =$$

$$= 1 - \frac{\pi a^2}{\sqrt{3}} + \frac{8\pi^2 - 6\pi\sqrt{3} - 9}{24} a^4 + O\left(a^6\right).$$

3) For arbitrary finite values of the inclusions conductivity λ and their small size $(a \to 0)$, the ThPhM solution is

$$q_{ThPhM}^{hex} = \frac{1 - \frac{\pi a^2}{2\sqrt{3}} + \lambda\left(1 + \frac{\pi a^2}{2\sqrt{3}}\right)}{1 + \frac{\pi a^2}{2\sqrt{3}} + \lambda\left(1 - \frac{\pi a^2}{2\sqrt{3}}\right)} =$$

$$= 1 + \frac{\lambda - 1}{\lambda + 1} \frac{\pi a^2}{\sqrt{3}} + \frac{(\lambda - 1)^2}{(\lambda + 1)^2} \frac{\pi^2 a^4}{6} + O\left(a^6\right) \approx \qquad (4.95)$$

$$\approx 1 + 1.813799365 \frac{\lambda - 1}{\lambda + 1} a^2 + 1.644934068 \left(\frac{\lambda - 1}{\lambda + 1}\right)^2 a^4 + O\left(a^6\right).$$

This solution coincids up to terms of order a^2, inclusive, with expansion in a series of expressions (4.63), (4.69), which describe the LA solution for medium-sized inclusions

$$q_{m.\,incl.}\bigg|_{\lambda>1} = 1 + \frac{\sqrt{3}\,\gamma a^2}{\sqrt{1-\gamma a^2}}\arctan\frac{\sqrt{3}}{3\sqrt{1-\gamma a^2}} +$$

$$+\frac{\sqrt{3}\,\gamma a^2}{3}\left[\frac{2}{\sqrt{1-\gamma a^2}}\left(2\arctan\frac{\sqrt{1-\sqrt{\gamma}a}}{\sqrt{1+\sqrt{\gamma}a}}+\arctan\left(\frac{\sqrt{\gamma}a}{\sqrt{1-\gamma a^2}}\right)\right)-\frac{\pi}{2}\right]=$$

$$= 1 + \frac{\lambda-1}{\lambda+1}\frac{\pi a^2}{\sqrt{3}} + \frac{(\lambda-1)^2}{(\lambda+1)^2}\frac{\left(3+2\pi\sqrt{3}\right)a^4}{8} + O\left(a^6\right) \approx \qquad (4.96)$$

$$\approx 1 + 1.813799365\,\frac{\lambda-1}{\lambda+1}\,a^2 + 1.735349524\left(\frac{\lambda-1}{\lambda+1}\right)^2 a^4 + O\left(a^6\right),$$

$$q_{m.\,incl.}\bigg|_{\lambda<1} = \left\{1 - \frac{\sqrt{3}\,\gamma a^2}{\sqrt{1+\gamma a^2}}\arctan\frac{\sqrt{3}}{3\sqrt{1+\gamma a^2}} - \right.$$

$$-\frac{\sqrt{3}\,\gamma a^2}{3}\left[\frac{2}{\sqrt{1+\gamma a^2}}\left(2\arctan\frac{\sqrt{1-\sqrt{-\gamma}a}}{\sqrt{1+\sqrt{-\gamma}a}}+\arctan\left(\frac{\sqrt{-\gamma}a}{\sqrt{1+\gamma a^2}}\right)\right)-\frac{\pi}{2}\right]\left.\right\}^{-1}=$$

$$= 1 + \frac{\lambda-1}{\lambda+1}\frac{\pi a^2}{\sqrt{3}} + \frac{(\lambda-1)^2}{(\lambda+1)^2}\frac{\left(8\pi^2-6\pi\sqrt{3}-9\right)a^4}{24} + O\left(a^6\right) \approx \qquad (4.97)$$

$$\approx 1 + 1.813799365\,\frac{\lambda-1}{\lambda+1}\,a^2 + 1.554518610\left(\frac{\lambda-1}{\lambda+1}\right)^2 a^4 + O\left(a^6\right).$$

Moreover, expansions (4.95)–(4.97) coincide up to terms of the order a^2 inclusive with the expansion in a series of the approximate analytical solution

$$q_P = 1 + \frac{2\gamma c}{1-\gamma c-\frac{0.075422\,\gamma^2 c^6}{1-1.060283\,\gamma^2 c^{12}}-0.000076\,\gamma^2 c^{12}}, \qquad (4.98)$$

obtained in [194]:

$$q_P = 1 + \frac{2\gamma c}{1-\gamma c-\frac{0.075422\,\gamma^2 c^6}{1-1.060283\,\gamma^2 c^{12}}-0.000076\,\gamma^2 c^{12}} \approx$$

$$\approx 1 + 1.813799365\,\frac{\lambda-1}{\lambda+1}\,a^2 + 1.644934069\left(\frac{\lambda-1}{\lambda+1}\right)^2 a^4 + O\left(a^6\right)'$$

where $c = \dfrac{|\Omega_i^-|}{|\Omega_i^+|} = \dfrac{\pi a^2}{2\sqrt{3}}$.

Very close to formula (4.98) in structure and results of numerical calculations is the solution obtained in [104] using the method of a complex variable and the properties of Weierstrass elliptic functions. This solution, presented in [104] in infinite series, taking into account the first correction of the asymptotics, can be transformed to the form:

$$q_G = \frac{(1+\lambda)^2 - (1-\lambda^2)\frac{\pi a^2}{2\sqrt{3}} - 171.873845(1-\lambda)^2\frac{a^{12}}{2^{12}}}{(1+\lambda)^2 + (1-\lambda^2)\frac{\pi a^2}{2\sqrt{3}} - 171.873845(1-\lambda)^2\frac{a^{12}}{2^{12}}}. \tag{4.99}$$

The effective parameter p is determined in [104] by the relations:

$$p = p_1 \frac{1 - 2V_2 P \, ||p_\gamma||}{p_1 + p_2}, \tag{4.100}$$

where p_1, p_2 are physical characteristics of the matrix and inclusions, respectively;

$$||p_\gamma|| = p_1 - p_2; \tag{4.101}$$

$$V_2 = \frac{\pi R^2}{\sin \mu}, \qquad \mu = \frac{\pi}{3}; \tag{4.102}$$

$$P = \frac{1}{1 + \chi V_2 - \chi^2 v_p^T M_p^{-1} \tilde{v}_p}; \tag{4.103}$$

$$\chi = \frac{||p_\gamma||}{p_1 + p_2}; \tag{4.104}$$

v_p (v_s), $\tilde{v}_p$ $(\tilde{v}_t)$ are the vectors (superscript T in (4.103) means the transposition operation) and M_p (m_{ts}) is the matrix with components:

$$v_s = R^{12s}\eta_{1\ 6s-1}; \tag{4.105}$$

$$\tilde{v}_t = \eta_{6t-1\ 1}; \tag{4.106}$$

$$m_{ts} = \delta_{6t-1\ 6s-1} - \chi^2 R^{12s} \sum_{i=1}^{\infty} R^{12i}\eta_{6t-1\ 6i+1} \eta_{6i+1\ 6s-1}, \tag{4.107}$$

$t, s = 1, 2, 3, \dots$

The parameters included in formulas (4.105)–(4.107) mean: R is the inclusion radius; $\delta_{6t-1\ 6s-1}$ is the Kronecker's symbol;

$$\eta_{k\ \ell} = -C^\ell_{k+\ell-1} \cdot S_{k+\ell}, \tag{4.108}$$

where

$$C^\ell_k = \frac{k!}{\ell!\,(k-\ell)!}, \tag{4.109}$$

$S_{k+\ell}$ is the series related to elliptic functions:

$$S_{k+\ell} = \sum_{m,\ n} {}' \beta_{mn}^{-k-\ell} \quad \text{for} \quad k+\ell \geq 2, \tag{4.110}$$

where the "prime" next to the sum sign means that the summation excludes case $m = n = 0$;

$$\beta_{mn} = m\omega_1 + m\omega_2 \quad \text{for} \quad m, n = 0, \pm 1, \pm 2, \ldots; \tag{4.111}$$

$$\omega_1 = (1, 0), \quad \omega_2 = (\cos \mu, \sin \mu). \tag{4.112}$$

From relations (4.100)–(4.112), we restricted ourselves to taking into account the first term of the asymptotics, we find:

$$\eta_{1\,5} = -S_6. \tag{4.113}$$

By virtue of the calculations given in [99] for $S_{k+\ell}$, we have

$$S_6 = 5.8630. \tag{4.114}$$

Then, taking into account (4.113), (4.114), we obtain

$$v_1 = R^{12} \eta_{1\,5} = -5.8630 R^{12}; \tag{4.115}$$

$$\tilde{v}_1 = \eta_{5\,1} = -29.3150; \tag{4.116}$$

$$m_{11} = 1 + O\left(R^{24}\right). \tag{4.117}$$

Finally, the expression for the effective parameter $q_G = p/p_1$, taking into account the first terms of the asymptotics and provided that the inclusion radius $a = 2R$ $(0 \le a \le 1)$, takes the form:

$$q_G = \frac{(1+\lambda)^2 - (1-\lambda^2) \cdot \frac{\pi a^2}{2\sqrt{3}} - 171.873845 \cdot (1-\lambda)^2 \frac{a^{12}}{2^{12}}}{(1+\lambda)^2 + (1-\lambda^2) \cdot \frac{\pi a^2}{2\sqrt{3}} - 171.873845 \cdot (1-\lambda)^2 \frac{a^{12}}{2^{12}}}, \tag{4.118}$$

where $\lambda = \frac{p_2}{p_1}$.

At zero approximately, the relationship (4.118) comes down to Maxwell's formula:

$$\frac{p}{p_1} = \frac{1 - c + \lambda\,(1+c)}{1 + c + \lambda\,(1-c)},$$

where $c = \frac{\pi a^2}{2\sqrt{3}}$ stands for the inclusion concentration.

Taking into account the first term of the asymptotics, solution (4.118) gives a correction to the Maxwell's solution for the case of finite values of the physical characteristics of inclusions, and this solution is within the Hashin-Shtrikman bounds.

Expression q_G is identical in structure and very close, according to the results of numerical calculations, to solution q_P obtained earlier in [194]:

- expansions into series of q_G (4.118) and q_P (4.98) for $a \to 0$ coincide up to terms of order a^{12} inclusive:

$$q_G = 1 + \frac{1.813799365\,(\lambda - 1)}{\lambda + 1}\,a^2 + \frac{1.644934069\,(\lambda - 1)^2}{(\lambda + 1)^2}\,a^4 +$$

$$+ \frac{1.491790185\,(\lambda - 1)^3}{(\lambda + 1)^3}\,a^6 + \frac{1.35294045\,(\lambda - 1)^4}{(\lambda + 1)^4}\,a^8 +$$

$$+ \frac{1.226948249\,(\lambda - 1)^5}{(\lambda + 1)^5}\,a^{10} + \frac{1.112718978\,(\lambda - 1)^6}{(\lambda + 1)^6}\,a^{12} +$$

$$+ \frac{(\lambda - 1)^3}{(\lambda + 1)^3}\left(0.07610953880 + \frac{1.009124488\,(\lambda - 1)^4}{(\lambda + 1)^4}\right) a^{14} + O\left(a^{16}\right);$$

$$q_P = 1 + \frac{1.813799365\,(\lambda - 1)}{\lambda + 1}\,a^2 + \frac{1.644934068\,(\lambda - 1)^2}{(\lambda + 1)^2}\,a^4 +$$

$$+ \frac{1.491790184\,(\lambda - 1)^3}{(\lambda + 1)^3}\,a^6 + \frac{1.35294044\,(\lambda - 1)^4}{(\lambda + 1)^4}\,a^8 +$$

$$+ \frac{1.226948249\,(\lambda - 1)^5}{(\lambda + 1)^5}\,a^{10} + \frac{1.112718977\,(\lambda - 1)^6}{(\lambda + 1)^6}\,a^{12} +$$

$$+ \frac{(\lambda - 1)^5}{(\lambda + 1)^5}\left(0.07611018688 + \frac{1.009124487\,(\lambda - 1)^2}{(\lambda + 1)^2}\right) a^{14} + O\left(a^{16}\right);$$

- expansions into series of q_G (4.118) and q_P (4.98) for $\lambda \to 1$ coincide up to terms of order $(\lambda - 1)^2$ inclusive:

$$q_G = 1 + 0.9068996824\,a^2\,(\lambda - 1) + (-0.4534498414 +$$

$$+ 0.4112335170\,a^2)\,a^2\,(\lambda - 1)^2 + (0.226724920 -$$

$$- 0.4112335171\,a^2 + 0.1864737730\,a^4 + 0.009513692344\,a^{12}) \times$$

$$\times a^2\,(\lambda - 1)^3 + O\left((\lambda - 1)^4\right),$$

$$q_P = 1 + 0.9068996823\,a^2\,(\lambda - 1) + (-0.4534498413 +$$

$$+ 0.4112335171\,a^2)\,a^2\,(\lambda - 1)^2 + (0.2267249206 -$$

$$- 0.4112335171\,a^2 + 0.1864737730\,a^4 + 0.000005333641822\,a^{24}) \times$$

$$\times a^2\,(\lambda - 1)^3 + O\left((\lambda - 1)^4\right).$$

Expression q_G does not describe the effective properties of the structure in the case of large sizes of inclusions ($a \to 1$) with large physical characteristics ($\lambda \gg 1$).

From a mathematical point of view, this is the discrepancy obtained from (4.118) with $\lambda \to \infty$, $a = 1$ expressions

$$q_G = \frac{1 + \frac{\pi}{2\sqrt{3}} - \frac{171.873845}{2^{12}}}{1 - \frac{\pi}{2\sqrt{3}} - \frac{171.873845}{2^{12}}} = 36.47 = const$$

with asymptotic solution q_{as} (4.55) found in [58].

From a physical point of view, this means that expression q_G does not describe quantitatively and qualitatively the percolation threshold.

However, in the cited paper, the analysis of the applicability of obtained solution was not carried out.

4)   For inclusion conductivities close to the matrix one, i.e. $\lambda \sim 1$, expansions in series of solutions (4.63), (4.69), and (4.92) coincide with the expansions of the relations q_P and q_G to terms of order $(\lambda - 1)$ inclusive for any size of inclusions a:

$$q_{ThPhM}^{hex} = \frac{1 - \frac{\pi a^2}{2\sqrt{3}} + \lambda \left(1 + \frac{\pi a^2}{2\sqrt{3}}\right)}{1 + \frac{\pi a^2}{2\sqrt{3}} + \lambda \left(1 - \frac{\pi a^2}{2\sqrt{3}}\right)} = 1 + \frac{\pi a^2}{2\sqrt{3}} (\lambda - 1) -$$

$$-\frac{\pi^2 a^2}{4\sqrt{3}} \left(1 - \frac{\pi a^2}{2\sqrt{3}}\right) (\lambda - 1)^2 + O\left((\lambda - 1)^3\right) \approx 1 + 0.06899683 \, a^2 \, (\lambda - 1) -$$

$$-0.453449841 \, a^2 \left(1 - 0.906899683 \, a^2\right) (\lambda - 1)^2 + O\left((\lambda - 1)^3\right),$$

$$q_{m.incl.}\Big|_{\lambda > 1} = 1 + \frac{\sqrt{3}\,\gamma a^2}{\sqrt{1 - \gamma a^2}} \arctan \frac{\sqrt{3}}{3\sqrt{1 - \gamma a^2}} +$$

$$+\frac{\sqrt{3}\,\gamma a^2}{3} \left[\frac{2}{\sqrt{1 - \gamma a^2}} \left(2\arctan \frac{\sqrt{1 - \sqrt{\gamma} a}}{\sqrt{1 + \sqrt{\gamma} a}} + \arctan \left(\frac{\sqrt{\gamma} a}{\sqrt{1 - \gamma a^2}}\right)\right) - \frac{\pi}{2}\right] =$$

$$= 1 + \frac{\pi a^2}{2\sqrt{3}} (\lambda - 1) - \frac{\pi a^2}{4\sqrt{3}} \left(1 - \frac{3a^2}{4} \left(1 + \frac{\sqrt{3}}{2\pi}\right)\right) (\lambda - 1)^2 + O\left((\lambda - 1)^3\right) \approx$$

$$\approx 1 + 0.906899683 \, a^2 \, (\lambda - 1) - 0.453449841 \, a^2 \times$$

$$\times \left(1 - 0.956748336 \, a^2\right) (\lambda - 1)^2 + O\left((\lambda - 1)^{5/2}\right),$$

$$q_{m.incl.}\Big|_{\lambda < 1} = \left\{1 - \frac{\sqrt{3}\,\gamma a^2}{\sqrt{1 + \gamma a^2}} \arctan \frac{\sqrt{3}}{3\sqrt{1 + \gamma a^2}} -\right.$$

$$-\frac{\sqrt{3}\,\gamma a^2}{3} \left[\frac{2}{\sqrt{1 + \gamma a^2}} \left(2\arctan \frac{\sqrt{1 - \sqrt{-\gamma} a}}{\sqrt{1 + \sqrt{-\gamma} a}} + \arctan \left(\frac{\sqrt{-\gamma} a}{\sqrt{1 + \gamma a^2}}\right)\right) - \frac{\pi}{2}\right]\Big\}^{-1} =$$

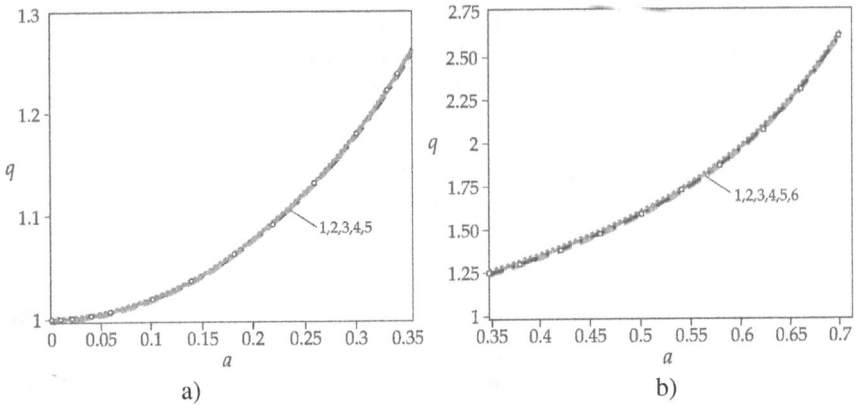

Figure 4.20 Inclusions of small size a with extremely large conductivity $\lambda \to \infty$: (a) $a \to 0$, (b) $a \ll 1$ (1– numerical solution [194], 2– q_{ThPhM}^{hex} (4.93), 3– $q_{m.incl}$ (4.62), 4– q_p (4.98) [194], 5– q_G (4.99) [104], 6– q (4.53)).

$$= 1 + \frac{\pi a^2}{2\sqrt{3}} (\lambda - 1) - \frac{\pi a^2}{4\sqrt{3}} \left(1 + \frac{3a^2}{4} \left(1 + \frac{\sqrt{3}}{2\pi} - \frac{4\pi}{3\sqrt{3}} \right) \right) (\lambda - 1)^2 +$$

$$+ O\left((\lambda - 1)^3 \right) \approx 1 + 0.906899683\, a^2\, (\lambda - 1) -$$

$$- 0.453449841\, a^2\, \left(1 - 0.857051030\, a^2 \right) (\lambda - 1)^2 + O\left((\lambda - 1)^{5/2} \right),$$

$$q_P \approx 1 + 0.9068996823\, a^2\, (\lambda - 1) - 0.4534498413\, a^2\, \times$$

$$\times\, \left(1 - 0.906899682\, a^2 \right) (\lambda - 1)^2 + O\left((\lambda - 1)^3 \right),$$

$$q_G = 1 + 0.9068996824\, a^2\, (\lambda - 1) - 0.4534498413\, a^2\, \times$$

$$\times\, \left(1 - 0.906899682\, a^2 \right) (\lambda - 1)^2 + O\left((\lambda - 1)^3 \right).$$

4.2.10 NUMERICAL RESULTS

Let us analyze the obtained solutions and comparing them with known analytical and numerical results.

1) Expressions (4.93), (4.62), (4.53), and (4.54) determine the asymptotic behavior of the effective parameter for extremely large physical characteristics ($\lambda \to \infty$) of a composite of hexagonal structure with inclusions of small, medium, large and close to limit value.
   Figures 4.20–4.24 show graphs of the effective parameter calculated by formulas (4.93), (4.62), (4.53) and (4.54), in comparison with the known results of analytical [104, 194] and asymptotic [58] solutions, as well as with numerical and experimental data [194].

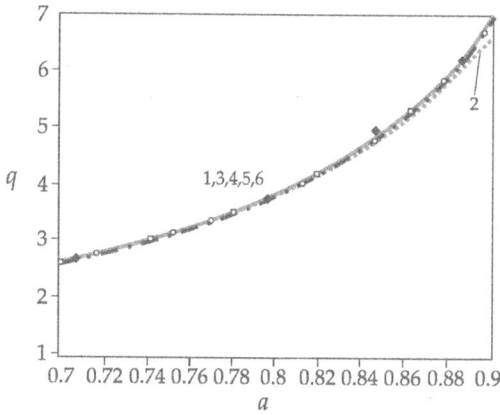

Figure 4.21 Inclusions of medium and large size with extremely high conductivity ($\lambda \to \infty$) (1– numerical solution [194], 2– q^{hex}_{ThPhM} (4.93), 3– q_p (4.98) [194], 4– q_G (4.99) [104], 5– q (4.53), 6– experimental data [194]).

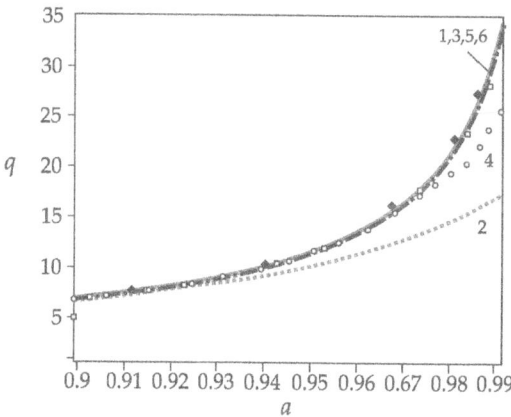

Figure 4.22 Large inclusions ($0 \ll a < 1$) with extremely high conductivity ($\lambda \to \infty$) (1– numerical solution [194], 2– q^{hex}_{ThPhM} (4.93), 3– q_p (4.98) [194], 4– q_G (4.99) [104], 5– q (4.53), 6– experimental data [194]).

2) The effective coefficient depending on the size of inclusions a for the case of large but finite conductivities of the inclusions is illustrated in Figs. 4.25, Fig. 4.26. Expressions (4.59), (4.63) and (4.92) are compared with the approximate analytical formula (4.98) [194], numerical and experimental results [194] and expression (4.99) [104].

3) Graphs of the effective parameter shown in Figs. 4.27–4.30 are built according to formulas (4.59), (4.63), (4.67), (4.69), (4.92) for different sizes of inclusions in the case of physical characteristics of inclusions and matrices of the same order, $\lambda \sim 1$.

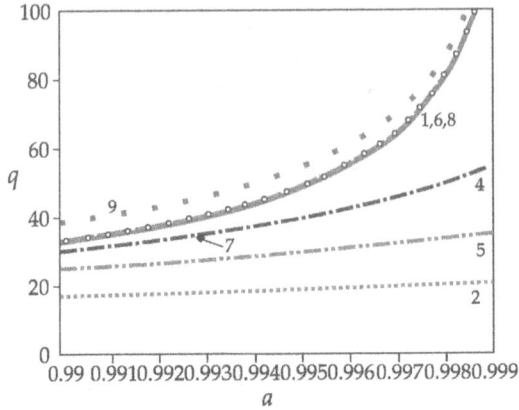

Figure 4.23 Inclusions of large and extremely large size with extremely large pconductivity ($\lambda \to \infty$) (1– numerical solution [194], 2– q_{ThPhM}^{hex} (4.93), 3– q_p (4.98) [194], 4– q_G (4.99) [104], 5– q (4.53), 6– experimental data [194], 7– q_{as}(4.54), 8– q_{as}(4.55)[58]).

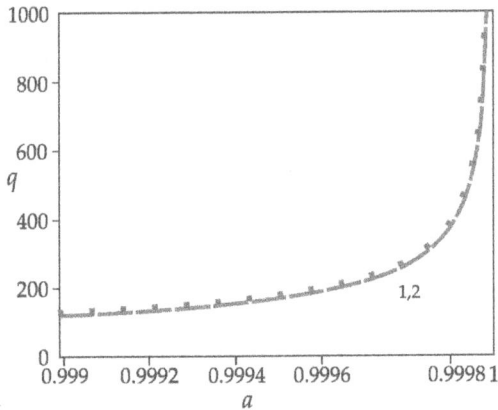

Figure 4.24 Inclusions of extremely large size ($a \to 1$) with extremely large conductivity ($\lambda \to \infty$) (1– q_{as}(4.54), 2– q_{as}(4.55)[58]).

4) In the case of small conductivity of inclusions, the effective coefficients for various a and $\lambda = 0.01$, calculated by formulas (4.67), (4.69), (4.92), are shown in Fig. 4.31 and Fig. 4.32.

5) Figures 4.33–4.38 present graphs of the effective parameters, described by relations (4.64)–(4.66), (4.94), in comparison with the asymptotic formula [58], written for $\lambda \to 0$ taking into account (4.55) and the Keller's theorem

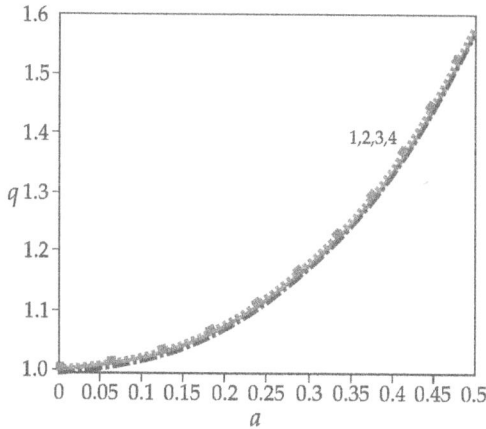

Figure 4.25 Inclusions of small and medium size with large conductivity ($\lambda = 100$) (1– q^{hex}_{ThPhM} (4.92), 2– $q_{m.incl}$ (4.63), 3– q_p (4.98) [194], 4– q_G (4.99) [104]).

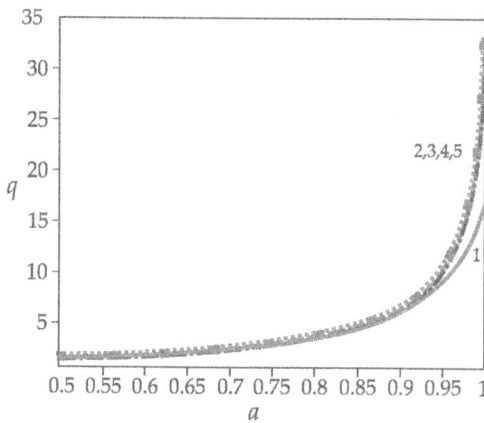

Figure 4.26 Inclusions of medium, large and extremely large size with high conductivity ($\lambda = 100$) (1– q^{hex}_{ThPhM} (4.92), 2– $q_{m.incl}$ (4.63), 3– q_p (4.98) [194], 4– q_G (4.99) [104], 5– q (4.59)).

(3.59), in the form

$$q_{as}\Big|_{\lambda \to 0} = \frac{\sqrt{1-a^2}}{\sqrt{3}\pi}. \tag{4.119}$$

In Tables 4.1–4.3 the effective thermal conductivity of a composite for various values of its physical (λ) and geometric (a) parameters are given in comparison with the results of other researchers.

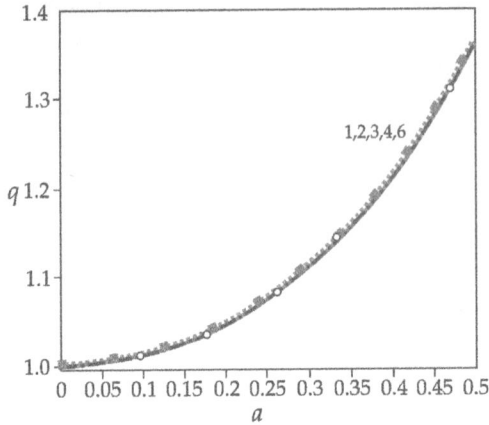

Figure 4.27 Inclusions of small and medium size with conductivity of inclusions and a matrix of the same order ($\lambda = 5$) ($1- q_{ThPhM}^{hex}$ (4.92), $2- q_{m.incl}$ (4.63), $3- q_p$ (4.98) [194], $4- q_G$ (4.99) [104], $6-$ numerical solution [194]).

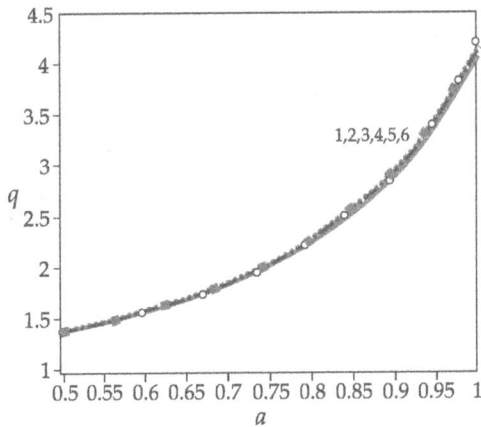

Figure 4.28 Inclusions of medium, large and extremely large size with conductivity of inclusions and a matrix of the same order ($\lambda = 5$) ($1- q_{ThPhM}^{hex}$ (4.92), $2- q_{m.incl}$ (4.63), $3- q_p$ (4.98) [194], $4- q_G$ (4.99) [104], $5- q$ (4.59), $6-$ numerical solution [194]).

4.2.11 TWO-SIDED ESTIMATES OF THE ASYMPTOTICS OF THE EFFECTIVE THERMAL CONDUCTIVITY PARAMETER FOR A HEXAGONAL STRUCTURE

Solution (4.62) for medium-size inclusions cannot be used in the case of their large sizes and does not give asymptotics (4.55) at $\lambda \to \infty$, $a \to 1$. Therefore, we transform relation (4.62) using AP.

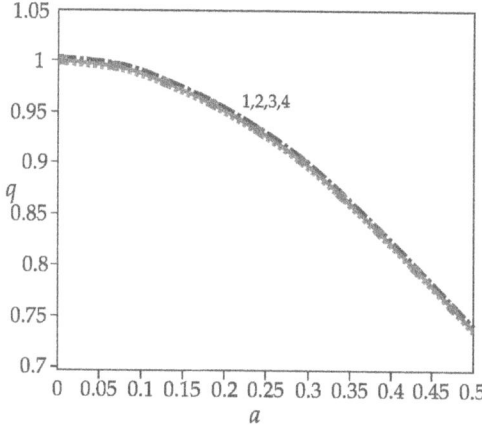

Figure 4.29 Inclusions of small and medium size with conductivity of inclusions and a matrix of the same order ($\lambda = 0.2$) ($1 - q^{hex}_{ThPhM}$ (4.92), $2 - q_{m.incl}$ (4.63), $3 - q_p$ (4.98) [194], $4 - q_G$ (4.99) [104]).

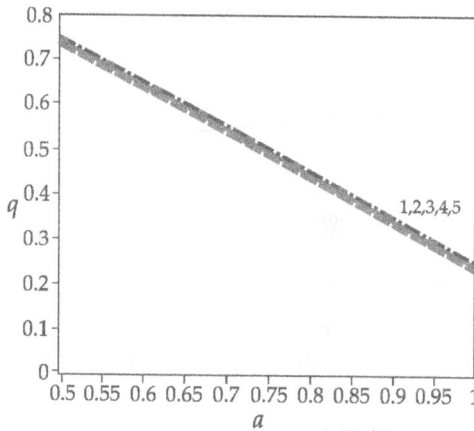

Figure 4.30 Inclusions of medium, large and extremely large size, with physical characteristics of inclusions and a matrix of the same order ($\lambda = 0.2$) ($1 - q^{hex}_{ThPhM}$ (4.92), $2 - q_{m.incl}$ (4.69), $3 - q_p$ (4.98) [194], $4 - q_G$ (4.99) [104], $5 - q$ (4.67)).

We write (4.62) as

$$q_{m.incl.} = 1 + \frac{2\sqrt{3}\,a^2}{\sqrt{1-a^2}} \left(\arctan \frac{\sqrt{3}}{3\sqrt{1-a^2}} + \frac{2}{3} \arctan \sqrt{\frac{1-a}{1+u}} \right) -$$

$$- \frac{\sqrt{3}\,\pi a^2}{6} + \frac{\sqrt{3}\,a^2}{\sqrt{1-a^2}} \left(\frac{2}{3} \arctan \frac{a}{\sqrt{1-a^2}} - \arctan \frac{\sqrt{3}}{3\sqrt{1-a^2}} \right) \qquad (4.120)$$

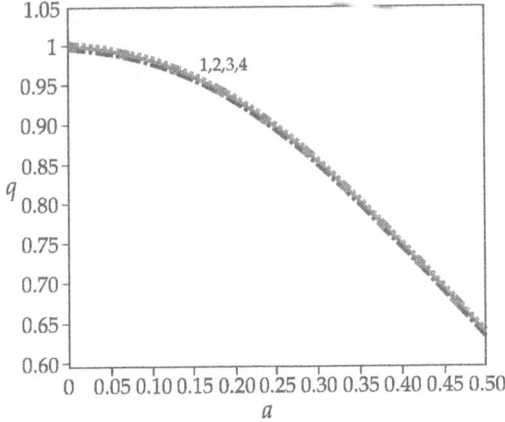

Figure 4.31 Inclusions of small and medium size with small conductivity of inclusions ($\lambda = 0.01$) ($1- q_{ThPhM}^{hex}$ (4.92), $2- q_{m.incl}$ (4.69), $3- q_p$ (4.98) [194], $4- q_G$ (4.99) [104]).

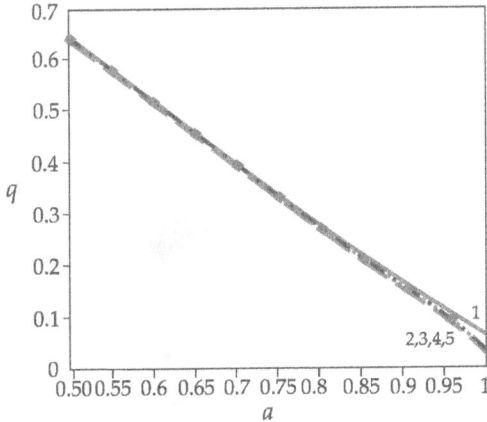

Figure 4.32 Inclusions of medium, large, and extremely large sizes with small conductivity of inclusions ($\lambda = 0.01$) ($1- q_{ThPhM}^{hex}$ (4.92), $2- q_{m.incl}$ (4.69), $3- q_p$ (4.98) [194], $4- q_G$ (4.99) [104], $5- q$ (4.67)).

and construct a sequence of APs $q_{[2/2]}^*, q_{[2/4]}^*, \ldots$ for the expression

$$q^* = \frac{\sqrt{3}\,a^2}{\sqrt{1-a^2}} \left(\frac{2}{3} \arctan \frac{a}{\sqrt{1-a^2}} - \arctan \frac{\sqrt{3}}{3\sqrt{1-a^2}} \right). \tag{4.121}$$

Then, we get

$$q_{[2/2]}^* = -\frac{1}{2} \frac{\pi^3 a^2}{\sqrt{3}\,(\pi+4)\,\pi a - \frac{1}{4}\left(2\sqrt{3}\,\pi^2 + 9\pi - 64\sqrt{3}\right) a^2},$$

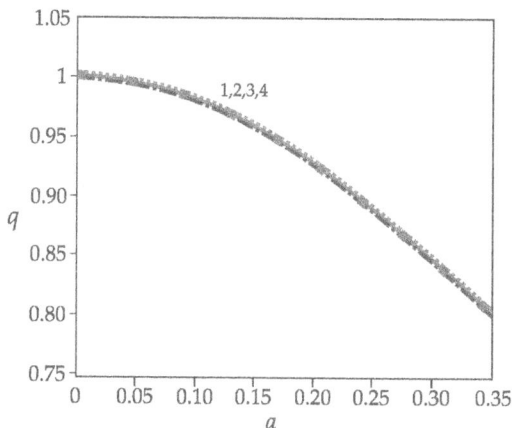

Figure 4.33 Small inclusions $a \to 0$, with extremely low conductivity $(\lambda \to 0)$ $(1 - q^{hex}_{ThPhM}$ (4.94), 2– $q_{m.incl}$ (4.66), 3– q_p (4.98) [194], 4– q_G (4.99) [104]).

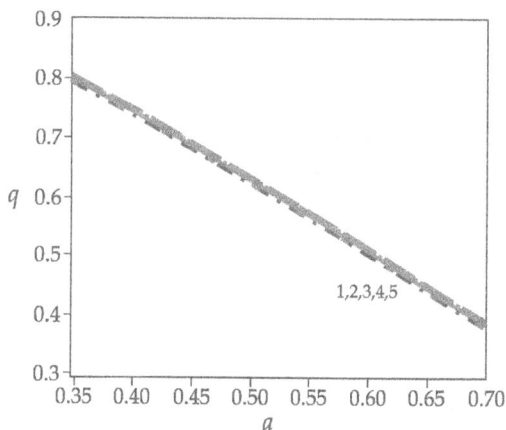

Figure 4.34 Inclusions of medium size $(a \ll 1)$ with extremely low conductivity $(\lambda \to 0)$ $(1 - q^{hex}_{ThPhM}$ (4.94), 2– $q_{m.incl}$ (4.66), 3– q_p (4.98) [194], 4– q_G (4.99) [104], 5– q (4.64)).

$$q^*_{[2/4]} = -\frac{\sqrt{3}}{2} \pi^5 a^2 / \left(3 \left(\pi + 4a \right) \pi^3 - \frac{3}{4} \left(2\pi^2 + 3\sqrt{3}\pi - 64 \right) \pi^2 a^2 - 2 \left(2\pi^2 + \right. \right.$$

$$\left. \left. +9\sqrt{3}\pi - 96 \right) \pi a^3 - \frac{1}{32} \left(12\pi^4 + 9\sqrt{3}\pi^3 + 94\pi^2 + 3456\sqrt{3}\pi - 24576 \right) a^4 \right),$$

$$q^*_{[2/6]} = -\frac{3}{2} \pi^7 a^2 / \left(3\sqrt{3} \left(\pi + 4a \right) \pi^5 - \frac{3}{4} \left(2\sqrt{3}\pi^2 + 9\pi - 64\sqrt{3} \right) \pi^4 a^2 - \right.$$

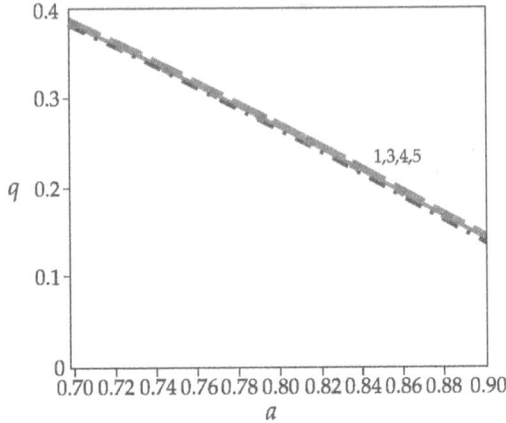

Figure 4.35 Inclusions of medium and large size with extremely small conductivity $(\lambda \to 0)$ $(1- q_{ThPhM}^{hex}$ (4.94), 3– q_p (4.98) [194], 4– q_G (4.99) [104], 5– q (4.64)).

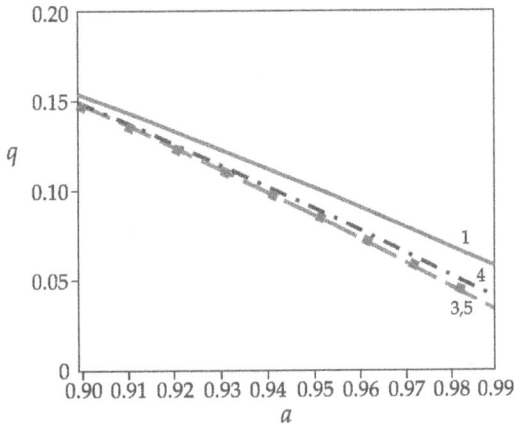

Figure 4.36 Inclusions of large size $(0 << a < 1)$ with extremely low conductivity $(\lambda \to 0)$ $(1- q_{ThPhM}^{hex}$ (4.94), 3– q_p (4.98) [194], 4– q_G (4.99) [104], 5– q (4.64)).

$$-2\left(2\sqrt{3}\,\pi^2 + 27\,\pi - 96\sqrt{3}\right)\pi^3 a^3 - \frac{1}{32}\left(12\sqrt{3}\,\pi^4 + 27\,\pi^3 + 94\sqrt{3}\,\pi^2 + \right.$$

$$\left. +10368\,\pi - 24576\sqrt{3}\right)\pi^2 a^4 - \frac{1}{20}\left(32\sqrt{3}\,\pi^4 + 315\,\pi^3 + 1215\sqrt{3}\,\pi^2 + \right.$$

$$\left. +34560\,\pi - 61440\sqrt{3}\right)\pi a^5 - \frac{1}{960}\left(180\,\sqrt{3}\,\pi^6 + 1603\sqrt{3}\,\pi^4 + \right.$$

$$\left. +153495\,\pi^3 - 589440\sqrt{3}\,\pi^2 + 8294400\,\pi - 11796480\sqrt{3}\right)a^6\right),$$

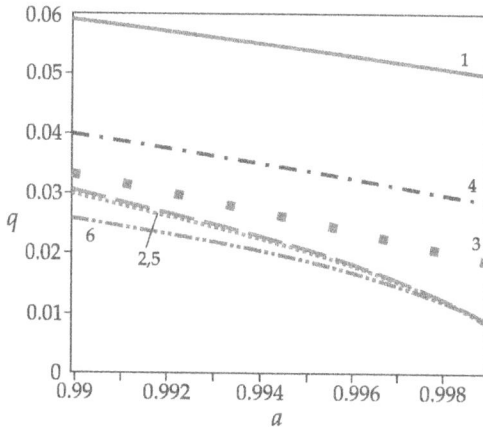

Figure 4.37 Inclusions of large and extremely large sizes with extremely small conductivity $(\lambda \to 0)$ $(1- q_{ThPhM}^{hex}$ (4.94), 2– q_{as} (4.65), 3– q_p (4.98) [194], 4– q_G (4.99) [104], 5– q (4.64), 6– q_{as} (4.119) [58, 127]).

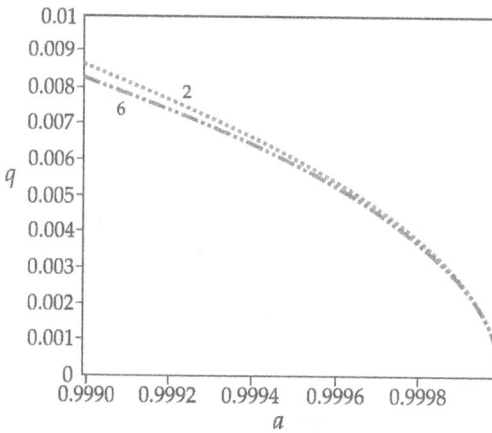

Figure 4.38 Inclusions of extremely large size $(a \to 1)$ with extremely small conductivity $(\lambda \to 0)$ $(2- q_{as}$ (4.65), 6– q_{as} (4.119) [58, 127]).

$$q_{[2/8]}^{*} = -\frac{3\sqrt{3}}{2}\,\pi^9 a^2 / \left(9\,(\pi+4a)\,\pi^7 - \frac{9}{4}\left(2\pi^2 + 3\sqrt{3}\,\pi - 64 \right)\pi^6 a^2 - \right.$$

$$-6\left(2\pi^2 + 9\sqrt{3}\,\pi \quad 96 \right)\pi^5 a^3 - \frac{1}{32}\left(36\pi^4 + 27\sqrt{3}\,\pi^3 + 282\pi^2 + \right.$$

$$+10368\sqrt{3}\,\pi - 73728 \right)\pi^4 a^4 - \frac{1}{20}\left(96\,\pi^4 + 315\sqrt{3}\,\pi^3 - 3645\pi^2 + \right.$$

$$+34560\sqrt{3}\,\pi - 184320\Big)\,\pi^3 a^5 - \frac{1}{320}\,(180\,\pi^6 + 1603\,\pi^4 + 51165\sqrt{3}\,\pi^3 -$$

$$-589440\,\pi^2 + 2764800\sqrt{3}\,\pi - 11796480\sqrt{3}\Big)\,\pi^2 a^6 -$$

$$-\frac{1}{2800}\Big(7680\,\pi^6 + 14490\sqrt{3}\,\pi^5 + 360255\,\pi^4 + 3534300\,\pi^3 -$$

$$-35817600\,\pi^2 + 116121600\sqrt{3}\,\pi - 412876800\Big)\,\pi a^7 -$$

$$-\frac{1}{35840}\,(12600\,\pi^8 - 8505\sqrt{3}\,\pi^7 + 125863\pi^6 + 3109554\sqrt{3}\,\pi^5 -$$

Table 4.1

The effective thermal conductivity coefficient for absolutely conducting inclusions ($\lambda \to \infty$) in comparison with the results of other authors.

Inclusion concentration c	Inclusion size a	ThPhM (4.93)	Formula (4.98) [194]	Formula (4.62)	Formula (4.53)	Numerical solution [194]	Asymptotic formula (4.54)	Asymptotic formula (4.55) [58]
0.001	0.0332	1.0020	1.0020	1.0020	–	–	–	–
0.005	0.0743	1.0101	1.0100	1.0101	–	–	–	–
0.01	0.1050	1.0202	1.0202	1.0202	–	–	–	–
0.05	0.2348	1.1053	1.1053	1.1055	–	–	–	–
0.1	0.3321	1.2222	1.2222	1.2232	–	1.2222	–	–
0.2	0.4696	1.5000	1.5000	1.5036	–	1.5000	–	–
0.3	0.5751	1.8571	1.8572	1.8639	–	1.8572	–	–
0.4	0.6641	2.3333	2.3340	–	2.3435	2.3340	–	–
0.5	0.7425	3.0000	3.0047	–	3.0202	3.0047	–	–
0.6	0.8134	4.0000	4.0267	–	4.0585	4.0267	–	–
0.65	0.8466	4.7143	4.7760	–	4.8240	4.7760	–	–
0.7	0.8786	5.6667	5.8111	–	5.8847	5.8111	–	–
0.73	0.8972	6.4074	6.6523	–	6.7482	6.6524	–	–
0.76	0.9154	7.3333	7.7595	–	7.8852	7.7600	8.4179	–
0.78	0.9274	8.0909	8.7210	–	8.8729	8.7223	9.4444	–
0.8	0.9392	–	9.9551	–	10.1409	9.9586	10.74683	–
0.82	0.9509	–	11.6099	–	11.8432	11.6207	12.4762	–
0.84	0.9624	–	13.9724	–	14.2845	14.0093	14.9323	–
0.85	0.9681	–	15.5933	–	15.9740	15.6656	16.6216	–
0.86	0.9738	–	17.6882	–	18.1860	17.8381	18.8258	–
0.87	0.9794	–	20.5205	–	21.2530	20.8570	21.8739	–
0.88	0.9851	–	24.6010	–	25.9080	25.4508	26.4927	–
0.89	0.9906	–	31.0680	–	34.2410	33.7007	34.7590	–
0.895	0.9934	–	–	–	41.9371	41.3407	42.4010	–
0.9	0.9962	..	–	–	56.9006	56.229	57.2822	62.3843
0.905	0.9990	–	–	–	113.5658	112.8	113.7890	118.8912
0.9055	0.9992	–	–	–	133.2113	–	133.4060	138.5082
0.906	0.9995	–	–	–	167.4997	–	167.6588	172.7610
0.9062	0.9996	–	–	–	190.6586	–	190.8002	195.902
0.9064	0.9997	–	–	–	226.5927	–	226.7136	231.8158
0.9066	0.9998	–	–	–	294.1396	–	294.2344	299.3365
0.9068	0.9999	–	–	–	513.8606	–	513.9163	519.0185

Table 4.2

The effective thermal conductivity coefficient for inclusions of high conductivity ($\lambda = 100$) in comparison with known results.

Inclusions concentration c	Inclusion size a	ThPhM (4.92)	Formula (4.98) [194]	Formula (4.63)	Formula (4.59)
0.001	0.0332	1.0020	1.0020	1.0020	–
0.005	0.0743	1.0099	1.0099	1.0099	–
0.01	0.1050	1.0198	1.0199	1.0198	–
0.05	0.2348	1.1031	1.1031	1.1033	–
0.1	0.3321	1.2173	1.2173	1.2183	–
0.2	0.4696	1.4877	1.4877	1.4912	–
0.3	0.5751	1.8331	1.8332	1.8397	–
0.4	0.6641	2.2899	2.2905	–	2.2997
0.5	0.7425	2.9223	2.9266	–	2.9408
0.6	0.8134	3.8557	3.8795	–	3.9065
0.7	0.8786	5.3722	5.4961	–	5.5514
0.8	0.9392	8.2661	9.0258	–	9.1097
0.9	0.9962	–	28.6827	–	27.2818
0.903	0.9978	–	31.1161	–	29.4538
0.906	0.9995	–	34.0740	–	32.1158
0.9062	0.9996	–	34.2942	–	32.3154
0.9064	0.9997	–	34.5176	–	32.5181
0.9066	0.9998	–	34.7443	–	32.7240
0.9068	0.9999	–	34.9744	–	32.9333

$$-58215948\,\pi^4 + 310464000\,\sqrt{3}\,\pi^3 - 2750791680\,\pi^2 +$$
$$+6936330240\sqrt{3}\,\pi - 21139292160\Big)\,a^8\Big),$$

$$q^*_{[2/10]} = -\frac{9}{2}\,\pi^{11}a^2 / \Big(9\sqrt{3}\,(\pi+4a)\,\pi^9 - \frac{9}{4}\left(2\sqrt{3}\pi^2 + 9\pi - 64\sqrt{3}\right)\pi^8 a^2 -$$

$$-6\left(2\sqrt{3}\pi^2 + 27\pi - 96\sqrt{3}\right)\pi^7 a^3 - \frac{1}{32}\left(36\sqrt{3}\pi^4 + 81\pi^3 + 282\sqrt{3}\,\pi^2 +\right.$$

$$+31104\pi - 73728\big)\,\pi^6 a^4 - \frac{1}{20}\,(96\,\sqrt{3}\,\pi^4 + 945\,\pi^3 - 3645\sqrt{3}\,\pi^2 +$$

$$+103680\pi - 184320\sqrt{3}\Big)\,\pi^5 a^5 - \frac{1}{320}\left(180\sqrt{3}\,\pi^6 + 1603\sqrt{3}\,\pi^4 +\right.$$

$$+153495\,\pi^3 - 589440\sqrt{3}\,\pi^2 + 8294400\,\pi - 11796480\sqrt{3}\Big)\,\pi^4 a^6 -$$

Table 4.3

The effective thermal conductivity for inclusions with a conductivity of the order of the matrix conductivity ($\lambda = 5$) in comparison with the results of other authors.

Conductivity of inclusions of order of matrix conductivity $\lambda = 5$						
Inclusions concentration c	Inclusion size a	ThPhM (4.92)	Formula (4.98) [194]	Formula (4.63)	Formula (4.59)	Numerical solution [194]
0.001	0.0332	1.0013	1.0013	1.0013	–	–
0.005	0.0743	1.0067	1.0067	1.0067	–	–
0.01	0.1050	1.0134	1.0134	1.0134	–	–
0.05	0.2348	1.0690	1.0690	1.0691	–	–
0.1	0.3321	1.1429	1.1429	1.1433	–	1.1429
0.2	0.4696	1.3077	1.3077	1.3094	–	1.3077
0.3	0.5751	1.5000	1.5000	1.5036	–	1.5000
0.4	0.6641	1.7273	1.7274	1.7330	–	1.7274
0.5	0.7425	2.0000	2.0008	–	2.0077	2.0008
0.6	0.8134	2.3333	2.3368	–	2.3435	2.3368
0.65	0.8466	2.5294	2.5363	–	2.5416	2.5363
0.7	0.8786	2.7500	2.7631	–	2.7654	2.7631
0.73	0.8972	2.8961	2.9152	–	2.9141	2.9152
0.76	0.9154	3.0541	3.0818	–	3.0755	3.0818
0.78	0.9274	3.1667	3.2022	–	3.1910	3.2022
0.8	0.9392	3.2857	3.3311	–	3.3135	3.3311
0.82	0.9509	3.4117	3.4698	–	3.4437	3.4699
0.84	0.9624	3.5455	3.6199	–	3.5823	3.6202
0.85	0.9681	3.6154	3.6998	–	3.6551	3.7002
0.86	0.9738	3.6875	3.7834	–	3.7303	3.7840
0.87	0.9794	3.7619	3.8710	–	3.8081	3.8719
0.88	0.9851	3.8387	3.9630	–	3.8887	3.9645
0.89	0.9906	3.9180	4.0600	–	3.9721	4.0623
0.895	0.9934	3.9587	4.1105	–	4.0149	4.1136
0.9	0.9962	4.0000	4.1625	–	4.0585	4.1665
0.905	0.9978	4.0420	4.2162	–	4.1029	4.2212
0.9068	0.9999	4.0573	4.2359	–	4.1191	4.2416
0.9068993	0.9999997891	4.0581	4.2369	–	4.1200	4.2427

$$-\frac{1}{2800} \left(7680\sqrt{3}\,\pi^6 + 43470\,\pi^5 - 360255\sqrt{3}\,\pi^4 + 10602900\,\pi^3 - \right.$$

$$\left. -35817600\sqrt{3}\,\pi^2 + 348364800\,\pi - 412876800\sqrt{3} \right) \pi^3 a^7 -$$

$$-\frac{1}{35840}\Big(12600\sqrt{3}\,\pi^8-25515\,\pi^7+125863\sqrt{3}\,\pi^6+9328662\,\pi^5-$$

$$-58215948\sqrt{3}\,\pi^4+931392000\,\pi^3-2740039680\sqrt{3}\,\pi^2+$$

$$+2079006720\,\pi-21139292160\sqrt{3}\Big)\,\pi^2 a^8-$$

$$-\frac{1}{26880}\Big(49152\sqrt{3}\,\pi^8+82134\,\pi^7-2489953\sqrt{3}\,\pi^6+74331432\,\pi^5-$$

$$-367895220\sqrt{3}\,\pi^4+4380687360\,\pi^3-11421204480\sqrt{3}\,\pi^2+$$

$$+71345111040\,\pi-63417876480\sqrt{3}\Big)\,\pi a^9- \qquad (4.122)$$

$$-\frac{1}{2150400}\Big(529200\sqrt{3}\,\pi^{10}-1990170\,\pi^9+6730776\sqrt{3}\,\pi^8+$$

$$+316328085\,\pi^7-2909134880\sqrt{3}\,\pi^6+50865513300\,\pi^5-$$

$$-210050507520\sqrt{3}\,\pi^4+2063016345600\,\pi^3-4802582937600\sqrt{3}\,\pi^2+$$

$$+25684239974400\,\pi-20293720473600\sqrt{3}\Big)\,a^{10}\Big).$$

A further increase in the order of the approximants is inappropriate, because starting from $q^*_{[2/12]}$, the AP sequence will diverge. This means that sequence $q^*_{[2/2]}$, $q^*_{[2/4]}, \cdots, q^*_{[2/2N]}, \cdots$ can be treated as an asymptotic, in which the optimal number of terms is preserved, which provides a close to exact solution of the problem.

Formula for effective parameter expression

$$q_{AP}=1+\frac{2\sqrt{3}\,a^2}{\sqrt{1-a^2}}\left(\arctan\frac{\sqrt{3}}{3\sqrt{1-a^2}}+\frac{2}{3}\arctan\sqrt{\frac{1-a}{1+a}}\right)-$$
$$-\frac{\sqrt{3}\,\pi a^2}{6}+q^*_{[2/10]} \qquad (4.123)$$

can be used for any size $(0\le a<1)$ of absolutely conducting inclusions and for $a\to 1$ gives the leading term of the asymptotics (4.55), which coincides with the well-known result obtained in [58]:

$$q_{AP}=\frac{\sqrt{3}\pi}{\sqrt{1-a^2}}-5+\frac{2\sqrt{3}}{3}-\frac{\pi\sqrt{3}}{6}-\frac{9\pi^{11}}{2}\left[\frac{567\sqrt{3}}{256}\,\pi^{10}-\right.$$

$$-\left(\frac{216513}{2048}-\frac{512\sqrt{3}}{7}\right)\frac{\pi^9}{5}-\left(1020681-\frac{22137707\sqrt{3}}{40}\right)\frac{\pi^8}{4480}-$$

$$-\left(\frac{266514867}{16}-\frac{26330161\sqrt{3}}{3}\right)\frac{\pi^7}{8960}-$$

$$-\left(1877769-\frac{191470627\sqrt{3}}{168}\right)\frac{\pi^6}{160}-\left(\frac{77374953}{16}-2284453\sqrt{3}\right)\frac{\pi^5}{64}-$$

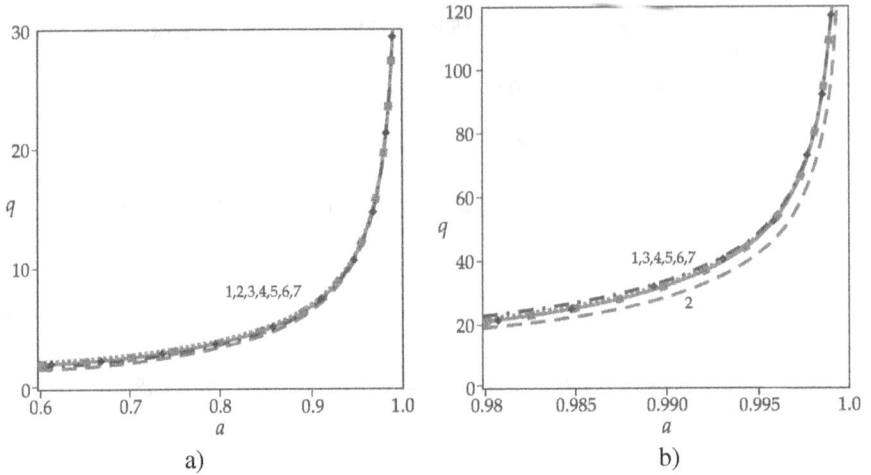

Figure 4.39 Graphs of the effective parameter described by the AP sequence: a) $0.6 \leq a < 1$; b) $0.98 \leq a < 1$ (1– q_{ThPhM} (4.53), 2– $q_{m.incl.}$ (4.62), 3– $q_{[2/2]}$, 4– $q_{[2/4]}$, 5– $q_{[2/6]}$, 6– $q_{[2/8]}$, 7– $q_{[2/10]}$).

$$-\left(287388 - \frac{16903659\sqrt{3}}{80}\right)\pi^4 - 3\left(513324 - 190784\sqrt{3}\right)\pi^3 -$$

$$-3072\left(864 - 919\sqrt{3}\right)\pi^2 - 147456\left(81 - 16\sqrt{3}\right)\pi +$$

$$+9437184\sqrt{3}\Big]^{-1} + O\left(\sqrt{1-a}\right)$$

or

$$q_{AP} = \frac{\sqrt{3}\pi}{\sqrt{1-a^2}} - 6.9345.$$

For comparison, according to the opinion of the authors of [96], the best result (formula (47) in [96]) was obtained in the following form

$$q = \frac{5.18766}{\sqrt{0.9069 - c}} - 6.2371 = \frac{\sqrt{3}\pi}{\sqrt{1-a^2}} - 6.2371,$$

where $c = \frac{\pi a^2}{2\sqrt{3}}$.

Figures 4.39a,b show the solutions described by the sequence AP $q^*_{[2/2]}$, $q^*_{[2/4]}$, ..., $q^*_{[2/10]}$.

Figures 4.40a,b present graphs of the effective parameter for absolutely conductive inclusions, shown in comparison with the results of calculation by formula (44) from [96], denoted by Q.

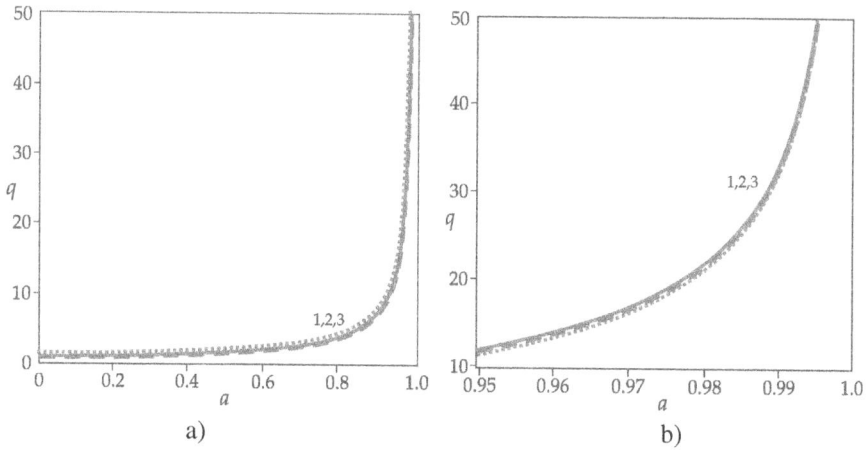

Figure 4.40 Graphs of the effective parameter for absolutely conducting inclusions: a) $0 \leq a < 1$; b) $0.95 \leq a < 1$ ($1- q_{ThPhM}$ (4.53), $2- q_{AP}$ (4.123), $3- q$ (44) [96]).

Table 4.4

Comparison of solutions obtained by various methods.

Inclusion size a	q_{AP} (4.122), (4.123)	Q	q_{LA} (4.53)	$\frac{Q-q_{AP}}{Q} \times 100\%$	$\frac{q_{ThPhM}-Q}{Q} \times 100\%$
0.92	8.0984	8.0996	8.2340	0.0151	1.6596
0.93	8.9538	8.9652	9.1225	0.1279	1.7539
0.94	10.0268	10.0536	10.2388	0.2670	1.8420
0.95	11.4224	11.4728	11.6929	0.4392	1.9179
0.96	13.3331	13.4206	13.6853	0.6522	1.9722
0.97	16.1618	16.3106	16.6347	0.9127	1.9868
0.98	20.9578	21.2155	21.6237	1.2146	1.9241
0.99	31.9262	32.4000	32.9428	1.4624	1.6752
0.995	47.6346	48.3041	48.9582	1.3861	1.3541
0.999	114.6774	115.5922	116.3830	0.7914	0.6841
0.9995	165.0648	166.0394	166.8318	0.5870	0.4772
0.9999	377.7790	378.9598	379.5979	0.3116	0.1684
0.99995	537.1656	538.5097	538.9913	0.2496	0.0894

Thus, the solutions according to the LA (4.53) and AP (4.122), (4.123) give, respectively, the upper and lower bounds of the asymptotics of the effective parameter in the case of absolute conductivity of the inclusions, and the "fork" between them is small at any size of inclusions, $0 \leq a < 1$, $a \to 1$ (Table 4.4).

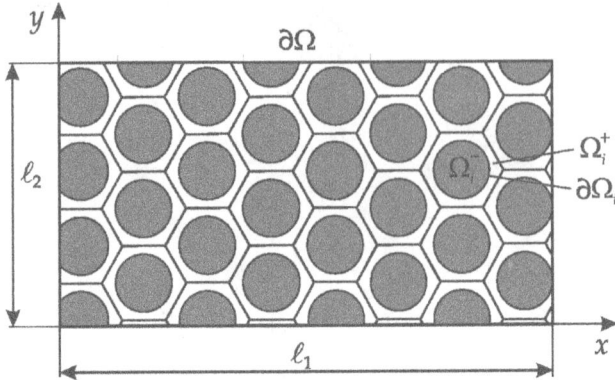

Figure 4.41 Rectangular composite membrane with hexagonal circular inclusion structure.

4.3 NATURAL VIBRATIONS OF COMPOSITE MEMBRANES

4.3.1 MEMBRANE WITH A HEXAGONAL ARRAY OF CIRCLE INCLUSIONS

The problem of transverse vibrations of a rectangular membrane rigidly clamped along the contour, which is a composite structure with periodically arranged in a hexagonal lattice of circular inclusions, is considered (Fig. 4.41).

In the general case, the problem of natural oscillations of such a membrane is written as

$$c^{+2}\left(\frac{\partial^2 u^+}{\partial x^2} + \frac{\partial^2 u^+}{\partial y^2}\right) = \frac{\partial^2 u^+}{\partial t^2} \quad \text{in} \quad \Omega_i^+, \tag{4.124}$$

$$c^{-2}\left(\frac{\partial^2 u^-}{\partial x^2} + \frac{\partial^2 u^-}{\partial y^2}\right) = \frac{\partial^2 u^-}{\partial t^2} \quad \text{in} \quad \Omega_i^-, \tag{4.125}$$

$$u^+ = u^-, \quad c^{+2}\frac{\partial u^+}{\partial \mathbf{n}} = c^{-2}\frac{\partial u^-}{\partial \mathbf{n}} \quad \text{at} \quad \partial\Omega_i, \tag{4.126}$$

$$u^\pm = 0 \quad \text{at} \quad \partial\Omega, \tag{4.127}$$

$$u^\pm = f^\pm(x, y), \quad \frac{\partial u^\pm}{\partial t} = F^\pm(x, y) \quad \text{at} \quad t = 0, \tag{4.128}$$

where: u^+, u^- are the transverse displacements of matrix and inclusions; $c^{\pm^2} = \frac{p^\pm}{\rho^\pm}$, $p^\pm$ is the tension in the membrane, $\rho^\pm$ are the surface densities; $\mathbf{n}$ is the outer normal to the inclusion contour.

We assume the following solution of problem (4.124)–(4.128):

$$u^\pm(x, y, t) = u^\pm(x, y) e^{i\omega t}, \tag{4.129}$$

where ω is the frequency.

Then, taking into account (4.129), relations (4.124)–(4.126) are transformed to the following form

$$\frac{\partial^2 u^+}{\partial x^2} + \frac{\partial^2 u^+}{\partial y^2} + \varpi u^+ = 0 \quad \text{in} \quad \Omega_i^+, \tag{4.130}$$

$$\lambda \left(\frac{\partial^2 u^-}{\partial x^2} + \frac{\partial^2 u^-}{\partial y^2} \right) + \varpi u^- = 0 \quad \text{in} \quad \Omega_i^-, \tag{4.131}$$

$$u^+ = u^-, \quad \frac{\partial u^+}{\partial \mathbf{n}} = \lambda \frac{\partial u^-}{\partial \mathbf{n}} \quad \text{at} \quad \partial \Omega_i, \tag{4.132}$$

where $\lambda = \frac{c^{-2}}{c^{+2}}$; $\varpi = \frac{\omega^2}{c^{+2}}$.

In accordance with the homohenization theory, we represent the solution of the problem (4.130)–(4.132), (4.127) in the form of series in powers of a dimensionless small parameter ε, characterizing the period of the structure:

$$u^\pm = u_0 (x, y) + \varepsilon \left[u_{10} (x, y) + u_1^\pm (x, y, \xi, \eta) \right] + \\ + \varepsilon^2 \left[u_{20} (x, y) + u_2^\pm (x, y, \xi, \eta) \right] + \dots \tag{4.133}$$

where ξ, η are the "fast" variables: $\xi = \frac{x}{\varepsilon}$, $\eta = \frac{y}{\varepsilon}$.

In a similar way, we represent the expansion of the frequency:

$$\varpi = \varpi_0 + \varepsilon \varpi_1 + \varepsilon^2 \varpi_2 + \dots \tag{4.134}$$

After splitting in ε relations (4.130), (4.131), we arrive at the following infinite sequence of equations

$$\varepsilon^{-1} : \quad \begin{aligned} \frac{\partial^2 u_1^+}{\partial \xi^2} + \frac{\partial^2 u_1^+}{\partial \eta^2} &= 0; \\ \frac{\partial^2 u_1^-}{\partial \xi^2} + \frac{\partial^2 u_1^-}{\partial \eta^2} &= 0; \end{aligned} \tag{4.135}$$

$$\varepsilon^0 : \quad \begin{aligned} \frac{\partial^2 u_0}{\partial x^2} + \frac{\partial^2 u_0}{\partial y^2} + 2\frac{\partial^2 u_1^+}{\partial x \partial \xi} + 2\frac{\partial^2 u_1^+}{\partial y \partial \eta} + \frac{\partial^2 u_2^+}{\partial \xi^2} + \frac{\partial^2 u_2^+}{\partial \eta^2} + \varpi_0 u_0 = 0, \\ \lambda \left(\frac{\partial^2 u_0}{\partial x^2} + \frac{\partial^2 u_0}{\partial y^2} + 2\frac{\partial^2 u_1^-}{\partial x \partial \xi} + 2\frac{\partial^2 u_1^-}{\partial y \partial \eta} + \frac{\partial^2 u_2^-}{\partial \xi^2} + \frac{\partial^2 u_2^-}{\partial \eta^2} \right) + \varpi_0 u_0 = 0, \end{aligned} \tag{4.136}$$

$$\varepsilon^1 : \quad \begin{aligned} \frac{\partial^2 u_1^+}{\partial x^2} + \frac{\partial^2 u_1^+}{\partial y^2} + \frac{\partial^2 u_{10}}{\partial x^2} + \frac{\partial^2 u_{10}}{\partial y^2} + 2\frac{\partial^2 u_2^+}{\partial x \partial \xi} + 2\frac{\partial^2 u_2^+}{\partial y \partial \eta} + \\ + \frac{\partial^2 u_3^+}{\partial \xi^2} + \frac{\partial^2 u_3^+}{\partial \eta^2} + \varpi_1 u_0 + \varpi_0 (u_1^+ + u_{10}) = 0, \\ \lambda \left(\frac{\partial^2 u_1^-}{\partial x^2} + \frac{\partial^2 u_1^-}{\partial y^2} + \frac{\partial^2 u_{10}}{\partial x^2} + \frac{\partial^2 u_{10}}{\partial y^2} + 2\frac{\partial^2 u_2^-}{\partial x \partial \xi} + 2\frac{\partial^2 u_2^-}{\partial y \partial \eta} + \right. \\ \left. \frac{\partial^2 u_3^-}{\partial \xi^2} + \frac{\partial^2 u_3^-}{\partial \eta^2} \right) + \varpi_1 u_0 + \varpi_0 (u_1^- + u_{10}) = 0, \end{aligned} \tag{4.137}$$

....

Accordingly, the conditions (4.132) take the form:

$$\varepsilon^1: \quad u_1^+ = u_1^- \quad \text{at} \quad \partial\Omega_i, \tag{4.138}$$

$$\varepsilon^2: \quad u_2^+ = u_2^- \quad \text{at} \quad \partial\Omega_i, \tag{4.139}$$

....

$$\varepsilon^0: \quad \frac{\partial u_1^+}{\partial\bar{\mathbf{n}}} + \frac{\partial u_0}{\partial\mathbf{n}} = \lambda\left(\frac{\partial u_1^-}{\partial\bar{\mathbf{n}}} + \frac{\partial u_0}{\partial\mathbf{n}}\right) \quad \text{at} \quad \partial\Omega_i, \tag{4.140}$$

$$\varepsilon^1: \quad \frac{\partial u_2^+}{\partial\bar{\mathbf{n}}} + \frac{\partial u_1^+}{\partial\mathbf{n}} + \frac{\partial u_{10}}{\partial\mathbf{n}} = \lambda\left(\frac{\partial u_2^-}{\partial\bar{\mathbf{n}}} + \frac{\partial u_1^-}{\partial\mathbf{n}} + \frac{\partial u_{10}}{\partial\mathbf{n}}\right) \quad \text{at} \quad \partial\Omega_i, \tag{4.141}$$

....

where derivatives with respect to the outer normal to the inclusion contour $\frac{\partial}{\partial\bar{\mathbf{n}}}, \frac{\partial}{\partial\mathbf{n}}$, written, respectively, in "fast" and "slow" variables are determined by relation (3.17).

Thus, the original eigenvalue problem in a multiply connected domain splits into a number of problems in domains with a much simpler geometry.

4.3.2 CONSTRUCTION OF ANALYTICAL RELATIONS FOR EIGENMODES AND OSCILLATION FREQUENCIES

The solution of the problem (4.130)–(4.132), (4.127) further decomposes into three stages:

1) At the first stage, the solution of the local problem is determined, i.e., the problem for the periodically repeating cell of the composite (Fig. 4.12)

$$\frac{\partial^2 u_1^\pm}{\partial\xi^2} + \frac{\partial^2 u_1^\pm}{\partial\eta^2} = 0 \quad \text{in} \quad \Omega_i^\pm, \tag{4.142}$$

$$u_1^+ = u_1^-, \quad \frac{\partial u_1^+}{\partial\bar{\mathbf{n}}} - \lambda\frac{\partial u_1^-}{\partial\bar{\mathbf{n}}} = (\lambda - 1)\frac{\partial u_0}{\partial\mathbf{n}} \quad \text{in} \quad \partial\Omega_i, \tag{4.143}$$

$$u_1^+ = 0 \quad \text{at} \quad \partial\Omega_i^*. \tag{4.144}$$

2) The second stage is finding the eigenfunctions and oscillation frequencies from the homogenized problem. The homogenized equation is obtained by applying the averaging operator (3.25) to expressions (4.136) and is represented by the relation

$$\bar{q}\left(\frac{\partial^2 u_0}{\partial x^2} + \frac{\partial^2 u_0}{\partial y^2}\right) + \frac{1}{|\Omega_i^*|}\left[\iint\limits_{\Omega_i^+}\left(\frac{\partial^2 u_1^+}{\partial x\partial\xi} + \frac{\partial^2 u_1^+}{\partial y\partial\eta}\right)d\xi\partial\eta +\right.$$

$$\left. + \lambda\iint\limits_{\Omega_i^-}\left(\frac{\partial^2 u_1^-}{\partial x\partial\xi} + \frac{\partial^2 u_1^-}{\partial y\partial\eta}\right)d\xi\partial\eta\right] + \varpi_0 u_0 = 0, \tag{4.145}$$

where $\Omega_i^* = \Omega_i^+ \bigcup \Omega_i^-$, $\bar{q} = \dfrac{|\Omega_i^+| + \lambda |\Omega_i^-|}{|\Omega_i^*|} = 1 - \dfrac{\pi a^2}{2\sqrt{3}} + \lambda \dfrac{\pi a^2}{2\sqrt{3}}$ – Voigt aver-
aged parameter.

Taking into account the expressions u_1^+, u_1^-, determined by the solution of the cell probleml (4.142)–(4.144), the homogenized problem is

$$q_x \frac{\partial^2 u_0}{\partial x^2} + q_y \frac{\partial^2 u_0}{\partial y^2} + \bar{q}\, \varpi_0 u_0 = 0 \quad \text{in} \quad \Omega^*, \tag{4.146}$$

$$u_0 = 0 \quad \text{at} \quad \partial\Omega, \tag{4.147}$$

where Ω^* : $\begin{Bmatrix} 0 \le x \le \ell_1 \\ 0 \le y \le \ell_2 \end{Bmatrix}$ and q_x, q_y are the homogenized parameters:

$$q_x = \bar{q} + \frac{1}{|\Omega_i^*|} \left(\iint\limits_{\Omega_i^+} \frac{\partial u_{1\,(1)}^+}{\partial \xi} d\xi d\eta + \lambda \iint\limits_{\Omega_i^-} \frac{\partial u_{1\,(1)}^-}{\partial \xi} d\xi d\eta \right), \tag{4.148}$$

$$q_y = \bar{q} + \frac{1}{|\Omega_i^*|} \left(\iint\limits_{\Omega_i^+} \frac{\partial u_{1\,(2)}^+}{\partial \eta} d\xi d\eta + \lambda \iint\limits_{\Omega_i^-} \frac{\partial u_{1\,(2)}^-}{\partial \eta} d\xi d\eta \right), \tag{4.149}$$

$u_{1\,(i)}^{\pm}$ ($i = 1$, 2) are the solutions of the local problem (4.142)–(4.144), written up to constant factors corresponding to the slow component of the solution:

$$u_1^{\pm} = u_{1\,(1)}^{\pm} (\xi,\, \eta) \frac{\partial u_0}{\partial x} + u_{1\,(2)}^{\pm} (\xi,\, \eta) \frac{\partial u_0}{\partial y}.$$

The solution of the homogenized eigenvalue problem (4.146), (4.147) can be represented as follows

$$u_0 = \sum_{m=1}^{\infty} \sum_{n=1}^{\infty} S_{mn} \sin \frac{m\pi x}{\ell_1} \sin \frac{n\pi y}{\ell_2}, \tag{4.150}$$

$$\varpi_0 = \frac{\pi^2}{\bar{q}} \left(q_x \left(\frac{m}{\ell_1} \right)^2 + q_y \left(\frac{n}{\ell_2} \right)^2 \right), \tag{4.151}$$

where the constants S_{mn} (m, $n = 1$, 2, ...) determined from the initial conditions (4.128).

3) At the third stage, the first correction ϖ_1 to the frequencies is determined.

To do this, one should define functions $u_2^{\pm}$ as a solution to the problem:

$$\frac{\partial^2 u_2^+}{\partial \xi^2} + \frac{\partial^2 u_2^+}{\partial \eta^2} =$$

$$= -\left(\frac{\partial^2 u_0}{\partial x^2} + \frac{\partial^2 u_0}{\partial y^2} + 2\frac{\partial^2 u_1^+}{\partial x \partial \xi} + 2\frac{\partial^2 u_1^+}{\partial y \partial \eta} \right) - \varpi_0 u_0 \quad \text{in} \quad \Omega_i^+, \tag{4.152}$$

$$
\lambda \left(\frac{\partial^2 u_2^-}{\partial \xi^2} + \frac{\partial^2 u_2^-}{\partial \eta^2} \right) =
$$
$$
= -\lambda \left(\frac{\partial^2 u_0}{\partial x^2} + \frac{\partial^2 u_0}{\partial y^2} + 2 \frac{\partial^2 u_1^-}{\partial x \partial \xi} + 2 \frac{\partial^2 u_1^-}{\partial y \partial \eta} \right) - \varpi_0 u_0 \quad \text{in} \quad \Omega_i^-,
$$
(4.153)

$$
u_2^+ = u_2^-,
$$
$$
\frac{\partial u_2^+}{\partial \bar{\mathbf{n}}} - \lambda \frac{\partial u_2^-}{\partial \bar{\mathbf{n}}} = (\lambda - 1) \frac{\partial u_{10}}{\partial \mathbf{n}} + \lambda \frac{\partial u_1^-}{\partial \mathbf{n}} - \frac{\partial u_1^+}{\partial \mathbf{n}} \quad \text{at} \quad \partial \Omega_i,
$$
(4.154)

$$
u_2^+ = 0 \quad \text{at} \quad \partial \Omega_i^*.
$$
(4.155)

The structure of the cell problems of the first and second approximations is identical and differs from the latter by the presence of an even component, which does not contribute to averaging and, therefore, is not used to determine ϖ_1. The expressions for functions $u_2^\pm$ obtained by solving problem (4.152)–(4.155) are generally represented by the relation

$$
u_2^\pm = u_1^\pm \left(u_0 \to u_{10} \right) + U_2^\pm \left(\xi, \eta \right),
$$

where

$$
U_2^\pm \left(-\xi, -\eta \right) = U_2^\pm \left(\xi, \eta \right),
$$

i.e., the average part of this solution is equal to zero.

By virtue of the last remark, the homogenized problem of the second approximation is obtained by applying the averaging operator (3.25) to expressions (4.137):

$$
q_x \frac{\partial^2 u_{10}}{\partial x^2} + q_y \frac{\partial^2 u_{10}}{\partial y^2} + \bar{q} \left(\varpi_0 u_{10} + \varpi_1 u_0 \right) = 0 \quad \text{in} \quad \Omega^*.
$$
(4.156)

The boundary condition is

$$
u_{10} = -\tilde{u}_1 \quad \text{at} \quad \partial \Omega,
$$
(4.157)

where $\tilde{u}_1$ is the averaged part of functions $u_1^\pm \left(x, y, \xi, \eta \right)$ at the membrane boundary.

In words, for edge $(0, y)$ we get

$$
\tilde{u}_1 = \frac{1}{\ell_2} \int_0^{\ell_2} u_1^\pm \left(0, y, \xi_0, \eta \right) d\eta,
$$
(4.158)

where ξ_0 is the value of ξ at the membrane boundary $x = 0$.

Similar relationships take place for the rest of the membrane boundaries. Expression (4.156) contains two unknown functions: u_{10} – homogenized solution of order ε and first frequency correction ϖ_1.

For determining ϖ_1, we use the well-known scheme [182, 183], namely, multiply equation by u_0 and integrate twice by parts over the region Ω^*. Taking into account the expression (4.146) and the boundary condition (4.147), one obtains

$$\varpi_1 \bar{q} \int_0^{\ell_1} \int_0^{\ell_2} u_0^2 \, dx dy - q_x \int_0^{\ell_2} \frac{\partial u_0}{\partial x} u_{10} \Big|_{x=0}^{x=\ell_1} dy - q_y \int_0^{\ell_1} \frac{\partial u_0}{\partial y} u_{10} \Big|_{y=0}^{y=\ell_2} dx = 0. \quad (4.159)$$

If

$$u_{10} = -\tilde{u}_1 = 0 \quad \text{on} \quad \partial\Omega,$$

then $\varpi_1 = 0$, and the expansion of natural oscillation frequencies can be written as follows:

$$\varpi = \varpi_0 + O\left(\varepsilon^2\right) + ...$$

If u_{10} does not satisfy the boundary conditions on the outer contour of the membrane, i.e. $u_{10} \neq 0$ on $\partial\Omega$, then we get a non-zero first correction to natural frequency

$$\varpi = \varpi_0 + O\left(\varepsilon\right) + ...,$$

which, by virtue of (4.159), is defined as

$$\varpi_1 = \frac{q_x \int_0^{\ell_2} \varphi(y)\, dy + q_y \int_0^{\ell_1} \phi(x)\, dx}{\bar{q} \int_0^{\ell_1} \int_0^{\ell_2} u_0^2 \, dx dy}, \quad (4.160)$$

where

$$\varphi(y) = \frac{\partial u_0}{\partial x} u_{10} \Big|_{x=0}^{x=\ell_1}, \quad \phi(x) = \frac{\partial u_0}{\partial y} u_{10} \Big|_{y=0}^{y=\ell_2}. \quad (4.161)$$

The subsequent terms of the expansions (4.133), (4.134) are found according to a similar scheme, and their determination does not cause difficulties.

4.3.3 COMPOSITE MEMBRANE WITH A SQUARE LATTICE OF CIRCLE INCLUSIONS

Using above proposed approach, we study natural oscillations of a membrane with a square array of circle inclusions (Fig. 4.42).

In this case, the homogenized equation (4.156) is transformed to the form:

$$q\left(\frac{\partial^2 u_{10}}{\partial x^2} + \frac{\partial^2 u_{10}}{\partial y^2}\right) + \bar{q}\left(\varpi_0 u_{10} + \varpi_1 u_0\right) = 0 \quad \text{in} \quad \Omega^*, \quad (4.162)$$

where q is the reduced parameter obtained by solving the problem on a cell; $\bar{q}$ is the value of the parameter found by Voigt averaging, and it reads

$$\bar{q} = 1 + (\lambda - 1)\frac{\pi a^2}{4}. \quad (4.163)$$

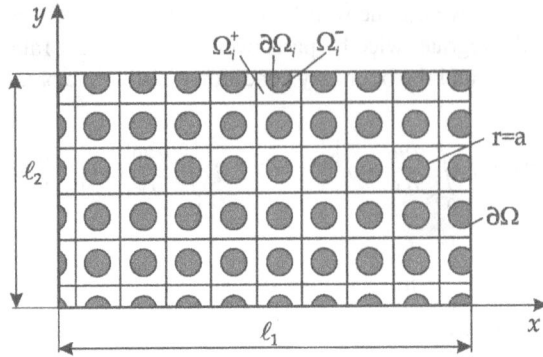

Figure 4.42 Rectangular membrane with a doubly periodic system of circle inclusions forming a square lattice.

Let us use the solution of the problem on a cell by the Schwarz alternating method [160] in the case of small inclusion sizes

$$u_1^{+\,(1)} = -\frac{\lambda-1}{\lambda+1}a^2\left(\frac{\partial u_0}{\partial x}\frac{\xi}{\xi^2+\eta^2} + \frac{\partial u_0}{\partial y}\frac{\eta}{\xi^2+\eta^2}\right), \tag{4.164}$$

$$u_1^{-\,(1)} = -\frac{\lambda-1}{\lambda+1}\left(\frac{\partial u_0}{\partial x}\xi + \frac{\partial u_0}{\partial y}\eta\right), \tag{4.165}$$

$$u_1^{(2)} = \frac{\lambda-1}{\lambda+1}\cdot\frac{\pi a^2}{4}\left(\frac{\partial u_0}{\partial x}\xi + \frac{\partial u_0}{\partial y}\eta\right) +$$

$$+\frac{\lambda-1}{\lambda+1}a^2\sum_{n=1}^{\infty}S_n\left[\frac{\partial u_0}{\partial x}\left(\sinh \pi n\xi \cos \pi n\eta - \cosh \pi n\eta \sin \pi n\xi\right) + \right. \tag{4.166}$$

$$\left. +\frac{\partial u_0}{\partial y}\left(\sinh \pi n\eta \cos \pi n\xi - \cosh \pi n\xi \sin \pi n\eta\right)\right],$$

$$q = 1 + \frac{\lambda-1}{\lambda+1}\frac{\pi a^2}{2}\left(1 + \frac{\lambda-1}{\lambda+1}\frac{\pi a^2}{4} + \frac{\lambda-1}{\lambda+1}\pi a^2\sum_{i=1}^{\infty}S_n n\right), \tag{4.167}$$

where

$$S_n = \frac{e^{-\pi n}\,\mathrm{Im}\,E_1\left(-\pi n + i\pi n\right) - e^{\pi n}\,\mathrm{Im}\,E_1\left(\pi n + i\pi n\right) + \pi e^{-\pi n}}{\sinh \pi n}, \tag{4.168}$$

$i = \sqrt{-1}$; $\mathrm{Im}\,E_1$ is imaginary part of the familiar exponential integral [2].

Then, taking into account expressions (4.167), (4.168), (4.163) for q and $\bar{q}$, eigenfunctions and oscillation frequencies in the zeroth approximation are determined by relations (4.150), (4.151), where $q_x = q_y = q$.

Table 4.5

Natural frequencies of a composite membrane with a square lattice of circle inclusions.

(m, n)		$\omega*$			
λ	a	(1, 1)	(1, 2)	(2, 2)	(1, 3)
0	0	19.7392	49.3480	78.9568	98.6960
	1/6	19.2898	48.2246	77.1593	96.4491
	1/3	18.0170	45.0425	72.0680	90.0850
0.1	1/6	19.3270	48.3175	77.3079	96.6349
	1/3	18.4954	46.2385	73.9816	92.4770
10	1/6	16.4258	41.0645	65.7032	82.1290
	1/3	12.6800	31.7225	50.7560	63.4450
100	1/6	6.2426	15.6066	24.9705	31.2131
	1/3	2.0595	5.1488	8.2381	10.2976

For a rectangular membrane rigidly clamped along the contour, the boundaries of which pass through the midpoints of the periodicity cells, i.e. we have

$$\text{at} \quad x = 0,\, \ell_1 : \quad \xi = 0,$$

$$\text{at} \quad y = 0,\, \ell_2 : \quad \eta = 0,$$

while from (4.158), (4.164)–(4.166) we get

$$\frac{1}{\ell_2} \int_0^{\ell_2} u_1^\pm (0, y, 0, \eta)\, d\eta = \frac{1}{\ell_2} \int_0^{\ell_2} u_1^\pm (\ell_1, y, 0, \eta)\, d\eta =$$

$$= \frac{1}{\ell_1} \int_0^{\ell_1} u_1^\pm (x, 0, \xi, 0)\, d\eta = \frac{1}{\ell_1} \int_0^{\ell_1} u_1^\pm (x, \ell_2, \xi, 0)\, d\eta = 0.$$

Consequently, this means that

$$\tilde{u}_1 = 0 \quad \text{at} \quad \partial\Omega,$$

and, therefore,

$$u_{10} = 0 \quad \text{at} \quad \partial\Omega.$$

In this case, the first correction to the natural oscillation frequencies $\varpi_1 = 0$. In Table 4.5 numerical values $\omega^* = \bar{\omega}\ell^2$ are given for square composite membrane, $\ell_1 = \ell_2 = \ell$, depending on the size of inclusions a and their physical characteristic λ.

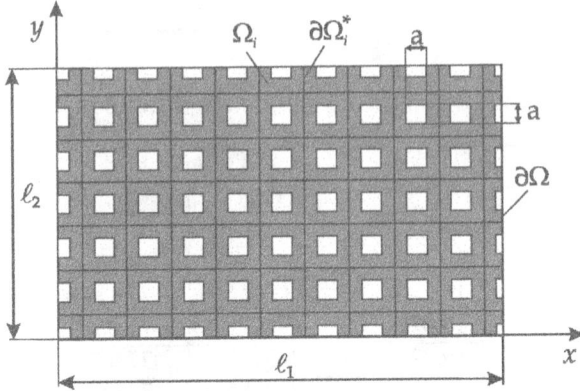

Figure 4.43 Rectangular membrane with a square lattice of square holes.

4.3.4 COMPOSITE MEMBRANE WITH SQUARE HOLES

To study the natural oscillations of a rectangular membrane rigidly clamped along the contour with a doubly periodic system of square perforations with side a, constituting a square lattice (Fig. 4.43), we use the solution of the problem on the cell

$$\frac{\partial^2 u_1}{\partial \xi^2} + \frac{\partial^2 u_1}{\partial \eta^2} = 0 \quad \text{in} \quad \Omega_i,$$

$$\frac{\partial u_1}{\partial \xi}\bigg|_{\xi=\pm a} = -\frac{\partial u_0}{\partial x}, \quad \frac{\partial u_1}{\partial \eta}\bigg|_{\eta=\pm a} = -\frac{\partial u_0}{\partial y},$$

$$u_1 = 0 \quad \text{at} \quad \partial \Omega_i^*,$$

obtained by the Bubnov-Galerkin variational method [25].

Then, we have

$$u_1 = \sum_{m=0}^{\infty} \sum_{n=0}^{\infty} (A_{1mn} \sin m\pi\xi \cos n\pi\eta + A_{2mn} \cos m\pi\xi \sin n\pi\eta +$$

$$+ A_{3mn} \cos m\pi\xi \cos n\pi\eta + A_{4mn} \sin m\pi\xi \sin n\pi\eta),$$

where A_{1mn} are determined from an infinite system of linear algebraic equations

$$\sum_{k=0}^{\infty} \sum_{\ell=0}^{\infty} A_{1k\ell} \left[\beta_{1k\ell} + \frac{k^2 + \ell^2}{\ell + n} \left(\frac{\sin \pi a (k-m)}{k-m} - \frac{\sin \pi a (k+m)}{k+m} \right) \times \right.$$

$$\times \sin \pi a (n+\ell) + \frac{2k}{\ell+n} \cos \pi ak \sin \pi am \sin \pi a (n+\ell) - \qquad (4.169)$$

$$\left. -2\ell \left(\frac{\sin \pi a (k-m)}{k-m} - \frac{\sin \pi a (k+m)}{k+m} \right) \sin \pi a\ell \cos \pi an \right] =$$

Table 4.6

Natural vibration frequencies of a composite membrane with a square lattice of square holes.

m, n	$\omega*$			
	(1, 1)	(1, 2)	(2, 2)	(1, 3)
a				
0	19.7392	49.3480	78.9568	98.6960
1/6	18.2805	45.7012	73.1219	91.4024
1/3	14.2513	35.6282	57.0053	71.2566

$$= -\frac{\partial u_0}{\partial x}\frac{2a}{\pi n}\sin \pi am \sin \pi an, \quad m = 1, 2, \ldots, \quad n = 0, 1, 2, \ldots$$

$$\beta_{1k\ell} = \begin{cases} -2\pi^2, & m = k, \ n = \ell = 0, \\ -\pi^2, & m = k, \ n = \ell \neq 0, \\ 0, & m \neq k. \end{cases} \tag{4.170}$$

A_{2mn} are found from a similar system of equations, taking into account the replacement: $\frac{\partial u_0}{\partial x} \to \frac{\partial u_0}{\partial y}$; $m \rightleftarrows n$; $k \rightleftarrows \ell$;

$$A_{3mn} = A_{4mn} = 0. \tag{4.171}$$

The effective coefficient q is written as

$$q = 1 - a^2 - \frac{1}{\pi}\sum_{m=0}^{\infty}\sum_{n=0}^{\infty} A_{mn}^* \sin m\pi a \sin n\pi a, \tag{4.172}$$

where $A_{1mn} = \frac{\partial u_0}{\partial x} a A_{mn}^*$, $A_{2mn} = \frac{\partial u_0}{\partial y} a A_{mn}^*$.
Voigt averaging leads to the relation

$$\bar{q} = 1 - a^2. \tag{4.173}$$

Further, similar to how it was done in the case of a composite membrane with a square lattice of circle inclusions, taking into account relations (4.169)–(4.173), we find the natural vibration frequencies $\omega^* = \bar{\omega}\ell^2$ of the square perforated membrane, the boundaries of which pass through the middle of the cells of periodicity and $\ell_1 = \ell_2 = \ell$, depending on hole size a (Table 4.6).

5 Using of Asymptotically Equivalent Functions for Analysis of Composite Structures

This chapter is focused on benefits of using asymptotically equivalent function for analysis of composite structures. It begins with determination of the effective thermal conductivity coeffcient by matching of asymptotic expansions. The latter thematic issue includes statement of the problem and constitutive relations, matching of asymptotic expansions in the case of high and low conductivity of inclusions, analysis of the effective thermal conductivity coefficient in limiting cases, and remarks on the effectiveness of the Hashin-Shtrikman bounds. Then, asymptotic solutions for a composite material with rhombic inclusions are derived and discussed. The analysis covers problems of non-intersecting rhombic inclusions of large sizes, intersecting of rhombic inclusions, rhombic inclusions in contact, asymptotic expressions for the effective coefficient, composites with equally represented phases and Dykhne structure, and physical equivalence of chess composite arrays.

Then, models of two-phase fibrous composites based on asymptotic approximations are studied, and equivalence of structures is addressed. The latter scope includes the following problems: advantages of the method of non-smooth transformation, high contrast densely packed composites, generalization to the case of non-conducting inclusions of large geometric sizes, analysis of the contact condition "matrix-inclusion", asymptotic representation of the solution by LA via employment of the non-smooth transformation, contact of non-conductive round inclusions in the presence of a thin layer at the phase boundary, models of composites with curvilinear rhombic inclusions, physical equivalence of composite structures, as well as hexagonal array of circle inclusions. Next, percolation threshold determination for elastic problems is investigated, including three main problems: self-consistency approximation, AP application for viral expansion, as well estimation of AP accuracy.

5.1 DETERMINATION OF THE EFFECTIVE THERMAL CONDUCTIVITY COEFFICIENT BY MATCHING OF ASYMPTOTIC EXPANSIONS

Determining the effective characteristics of inhomogeneous media is one of the main problems in the mechanics of composites. The variety of composite materials, a wide range of their practical application in engineering, features of the technology for

DOI: 10.1201/9781003391029-5

the production of structures from them yield a challenging problem for scientists to develop suitable methods for studying them.

One of the possible ways to solve the problem is to study the behavior of composites in the case of limiting values of their physical and geometric characteristics, followed by the application of a number of asymptotic simplifications. In this case, the following features of the asymptotic study of such structures should be noted:

(i) firstly, the inefficiency of using numerical methods for calculations of composite materials for the parameter limiting values;

(ii) secondly, the need to build appropriate models for various limiting values of the geometric and physical characteristics of the structure, when the physical essence of the processes occurring in the composite changes dramatically;

(iii) thirdly, the effective characteristics have different asymptotics in the various limiting cases.

In this regard, it seems natural to single out two limiting cases, which have their own characteristic features and, therefore, are described by different physical models and require different mathematical approaches:

(i) inclusions are small. In this case, the entire original composite material can be modeled by an infinite matrix with a single inclusion located in it (see Chapter 3).

(ii) inclusions are large. Then, the solution of the problem requires a different approach, because with an increase in the size of the inclusions, the physical processes occurring in the composite changes. In this case, it is advisable to use the lubrication theory (see Chapter 4).

5.1.1   STATEMENT OF THE PROBLEM AND CONSTITUTIVE RELATIONS

Consider the problem of heat conduction for a two phase composite structure consisting of a continuous matrix with a doubly periodic system (the period of the structure is the same in both directions) of cylindrical inclusions, square in cross section (Fig. 5.1). A characteristic structural cell of the composite is shown in Fig. 5.2.

To determine the effective thermal conductivity of a composite array with cylindrical inclusions of a square profile, the following models were used:

1) ThPhM, the use of which leads to the expression of the homogenized coefficient q_{ThPhM} (3.131), which can be used over the entire range of the conductivity of inclusions ($0 \leq \lambda < \infty$, $\lambda \to \infty$) and their geometric sizes ($0 \leq a \leq 1$):

$$q_{ThPhM} = \frac{1 - a^2 + \lambda \left(1 + a^2\right)}{1 + a^2 + \lambda \left(1 - a^2\right)}. \tag{5.1}$$

2) A two phase composite model (2PhM), constructed under the assumption of small geometric sizes of inclusions ($a \to 0$), according to which the

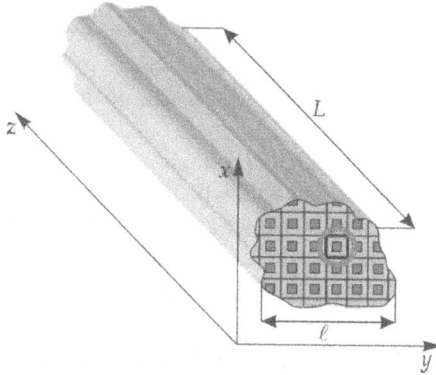

Figure 5.1 Composite structure with square cylindrical inclusions.

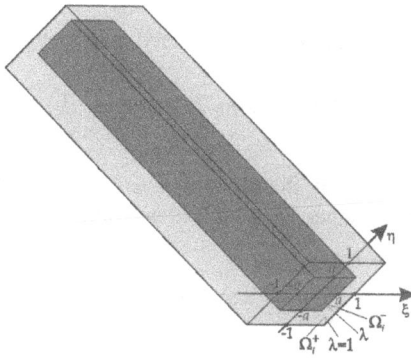

Figure 5.2 Characteristic structural cell of the composite.

homogenized parameter q_{2PhM} is determined by the ratio

$$q_{2PhM} = 1 + \frac{2(\lambda - 1)}{\lambda + 1}a^2. \tag{5.2}$$

3) Composite model for a structure with large inclusions ($a \to 1$) of high conductivity ($\lambda \gg 1$), the mathematical description of which is carried out on the basis of ThPhM and gives the expression for the homogenized thermal conductivity q_{LA} as (4.16):

$$q_{LA} = \frac{\lambda(1 - a^2 + a^3) + a^2(1 - a)}{\lambda(1 - a) + a}. \tag{5.3}$$

As was shown earlier, in the case of limiting values of the conductivity of inclusions ($\lambda \to 0$, $\lambda \to \infty$) and / or their geometric size ($a \to 0$, $a \to 1$), relations (5.1), (5.2) and (5.1), (5.3) give the same asymptotic representations of the effective thermal conductivity parameter, coinciding in special cases with known data.

Moreover, the analysis and comparison of the above obtained results confirm that ThPhM, in the case of cylindrical inclusions of a square profile, gives sufficiently accurate solutions for the effective thermal conductivity parameter over the entire range of the geometric size of the inclusions and of their physical characteristics, including limiting cases.

However, it should be noted there is no single analytical expression for the effective thermal conductivity coefficient with which comparison q_{ThPhM} could be made. In this regard, it is of interest to construct an analytical expression for the effective thermal conductivity by matching the asymptotic representations (5.2), (5.3). The nontriviality of the problem lies in the fact that each of the relations (5.2), (5.3) from a mathematical point of view is a function of two variables, and matching should be performed for each of them. Let us divide the solution of the problem into three steps:

(i) splitting of asymptotic representations (5.2), (5.3) by the inclusion size a for fixed conductivity $\lambda \gg 1$;

(ii) approximation obtained at $\lambda \gg 1$ effective parameter for cases $0 < \lambda \ll 1$;

(iii) splitting of the asymptotic relations $q^{(\infty)}$ and $q^{(0)}$ by conductivity of inclusions λ.

In order to solve the problem, the method of asymptotically equivalent functions [8, 11, 215] is used, the essence of which is as follows: let the function $f(z)$ for $z \to \infty$ has asymptotics $F(z)$:

$$f(z) \sim F(z) \quad \text{at} \quad z \to \infty, \tag{5.4}$$

and at $z \to 0$:

$$f(z) \sim \sum_{i=0}^{\infty} c_i z^i \quad \text{at} \quad z \to 0. \tag{5.5}$$

An asymptotically equivalent function can be defined as follows:

$$f(z) = \frac{\sum\limits_{i=0}^{m} \alpha_i(z) z^i}{\sum\limits_{i=0}^{m} \beta_i(z) z^i}. \tag{5.6}$$

Functions $\alpha_i(z)$ and $\beta_i(z)$ are chosen in such a way that the expansion of expression (5.6) in powers z for $z \to 0$ corresponded to expansion (11.1.5), while the asymptotic representation (5.6) for $z \to \infty$ matched the expression $F(z)$ (5.4).

5.1.2   MATCHING ASYMPTOTIC EXPANSIONS IN THE CASE OF HIGH CONDUCTIVITY OF INCLUSIONS

Consider the case of inclusions of high conductivity, $\lambda \gg 1$. As the splitting parameter ε_a, we take the value

$$\varepsilon_a = \frac{1}{2}a^2(1+a) + \frac{5}{2}a^2(1-a) - 3a^2(1-a)^2 + 12a^3(1-a)^2 -$$

$$-16a^3(1-a)^3 - 6a^4(1-a)^2 + 20a^4(1-a)^3 + 35a^4(1-a)^4 +$$
$$+56a^5(1-a)^5 - 70a^4(1-a)^6 + 70a^5(1-a)^6 + 140a^6(1-a)^6 - \quad (5.7)$$
$$-322a^5(1-a)^7 - 924a^6(1-a)^7 + 1716a^7(1-a)^7 - 3432a^7(1-a)^8 +$$
$$+6435a^8(1-a)^8 - 12870a^8(1-a)^9 + 24310a^9(1-a)^9 -$$
$$-48620a^9(1-a)^{10} + 92378a^{10}(1-a)^{10}.$$

The expression of parameter ε_a is obtained from the conditions that the expansion into a series of the expression of the homogenized parameter

$$q^{(\infty)} = (1 - \varepsilon_a) \, q_{2PhM}^{(\infty)} + \varepsilon_a \, q_{LA}^{(\infty)} \quad (5.8)$$

matches:

1) with series expansion at $a \to 0$ expressions $q_{2PhM}^{(\infty)} \equiv q_{2PhM}$ (5.2) up to terms of order a^{10} inclusive

$$q^{(\infty)} \sim q_{2PhM}^{(\infty)} \sim 1 + \frac{2\,(\lambda - 1)}{\lambda + 1}\, a^2 + o\left(a^{10}\right) \quad \text{at} \quad a \to 0, \quad (5.9)$$

2) with series expansion for $a \to 1$ expressions $q_{LA}^{(\infty)} \equiv q_{LA}$ (5.3) up to terms of order $(1-a)^{10}$ inclusive

$$q^{(\infty)} \sim q_{LA}^{(\infty)} \sim \lambda - (\lambda^2 - 1)\,(1-a) + (\lambda^3 - \lambda^2 + \lambda - 1)\,(1-a)^2 -$$
$$-\lambda\,(\lambda^3 - 2\lambda^2 + 2\lambda - 1)\,(1-a)^3 + \lambda\,(\lambda^4 - 3\lambda^3 + 4\lambda^2 - 3\lambda + 1)\,(1-a)^4 -$$
$$-\lambda\,\left(\lambda^5 - 4\lambda^4 + 7\lambda^3 - 7\lambda^2 + 4\lambda - 1\right)(1-a)^5 + \quad (5.10)$$
$$+\lambda\,\left(\lambda^6 - 5\lambda^5 + 11\lambda^4 - 14\lambda^3 + 11\lambda^2 - 5\lambda + 1\right)(1-a)^6 -$$
$$-\lambda\,\left(\lambda^7 - 6\lambda^6 + 16\lambda^5 - 25\lambda^4 + 25\lambda^3 - 16\lambda^2 + 6\lambda - 1\right)(1-a)^7 +$$
$$+\lambda\,\left(\lambda^8 - 7\lambda^7 + 22\lambda^6 - 41\lambda^5 + 50\lambda^4 - 41\lambda^3 + 22\lambda^2 - 7\lambda + 1\right)(1-a)^8 -$$
$$-\lambda\,\left(\lambda^9 - 8\lambda^8 + 29\lambda^7 - 63\lambda^6 + 91\lambda^5 - 91\lambda^4 + 63\lambda^3 - 29\lambda^2 + 8\lambda - 1\right) \times$$
$$\times (1-a)^9 + \lambda\,\left(\lambda^{10} - 9\lambda^9 + 37\lambda^8 - 92\lambda^7 + 154\lambda^6 - 182\lambda^5 +\right.$$
$$\left.+154\lambda^4 - 92\lambda^3 + 37\lambda^2 - 9\lambda + 1\right)(1-a)^{10} + o\left((1-a)^{10}\right) \quad \text{for} \quad a \to 1.$$

It is also obvious that relation (5.8) also holds in the limiting cases: $a - 0$ and $a = 1$. Really, observe that

$$\begin{aligned}
\text{at} \quad a = 0: \quad & \varepsilon_a = 0 \quad \Rightarrow \quad q^{(\infty)} = q_{2PhM}^{(\infty)} \\
\text{at} \quad a = 1: \quad & \varepsilon_a = 1 \quad \Rightarrow \quad q^{(\infty)} = q_{LA}^{(\infty)}
\end{aligned} \quad (5.11)$$

Consequently, taking into account (5.2), (5.3), (5.8), expression $q^{(\infty)}$ is defined as follows

$$q^{(\infty)} = (1 - \varepsilon_a)\left(1 + \frac{2\,(\lambda - 1)\,a^2}{\lambda + 1}\right) + \varepsilon_a\left(1 + \frac{(\lambda - 1)\,a\,(1 - a + a^2)}{\lambda\,(1 - a) + a}\right), \quad (5.12)$$

where ε_a has the form (5.7).

5.1.3 MATCHING ASYMPTOTIC EXPANSIONS IN THE CASE OF LOW CONDUCTIVITY OF INCLUSIONS

Consider the case of inclusions of low conductivity, $0 < \lambda \ll 1$. The difference here from case $\lambda \gg 1$ is that expression (5.3), found using LA, is valid only for $\lambda \gg 1$. We are aimed on constructing an expression valid for $0 < \lambda \ll 1$ and $a \to 1$. Let us use the Keller's theorem [127], due to which the given parameters for the reciprocal values of the conductivity of inclusions are coupled by relation (3.59). Then, using the Keller's theorem, we obtain for $0 < \lambda \ll 1$ and $a \to 1$ the following expression for the effective thermal conductivity

$$q_{LA}^{(0)} = \frac{(1 - a) + \lambda\,a}{1 - a^2 + a^3 + \lambda\,a^2\,(1 - a)}. \quad (5.13)$$

Expression for effevtive parameter q_{2PhM} (5.2), obtained using 2PhM, is generally applicable for any values of the conductivity of the inclusions $0 \le \lambda < \infty$ and asymptotically (up to terms of order a^3 inclusive) satisfies the Keller's theorem. Therefore, due to the identity of the relations used in the splitting, the expression $q_{2PhM}^{(0)}$ takes the following form

$$q_{2PhM}^{(0)} = 1 + \frac{2\,(\lambda - 1)\,a^2}{\lambda + 1 - 2\,(\lambda - 1)\,a^2}. \quad (5.14)$$

Splitting expressions $q_{2PhM}^{(0)}$ (5.14) and $q_{LA}^{(0)}$ (5.13) using the parameter ε_a (5.7) yields a beautiful result: expansion into a series of the expression of the given parameter

$$q^{(0)} = (1 - \varepsilon_a)\,q_{2PhM}^{(0)} + \varepsilon_a\,q_{LA}^{(0)} \quad (5.15)$$

also matches:

1) with series expansion at $a \to 0$ expressions $q_{2PhM}^{(0)}$ (5.14) up to terms of order a^{10} inclusive:

$$q^{(0)} \sim q_{2PhM}^{(0)} \sim 1 + \frac{2\,(\lambda - 1)}{\lambda + 1}\,a^2 + \frac{4\,(\lambda - 1)^2}{(\lambda + 1)^2}\,a^4 + \frac{8\,(\lambda - 1)^3}{(\lambda + 1)^3}\,a^6 +$$
$$+ \frac{16\,(\lambda - 1)^4}{(\lambda + 1)^4}\,a^8 + \frac{32\,(\lambda - 1)^5}{(\lambda + 1)^5}\,a^{10} + o\left(a^{10}\right) \quad \text{at} \quad a \to 0, \quad (5.16)$$

2) with series expansion at $a \to 1$ expression $q_{LA}^{(0)}$ (5.13) up to the terms of order $(1-a)^{10}$ inclusive:

$$q^{(0)} \sim q_{LA}^{(0)} \sim \lambda - (\lambda^2 - 1)(1-a) + (\lambda^3 + \lambda^2 - 3\lambda + 1)(1-a)^2 -$$

$$- (\lambda^4 + 2\lambda^3 - 5\lambda^2 + \lambda + 1)(1-a)^3 + \left(\lambda^5 + 3\lambda^4 - 6\lambda^3 - 3\lambda^2 + 7\lambda - 2\right) \times$$

$$\times (1-a)^4 - \left(\lambda^6 + 4\lambda^5 - 6\lambda^4 - 11\lambda^3 + 18\lambda^2 - 5\lambda - 1\right)(1-a)^5 +$$

$$+ \left(\lambda^7 + 5\lambda^6 - 5\lambda^5 - 22\lambda^4 + 28\lambda^3 + 3\lambda^2 - 14\lambda + 4\right)(1-a)^6 - \quad (5.17)$$

$$- \left(\lambda^8 + 6\lambda^7 - 3\lambda^6 - 35\lambda^5 + 31\lambda^4 + 36\lambda^3 - 53\lambda^2 + 17\lambda\right)(1-a)^7 +$$

$$+ \left(\lambda^9 + 7\lambda^8 - 49\lambda^6 + 22\lambda^5 + 100\lambda^4 - 110\lambda^3 + 13\lambda^2 + 23\lambda - 7\right)(1-a)^8 -$$

$$- \left(\lambda^{10} + 8\lambda^9 + 4\lambda^8 - 63\lambda^7 - 3\lambda^6 + 193\lambda^5 - 150\lambda^4 - 80\lambda^3 + 133\lambda^2 -\right.$$

$$\left. - 46\lambda + 3\right)(1-a)^9 + \left(\lambda^{11} + 9\lambda^{10} + 9\lambda^9 - 76\lambda^8 - 47\lambda^7 + 306\lambda^6 - 121\lambda^5 -\right.$$

$$\left. - 345\lambda^4 + 370\lambda^3 - 89\lambda^2 - 28\lambda + 11\right)(1-a)^{10} + o\left((1-a)^{10}\right) \quad \text{at} \quad a \to 1.$$

In limiting cases $a = 0$ and $a = 1$, we have

$$
\begin{aligned}
&\text{for} \quad a = 0: \quad \varepsilon_a = 0 \quad \Rightarrow \quad q^{(0)} = q_{2PhM}^{(0)}, \\
&\text{for} \quad a = 1: \quad \varepsilon_a = 1 \quad \Rightarrow \quad q^{(0)} = q_{LA}^{(0)}.
\end{aligned}
\quad (5.18)
$$

Therefore, due to (5.13)–(5.15), expression $q^{(0)}$ is obtained in the following form

$$q^{(0)} = (1 - \varepsilon_a)\left(1 + \frac{2(\lambda - 1)a^2}{\lambda + 1 - 2(\lambda - 1)a^2}\right) +$$

$$+ \varepsilon_a \left(1 + \frac{(\lambda - 1)a(1 - a + a^2)}{1 - a^2 + a^3 + \lambda a^2(1-a)}\right) \quad (5.19)$$

where ε_a is determined by the ratio (5.7).

5.1.4 MATCHING ASYMPTOTIC EXPANSIONS ON THE CONDUCTIVITY OF INCLUSIONS

We proceed further to the second step of solving the problem: we match the asymptotic expansions $q^{(\infty)}$ (5.12) and $q^{(0)}$ (5.19) by conductivity of inclusions λ. In this case, the expression for the effective thermal conductivity parameter is represented as

$$q = (1 - \varepsilon_\lambda) q^{(0)} + \varepsilon_\lambda q^{(\infty)}. \quad (5.20)$$

As the splitting parameter, we take

$$\varepsilon_\lambda = \frac{\lambda^2 + \frac{\lambda}{2}}{\lambda^2 - 1}. \tag{5.21}$$

With this choice of parameter ε_λ, in the limiting cases $\lambda = 0$ and $\lambda \to \infty$, one gets

$$\begin{aligned}
\text{at} \quad \lambda = 0: \quad & \varepsilon_\lambda = 0 \quad \Rightarrow \quad q = q^{(0)}, \\
\text{at} \quad \lambda \to \infty: \quad & \varepsilon_\lambda = 1 \quad \Rightarrow \quad q = q^{(\infty)}.
\end{aligned} \tag{5.22}$$

Thus, by virtue of relations (5.12), (5.19), (5.20), (5.21), the expression for the homogenized thermal conductivity coefficient q is finally determined as follows:

$$\begin{aligned}
q = & \frac{\lambda + 1}{\lambda + 1 - 2(\lambda - 1)a^2} + \varepsilon_a (\lambda - 1) a \left(\frac{1 - a + a^2}{1 - a^2 + a^3 + \lambda a^2 (1 - a)} - \right. \\
& \left. - \frac{2a}{\lambda + 1 - 2(\lambda - 1)a^2} \right) + \varepsilon_\lambda 2(\lambda - 1) a^2 \left(\frac{1}{\lambda + 1} - \frac{1}{\lambda + 1 - 2(\lambda - 1)a^2} \right) + \\
& + \varepsilon_a \varepsilon_\lambda (\lambda - 1) a \left(\frac{2a}{\lambda + 1 - 2(\lambda - 1)a^2} - \frac{2a}{\lambda + 1} + \frac{1 - a + a^2}{\lambda(1 - a) + a} - \right. \\
& \left. - \frac{1 - a + a^2}{1 - a^2 + a^3 + \lambda a^2 (1 - a)} \right)
\end{aligned} \tag{5.23}$$

where ε_a, ε_λ, respectively, have the form (5.7), (5.21).

5.1.5 ANALYSIS OF THE EFFECTIVE THERMAL CONDUCTIVITY COEFFICIENT IN LIMITING CASES

Let us show that the use of representations (5.23), (5.7), (5.21) makes it possible to perform correctly all limiting cases:

1) $\lambda \to 0$ – inclusions of low conductivity, in the limit – a composite with inclusions – heat insulators:

1.1. $a \to 0$ – small sizes of inclusions:

$$q \sim q_{2PhM}^{(0)} \sim 1 - 2 \left(1 - 2\lambda + 2\lambda^2 \right) a^2 + \dots \tag{5.24}$$

1.2. $a \to 1$ – large inclusion sizes:

$$q \sim q_{LA}^{(0)} \sim \lambda + \left(1 - \lambda^2 \right) (1 - a) + (1 - a)^2 + \dots \tag{5.25}$$

2) $\lambda \to 1$ – composite, in which the conductivities of the matrix and inclusions are of the same order:

2.1. $a \to 0$ – inclusions are small:

$$q \sim q_{2PhM}^{(0)} \sim q_{2PhM}^{(\infty)} \sim 1 + (\lambda - 1) \left(1 - \frac{\lambda - 1}{2} + \frac{(\lambda - 1)^2}{4} \right) a^2 + \dots \tag{5.26}$$

2.2. $a \rightarrow 1$ – large inclusion sizes:

$$q \sim q_{LA}^{(0)} \sim q_{LA}^{(\infty)} \sim 1 + (\lambda - 1) - (\lambda - 1)\left(2 + (\lambda - 1)\right)(1 - a) + ...$$
(5.27)

3) $\lambda \rightarrow \infty$ – inclusions of high conductivity, in the limit – a composite with absolutely conductive inclusions:

[3.1. $a \rightarrow 0$ – small inclusions:

$$q \sim q_{2PhM}^{(\infty)} \sim 1 + 2\left(1 - \frac{2}{\lambda} + \frac{2}{\lambda^2}\right)a^2 + ...$$
(5.28)

3.2. $a \rightarrow 1$ – large inclusion sizes:

$$q \sim q_{LA}^{(\infty)} \sim \frac{1}{1-a}\left(1 + \frac{2}{\lambda}\right) - 1 + 2(1-a) - (1-a)^2 + ...$$
(5.29)

Similarly, asymptotic relations can be obtained with a change in the order of limit transitions: for a composite with small (large) inclusions, and a change in their conductivity:

4) $a \rightarrow 0$ – composite structure with inclusions of small sizes:

4.1. $\lambda \rightarrow 0$ – passage to low conductivity case:

$$q \sim q_{2PhM}^{(0)} \sim \left(1 - 2a^2 + 4a^4\right) + 4a^2\lambda - 4a^2\lambda^2 + 4a^2\lambda^3 - 4a^2\lambda^4 + ...$$
(5.30)

4.2. $\lambda \rightarrow 1$ – conductivity of the matrix and inclusions of the same order:

$$q \sim q_{2PhM}^{(0)} \sim q_{2PhM}^{(\infty)} \sim 1 + a^2(\lambda - 1) - \frac{a^2}{2}(\lambda - 1)^2 +$$
$$+ \frac{a^2}{4}(\lambda - 1)^3 - \frac{a^2}{8}(\lambda - 1)^4 + ...$$
(5.31)

4.3. $\lambda \rightarrow \infty$ – inclusion of high conductivity:

$$q \sim q_{2PhM}^{(\infty)} \sim 1 + 2a^2 - 4a^2\frac{1}{\lambda} + 4a^2\frac{1}{\lambda^2} - 4a^2\frac{1}{\lambda^3} + 4a^2\frac{1}{\lambda^4} + ... \quad (5.32)$$

5) $a \rightarrow 1$ – composite array with inclusions of large sizes:

5.1. $\lambda \rightarrow 0$ – passage to low conductivity:

$$q \sim q_{LA}^{(0)} \sim (1 - a)\left(1 + (1 - a) - (1 - a)^2 - 2(1 - a)^3\right) +$$
$$+ \lambda - (1 - a)\left(1 - (1 - a)\right)\lambda^2 + ...$$
(5.33)

5.2. $\lambda \rightarrow 1$ conductivity of the matrix and inclusions of the same order:

$$q \sim q_{LA}^{(0)} \sim q_{LA}^{(\infty)} \sim 1 + \left(1 - 2(1 - a)\right)(\lambda - 1) - (1 - a)(\lambda - 1)^2 + ...$$
(5.34)

5.3. $\lambda \to \infty$ – inclusion of high conductivity:

$$q \sim q_{LA}^{(\infty)} \sim \lambda + (1-a)\left(1-(1-a)\right) + \dots \qquad (5.35)$$

Comparison of expressions (5.24) and (5.30), (5.25) and (5.33), (5.26) and (5.31), (5.27) and (5.34), (5.28) and (5.32) testifies to the obvious coincidence of the corresponding terms of the asymptotic expansions.

In the case of a composite with inclusions of large sizes of high conductivity, we have different asymptotic representations for $\lambda \to \infty$ (5.35) and for $a \to 1$ (5.29), which from a physical point of view corresponding to different structures:

(i) in the first case ($\lambda \to \infty$) in the limit, we have a homogeneous material with ideal conductivity equal to the conductivity of inclusions;

(ii) in the second case ($a \to 1$), the structure is absolutely conductive cylindrical inclusions of a square profile, rigidly connected to each other along the side surfaces.

Comparison of relation (5.23), obtained by matching the asymptotic representations 2PhM (5.2) and LA (5.3), with the homogenized thermal conductivity coefficient (5.1), determined in accordance with ThPhM, yields

1) $\lambda \to 0$:

$$\left.\begin{aligned} q &\sim 1 - 2\left(1-2\lambda+2\lambda^2\right)a^2 + \\ &+ 2\left(2-7\lambda+14\lambda^2\right)a^4 + \dots \\ q_{ThPhM} &\sim 1 - 2\left(1-2\lambda+2\lambda^2\right)a + \\ &+ 2\left(1-4\lambda+8\lambda^2\right)a^4 + \dots \end{aligned}\right\} \quad \text{at} \quad a \to 0, \qquad (5.36)$$

$$\left.\begin{aligned} q &\sim \lambda - \left(\lambda^2-1\right)\left(1-a\right) + \\ &+ \left(1-2\lambda+\lambda^2\right)\left(1-a\right)^2 + \dots \\ q_{ThPhM} &\sim \lambda - \left(\lambda^2-1\right)\left(1-a\right) + \\ &+ \left(\frac{1}{2}-\lambda-\frac{\lambda^2}{2}\right)\left(1-a\right)^2 + \dots \end{aligned}\right\} \quad \text{at} \quad a \to 1. \qquad (5.37)$$

2) $\lambda \to 1$:

$$\left.\begin{aligned} q &\sim 1 + (\lambda-1)\left(1-\frac{\lambda-1}{2}+\frac{(\lambda-1)^2}{4}\right)a^2 - \\ &- \frac{\lambda-1}{2}\left(\frac{3}{2}-\frac{7(\lambda-1)}{4}+\frac{3(\lambda-1)^2}{2}\right)a^4 + \dots \\ q_{ThPhM} &\sim 1 + (\lambda-1)\left(1-\frac{\lambda-1}{2}+\frac{(\lambda-1)^2}{4}\right)a^2 + \\ &+ \frac{\lambda-1}{2}\left((\lambda-1)-(\lambda-1)^2\right)a^4 + \dots \end{aligned}\right\} \quad \text{at} \quad a \to 0,$$

$$\qquad (5.38)$$

$$\left.\begin{aligned}
& q \sim 1+(\lambda-1)-(\lambda-1)\,(2+(\lambda-1))\,(1-a)+ \\
& +(\lambda-1)\left(\frac{1}{2}+\frac{9\,(\lambda-1)}{4}+\frac{7(\lambda-1)^2}{8}\right)(1-a)^2+... \\
& q_{ThPhM} \sim 1+(\lambda-1)-(\lambda-1)\,(2+(\lambda-1))\,(1-a)+ \\
& +(\lambda-1)\left(1+\frac{5\,(\lambda-1)}{2}+(\lambda-1)^2\right)(1-a)^2+...
\end{aligned}\right\} \quad \text{at}\quad a \to 1.$$

$$(5.39)$$

3) $\lambda \to \infty$:

$$\left.\begin{aligned}
& q \sim 1+2\left(1-\frac{2}{\lambda}+\frac{2}{\lambda^2}\right)a^2- \\
& -2\left(\frac{1}{\lambda}-\frac{2}{\lambda^2}\right)a^4+... \\
& q_{ThPhM} \sim 1+2\left(1-\frac{2}{\lambda}+\frac{2}{\lambda^2}\right)a^2+ \\
& +2\left(1-\frac{4}{\lambda}+\frac{8}{\lambda^2}\right)a^4+...
\end{aligned}\right\} \quad \text{at}\quad a \to 0, \qquad (5.40)$$

$$\left.\begin{aligned}
& q \sim \frac{1}{1-a}-\left(1+\frac{3}{\lambda}+\frac{4}{\lambda^2}\right)+ \\
& +\left(2+\frac{3}{2\lambda}+\frac{6}{\lambda^2}\right)(1-a)+... \\
& q_{ThPhM} \sim \frac{1}{1-a}-\left(\frac{1}{2}-\frac{1}{4\lambda}-\frac{1}{16\lambda^3}\right)+ \\
& +\left(\frac{1}{4}-\frac{1}{16\lambda^2}\right)(1-a)+...
\end{aligned}\right\} \quad \text{at}\quad a \to 1. \qquad (5.41)$$

Thus, from relations (5.36)–(5.41) it follows: the effective thermal conductivity parameter (5.23), (5.7), (5.21), obtained by matching the asymptotic representations 2PhM (5.2) and LA (5.3), gives for the homogenized thermal conductivity (5.1) found from ThPhM: upper bound for $\lambda \le 1$, and lower bound at $\lambda \ge 1$.

Figures 5.3–5.4 show graphs of the effective thermal conductivity calculated using the representation (5.23) and ThPhM (5.1); the Hashin–Shtrikman bounds are also shown for comparison:

for $1 \le \lambda < \infty$:

$$\frac{1-a^2+\lambda\,(1+a^2)}{1+a^2+\lambda\,(1-a^2)}=\underline{q}_{HS} \le q \le \bar{q}_{HS}=\lambda\,\frac{2-a^2+\lambda\,a^2}{a^2+\lambda\,(2\ a^2)}, \qquad (5.42)$$

for $0 \le \lambda \le 1$:

$$\lambda\,\frac{2-a^2+\lambda\,a^2}{a^2+\lambda\,(2-a^2)}=\underline{q}_{HS} \le q \le \bar{q}_{HS}=\frac{1-a^2+\lambda\,(1+a^2)}{1+a^2+\lambda\,(1-a^2)}. \qquad (5.43)$$

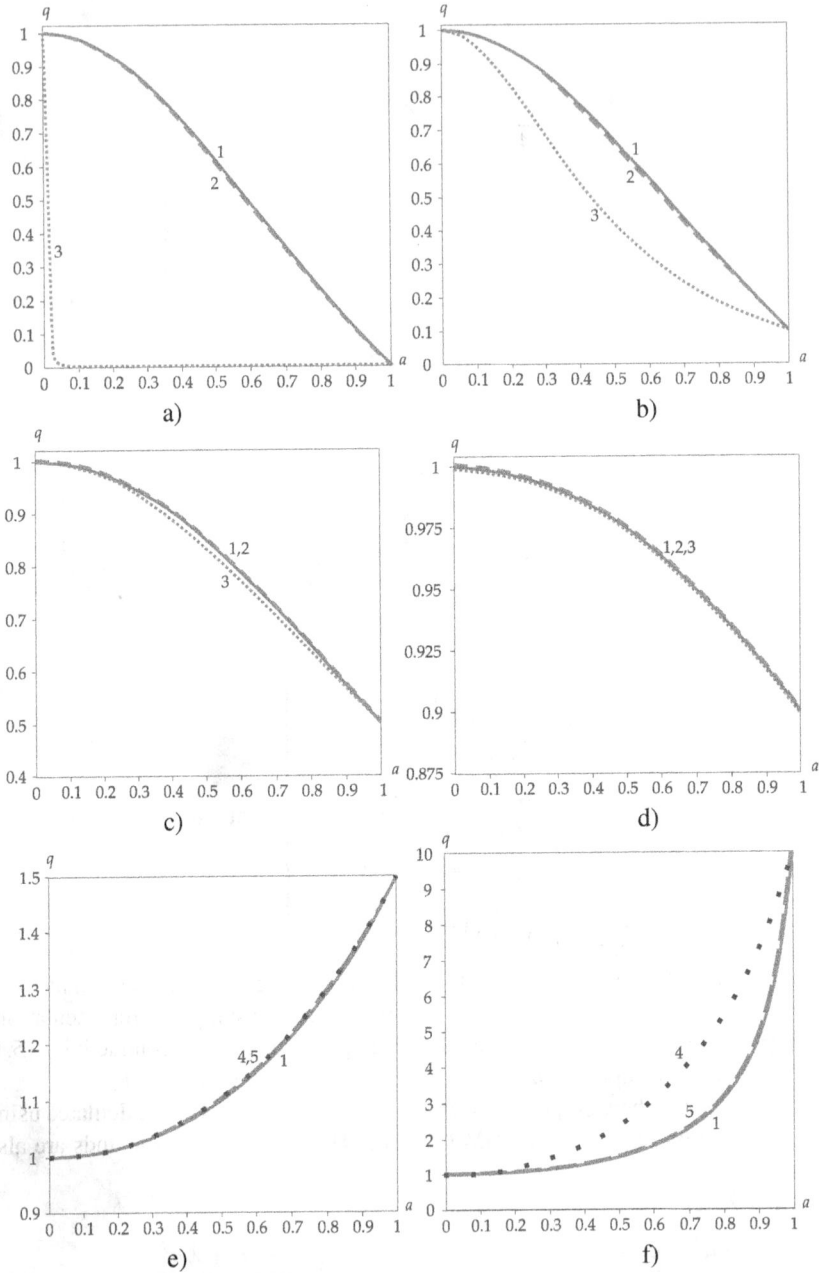

Figure 5.3 Effective thermal conductivity at: a) $\lambda = 10^{-5}$, b) $\lambda = 0.1$, c) $\lambda = 0.5$, d) $\lambda = 0.9$, e) $\lambda = 1.5$, f) $\lambda = 10$ (1– q (5.23), 2– upper Hashin-Shtrikman bound (q_{ThPhM}), 3– lower Hashin-Shtrikman bound, 4– upper Hashin-Shtrikman bound, 5– lower Hashin-Shtrikman bound (q_{ThPhM})).

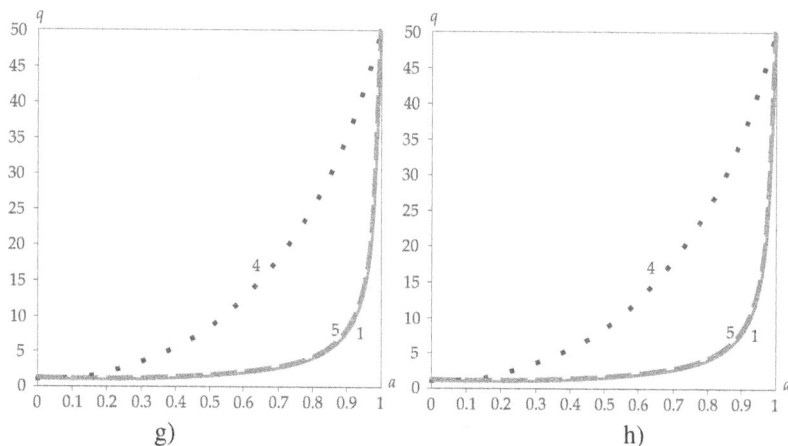

Figure 5.3 **(continued)**: g) $\lambda = 50$, h) $\lambda = 10^3$ (1– q (5.23), 4– upper Hashin-Shtrikman bound, 5– lower Hashin-Shtrikman bound (q_{ThPhM})).

Figures 5.3a-h present the given parameter as a function of the size of inclusions a at their fixed conductivity λ. The choice of values λ is due to the following considerations:

(i) $\lambda = 10^{-5}$ (Fig. 5.3a) detects practically non-conductive inclusions ($\lambda \to 0$);
(ii) $\lambda = 0.1$, $\lambda = 0.5$ (Fig. 5.3b and Fig. 5.3c, respectively) there is a case of inclusions of low conductivity ($\lambda << 1$);
(iii) values $\lambda = 0.9$, $\lambda = 1.5$ (Fig. 5.3d and Fig. 5.3e, respectively) correspond to the case when the conductivities of the matrix and inclusions are of the same order ($\lambda \sim 1$);
(iv) if $\lambda = 10$, $\lambda = 50$ (Fig. 5.3f and Fig. 5.3g, respectively), then we have a composite with inclusions of high conductivity ($\lambda >> 1$);
(v) values $\lambda = 10^3$ (Fig. 5.3h) corresponds to almost absolutely conductive inclusions ($\lambda \to \infty$).

Graphs in Figs. 5.4a–e present the homogenized coefficient depending on the conductivity of inclusions λ with their fixed geometric size a. The calculated values of the inclusion size a correspond to the cases:

(i) $a = 0.01$, $a = 0.5$ (Fig. 5.4a and Fig. 5.4b, respectively) – small inclusions ($a \to 0$);
(ii) $a = 0.5$ (Fig. 5.4c) – inclusions of medium size;
(iii) $a = 0.75$, $a = 0.99$ (Fig. 5.4d and Fig. 5.4e, respectively) – large inclusions ($a \to 1$).

Fig. 5.5 illustrates a qualitative picture of the dependence of the effective thermal conductivity coefficient q, found using the relation (5.23) with regard to two parameters: geometric with respect to the size of the inclusions a, and physical with regard to the conductivity of the inclusions λ.

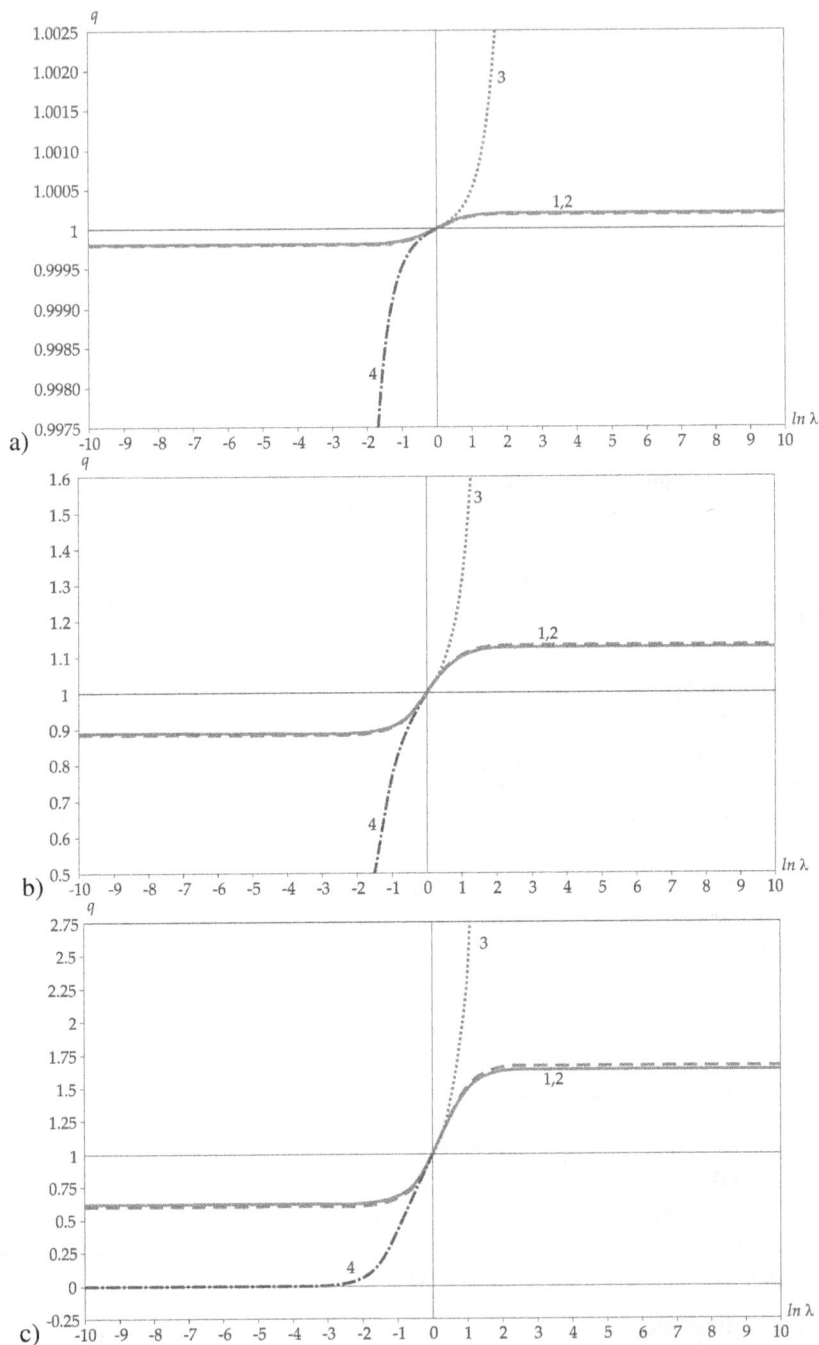

Figure 5.4 Effective thermal conductivity at: a) $a = 0.01$, b) $a = 0.25$, c) $a = 0.5$ (1– q (5.23), 2– lower (upper) Hashin-Shtrikman bound (q_{ThPhM}) for $1 \leq \lambda < \infty$ $(0 \leq \lambda < 1)$, 3– upper Hashin-Shtrikman bound, 4– lower Hashin-Shtrikman bound).

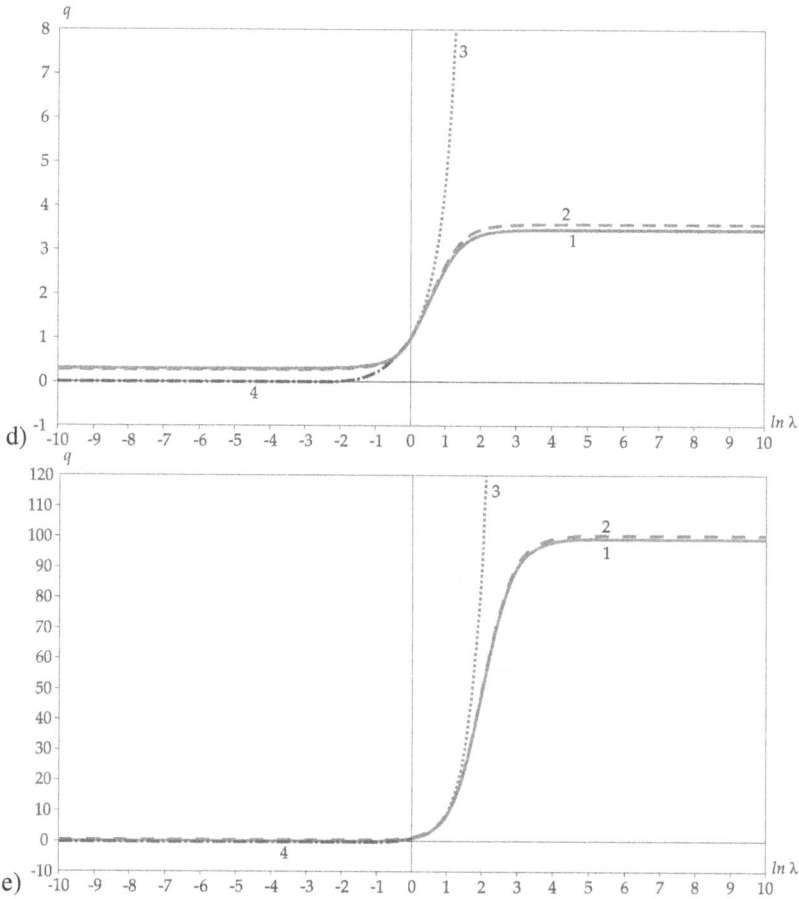

Figure 5.4 (**continued**): d) $a = 0.75$, e) $a = 0.99$.

Fig. 5.6 presents the regions schematically defined in which the main contribution to the expression of the effective thermal conductivity coefficient is made by the corresponding asymptotic representations: q – (5.23), (5.7), (5.21); $q^{(0)}$ – (5.19), (5.7); $q^{(\infty)}$ – (5.12), (5.7); q_{2PhM} – (5.2); $q_{2PhM}^{(0)}$ (5.14); $q_{2PhM}^{(\infty)}$ – (5.2); q_{LA} – (5.3); $q_{LA}^{(0)}$ – (5.13); $q_{LA}^{(\infty)}$ – (5.3).

The analysis of the presented asymptotic expansions and graphical results confirm the sufficient accuracy of the calculations of the effective thermal conductivity parameter, constructed by matching the asymptotic relations, in comparison with the calculations using ThPhM. The maximum descripancies in comparison with that calculated from ThPhM are less than 5%.

5.1.6 A NOTE ON THE EFFECTIVENESS OF HASHIN-SHTRIKMAN BOUNDS

Unfortunately, the Hashin-Shtrikman bounds are not always effective for estimating the effective parameters. For example, let there be two composites with cylindrical

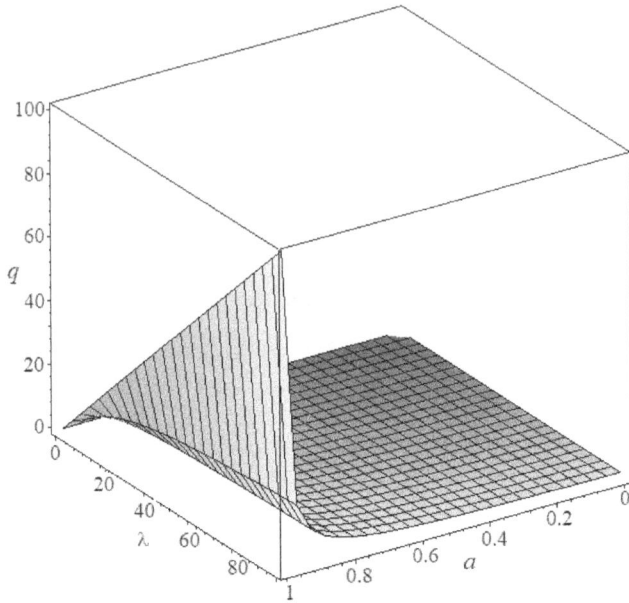

Figure 5.5 Dependence of the effective thermal conductivity coefficient q on the size of inclusions a and their conductivity λ.

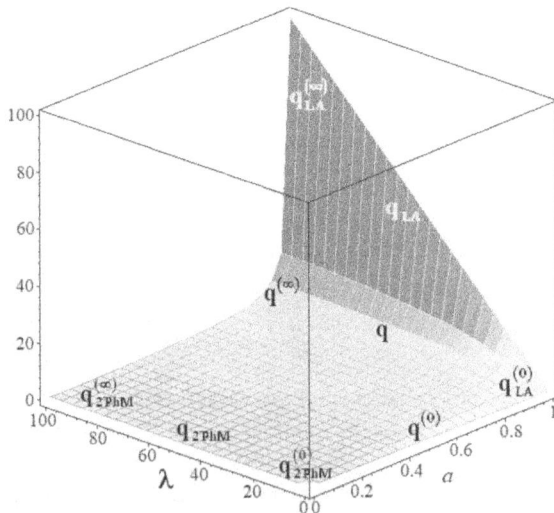

Figure 5.6 Qualitative picture of the distribution of the regions of change of the asymptotic relations of the effective coefficient of thermal conductivity.

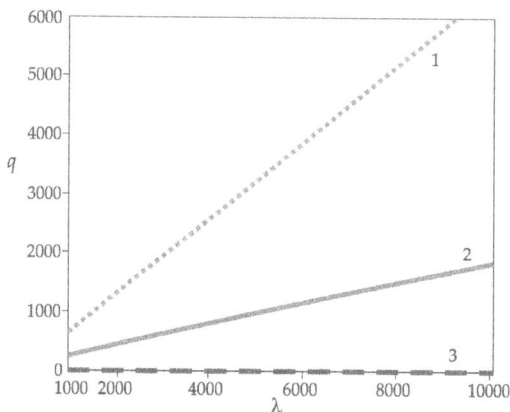

Figure 5.7 Effective parameters of composites with square and circle inclusions, $c = \pi/4$ ($1 - \bar{q}_{HS}$, $2 - q^{(b)}$, $3 - q^{(a)} \equiv \underline{q}_{HS}$).

inclusions, the same concentration of inclusions and the same conductivity – composites (a) and (b). Obviously, for the given parameters of these composites, the Hashin-Shtrikman bounds are the same.

In what follows, we take

$$c = \frac{\pi}{4}, \quad 10^3 \le \lambda \le 10^4;$$

(i) composite (a) with square inclusions and composite (b) with circle inclusions (see Fig. 5.7);
(ii) effective thermal conductivity parameter of the composite (a), found by the ThPhM (5.1): $q^{(a)} = q_{ThPhM}$;
(iii) effective thermal conductivity parameter $q^{(b)}$ of the composite (b) found from the results of [104];
(iv) Hashin–Shtrikman bounds (5.42): upper bound $\bar{q}_{HS}$, lower bound $\underline{q}_{HS}$.

We obtain

$$q^{(a)}\Big|_{\lambda=10^3} = 8.2519, \qquad q^{(a)}\Big|_{\lambda=10^4} = 8.3128,$$

$$q^{(b)}\Big|_{\lambda=10^3} = 244.1699, \qquad q^{(b)}\Big|_{\lambda=10^4} = 1816.7255,$$

$$\underline{q}_{HS}\Big|_{\lambda=10^3} = 8.2519, \qquad \underline{q}_{HS}\Big|_{\lambda=10^4} = 8.3128,$$

$$\bar{q}_{HS}\Big|_{\lambda=10^3} = 647.2116, \qquad \bar{q}_{HS}\Big|_{\lambda=10^4} = 6466.8833.$$

i.e., the effective parameters of the composites $q^{(a)}$ and $q^{(b)}$ differ by more than 30-200 times, while the Hashin-Shtrikman bounds are the same, and difference between them is very large.

The expression above for the effective thermal conductivity coefficient q (5.23) was constructed for square inclusions and contains all possible asymptotics, for all limiting values of geometric and physical parameters. Comparison of solutions q_{ThPhM} (5.1) and q (5.23) for any parameter values a and λ, including the limit ones, shows that all cases the discrepancy between the results is insignificant (this, in particular, can be seen in Figs. 5.3–5.4). In fact, the matched expression for the effective coefficient q (5.23) allows us to show that for the case of square inclusions, the exact solution is extremely close to the Hashin-Shtrikman upper bound for $\lambda \leq 1$ and lower bound at $\lambda \geq 1$.

5.2 ASYMPTOTIC SOLUTIONS FOR A COMPOSITE MATERIAL WITH RHOMBIC INCLUSIONS

Extensive literature is devoted to the study of fibrous composites with inclusions of a circular cross section [53, 71, 161], and the issues of determining their effective characteristics have been studied quite deeply. A much smaller number of works are devoted to inclusions with corner points [52, 112, 162, 223].

At a low concentration of inclusions, their shape does not have a fundamental effect on the effective properties of the composite. However, when the sizes of inclusions are large and they tend to contact, the solution of the problem is far from trivial. Depending on the ratio of the parameters that determine the system, i.e., the size of the inclusions and the characteristics of the constituent phases of the composite, there are different asymptotics of the effective coefficient.

We consider a two phase micro-inhomogeneous material consisting of a continuous matrix and inclusions periodically located in it in the form of rhombic-section fibers that make up a simple square lattice. The problem of determining the effective thermal conductivity of the structure is solved under the assumption of large sizes and high conductivity of the inclusions. The solution of the problem is based on the use of the homogenization method followed by the application of asymptotic representations.

The problem is solved for composite materials of different structure. We consider:

1) rhombic inclusions of large sizes, approaching the contact, but not reaching it (matrix contact).
2) intersecting rhombic inclusions (presence of a small contact area of the inclusions).
3) case of contact of rhombic inclusions (existence of a point of contact of inclusions).
4) composites with equally represented phases.

At first glance, in the presence of a contact point of inclusions in the case under consideration, we arrive at a composite structure with equally represented phases

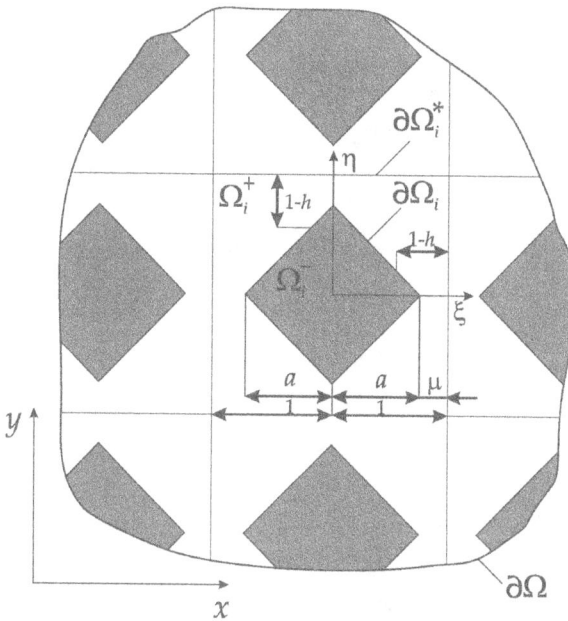

Figure 5.8 Composite with rhombic fibers.

[49], the effective parameter of which is determined by the well-known Dykhne formula [80]. A detailed analysis, however, shows that the situation here is more complicated, and the very concept of equally represented phases needs to be precisely defined.

On the example of the composite structure under consideration, an analysis of the possible limiting forms of a composite with rhombic inclusions was carried out, limit transitions were made, leading to the Dykhne solution [80], and the limits of applicability of the obtained expressions were determined.

5.2.1 NON-INTERSECTING RHOMBIC INCLUSIONS OF LARGE SIZES

Let us consider the problem of determining the effective thermal conductivity coefficient of a two-phase micro-inhomogeneous material with periodically arranged large-sized rhombic-section fibers forming a square lattice (Fig. 5.8).

Due to the symmetry of the structure, it is possible to separate the heat fluxes in the direction of the coordinate axes ξ, η, and consider one of them, for example, in the direction of the axis $O\eta$. Assuming large sizes $(a \gg 0)$ and high conductivity $(\lambda \gg 1)$ of inclusions we look for a solution to the problem using the lubrication approach (see Chapter 4). The calculation model for 1/2 cell is shown in Fig. 5.9. Let us denote the value of the gap between the inclusions through $2(1 - h(\xi))$ (Fig. 5.8). Then, problems (3.22)–(3.24) on the periodicity cell (Fig. 5.8) in "fast" variables

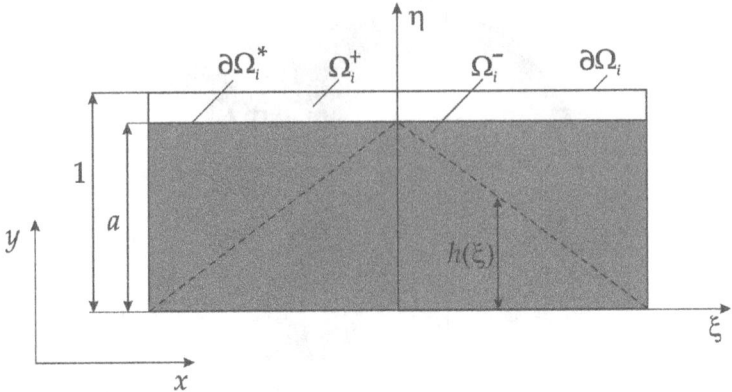

Figure 5.9 Calculation model of the lubrication approach for rhombic inclusions.

ξ, η will be written as follows:

$$\frac{\partial^2 u_1^+}{\partial \eta^2} = 0 \quad \text{in} \quad \Omega_i^+, \qquad \frac{\partial^2 u_1^-}{\partial \xi^2} + \frac{\partial^2 u_1^-}{\partial \eta^2} = 0 \quad \text{in} \quad \Omega_i^-, \tag{5.44}$$

$$u_1^+ = u_1^-, \qquad \frac{\partial u_1^+}{\partial \bar{n}} - \lambda \frac{\partial u_1^-}{\partial \bar{n}} = (\lambda - 1) \frac{\partial u_0}{\partial n} \quad \text{at} \quad \eta = h, \tag{5.45}$$

$$u_1^- = 0 \quad \text{at} \quad \eta = 0, \tag{5.46}$$

$$u_1^+ = 0 \quad \text{at} \quad \eta = 1, \tag{5.47}$$

where

$$h(\xi) = a - \xi. \tag{5.48}$$

The solution of the boundary value problem (5.44)–(5.48) yields

$$u_1^+ = \frac{(\lambda - 1) h (\eta - 1)}{\lambda - (\lambda - 1) h} \frac{\partial u_0}{\partial y}, \qquad u_1^- = -\frac{(\lambda - 1) (1 - h) \eta}{\lambda - (\lambda - 1) h} \frac{\partial u_0}{\partial y}. \tag{5.49}$$

Then, taking into account relations (5.48), (5.49), after performing the necessary transformations, the homogenized coefficient is obtained from the relation

$$q = \frac{1}{|\Omega_i^*|} \left[|\Omega_i^+| + \lambda |\Omega_i^-| + (\lambda - 1) \iint\limits_{\Omega_i^+} \frac{h(\xi)}{\lambda - (\lambda - 1) h(\xi)} d\xi d\eta + \right.$$

$$\left. + \lambda (\lambda - 1) \iint\limits_{\Omega_i^-} \frac{h(\xi) - 1}{\lambda - (\lambda - 1) h(\xi)} d\xi d\eta \right] = F, \tag{5.50}$$

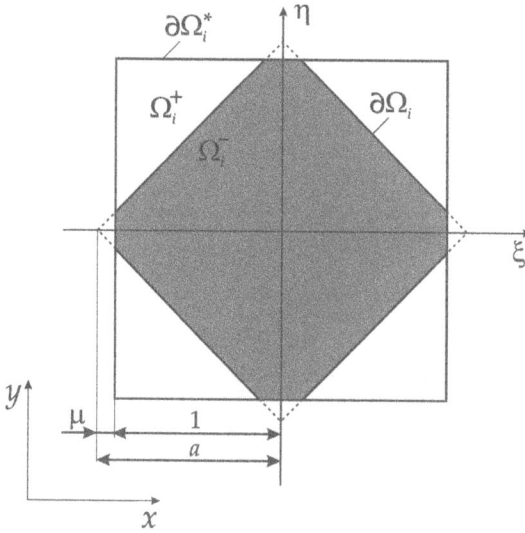

Figure 5.10 Intersecting rhombic inclusions: the presence of a small contact area of the inclusions.

as

$$q = 1 - a + \frac{\lambda}{\lambda - 1} \ln \left(\frac{\lambda}{\lambda - (\lambda - 1)a} \right). \tag{5.51}$$

By introducing a small parameter

$$\mu = |1 - a| \to 0 \quad \text{at} \quad a \to 1, \tag{5.52}$$

we rewrite (5.51) as an expression

$$q_{cm} = \mu + \frac{\lambda}{\lambda - 1} \ln \left(\frac{\lambda}{1 + (\lambda - 1)\mu} \right), \tag{5.53}$$

which can be considered as an approach to the contact point from the inside of the cell (at $a \to 1 - 0$), i.e., during the contact of the matrices.

5.2.2   INTERSECTING RHOMBIC INCLUSIONS: THE PRESENCE OF A SMALL CONTACT AREA OF THE INCLUSIONS

To approach the point of contact from the other side, i.e. the outer side of the cell (at $a \to 1 + 0$) in the presence of contact of inclusions, we consider the case of intersecting rhombic inclusions.

The periodicity cell for this case is shown in Fig. 5.10, and the value $h(\xi)$ is determined by formula

$$|h(\xi)| = \begin{cases} a - |\xi| & \text{at} \quad |\xi| \geq -\mu \\ 1 & \text{at} \quad |\xi| \leq -\mu. \end{cases} \tag{5.54}$$

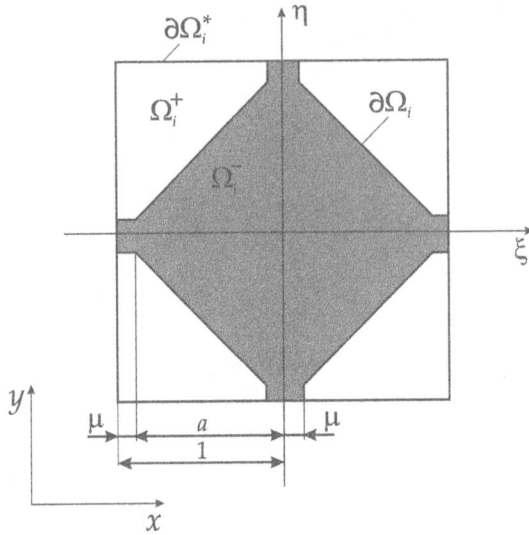

Figure 5.11 Rhombic inclusions contact: existence of an inclusion contact point.

According to formula (5.50), taking into account relation (5.54), the effective coefficient is determined by the expression

$$q_{cd} = \lambda\mu + \frac{\lambda}{\lambda - 1} \ln\left(\lambda - (\lambda - 1)\mu\right). \tag{5.55}$$

5.2.3 RHOMBIC INCLUSION CONTACT: EXISTENCE OF AN INCLUSION CONTACT POINT

To study the point contact between the inclusions, we approximate the rhombic shape of the inclusion by the region shown in Fig. 5.11. In this case, the value $h(\xi)$ is defined as follows

$$|h(\xi)| = \begin{cases} 1 & \text{at} \quad |\xi| \leq \mu, \\ 1 - |\xi| & \text{at} \quad \mu < |\xi| < 1 - \mu, \\ \mu & \text{at} \quad |\xi| \geq 1 - \mu. \end{cases} \tag{5.56}$$

By virtue of formula (5.50) and dependencies (5.56), we obtain the following expression for the effective coefficient:

$$q_{av} = \mu + \lambda\mu + \frac{\lambda}{\lambda - 1} \ln\left(\frac{\lambda - (\lambda - 1)\mu}{1 + (\lambda - 1)\mu}\right). \tag{5.57}$$

5.2.4 ASYMPTOTIC EXPRESSIONS FOR THE EFFECTIVE COEFFICIENT

We study the asymptotics of the effective thermal conductivity coefficient depending on the ratio of the quantities μ and λ. We introduce parameter α characterizing this

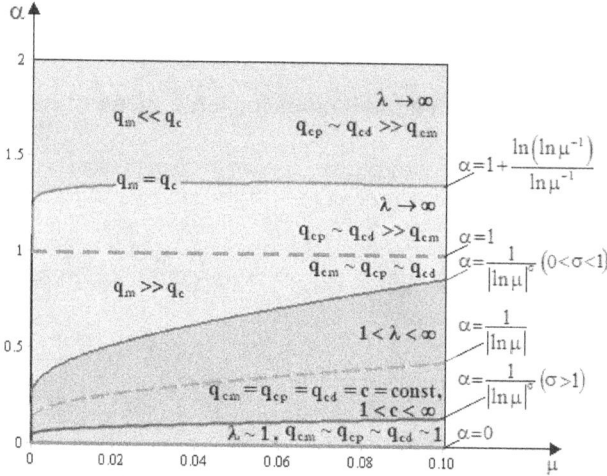

Figure 5.12 Illustration of the areas of applicability of asymptotic representations of the effective parameter.

relation as $\lambda \sim \mu^{-\alpha}$. By varying the value α, one can obtain different asymptotics of the solution in the entire range of μ and λ for $\mu \to 0$ (Table 5.1).

As can be seen from Fig. 5.12, all possible limit values are covered by μ and λ in the case of inclusions of large size and high conductivity.

Data in Table 5.1 shows that approximation (5.55) makes it possible to pass to the contact point and obtain the correct asymptotics of the solution for $a = 1$ for any values of quantity α. Approximation (5.53) at the point of contact correctly describes the behavior of the effective coefficient only at $\alpha \in [0; 1]$. For $\alpha > 1$ formula (5.53) cannot be used at the contact point, since the formulation of the original problem assumes the presence of an infinitely thin matrix layer between the inclusions.

Let us analyze the physical meaning of the resulting asymptotics. In the general case, the effective thermal conductivity coefficient consists of two components:

$$q = q_m + q_c, \tag{5.58}$$

where q_m is the contribution due to the heat flux through the matrix material; $q_c = \lambda \mu$ is the point of contact contribution.

Comparison given in Table 5.1 asymptotic relations shows (Fig. 5.12) that

(i) for $0 \le \alpha \le 1$ the contribution of the second term in (5.58) is insignificant, and the effective conductivity of the medium is determined mainly by the quantity q_m, i.e. due to the matrix:

$$q_m \gg q_c, \quad q \sim q_m;$$

Table 5.1
Asymptotics of the effective coefficient for large sizes of inclusions with high conductivity.

Asymptotics of the effective coefficient for composites with rhombic inclusions		
	Parameter order α	The value of the conductivity of inclusions $\lambda\sim\mu^{-\alpha}$
Matrix contact $a\to1-0$; q_{cm} (5.53)	Contact point of inclusions $a=1$; q_{cp} (5.57)	Contact area of inclusions $a\to1+0$; q_{cd} (5.55)
1.	$0\le\alpha\le\dfrac{1}{\|\ln\mu\|^\sigma},\sigma>1$	$\lambda\sim1$
1.1.	$\alpha=\mu$	$\lambda=\mu^{-\mu}$
$q_{cm}=1+\left(\frac12-\mu\right)(\lambda-1)$	$q_{cp}=1+2\mu+\frac12(\lambda-1)$	$q_{cd}=1+2\mu+\left(\frac12+\mu\right)(\lambda-1)$
1.2.	$\alpha=\dfrac{1}{\|\ln\mu\|^3}$	$\lambda=e^{\frac{1}{\ln^2\mu}}$
$q_{cm}=1+\left(\frac12-\mu\right)(\lambda-1)$	$q_{cp}=1+\frac12(\lambda-1)$	$q_{cd}=1+\left(\frac12+\mu\right)(\lambda-1)$
2.	$\alpha\sim\dfrac{1}{\|\ln\mu\|}$	$1<\lambda<\infty$
2.1.	$\alpha\sim\dfrac{1}{3\|\ln\mu\|}$	$\lambda=\sqrt[3]{e}$
$q_{cm}=\dfrac{\sqrt[3]{e}}{3(\sqrt[3]{e}-1)}$	$q_{cp}=\dfrac{\sqrt[3]{e}}{3(\sqrt[3]{e}-1)}$	$q_{cd}=\dfrac{\sqrt[3]{e}}{3(\sqrt[3]{e}-1)}$
2.2.	$\alpha=\dfrac{1}{\|\ln\mu\|}$	$\lambda=e$
$q_{cm}=\dfrac{e}{e-1}$	$q_{cp}=\dfrac{e}{e-1}$	$q_{cd}=\dfrac{e}{e-1}$
2.3.	$\alpha=\dfrac{3}{\|\ln\mu\|}$	$\lambda=e^3$
$q_{cm}=\dfrac{3e^3}{e^3-1}$	$q_{cp}=\dfrac{3e^3}{e^3-1}$	$q_{cd}=\dfrac{3e^3}{e^3-1}$
3.	$\dfrac{1}{\|\ln\mu\|^\sigma}\le\alpha\le1,\ 0<\sigma<1$	$\lambda\to\infty$
3.1.	$\alpha=\dfrac{1}{\sqrt[3]{\|\ln\mu\|}}$	$\lambda=\mu^{-\frac{1}{\sqrt[3]{\|\ln\mu\|}}}$
$q_{cm}=\dfrac{\lambda}{\lambda-1}\ln\lambda-\lambda\mu+\mu$	$q_{cp}=\dfrac{\lambda}{\lambda-1}\ln\lambda$	$q_{cd}=\dfrac{\lambda}{\lambda-1}\ln\lambda+\lambda\mu-\mu$
3.2.	$\alpha=1$	$\lambda=\dfrac{1}{\mu}$
$q_{cm}=\dfrac{\lambda}{\lambda-1}\ln\left(\dfrac{\lambda}{2}\right)$	$q_{cp}=1+\dfrac{\lambda}{\lambda-1}\ln\left(\dfrac{\lambda-1}{2}\right)$	$q_{cd}=1+\dfrac{\lambda}{\lambda-1}\ln(\lambda-1)$
4.	$\alpha=1+\sigma,\ \sigma\to+0$	$\lambda=\dfrac{1}{\mu^{1+\sigma}}$
4.1.	$\sigma=\dfrac{\ln(\ln\mu^{-1})}{\ln\mu^{-1}}$	$\lambda=\dfrac{\ln\mu^{-1}}{\mu}$
$q_{cm}=\dfrac{\lambda}{\lambda-1}\ln\dfrac{\lambda}{1+\lambda^\sigma}$	$q_{cp}=\lambda^\sigma+\dfrac{\lambda}{\lambda-1}\ln\dfrac{\lambda-\lambda^\sigma}{1+\lambda^\sigma}$	$q_{cd}=\lambda^\sigma+\dfrac{\lambda}{\lambda-1}\ln(\lambda-\lambda^\sigma)$
5.	$1<\alpha<\infty$	$\lambda\to\infty$
$q_{cm}=\dfrac{1}{\alpha}\dfrac{\lambda}{\lambda-1}\ln\lambda$	$q_{cp}=\lambda^{1-1/\alpha}+\dfrac{1}{\alpha}\dfrac{\lambda}{\lambda-1}\ln\lambda$	$q_{cd}=\lambda^{1-1/\alpha}+\dfrac{\lambda}{\lambda-1}\ln\lambda$

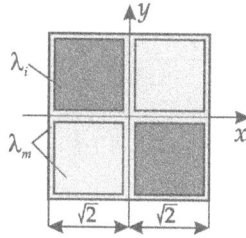

Figure 5.13 Composite (m), whose phase contact lines have matrix conductivity.

(ii) for $\alpha \sim 1 + \frac{\ln(\ln\mu^{-1})}{\ln\mu^{-1}}$ contribution to the effective conductivity of the matrix q_m and points of contact q_c is equivalent:

$$q_m \sim q_c, \qquad q \sim q_m \sim q_c;$$

(iii) for $1 < \alpha < \infty$ the main term of the asymptotics is determined by the presence of the contact point of the inclusions:

$$q_m \ll q_c, \qquad q \sim q_c.$$

The results obtained have a clear physical interpretation. At finite values of the conductivity of the inclusions, the main value of the heat flux is directed along the areas, including the matrix material, due to their much larger geometric sizes compared to the contact area. On the other hand, when the conductivity of the inclusions increases ($\lambda \to \infty$), ideally conducting clusters are formed in the composite, passing through the contact point of the inclusions, through which the heat fluxes are directed, which determine the overall effective conductivity of the structure.

5.2.5 COMPOSITES WITH EQUALLY REPRESENTED PHASES AND DYKHNE STRUCTURE

The composite structures considered above in the case of a point contact between inclusions are composites with two phases and equal concentration [49, 80]. The seeming contradiction between the obtained expressions and the well-known Dykhne formula [80] actually points to the different physical nature of the composites and, accordingly, different mathematical description of the limit transitions leading to chess-like structures.

Due to the symmetry of chess-like structures, three different limiting forms are possible for them (unlike any other types of inclusions), when there are only two such forms, i.e. an inclusion contact or a matrix contact. In what follows, consider different types of chess structures.

1) Phase contact lines have matrix conductivity (Fig. 5.13). The conductivity of such a structure (let us call it the structure (m)) in the direction, for example, of the axis x $\left(y = const, \ 0 < y < \sqrt{2} \right)$ within the selected element

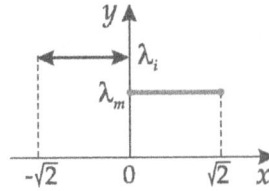

Figure 5.14 Geometric representation of the function (5.59) of the conductivity of the composite.

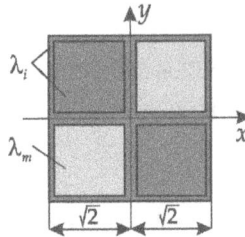

Figure 5.15 Composite (i), the phase contact lines of which have an inclusion conductivity.

is described by a discontinuous function of the form (Fig. 5.14):

$$\lambda = \begin{cases} \lambda_i, & -\sqrt{2} < x < 0, \\ \lambda_m, & 0 \leq x \leq \sqrt{2}, \end{cases} \tag{5.59}$$

where λ_i, λ_m are conductivity of inclusions and matrix, respectively.
For the composite structure under consideration, in the limiting case $\lambda_i = 0$ (non-conductive inclusions), the effective conductivity is not equal to zero due to the possibility of heat flux along the phase contact lines. The effective coefficient of such a structure, which we denote by q_m, according to the relation (5.53), is determined by an expression of the form

$$q_m = \lambda_m (1-a) + \frac{\lambda_i \lambda_m}{\lambda_i - \lambda_m} \ln \frac{\lambda_i}{\lambda_m a + \lambda_i (1-a)}. \tag{5.60}$$

2) The phase contact lines have an inclusion conductivity (structure (i), Fig. 5.15). The conductivity function corresponding to such a structure has the form (Fig. 5.16):

$$\lambda = \begin{cases} \lambda_i, & -\sqrt{2} \leq x \leq 0, \\ \lambda_m, & 0 < x < \sqrt{2}. \end{cases} \tag{5.61}$$

Obviously, in the limiting case $\lambda_i = 0$, due to the absence of contact between the elements of the matrix, such a structure as a whole will be non-conductive. The effective coefficient q_i of the described composite array

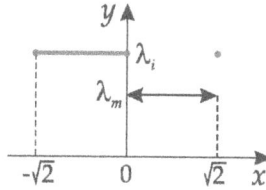

Figure 5.16   Geometric representation of the function (5.61) of the conductivity of the composite.

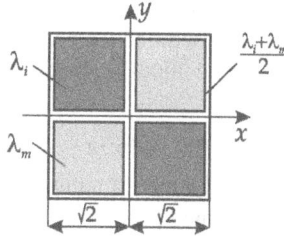

Figure 5.17   Dykhne structure (D): matrix and inclusions are separated by a phase line.

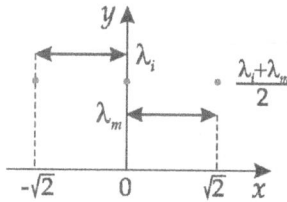

Figure 5.18   Geometric representation of the function (5.63) of the conductivity of the composite.

according to relations (5.55), (5.52) is determined by the formula

$$q_i = \lambda_i (a - 1) + \frac{\lambda_i \lambda_m}{\lambda_i - \lambda_m} \ln \frac{\lambda_i (2 - a) + \lambda_m (a - 1)}{\lambda_m}. \qquad (5.62)$$

Note that the considered composite systems transform into one another when replacing $\lambda_i \rightleftarrows \lambda_m$, $a \rightleftarrows (2 - a)$. However, in both cases it is impossible to speak of equal representation of the phases, since the replacement $\lambda_i \rightleftarrows \lambda_m$, $a \rightleftarrows (2 - a)$ leads to a fundamental change in the structure of the composite material. Thus, in these cases, the Dykhne formula is not applicable.

3) Dykhne (D) structure: the matrix and inclusions are separated from each other by a phase separation line – this is the physical interpretation of the composite structure presented in Dykhne's work [80] (Fig. 5.17).

The conductivity function of such a structure (let us designate it as the structure (D)) is shown in Fig. 5.18 Moreover, on the phase separation lines, this function is mathematically determined by the average value of its limits on the left and right

$$
\lambda =
\begin{cases}
\lambda_i, & -\sqrt{2} < x < 0, \\
\dfrac{\lambda_i + \lambda_m}{2}, & x = \pm\sqrt{2}, \ x = 0, \\
\lambda, & 0 < x < \sqrt{2}.
\end{cases}
\tag{5.63}
$$

The fundamental difference between the Dykhne structure and the cases considered above is the equal representation of the phases, which means that the replacement $\lambda_i \rightleftarrows \lambda_m$ does not lead to a change in either the geometric or physical characteristics of the composite, and the system then transforms itself (self-dual media).

The effective conductivity of q_D composite structures of this type is determined by the Dykhne formula [80]

$$
q_D = \sqrt{\lambda_i \lambda_m}.
\tag{5.64}
$$

In the limiting case $\lambda_i = 0$ in the Dykhne structure, some infinitely thin framework of nonzero conductivity remains, which preserves the conductivity of the structure. Therefore, although $\lambda_i \to 0$ we get $q_m \to 0$ and $q_D \to 0$, but asymptotically we have the estimate

$$
q_D \gg q_m \quad \text{at} \quad \lambda_i \to 0.
$$

The above considerations allow us to conclude that this Dykhne structure can be considered as the limiting case of a composite material with rhombic inclusions. Thus, the Dykhne formula can be derived from the previously obtained relations (5.60), (5.62) with the appropriate passage to the limit.

To carry out the passage to the limit on the Dykhne structure (Fig. 5.17), we use relations (5.60), (5.62) with $a = 1$ with replacing λ_i, λ_m to $\tilde{\lambda}_i$, $\tilde{\lambda}_m$, in the following way

$$
\begin{aligned}
\tilde{\lambda}_i &= \frac{\lambda_i + \lambda_m}{2} + sign\,(1 - a)\,\frac{\lambda_i - \lambda_m}{2}, \\
\tilde{\lambda}_m &= \frac{\lambda_i + \lambda_m}{2} - sign\,(1 - a)\,\frac{\lambda_i - \lambda_m}{2},
\end{aligned}
\tag{5.65}
$$

and hence near the contact line of the phases of the composite, we have

$$
\tilde{\lambda}_i = \tilde{\lambda}_m = \frac{\lambda_i + \lambda_m}{2} \quad \text{at} \quad a = 1.
$$

Then, taking into account expressions (5.65), from formulas (5.60), (5.62) for the effective conductivity of the Dykhne structure, we obtain:

$$
q = q_D = \frac{\tilde{\lambda}_i \tilde{\lambda}_m}{\tilde{\lambda}_i - \tilde{\lambda}_m} \ln \frac{\tilde{\lambda}_i}{\tilde{\lambda}_m}
\tag{5.66}
$$

Due to the asymptotic representation [2]

$$
\ln z \sim \sqrt{z} - \frac{1}{\sqrt{z}} \quad \text{at} \quad z \to 1,
$$

relation (5.66) is transformed to the following form

$$q_D = \frac{\tilde{\lambda}_i \tilde{\lambda}_m}{\tilde{\lambda}_i - \tilde{\lambda}_m} \ln \frac{\tilde{\lambda}_i}{\tilde{\lambda}_m} = \frac{\tilde{\lambda}_i \tilde{\lambda}_m}{\tilde{\lambda}_i - \tilde{\lambda}_m} \frac{\frac{\tilde{\lambda}_i}{\tilde{\lambda}_m} - 1}{\sqrt{\frac{\tilde{\lambda}_i}{\tilde{\lambda}_m}}} = \frac{\tilde{\lambda}_i \tilde{\lambda}_m}{\tilde{\lambda}_i - \tilde{\lambda}_m} \frac{\tilde{\lambda}_i - \tilde{\lambda}_m}{\sqrt{\tilde{\lambda}_i \tilde{\lambda}_m}} = \sqrt{\tilde{\lambda}_i \tilde{\lambda}_m} =$$

$$= \sqrt{\left(\frac{\lambda_i + \lambda_m}{2}\right)^2 - \left(\frac{\lambda_i - \lambda_m}{2}\right)^2} = \sqrt{\lambda_i \lambda_m},$$

i.e. coincides with the Dykhne formula (5.64).

5.2.6 PHYSICAL EQUIVALENCE OF CHESS COMPOSITE ARRAYS

It has been established that for chess-like composite materials there are three different limiting shapes determined by the different geometric structure of the composites.

In this regard, the question arises about the possible physical equivalence of these forms and the conditions under which such dependences take place. In other words, it is necessary to find the ratios of physical and geometrical parameters for which the effective coefficients q_m (5.60) and q_i (5.62) corresponding structures will asymptotically coincide with the effective Dykhne coefficient (5.64). Further, without loss of generality, we assume

$$\lambda_m = 1, \qquad \lambda = \frac{\lambda_i}{\lambda_m} = \lambda_i,$$

and therefore, relations (5.60), (5.62) take the form, respectively

$$q_m = 1 - a + \frac{\lambda}{\lambda - 1} \ln \frac{\lambda}{a + \lambda (1 - a)}, \qquad (5.67)$$

$$q_i = \lambda (a - 1) + \frac{\lambda}{\lambda - 1} \ln (\lambda (2 - a) + (a - 1)). \qquad (5.68)$$

Then we find:

1) $\lambda \sim 1$ – the structure is close to homogeneous. In this case, the result is obvious:

$$q_m \sim q_i \sim q_D \sim \sqrt{\lambda} \sim 1 \quad \text{at} \quad \lambda \sim 1, \qquad (5.69)$$

because at $a \to 1$, we have

$$q_m \sim q_i \sim \frac{\lambda}{\lambda - 1} \ln \lambda = \frac{\lambda}{\lambda - 1} \left(\sqrt{\lambda} - \frac{1}{\sqrt{\lambda}}\right) = \frac{\lambda}{\lambda - 1} \frac{\lambda - 1}{\sqrt{\lambda}} = \sqrt{\lambda}.$$

2) $\lambda \sim \frac{1}{(a-1)^2}$ – the conductivity of the inclusions is large, but finite. Therefore, from (5.67), (5.68) we have

$$q_m \sim \frac{1}{2} \frac{\lambda}{\lambda - 1} \ln \lambda, \qquad q_i \sim \sqrt{\lambda} + \frac{\lambda}{\lambda - 1} \ln \lambda,$$

i.e.

$$q_i \sim q_D, \quad q_m \ll q_D, \quad q_m \ll q_i \quad \text{for} \quad \lambda \sim \frac{1}{(a-1)^2}, \tag{5.70}$$

3) $\lambda \gg \frac{1}{(a-1)^2}$ – perfectly conducting inclusions. In this case, heat propagation along the phase contact lines becomes the main factor determining the effective conductivity of the composite, i.e., the following estimates take place

$$q_i \gg q_D \gg q_m \quad \text{at} \quad \lambda \gg \frac{1}{(a-1)^2}. \tag{5.71}$$

For example, at $\lambda \sim (1-a)^{-3}$ we have

$$q_i \sim \sqrt[3]{\lambda^2}, \quad q_D = \sqrt{\lambda}, \quad q_m \sim \frac{1}{3} \ln \lambda.$$

The results (5.69)–(5.71) indicate that with an increase in the conductivity of inclusions, the structure of chess-like composites has an even more significant effect on their effective conductivity:

(i)   when the conductivity of the inclusions is close to the conductivity of the matrix, the conductivity of the contact lines of the phases of the composite is not fundamental: the effective conductivities of the composites (m), (i), and (D) are asymptotic of the same order;

(ii)   at a high but finite conductivity of the inclusions, the contact of the phases of the composite over the matrix material significantly reduces its effective characteristic, while the main asymptotic term in the expressions for the effective parameters of the structures (i) and (D) is the same;

(iii)   for absolutely conducting inclusions, the main part of the heat flux is directed precisely along the phase contact lines, which continuously pass through the entire composite array, which causes a higher order of asymptotics in the expression for the effective conductivity of the composite (i) compared to the structure (D), and even more so compared to structure (m).

5.3   MODELS OF TWO-PHASE FIBROUS COMPOSITES: ASYMPTOTIC APPROXIMATIONS AND EQUIVALENCE OF STRUCTURES

For various models of layered composites, an effective mathematical tool for analytical study is the technique of non-smooth transformation of the argument (τ-transformations). Initially, this method was developed for the case of a symmetric saw-tooth function [195, 197, 221] and was used to solve ODEs describing periodic processes with regularly localized singularities, for nonlinear systems with periodic pulsed excitation, in the case of excitations with a periodic series of discontinuities of the first kind, and more.

The generalization of the method, which takes into account the asymmetry of the transformation [197], enables one to significantly expand the range of its practical applications both in the theory of oscillations and waves [195], and in the study of the statics of elastic systems of a periodic structure (in particular, in solving periodic problems of the theory of elasticity for layered composites [196, 197]).

Further, within the framework of the homogenization theory, using the technique of τ- transformations [195, 197] and the apparatus of asymptotically equivalent functions (page 192), a number of problems are considered for fibrous composites of various structures, with inclusions of various shapes, conductivity, and phase contact conditions:

(i) determination of the effective thermal conductivity of high contrast densely packed composites with circle inclusions;
(ii) study of a composite material with non-conductive inclusions of large geometric sizes;
(iii) analysis of "matrix-inclusion" contact conditions in composite structures;
(iv) study of composites with a thin interlayer at the phase boundary;
(v) modeling of composites with curvilinear rhombic inclusions;
(vi) analysis of physically equivalent composite structures;
(vii) study of composites with a hexagonal array of circle inclusions.

5.3.1 MAIN PROVISIONS OF THE METHOD OF NON-SMOOTH TRANSFORMATION OF THE ARGUMENT

The essence of the method of non-smooth transformation of the argument (τ-transformations) consists in the possibility of transition for any 4ℓ – periodic function $f(x)$ from the original argument x to the new, $\tau = \tau\left(\frac{x}{\ell}\right)$, through relations that are identically valid for any values of x [197]:

$$f(x) = P(\tau) + Q(\tau)\tau',$$ (5.72)

where

$$P(\tau) = \frac{1+\theta}{2}\left\{(1+\theta)\, f\left[(1+\theta)\,\tau\ell\right] + (1-\theta)\, f\left[(2-(1-\theta)\,\tau)\,\ell\right]\right\},$$

$$Q(\tau) = \frac{1-\theta^2}{2}\left\{(1-\theta^2)\, f\left[(1+\theta)\,\tau\ell\right] - f\left[(2-(1-\theta)\,\tau)\,\ell\right]\right\}.$$ (5.73)

Here $\tau = \tau(x)$ is a piece-wise linear periodic function with period 4, defined on a period in the form (see Fig. 5.19):

$$\tau = \begin{cases} \dfrac{x}{1+\theta} & \text{at} \quad -(1+\theta) \leq x \leq 1+\theta, \\[2mm] -\dfrac{x-2}{1-\theta} & \text{at} \quad 1+\theta \leq x \leq 3-\theta, \end{cases}$$ (5.74)

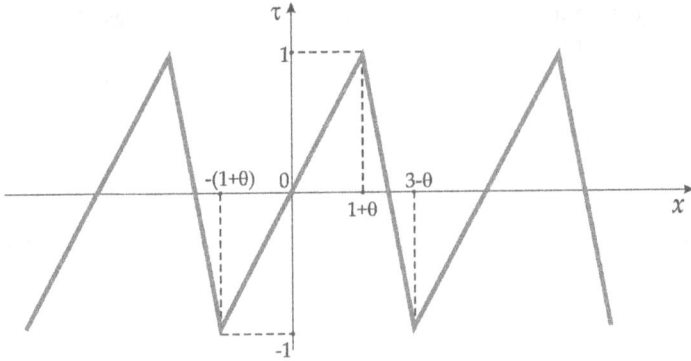

Figure 5.19 Saw-tooth periodic function $\tau = \tau(x)$.

and θ $(-1 \leq \theta \leq 1)$ stands for a parameter characterizing the inclination of the saw teeth.

The value $\theta = 0$ in expression (5.74) corresponds to the case of a saw-tooth function, symmetric with respect to a quarter of period $(x = 1)$.

We present the main properties of the sawtooth function $\tau = \tau(x)$ [197]:

1) The derivative of function τ' is defined by the relation:

$$\tau' = \begin{cases} \dfrac{1}{1+\theta} & \text{at} \quad -(1+\theta) < x < 1+\theta, \\[2mm] -\dfrac{\theta}{1-\theta^2} & \text{at} \quad x = \pm(1+\theta), \; x = 3-\theta, \\[2mm] -\dfrac{1}{1-\theta} & \text{at} \quad 1+\theta < x < 3-\theta \end{cases} \qquad (5.75)$$

and is graphically represented as shown in Fig. 5.20.

2) The square of the derivative of at $\theta \neq 0$ is a piecewise constant function with a periodic series of discontinuities of the first kind and is defined by the relation:

$$\tau'^2 = \frac{1}{1-\theta^2} - \frac{2\theta}{1-\theta^2}\tau' \qquad (5.76)$$

or (see Fig. 5.21):

$$\tau'^2 = \begin{cases} \dfrac{1-\theta}{(1+\theta)^2} & \text{at} \quad -(1+\theta) \leq x \leq 1+\theta \\[2mm] \dfrac{1+\theta^2}{(1-\theta^2)^2} & \text{at} \quad x = \pm(1+\theta), \quad x = 3-\theta \\[2mm] \dfrac{1+\theta}{(1-\theta)^2} & \text{at} \quad 1+\theta \leq x \leq 3-\theta \end{cases} \qquad (5.77)$$

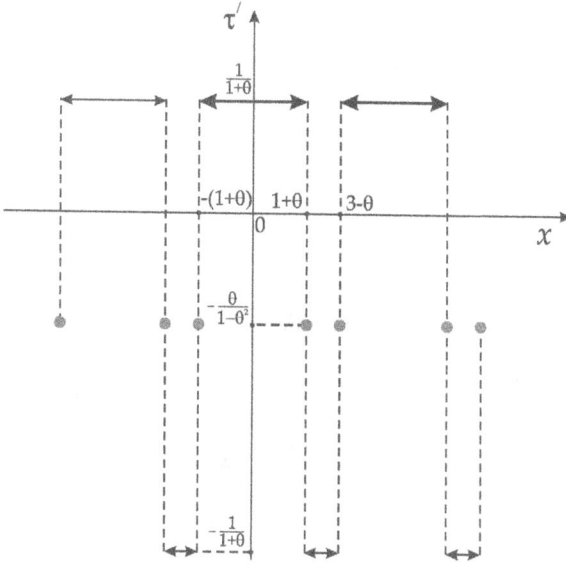

Figure 5.20   Derivative $\tau'(x)$ of the saw-tooth function.

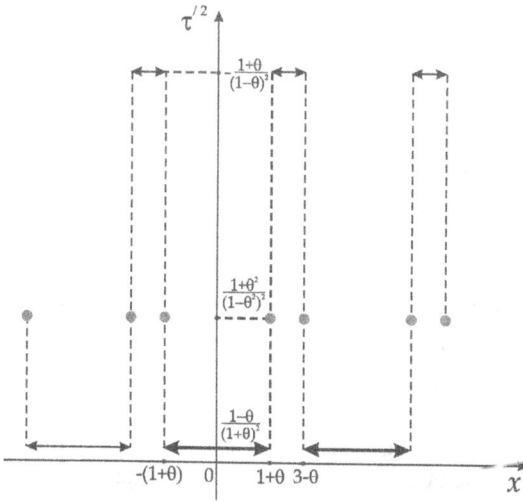

Figure 5.21   The square of the derivative of the saw-tooth function at $\theta \neq 0$.

Second derivative τ'' has singularity in the form of a Dirac delta functions

$$\tau'\tau'' = -\frac{\theta}{1-\theta^2}\tau'' \qquad\qquad (5.78)$$

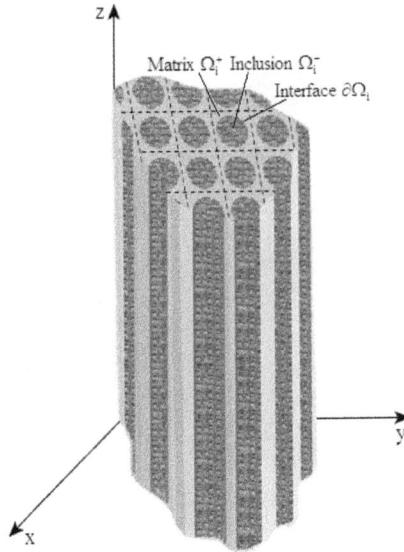

Figure 5.22 Composite with circle cylindrical inclusions forming a square lattice.

In the particular case of a symmetric saw-tooth function, we have $\theta = 0$, and expressions (5.73)–(5.76), (5.77) coincide with the corresponding relations given in [195].

In particular, the expressions for the square of the derivative τ'^2 (5.76), (5.77) take the form

$$\tau'^2 = 1 \quad \text{for all values of} \quad x.$$

5.3.2 HIGH CONTRAST DENSELY PACKED COMPOSITES

A two phase composite structure is considered, doubly periodic with a period that is the same in both directions, which is a matrix with cylindrical inclusions of a circle profile arranged in a square lattice (Fig. 5.22).

The problem of determining the effective thermal conductivity of such a structure is reduced to solving the Poisson's equation with conjugation conditions at the phase boundaries and the boundary condition at the outer contour of the composite:

$$\lambda_m \Delta u^+ = F \quad \text{in} \quad \Omega_i^+, \qquad \lambda_i \Delta u^- = F \quad \text{in} \quad \Omega_i^-, \tag{5.79}$$

$$u^+ = u^-, \quad \lambda_m \frac{\partial u^+}{\partial \mathbf{n}} = \lambda_i \frac{\partial u^-}{\partial \mathbf{n}} \quad \text{at} \quad \partial \Omega_i, \tag{5.80}$$

$$C_1 u^+ + C_2 \frac{\partial u^+}{\partial \mathbf{n}} = 0 \quad \text{at} \quad \partial \Omega, \tag{5.81}$$

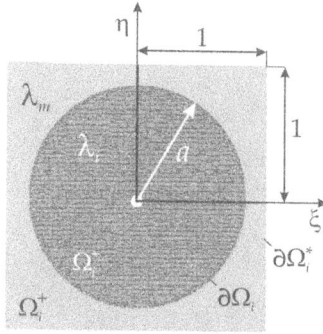

Figure 5.23 Typical cell of a composite with circle inclusions.

where u^+, u^- are the temperature distribution functions in the matrix Ω_i^+ and inclusions Ω_i^-; λ_m and λ_i are the thermal conductivity of the phases of the composite array; F is the heat source density; $\partial\Omega_i$, $\partial\Omega$ are the contours of inclusions and the outer contour of the structure, respectively; n is the outer normal to the contour; C_1, C_2 are the constants.

The application of the homogenization method, based on the technique of multiple scale expansions, allows us to reduce the original problem (5.79)–(5.81) to the problem for the characteristic cell of the composite (Fig. 5.23):

$$\frac{\partial^2 u_1^+}{\partial \xi^2} + \frac{\partial^2 u_1^+}{\partial \eta^2} = 0 \quad \text{in} \quad \Omega_i^+, \qquad \frac{\partial^2 u_1^-}{\partial \xi^2} + \frac{\partial^2 u_1^-}{\partial \eta^2} = 0 \quad \text{in} \quad \Omega_i^-, \qquad (5.82)$$

$$u_1^+ = u_1^-, \qquad \frac{\partial u_1^+}{\partial \bar{n}} - \lambda \frac{\partial u_1^-}{\partial \bar{n}} = (\lambda - 1) \frac{\partial u_0}{\partial n} \quad \text{on} \quad \partial\Omega_i, \qquad (5.83)$$

$$u_1^+ \Big|_{\xi=-1} = u_1^+ \Big|_{\xi=1}, \qquad \frac{\partial u_1^+}{\partial \xi} \Big|_{\xi=-1} = \frac{\partial u_1^+}{\partial \xi} \Big|_{\xi=1}, \qquad (5.84)$$

$$u_1^+ \Big|_{\eta=-1} = u_1^+ \Big|_{\eta=1}, \qquad \frac{\partial u_1^+}{\partial \eta} \Big|_{\eta=-1} = \frac{\partial u_1^+}{\partial \eta} \Big|_{\eta=1}, \qquad (5.85)$$

where $\lambda = \frac{\lambda_i}{\lambda_m}$.

Let us obtain the solution of the problem on the cell (5.82)–(5.85) for composites with inclusions of large sizes ($a \to 1$) and high conductivity ($\lambda \gg 1$), using the lubrication approach.

We separate the heat fluxes in the direction of the coordinate axes ξ, η and consider one of them, for example, in the direction of the axis $O\eta$ (obviously, due to the symmetry of the problem, this is not important).

For inclusion sizes close to the maximum ($a = 1$), inclusions almost touch each other, i.e. $1 - a \to 0$. Consequently, within this region (nears almost touching inclusions of high conductivity), the maximum heat fluxes take place.

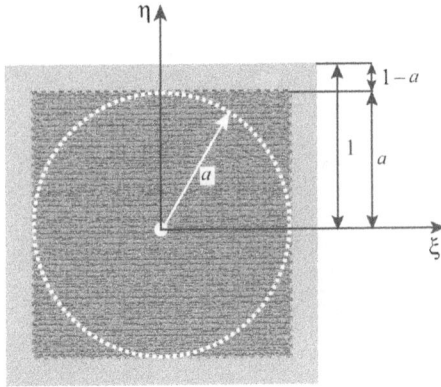

Figure 5.24 Circle inclusion transformation within LA.

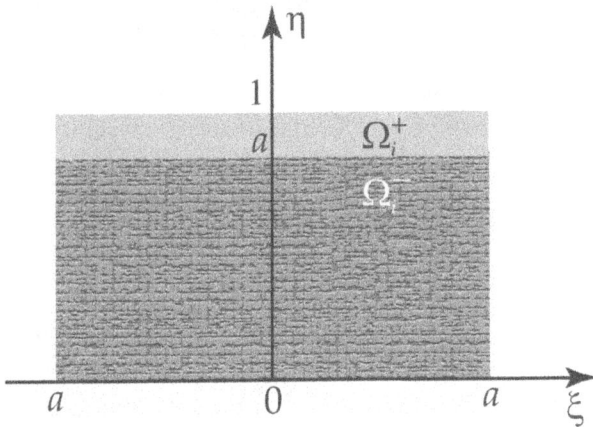

Figure 5.25 Calculation model of a cell according to LA.

These physical considerations determine the legitimacy and expediency of using LA, which allows us to consider a geometrically simpler problem, namely, to replace the circle contour of the inclusion with a square one with side a (Fig. 5.24). Due to the symmetry of the structure, we consider 1/2 of the transformed cell (for $\eta \geq 0$), denoting, respectively, the areas of the matrix and inclusions Ω_i^+ and Ω_{i1}^-. In accordance with the general idea of the LA, described in detail in Section 4.1.2, as the calculation model of a characteristic cell, we take its transformed domain, with the exception of the edge zones (Fig. 5.25). Thus, using the homogenization approach and the LA, the solution of the original problem is divided into two main stages.

1) At the first stage, the solution of the local problem is determined in the transformed area of the cell:

$$\frac{\partial^2 u_{11}^+}{\partial \eta^2} = 0 \quad \text{in} \quad \Omega_{i1}^+, \qquad \frac{\partial^2 u_{11}^-}{\partial \xi^2} + \frac{\partial^2 u_{11}^-}{\partial \eta^2} = 0 \quad \text{in} \quad \Omega_{i1}^-, \qquad (5.86)$$

$$u_{11}^+ = u_{11}^-, \qquad \frac{\partial u_1^+}{\partial \eta} - \lambda \frac{\partial u_1^-}{\partial \eta} = (\lambda - 1) \frac{\partial u_0}{\partial \eta} \quad \text{at} \quad \eta = a, \qquad (5.87)$$

$$u_{11}^- = 0 \quad \text{at} \quad \eta = 0, \qquad (5.88)$$

$$u_{11}^- = 0 \quad \text{at} \quad \eta = 1. \qquad (5.89)$$

The general solutions of equations (5.86) are:

$$u_{11}^+ = A_0 + B_0 \eta, \qquad (5.90)$$

$$u_{11}^- = C_0 + D_0 \eta + \sum_{n=1}^{\infty} \left[\left(C_n \cosh \pi n \eta + D_n \sinh \pi n \eta \right) \cos \pi n \xi + \right.$$
$$\left. + \left(\bar{C}_n \cosh \pi n \eta + \bar{D}_n \sinh \pi n \eta \right) \sin \pi n \xi \right], \qquad (5.91)$$

where A_0, B_0, C_n, D_n, $\bar{C}_n$, $\bar{D}_n$ $(n = 0, 1, 2, ...)$ are the constants determined from conditions (5.87) and (5.88), (5.89).
From the conjugation conditions (5.87), it follows that

$$C_n = D_n = \bar{C}_n = \bar{D}_n = 0, \qquad n = 1, 2, ... ,$$

since the expression (5.90) contains only zero terms in the expansion of the function u_{11}^+ in the Fourier series with respect to the variable ξ.
Further, due to the symmetry condition (5.88), we have

$$C_0 = 0.$$

The three arbitrary constants A_0, B_0, D_0 in expressions (5.90), (5.91) are determined from the three conditions (5.87), (5.88) as follows

$$A_0 = (1 - \lambda \Delta) \frac{\partial u_0}{\partial y}, \quad B_0 = -(1 - \lambda \Delta) \frac{\partial u_0}{\partial y}, \quad D_0 = -(1 - \Delta) \frac{\partial u_0}{\partial y}, \qquad (5.92)$$

where

$$\Delta = (a + \lambda (1 - a))^{-1}.$$

Expressions of functions u_{12}^+, u_{12}^- that determine the heat flux in the direction of axis $O\xi$ can be obtained in a similar way in the form

$$u_{12}^+ = u_{11}^+, \qquad u_{12}^- = u_{11}^- \left(\frac{\partial u_0}{\partial y} \to \frac{\partial u_0}{\partial x} \right).$$

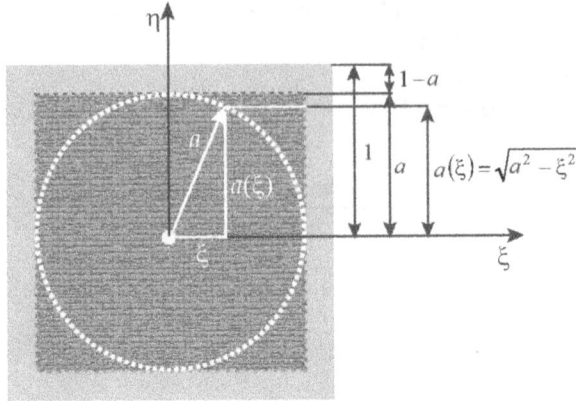

Figure 5.26 Approximation of a circle inclusion contour.

2) At the second stage, we consider the inclusion size a as a variable function of the coordinate ξ (Fig. 5.26):

$$a(\xi) = \sqrt{a^2 - \xi^2}. \tag{5.93}$$

Applying the averaging operator (3.25) to the equation

$$\frac{\partial^2 u_0}{\partial x^2} + \frac{\partial^2 u_0}{\partial y^2} + 2\frac{\partial^2 u_1^+}{\partial x \partial \xi} + 2\frac{\partial^2 u_1^+}{\partial y \partial \eta} + \frac{\partial^2 u_2^+}{\partial \xi^2} + \frac{\partial^2 u_2^+}{\partial \eta^2} +$$

$$+ \lambda \left(\frac{\partial^2 u_0}{\partial x^2} + \frac{\partial^2 u_0}{\partial y^2} + 2\frac{\partial^2 u_1^-}{\partial x \partial \xi} + 2\frac{\partial^2 u_1^-}{\partial y \partial \eta} + \frac{\partial^2 u_2^-}{\partial \xi^2} + \frac{\partial^2 u_2^-}{\partial \eta^2} \right) = F,$$

where

$$u_1^+ = u_{11}^+ + u_{12}^+, \qquad u_1^- = u_{11}^- + u_{12}^-,$$

we obtain the following homogenized equation

$$\frac{1}{|\Omega_i^*|} \left[\iint_{\Omega_i^+} \left(\frac{\partial^2 u_0}{\partial x^2} + \frac{\partial^2 u_0}{\partial y^2} + \frac{\partial^2 u_1^+}{\partial x \partial \xi} + \frac{\partial^2 u_1^+}{\partial y \partial \eta} \right) d\xi \, d\eta + \right.$$

$$\left. + \lambda \iint_{\Omega_i^-} \left(\frac{\partial^2 u_0}{\partial x^2} + \frac{\partial^2 u_0}{\partial y^2} + \frac{\partial^2 u_1^-}{\partial x \partial \xi} + \frac{\partial^2 u_1^-}{\partial y \partial \eta} \right) d\xi \, d\eta \right] = F. \tag{5.94}$$

Integration is performed over the original area of the cell, $\Omega_i^* = \Omega_i^+ \bigcup \Omega_i^-$, taking into account relation (5.93), i.e., considering the constants A_0, B_0, D_0 (5.92) as the functions of a variable ξ.

After performing the necessary transformations, we find expression for the effective thermal conductivity coefficient q (on condition $\lambda > \frac{a}{1+a}$)

$$q = 1 - a + \frac{\lambda}{\lambda - 1}\left(\frac{2\lambda \arctan \sqrt{\frac{\lambda(1+a)-a}{\lambda(1-a)+a}}}{\sqrt{(\lambda(1+a)-a)(\lambda(1-a)+a)}} - \frac{\pi}{2}\right) \quad (5.95)$$

at $\lambda \gg 1, 0 \ll a < 1$.

From relation (5.95), the following asymptotic representations can be obtained:

(i) for absolutely conducting inclusions of large geometric sizes, close to the limit

$$q = \frac{2 \arctan \sqrt{\frac{1+a}{1-a}}}{\sqrt{1-a^2}} + 1 - a - \frac{\pi}{2} \quad \text{at} \quad \lambda \to \infty, \quad 0 \ll a < 1, \ (5.96)$$

and in particular,

$$q = \frac{\pi}{\sqrt{1-a^2}} - \frac{\pi}{2} - 1 \quad \text{at} \quad \lambda \to \infty, \quad a \to 1; \quad (5.97)$$

(ii) for inclusions of almost maximal geometric size and high conductivity $\left(\lambda > \frac{1}{2}\right)$:

$$q = \frac{\lambda}{\lambda - 1}\left(\frac{2\lambda \arctan \sqrt{2\lambda - 1}}{\sqrt{2\lambda - 1}} - \frac{\pi}{2}\right) \quad \text{at} \quad a = 1, \quad \lambda \gg 1,$$

$$(5.98)$$

which for absolutely conducting inclusions is transformed to the form

$$q_{ci}^{(\infty)} = \pi\sqrt{\frac{\lambda}{2}} - \frac{\pi}{2} - 1 \quad \text{at} \quad a = 1, \quad \lambda \to \infty. \quad (5.99)$$

5.3.3 GENERALIZATIONS TO THE CASE OF NON-CONDUCTING INCLUSIONS OF LARGE GEOMETRIC SIZES

Using the Kellers' theorem (3.59), from relation (5.99), an expression for the effective thermal conductivity in the case of inclusions of low conductivity can be written as follows (on condition $\lambda < \frac{1+a}{a}$):

$$q = \frac{2(1-\lambda)\sqrt{(1+a-\lambda a)(1-a+\lambda a)}}{4 \arctan \sqrt{\frac{1+a-\lambda a}{1-a+\lambda a}} + (2(1-a)(1-\lambda)-\pi)\sqrt{(1+a-\lambda a)(1-a+\lambda a)}}$$

$$(5.100)$$

at $\lambda \ll 1, 0 \ll a < 1$.

Asymptotic expressions (5.96)–(5.99) are transformed to the following form

(i) non-conductive inclusions of large geometric sizes, close to the maximal possible:

$$q = \frac{\sqrt{1-a^2}}{2 \arctan \sqrt{\frac{1+a}{1-a}} + \left(1-a-\frac{\pi}{2}\right)\sqrt{1-a^2}} \qquad (5.101)$$

at $\lambda \to 0$, $0 << a < 1$;
in particular,

$$q = \frac{\sqrt{1-a^2}}{\pi - \left(\frac{\pi}{2}+1\right)\sqrt{1-a^2}} \qquad \text{at} \quad \lambda \to 0, \quad a \to 1; \qquad (5.102)$$

(ii) inclusions with low conductivity of large geometric sizes, close to the maximal possible (on condition $\lambda < 2$):

$$q = \frac{2(1-\lambda)\sqrt{\lambda(2-\lambda)}}{4 \arctan \sqrt{\frac{2-\lambda}{\lambda}} - \pi\sqrt{\lambda(2-\lambda)}} \qquad \text{at} \quad a = 1, \quad \lambda << 1 \quad (5.103)$$

or for practically non-conductive inclusions:

$$q_{ci}^{(0)} = \frac{\sqrt{2\lambda}}{\pi - \left(\frac{\pi}{\sqrt{2}}+\sqrt{2}\right)\sqrt{\lambda}} \qquad \text{at} \quad a = 1, \quad \lambda \to 0. \qquad (5.104)$$

5.3.4   ANALYSIS OF THE CONTACT CONDITIONS "MATRIX-INCLUSION"

We note that the leading term of the asymptotics (5.87), corresponding to the case of large absolutely conducting inclusions, coincides with the result obtained in [104]. However, the asymptotics (5.99) for the case of contact of inclusions does not coincide with that given in the paper [158]:

$$q = \frac{\pi\lambda}{2(\ln\lambda - \gamma)} \qquad \text{at} \quad a = 1, \quad \lambda \to \infty, \qquad (5.105)$$

where γ is Euler constant; $\gamma \approx 0.58$ [2].

In our opinion, such a discrepancy between the results is due to the difference in the geometric structure of the composites considered in both cases, which was originally included in the formulation of the problem. In [158], it is assumed that in the limiting case of inclusion sizes, they are in contact, i.e., there is a contact point of inclusions.

However, we consider an elementary cell of a composite (a matrix with an inclusion embedded in it), and the periodicity conditions for a matrix assume its repeatability. This means that in the limit, when the inclusion diameter is equal to the size of the side of the cell, some infinitely thin matrix layer remains at the contact of the inclusion. Generally speaking, due to the structural, technological or other features of the composite, three different limiting cases of their structure are possible:

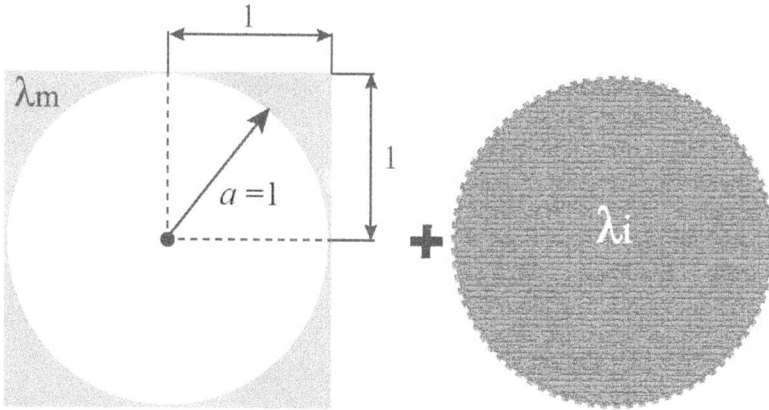

Figure 5.27 Composite structure: frame made of matrix material with embedded inclusions.

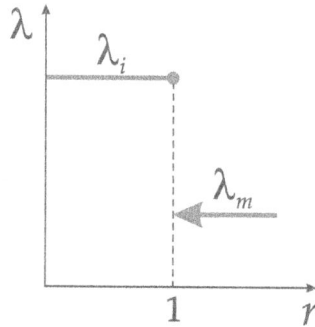

Figure 5.28 Conductivity function in the case of inclusions separated by a matrix interlayer.

(i) there is a frame made of the material with conductivity of the matrix λ_m, in which inclusions of conductivity λ_i are inserted. In this case, there is no contact between the inclusions: a thin layer of matrix material remains between them (Fig. 5.27);

(ii) frame made of material conductivity λ_i with inserted curvilinear rhombic inclusions with conduction λ_m. Then the contact of the inclusions is carried out (Fig. 5.29);

(iii) there is a thin layer (interface), the conductivity of which is equal to the average value of the conductivities of the materials λ_i and λ_m, i.e. $\frac{\lambda_i + \lambda_m}{2}$ (Fig. 5.31).

Let us explain the interface model used by us. In the manufacture of fibrous composites, due to the chemical, thermal, and mechanical interaction between the fibers and the matrix, an interfacial transition zone arises between them with characteristics that vary in the direction from the inclusion to the matrix [90, 91]. It is

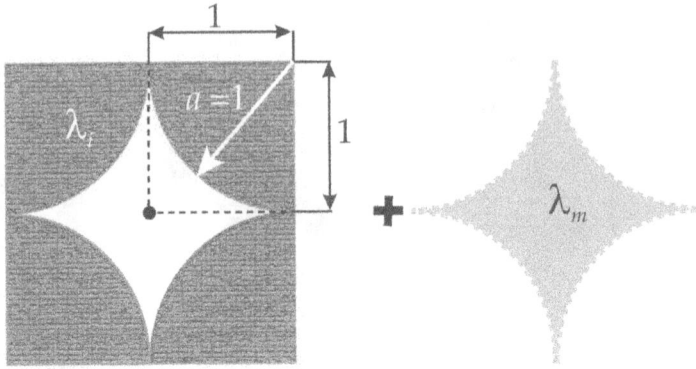

Figure 5.29 Composite structure: frame made of inclusion material with curvilinear rhombic inserts.

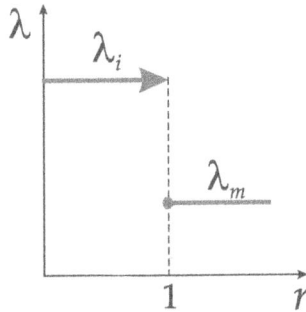

Figure 5.30 Conductivity function in case of contact of inclusions.

difficult to determine these characteristics experimentally; therefore, various approximations are used in calculations: linear, parabolic, hyperbolic, and logarithmic [1]. Such approximations significantly complicate the calculations; therefore, a number of authors proposed the idea of approximating radially variable properties by several layers (piecewise constant change in properties). This approximation works well in practice.

We use a single-layer variable interface approximation. In the latter case, it is natural to consider its properties as the arithmetic mean between the properties of the matrix and the fiber.

From a mathematical point of view, shown in Figs. 5.27, Fig. 5.29, and Fig. 5.31, cases of contact "matrix - inclusion" are described, respectively, by various generalized conductivity functions. The form of such dependences in an arbitrary radial section of the cell is schematically illustrated in Fig. 5.28, Fig. 5.30, and Fig. 5.32. Obviously, when determining the effective parameters of composites, the question of the presence or absence of an infinitely thin interlayer at the boundary of its phases is not fundamental until the sizes of the inclusions reach their limiting value, and

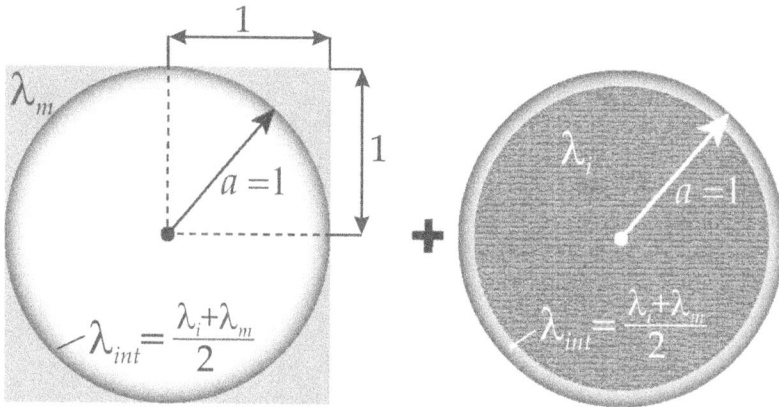

Figure 5.31 Composite structure: a thin interlayer at the phase boundary, characterized by an average value of matrix and inclusion conductivities.

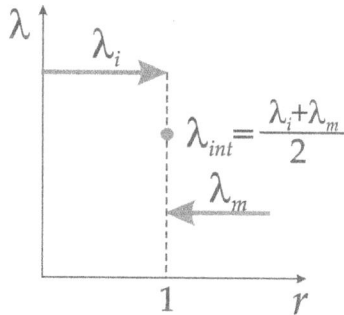

Figure 5.32 Conductivity function in the case of contact of phases along the interlayer with an average value of conductivities.

their conductivity does not approach infinitesimal or infinitely large. In this regard, it is of interest to study the asymptotics and the relationship between the effective parameters presented in Fig. 5.27, Fig. 5.29, and Fig. 5.31 composite structures.

Further, we use the technique of non-smooth transformation of the argument [195, 197], which makes it possible correctly describe local properties of inhomogeneous structures.

5.3.5 ASYMPTOTIC REPRESENTATION OF THE SOLUTION BY LA USING THE NON-SMOOTH TRANSFORMATION OF THE ARGUMENT

Consider a composite material shown in Fig. 5.33: inclusions of high conductivity ($\lambda \gg 1$) and large size ($a \to 1$) have a thin interlayer at the phase boundary, which is characterized by an average value of the conductivity of the matrix and inclusions, $\frac{\lambda_i + \lambda_m}{2}$.

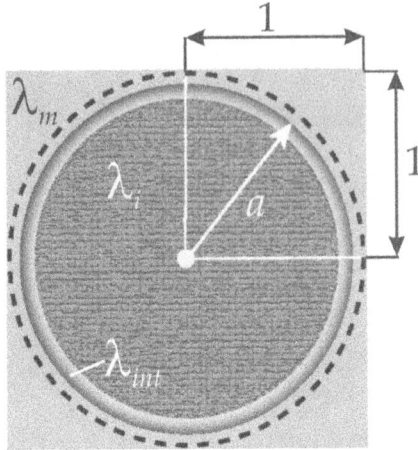

Figure 5.33 Composite with a thin interlayer at the phase boundary with an average conductivity of the matrix and inclusions.

Let us consider the conductivities of the inclusions and the matrix as functions of the size of the inclusions a, describing them by analytical relations of the form

$$\tilde{\lambda}_i = a\lambda_i + (1-a)\lambda_m + 2a(1-a)(\lambda_i - \lambda_m)\tau'_i, \qquad (5.106)$$

$$\tilde{\lambda}_m = a\lambda_i + (1-a)\lambda_m - 2a(1-a)(\lambda_i - \lambda_m)\tau'_m, \qquad (5.107)$$

where $\tau_i = \tau_i(\tilde{x})$, $\tau_m = \tau_m(\tilde{x})$, $\tilde{x} = 2x$ are introduced in accordance with expression (5.74) (saw-tooth functions of inclusions) and matrix (Fig. 5.34), transformed, taking into account the parameters of the inclination of the saw teeth

$$\theta_i = 2a - 1, \qquad \theta_m = 1 - 2a. \qquad (5.108)$$

The following notation has been employed

$$\tau_i(\tilde{x}) = \begin{cases} \dfrac{\tilde{x}}{2a} & \text{at} \quad -2a \le \tilde{x} \le 2a, \\[2ex] -\dfrac{\tilde{x}-2}{2(1-a)} & \text{at} \quad 2a \le \tilde{x} \le 2(2-a), \end{cases} \qquad (5.109)$$

$$\tau_m(\tilde{x}) = \begin{cases} \dfrac{\tilde{x}-2}{2(1-a)} & \text{at} \quad 2a \le \tilde{x} \le 2(2-a), \\[2ex] -\dfrac{\tilde{x}-4}{2a} & \text{at} \quad 2(2-a) \le \tilde{x} \le 2(2+a), \end{cases} \qquad (5.110)$$

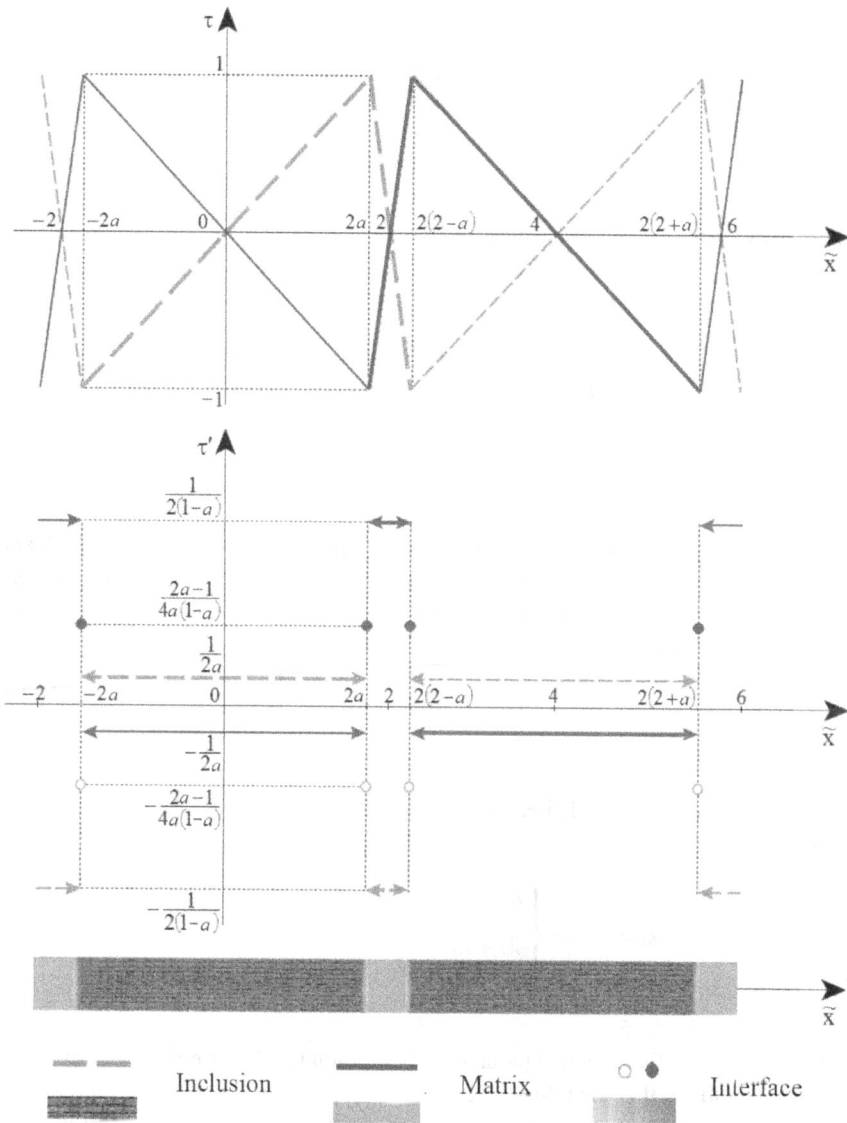

Figure 5.34   Saw-tooth functions $\tau_i(\tilde{x})$, $\tau_m(\tilde{x})$ and their derivatives $\tau'_i(\tilde{x})$, $\tau'_m(\tilde{x})$.

$$\tau'_i(\tilde{x}) = \begin{cases} \dfrac{1}{2a} & \text{at} \quad -2a < \tilde{x} < 2a, \\[2mm] -\dfrac{2a-1}{4a\,(1-a)} & \text{at} \quad \tilde{x} = \pm 2a,\ \tilde{x} = 2\,(2-a), \\[2mm] -\dfrac{1}{2\,(1-a)} & \text{at} \quad 2a \le \tilde{x} \le 2\,(2-a), \end{cases} \qquad (5.111)$$

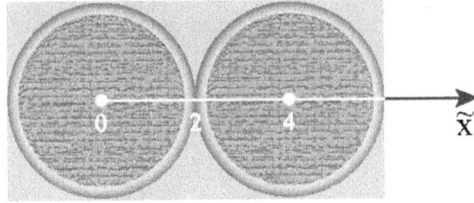

Figure 5.35 Inclusions adjoining along a thin layer at the phase boundary.

$$\tau'_m(\tilde{x}) = \begin{cases} \dfrac{1}{2(1-a)} & \text{at}\quad 2a \le \tilde{x} \le 2(2-a), \\[2mm] \dfrac{2a-1}{4a(1-a)} & \text{at}\quad \tilde{x} = 2(1\pm a),\ \tilde{x} = 2a, \\[2mm] -\dfrac{1}{2a} & \text{at}\quad 2(2-a) \le \tilde{x} \le 2(2+a). \end{cases} \qquad (5.112)$$

Consider the expression (5.95) of the effective parameter found from the LA and present it for the case $\lambda \gg 1$, $a \to 1$, taking into account the expressions for conductivities (5.106), (5.107), in the following form

$$q = \frac{\pi \tilde{\lambda}}{\sqrt{2}} \frac{\sqrt{\tilde{\lambda}}}{\tilde{\lambda}-1}, \qquad (5.113)$$

where $\tilde{\lambda} = \frac{\lambda_i}{\lambda_m}$.

In the limiting case $a = 1$, i.e. for inclusions touching along a thin layer, the following relations hold

$$\lambda_i = \lambda_m = \begin{cases} \lambda_i & \text{at}\quad 0 \le \tilde{x} < 2,\ 2 < \tilde{x} \le 4, \\[2mm] \dfrac{\lambda_i + \lambda_m}{2} & \text{at}\quad \tilde{x} = 2, \end{cases}$$

on the period (Fig. 5.35).

This means that the effective parameter expression (5.113) is not defined at $a = 1$. In order to remove this uncertainty and correctly pass from saw-tooth functions to their smooth analogs, we use the method of asymptotically equivalent functions. From (5.113) at $\tilde{\lambda} \to 1$, we obtain

$$q = \frac{\pi \tilde{\lambda}}{\sqrt{2}} \frac{\sqrt{\tilde{\lambda}}}{\tilde{\lambda}-1}. \qquad (5.114)$$

At the same time, we have

(i) expansions into series of functions $f_1\left(\tilde{\lambda}\right)$ and $f_0\left(\tilde{\lambda}\right)$, standing respectively in the left and right parts of relation (5.114), coincide up to terms of

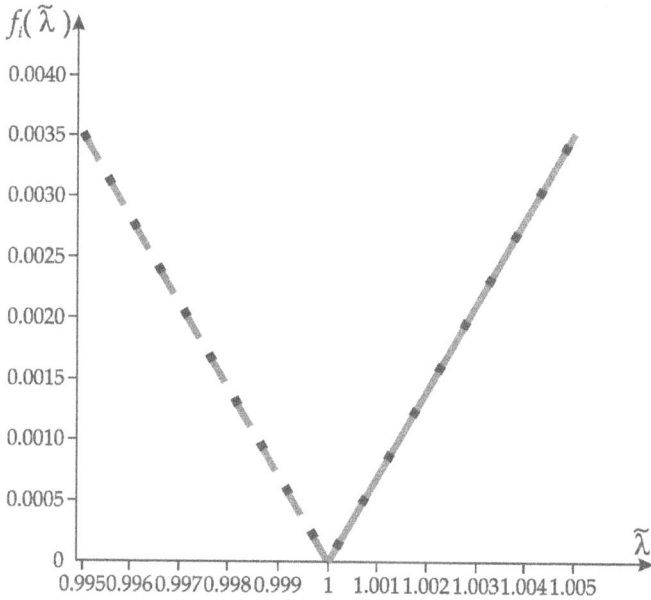

Figure 5.36 Graphs of asymptotically equivalent at $\tilde{\lambda} \to 1$ functions ($f_1(\tilde{\lambda})$ – solid line $f_2(\tilde{\lambda})$ – dashed line, $f_0(\tilde{\lambda})$ – dotted line).

the order $\left(\tilde{\lambda} - 1\right)^4$ inclusive:

$$\frac{\tilde{\lambda} - 1}{\sqrt{2\tilde{\lambda}}} = \frac{\sqrt{2}}{2}\left(\tilde{\lambda} - 1\right) - \frac{\sqrt{2}}{4}\left(\tilde{\lambda} - 1\right)^2 + \frac{3\sqrt{2}}{16}\left(\tilde{\lambda} - 1\right)^3 -$$

$$- \frac{5\sqrt{2}}{32}\left(\tilde{\lambda} - 1\right)^4 + \frac{35\sqrt{2}}{256}\left(\tilde{\lambda} - 1\right)^5 + O\left(\tilde{\lambda} - 1\right)^6;$$

$$\frac{\sqrt{\frac{\tilde{\lambda}}{2} + \frac{1}{2\tilde{\lambda}}}\ \ln\left(\frac{\tilde{\lambda}}{2} + \frac{1}{2\tilde{\lambda}}\right)}{\sqrt{\frac{\tilde{\lambda}}{2} + \frac{1}{2\tilde{\lambda}} - 1}} = \frac{\sqrt{2}}{2}\left(\tilde{\lambda} - 1\right) - \frac{\sqrt{2}}{4}\left(\tilde{\lambda} - 1\right)^2 +$$

$$+ \frac{3\sqrt{2}}{16}\left(\tilde{\lambda} - 1\right)^3 - \frac{5\sqrt{2}}{32}\left(\tilde{\lambda} - 1\right)^4 + \frac{101\sqrt{2}}{768}\left(\tilde{\lambda} - 1\right)^5 + O\left(\tilde{\lambda} - 1\right)^6;$$

(ii) expression $f_0\left(\tilde{\lambda}\right) = \dfrac{\sqrt{\frac{\tilde{\lambda}}{2} + \frac{1}{2\tilde{\lambda}}}\ \ln\left(\frac{\tilde{\lambda}}{2} + \frac{1}{2\tilde{\lambda}}\right)}{\sqrt{\frac{\tilde{\lambda}}{2} + \frac{1}{2\tilde{\lambda}} - 1}}$ does not change by transformation

$\tilde{\lambda} \rightleftarrows \frac{1}{\tilde{\lambda}}$, i.e., it also adequately describes the function $f_2\left(\tilde{\lambda}\right) = \dfrac{\tilde{\lambda}^{-1} - 1}{\sqrt{2\tilde{\lambda}^{-1}}}$ for $\tilde{\lambda} \to 1 - 0$ (Fig. 5.36, Fig. 5.37).

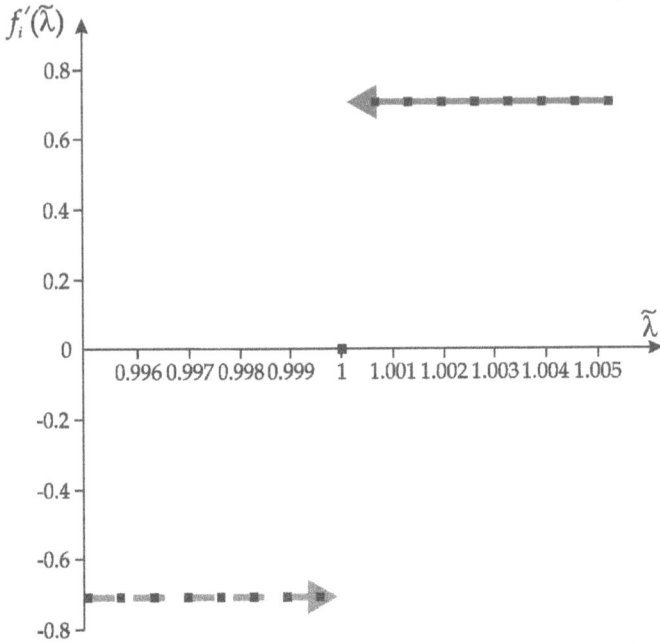

Figure 5.37 Graphs of asymptotically equivalent at $\tilde{\lambda} \to 1$ derivatives of functions under consideration ($f_1'(\tilde{\lambda})$ – solid line, $f_0'(\tilde{\lambda})$ – dotted line, $f_2'(\tilde{\lambda})$ – dashed line).

Observe that function $f_0\left(\tilde{\lambda}\right)$ has the following characteristic feature: on the one hand, the form of the function near point $\tilde{\lambda} = 1$ is similar to the symmetric saw-tooth function on the half-period; on the other hand, this function allows, in the case of inclusions touching along a thin layer, to replace a saw-tooth to a smooth argument for all values of $\tilde{x}$, i.e., there is a kind of transfer of the properties of the argument – the function.

Indeed, for inclusions touching along a thin layer ($a = 1$), the function

$$f_0\left(\tilde{\lambda}\right) = \frac{\sqrt{\frac{\tilde{\lambda}}{2} + \frac{1}{2\tilde{\lambda}}} \ln\left(\frac{\tilde{\lambda}}{2} + \frac{1}{2\tilde{\lambda}}\right)}{\sqrt{\frac{\tilde{\lambda}}{2} + \frac{1}{2\tilde{\lambda}} - 1}} = \frac{\sqrt{\frac{\tilde{\lambda}_i^2 + \tilde{\lambda}_m^2}{2\tilde{\lambda}_i\tilde{\lambda}_m}} \ln\left(\frac{\tilde{\lambda}_i^2 + \tilde{\lambda}_m^2}{2\tilde{\lambda}_i\tilde{\lambda}_m}\right)}{\sqrt{\frac{\tilde{\lambda}_i^2 + \tilde{\lambda}_m^2}{2\tilde{\lambda}_i\tilde{\lambda}_m} - 1}} \qquad (5.115)$$

can be transformed, taking into account the relations (5.106)–(5.108), (5.111), (5.112) and the properties of the saw-tooth function (5.76). Finally, we find that

$$\tilde{\lambda}_i^2 = \lambda_i^2 + 4\left(1 - a\right)\lambda_i\left(\lambda_i - \lambda_m\right)\tau'_i -$$

$$-4\left(1 - a\right)^2\left(\lambda_i - \lambda_m\right)^2 \left(\frac{1}{4\left(1 - a\right)} + \frac{\tau'_i}{2\left(1 - a\right)}\right) =$$

$$= \lambda_i^2 + 2\,(1-a)\,(\lambda_i^2 - \lambda_m^2)\,\tau'_i = \begin{cases} \lambda_i^2 & \text{at} \quad \tilde{x} < 2, \\[6pt] \dfrac{\lambda_i^2 + \lambda_m^2}{2} & \text{at} \quad \tilde{x} = 2, \\[6pt] \lambda_m^2 & \text{at} \quad \tilde{x} > 2, \end{cases}$$

$$\tilde{\lambda}_m^2 = \lambda_i^2 - 4\,(1-a)\,\lambda_i\,(\lambda_i - \lambda_m)\,\tau'_i +$$

$$+ 4\,(1-a)^2\,(\lambda_i - \lambda_m)^2 \left(\dfrac{1}{4\,(1-a)} + \dfrac{\tau'_m}{2\,(1-a)} \right) =$$

$$= \lambda_i^2 - 2\,(1-a)\,(\lambda_i^2 - \lambda_m^2)\,\tau'_m = \begin{cases} \lambda_m^2 & \text{at} \quad \tilde{x} < 2, \\[6pt] \dfrac{\lambda_i^2 + \lambda_m^2}{2} & \text{at} \quad \tilde{x} = 2, \\[6pt] \lambda_i^2 & \text{at} \quad \tilde{x} > 2. \end{cases}$$

i.e. for any values $\tilde{x}$, we have

$$\tilde{\lambda}_i^2 + \tilde{\lambda}_m^2 = \lambda_i^2 + \lambda_m^2. \tag{5.116}$$

Similarly, for all values $\tilde{x}$, we obtain

$$\tilde{\lambda}_i \tilde{\lambda}_m = \lambda_i^2 + 2\,(1-a)\,\lambda_i\,(\lambda_i - \lambda_m)\,\tau'_i - 2\,(1-a)\,\lambda_i\,(\lambda_i - \lambda_m)\,\tau'_m -$$

$$- 4\,(1-a)^2\,(\lambda_i - \lambda_m)^2\,\tau'_i \tau'_m = \lambda_i^2 - 2\,(1-a)\,\lambda_i\,(\lambda_i - \lambda_m)\cdot \dfrac{1}{2\,(1-a)}$$

i.e.

$$\tilde{\lambda}_i \tilde{\lambda}_m = \lambda_i \lambda_m. \tag{5.117}$$

Then, taking into account (5.116), (5.117), expression (5.115) is transformed as follows:

$$\dfrac{\sqrt{\dfrac{\tilde{\lambda}}{2} + \dfrac{1}{2\tilde{\lambda}}}\,\ln\left(\dfrac{\tilde{\lambda}}{2} + \dfrac{1}{2\tilde{\lambda}}\right)}{\sqrt{\dfrac{\tilde{\lambda}}{2} + \dfrac{1}{2\tilde{\lambda}} - 1}} = \dfrac{\sqrt{\dfrac{\lambda_i^2 + \lambda_m^2}{2\lambda_i \lambda_m}}\,\ln\left(\dfrac{\lambda_i^2 + \lambda_m^2}{2\lambda_i \lambda_m}\right)}{\sqrt{\dfrac{\lambda_i^2 + \lambda_m^2}{2\lambda_i \lambda_m} - 1}} =$$

$$= \dfrac{\sqrt{\dfrac{\lambda}{2} + \dfrac{1}{2\lambda}}\,\ln\left(\dfrac{\lambda}{2} + \dfrac{1}{2\lambda}\right)}{\sqrt{\dfrac{\lambda}{2} + \dfrac{1}{2\lambda} - 1}}$$

and

$$\tilde{\lambda} = \dfrac{\tilde{\lambda}}{2} + \dfrac{1}{2\tilde{\lambda}} + \sqrt{\left(\dfrac{\tilde{\lambda}}{2} + \dfrac{1}{2\tilde{\lambda}}\right)^2 - 1} = \dfrac{\lambda}{2} + \dfrac{1}{2\lambda} + \sqrt{\left(\dfrac{\lambda}{2} + \dfrac{1}{2\lambda}\right)^2 - 1} = \lambda.$$

Therefore, at $\lambda \to \infty$, we get

$$\frac{\sqrt{\frac{\tilde{\lambda}}{2} + \frac{1}{2\tilde{\lambda}}} \ \ln\left(\frac{\tilde{\lambda}}{2} + \frac{1}{2\tilde{\lambda}}\right)}{\sqrt{\frac{\tilde{\lambda}}{2} + \frac{1}{2\tilde{\lambda}} - 1}} \sim \ln\frac{\lambda}{2}. \tag{5.118}$$

Taking into account (5.118), we obtain that relation (5.113) for $a = 1$, $\lambda \to \infty$ takes the following form

$$q = \frac{\pi}{2} \frac{\lambda}{\ln\frac{\lambda}{2}}.$$

Thus, we finally have an asymptotic expression for the effective parameter in the case of contact of absolutely conducting inclusions with an interlayer, which reads

$$q_{ci\ \text{int}}^{(\infty)} = \frac{\pi\lambda}{2\,(\ln\lambda - \ln 2)}. \tag{5.119}$$

Note that the main part of the asymptotic formula (5.119)

$$q \sim \frac{\pi\lambda}{2\ln\lambda}.$$

coincides with the asymptotics [187] and differs from the result (5.105) [158] by the constant term $\ln 2 \approx 0.69$.

5.3.6 CONTACT OF NON-CONDUCTIVE ROUND INCLUSIONS IN THE PRESENCE OF A THIN LAYER AT THE PHASE BOUNDARY

In Section 5.3.3, expression (5.103) was obtained for the effective thermal conductivity parameter for inclusions with low conductivity of large geometric sizes.

Let us generalize this relation to the case of a composite material with inclusions of low conductivity $0 < \lambda << 1$, which is separated by a thin layer (interface) at the phase boundary. The interface conductivity λ_{int} is characterized by the average conductivity of the matrix and inclusions, $\lambda_{\text{int}} = \frac{\lambda_i + \lambda_m}{2}$ (Fig. 5.38).

Using the technique of non-smooth transformation of the argument described above, we assume that the conductivities of the inclusions and the matrix are represented by analytical relations (5.106), (5.107). Similar to the approach described above, we represent the ratio of the effective thermal conductivity parameter (5.103) in the case of inclusions of extremely large geometric sizes ($a = 1$) and conductivity ($\lambda << 1$) as

$$q = \frac{\left(1 - \tilde{\lambda}\right)\sqrt{2\tilde{\lambda}}}{\pi}, \tag{5.120}$$

and obtain the expression in the numerator by the asymptotically equivalent function.

Then, we have

$$\left(1 - \tilde{\lambda}\right)\sqrt{2\tilde{\lambda}} \sim \frac{2\tilde{\lambda}\sqrt{\frac{\tilde{\lambda}}{2} + \frac{1}{2\tilde{\lambda}}} \ \ln\left(\frac{\tilde{\lambda}}{2} + \frac{1}{2\tilde{\lambda}}\right)}{\sqrt{\frac{\tilde{\lambda}}{2} + \frac{1}{2\tilde{\lambda}} - 1}} \quad \text{at} \quad \tilde{\lambda} \to 1, \tag{5.121}$$

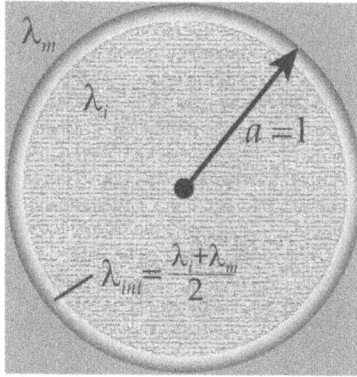

Figure 5.38 Characteristic cell of a composite with low-conductivity inclusions touching along a thin layer at the phase boundary.

where the expansions into series of functions on the left and right sides of (5.121) coincide up to terms of order $\left(1 - \tilde{\lambda}\right)^{4}$ inclusive:

$$
\left(1 - \tilde{\lambda}\right) \sqrt{2\tilde{\lambda}} = \sqrt{2}\left(1 - \tilde{\lambda}\right) - \frac{\sqrt{2}}{2}\left(1 - \tilde{\lambda}\right)^{2} - \frac{\sqrt{2}}{8}\left(1 - \tilde{\lambda}\right)^{3} -
$$
$$
- \frac{\sqrt{2}}{16}\left(1 - \tilde{\lambda}\right)^{4} - \frac{5\sqrt{2}}{128}\left(1 - \tilde{\lambda}\right)^{5} + o\left(1 - \tilde{\lambda}\right)^{6},
$$

$$
\frac{2\tilde{\lambda}\sqrt{\frac{\tilde{\lambda}}{2} + \frac{1}{2\lambda}}\,\ln\left(\frac{\tilde{\lambda}}{2} + \frac{1}{2\lambda}\right)}{\sqrt{\frac{\tilde{\lambda}}{2} + \frac{1}{2\lambda} - 1}} = \sqrt{2}\left(1 - \tilde{\lambda}\right) - \frac{\sqrt{2}}{2}\left(1 - \tilde{\lambda}\right)^{2} -
$$
$$
- \frac{\sqrt{2}}{8}\left(1 - \tilde{\lambda}\right)^{3} - \frac{\sqrt{2}}{16}\left(1 - \tilde{\lambda}\right)^{4} - \frac{19\sqrt{2}}{384}\left(1 - \tilde{\lambda}\right)^{5} + o\left(1 - \tilde{\lambda}\right)^{6}.
$$

Since, using (5.116), (5.117) at $\lambda \to 0$, we have

$$
\frac{\sqrt{\frac{\tilde{\lambda}}{2} + \frac{1}{2\lambda}}\,\ln\left(\frac{\tilde{\lambda}}{2} + \frac{1}{2\lambda}\right)}{\sqrt{\frac{\tilde{\lambda}}{2} + \frac{1}{2\lambda} - 1}} \sim \ln\frac{1}{2\lambda}, \tag{5.122}
$$

then the expression of the effective parameter (5.120) is converted to the form

$$
q = \frac{2\lambda\,\ln\frac{1}{2\lambda}}{\pi}.
$$

Consequently, we have an asymptotic expression for the effective parameter in the case of contact of non-conductive inclusions with an interlayer:

$$
q_{ci\,\text{int}}^{(0)} = \frac{2\lambda\left(\ln\lambda^{-1} - \ln 2\right)}{\pi}. \tag{5.123}
$$

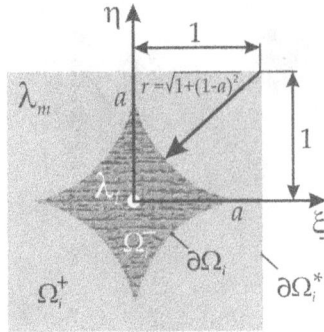

Figure 5.39 Typical cell of a composite with curvilinear rhombic inclusions.

It should be emphasized that the obtained asymptotic relations of the parameters (5.119), (5.123) in the case $a = 1$ for a thin layer of inclusions, for absolutely conductive ($\lambda \to \infty$) and non-conductive ($\lambda \to 0$) inclusions, satisfy the Keller's theorem:

$$\frac{1}{q^{(0)}_{ci\ int} (\lambda^{-1})} = q^{(\infty)}_{ci\ int} (\lambda)$$

because

$$\frac{1}{\frac{2\lambda^{-1}(\ln\lambda - \ln 2)}{\pi}} = \frac{\pi\lambda}{2(\ln\lambda - \ln 2)}.$$

5.3.7 MODELS OF COMPOSITES WITH CURVILINEAR RHOMBIC INCLUSIONS

Fig. 5.39 shows the cell of a composite with curvilinear rhombic inclusions. For an asymptotic study of composites with large inclusions ($0 << a < 1$) and high conductivity ($\lambda >> 1$), we apply LA.

The inclusion size a as a variable function of coordinate ξ will be described in this case as follows (Fig. 5.40):

$$a(\xi) = 1 - \sqrt{1 + (1-a)^2 - (1-\xi)^2}. \tag{5.124}$$

Taking into account the representation (5.124), the homogenized equation (5.94) is transformed to the following form

$$q = \frac{1}{|\Omega_i^*|} \left(\iint\limits_{\Omega_i^+} (1 + B_0^*)\, d\xi d\eta + \lambda \iint\limits_{\Omega_i^-} (1 + D_0^*)\, d\xi d\eta \right), \tag{5.125}$$

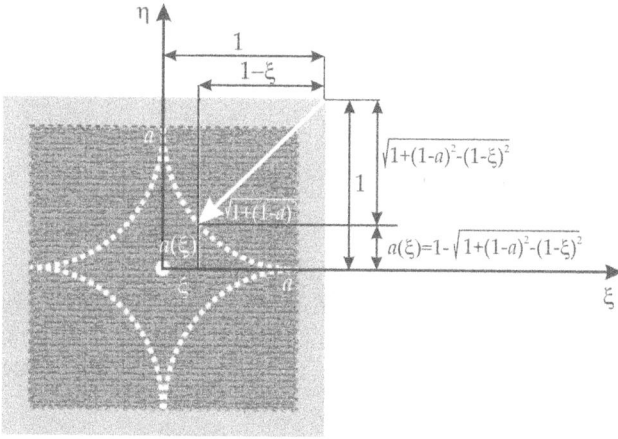

Figure 5.40   Approximation of a curvilinear rhombic inclusion in the LA model.

where integration is performed over the original cell area with a curvilinear rhombic inclusion, $\Omega_i^* = \Omega_i^+ \bigcup \Omega_i^-$; B_0^*, D_0^* are determined up to a factor $\frac{\partial u_0}{\partial y}$. The inclusion form is a variable function of the coordinate ξ, represented by the relation (5.124).

Performing integration in (5.125), we find the expression for the effective thermal conductivity coefficient at $\lambda \gg 1$, $0 \ll a < 1$ $\left(\lambda > 1 + \frac{1}{\sqrt{1+(1-a)^2}} \right)$ in the following form

$$q = 1 - a + \frac{\lambda}{\lambda - 1} \left(\frac{\pi}{2} - \frac{1}{\sqrt{\Delta_1 - 1}} \ln \frac{\sqrt{\Delta_1 - 1} + \sqrt{\Delta_1 - 1}}{\sqrt{\Delta_1 - 1} - \sqrt{\Delta_1 + 1}} \right), \qquad (5.126)$$

where

$$\Delta_1 = (\lambda - 1)^2 \left(1 + (1 - a)^2 \right).$$

In particular cases, from (5.126), the following asymptotic relations can be obtained:

(i) for absolutely conductive inclusions ($\lambda \to \infty$) of large geometric sizes close to the limiting possible value ($0 \ll a < 1$)

$$q = 1 - a + \frac{\pi}{2} - \frac{\ln \left(2\lambda \sqrt{1 + (1-a)^2} \right)}{\lambda \sqrt{1 + (1-a)^2}}. \qquad (5.127)$$

In particular, at $\lambda \to \infty$, $a \to 1$, we get

$$q = \frac{\pi}{2} - \frac{\ln 2\lambda}{\lambda}; \qquad (5.128)$$

(ii) for inclusions of extremely large sizes ($a = 1$) and high conductivity ($\lambda \gg 1$):

$$q = \frac{\lambda}{\lambda - 1}\left(\frac{\pi}{2} - \frac{1}{\sqrt{\lambda\,(\lambda - 2)}}\ln\frac{\sqrt{\lambda} + \sqrt{\lambda - 2}}{\sqrt{\lambda} - \sqrt{\lambda - 2}}\right) \qquad (5.129)$$

or for extremely large sizes ($a = 1$) absolutely conductive inclusions ($\lambda \to \infty$):

$$q_{cri}^{(\infty)} = \frac{\pi}{2} - \frac{\ln 2\lambda}{\lambda}. \qquad (5.130)$$

1) Asymptotic representations of the effective thermal conductivity for composites with large curvilinear rhombic inclusions ($0 \ll a < 1$) of low conductivity ($\lambda \ll 1$, $\lambda < 1 - \frac{1}{1+\sqrt{1+(1-a)^2}}$) can be written using expressions (5.127)–(5.130) and Keller's theorem:

$$q = \frac{(1 - \lambda)\sqrt{\Delta_2 - 1}}{\left((1 - a)(1 - \lambda) + \frac{\pi}{2}\right)\sqrt{\Delta_2 - 1} - \ln\frac{\sqrt{\Delta_2 - 1} + \sqrt{\Delta_2 - 1}}{\sqrt{\Delta_2 - 1} - \sqrt{\Delta_2 + 1}}} \qquad (5.131)$$

at $\lambda \ll 1$, $\quad 0 \ll a < 1$, where

$$\Delta_2 = \left(\frac{1 - \lambda}{\lambda}\right)^2\left(1 + (1 - a)^2\right).$$

Accordingly, formulas (5.127)–(5.130) are transformed to the form:

1.1. Non-conductive inclusions of large sizes, close to the limiting possible value

$$q = \frac{2}{2(1 - a) + \pi - \frac{2\lambda}{\sqrt{1+(1-a)^2}}\ln\left(\frac{2\sqrt{1+(1-a)^2}}{\lambda}\right)} \qquad (5.132)$$

at $\lambda \to 0$, $\quad 0 \ll a < 1$; in particular

$$q = \frac{2}{\pi - 2\lambda\ln\left(\frac{2}{\lambda}\right)} \qquad \text{at} \quad \lambda \to 0, a \to 1; \qquad (5.133)$$

1.2. Inclusions of extremely large sizes and low conductivity (on condition $\lambda < \frac{1}{2}$):

$$q = \frac{2(1 - \lambda)}{\pi - \frac{2\lambda}{\sqrt{1-2\lambda}}\ln\frac{1+\sqrt{1-2\lambda}}{1-\sqrt{1-2\lambda}}} \qquad \text{at} \quad a = 1, \lambda \ll 1, \qquad (5.134)$$

or for practically non-conductive inclusions

$$q_{cri}^{(0)} = \frac{2}{\pi - 2\lambda\ln\left(\frac{2}{\lambda}\right)} \qquad \text{at} \quad a = 1, \lambda \to 0. \qquad (5.135)$$

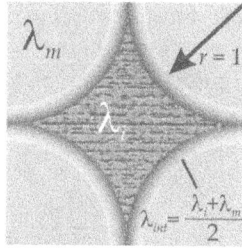

Figure 5.41 Composite with highly conductive curvilinear rhombic inclusions touching along a thin layer at the phase boundary.

2) Consider a composite with highly conductive curvilinear rhombic inclusions of large sizes, touching along a thin layer at the phase boundary (Fig. 5.126).

The generalization in this case of the relations obtained in 5.3.7, is nontrivial, because the asymptotic formula (5.129) holds at $\lambda > 2$ and cannot be used to estimate conductivities (5.106), (5.107) at $\tilde{\lambda} \sim 1$.

To find these estimates, we present the expression (5.126) in a transformed form under the condition $0 < \lambda < 1 + \frac{1}{\sqrt{1+(1-a)^2}}$ in the following way:

$$q = 1 - a + \frac{\lambda}{\lambda - 1}\left(\frac{\pi}{2} - \frac{2}{\sqrt{1-\Delta_1}}\arctan\sqrt{\frac{1-\sqrt{\Delta_1}}{1+\sqrt{\Delta_1}}}\right). \qquad (5.136)$$

Then from (5.136), using Keller's theorem, we have at $1 - \frac{1}{1+\sqrt{1+(1-a)^2}} < \lambda < \infty$:

$$q = \frac{1}{1 - a + \frac{1}{1-\lambda}\left(\frac{\pi}{2} - \frac{2}{\sqrt{1-\Delta_2}}\arctan\sqrt{\frac{1-\sqrt{\Delta_2}}{1+\sqrt{\Delta_2}}}\right)}. \qquad (5.137)$$

Expression (5.137) is transformed at $a = 1$, $\lambda > \frac{1}{2}$ to the following one

$$q = \frac{\lambda - 1}{\frac{2\lambda}{\sqrt{2\lambda-1}}\arctan\sqrt{2\lambda-1} - \frac{\pi}{2}},$$

or at $\lambda \gg 1$:

$$q = \frac{2(\lambda - 1)}{\pi\sqrt{2\lambda}}. \qquad (5.138)$$

The expression (5.138) is related to the formula (5.113) of the effective parameter found using LA for absolutely conductive inclusions ($\lambda \to \infty$) as follows:

$$q_{cri} = \frac{\tilde{\lambda}}{q_{ci}}. \qquad (5.139)$$

Consequently, the asymptotic estimates (5.114), (5.118) remain valid and can be used for the case under consideration.

Thus, for a composite with curvilinear rhombic inclusions of absolute conductivity ($\lambda \to \infty$), subdivided by a thin layer at the phase boundary, with conductivity $\lambda_{\text{int}} = \frac{\lambda_i + \lambda_m}{2}$, the effective thermal conductivity parameter has the following form

$$q_{cri \text{ int}}^{(\infty)} = \frac{2(\ln \lambda - \ln 2)}{\pi}. \tag{5.140}$$

3) Scheme for an asymptotic study of the effective conductivity coefficient of composites with large curvilinear rhombic inclusions $0 \ll a < 1$ and low conductivity $\lambda \ll 1$, touching along a thin layer at the phase boundary, is identical to that described in 5.3.7. Therefore, we briefly present only the final results:

(i) at condition $1 - \frac{1}{1+\sqrt{1+(1-a)^2}} < \lambda < \infty$ relation (5.131) is represented as (5.137);

(ii) using the Keller's theorem allows one to generalize (5.137) to the case $0 < \lambda < 1 + \frac{1}{\sqrt{1+(1-a)^2}}$, transforming it to the form (5.136);

(iii) for contacting inclusions $(a = 1)$ expression (5.136) is written as follows (at $\lambda < 2$):

$$q = \frac{\lambda}{\lambda - 1}\left(\frac{\pi}{2} - \frac{2}{\sqrt{\lambda(2-\lambda)}} \arctan \sqrt{\frac{2-\lambda}{\lambda}}\right); \tag{5.141}$$

(iv) respectively, at $\lambda \ll 1$ from (5.141), we have

$$q = \frac{\pi}{1-\lambda}\sqrt{\frac{\lambda}{2}}; \tag{5.142}$$

(v) the expression (5.142) written in terms of the saw-tooth transformation of the argument is related to the formula (5.120) for the effective thermal conductivity parameter of the composite with non-conductive inclusions ($\lambda \to 0$) by the relation (5.139).

We use the asymptotic formulas (5.121), (5.122) to evaluate the expression (5.142) represented by the saw-tooth argument transformation and generalized relations for the conductivities of the inclusions and the matrix (5.106), (5.107). Then, for a composite structure with curvilinear rhombic non-conducting inclusions ($\lambda \to 0$), contacting ($a = 1$) along a thin layer at the phase boundary, we obtain the following expression for the effective thermal conductivity parameter

$$q_{cri \text{ int}}^{(0)} = \frac{\pi}{2(\ln \lambda^{-1} - \ln 2)}. \tag{5.143}$$

5.3.8 PHYSICAL EQUIVALENCE OF COMPOSITE STRUCTURES

In theoretical studies of composite materials, effective conductivity is determined not in absolute, but in relative terms, as a rule, in relation to the conductivity of the

Figure 5.42 Composite material of regular structure.

matrix. However, for a significant class of composites with a regular structure, the concepts of a matrix and inclusions are very relative and depend on their choice by the researcher. For example, the composite shown in Fig. 5.42 can be interpreted as a structure with rhombic inclusions with sharp corners, and as an array with rhombic inclusions with rounded corners. The second important point of research is the description of the physical characteristics (in this case, thermal conductivity) of the matrix and inclusions of the composite.

It is obvious that the terms "absolutely conductive" and "non-conductive" inclusions, widely used in the mechanics of composites, are nothing but mathematical idealization. In such cases, it would be more accurate to describe the real properties of the phases through the relations

$$\lambda_i \gg \lambda_m \;\Rightarrow\; \lambda = \frac{\lambda_i}{\lambda_m} \gg 1 \tag{5.144}$$

and

$$\lambda_i \ll \lambda_m \;\Rightarrow\; \lambda = \frac{\lambda_i}{\lambda_m} \ll 1. \tag{5.145}$$

However, if we change the normalization in relations (5.144), (5.145), i.e., determine in relation not to the conductivity of the matrix, but relative to the conductivity of the inclusions, or (which is the same) change the roles of the "inclusions" – "matrix", then the structure with "absolutely conductive" inclusions will already be described as an array with "non-conductive" inclusions, and vice versa. The above reasoning remains valid not only in the limiting values of matrix and inclusions conductivities ($\lambda \to 0$ or $\lambda \to \infty$), but also for any of their finite values.

For example, if for the shown in Fig. 5.42 composite, the following holds

(i) phase I: rhombus with rounded corners;
(ii) phase II: rhombus with sharp corners;
(iii) $\lambda_I \ll \lambda_{II}$, such a structure can be described:
   – as a composite with absolutely conducting rhombic inclusions with sharp corners (Fig. 5.43a);

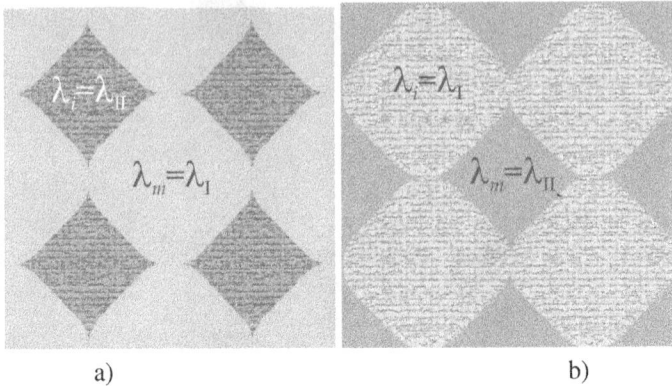

a)   b)

Figure 5.43 Physically equivalent composite structures: a) composite with highly conductive rhombic inclusions with sharp corners $\lambda_m = \lambda_I$, $\lambda_i = \lambda_{II}$, $\lambda^{(1)} = \frac{\lambda_i}{\lambda_m} = \frac{\lambda_{II}}{\lambda_I} \gg 1$; b) composite with low conductive rhombic inclusions with rounded corners: $\lambda_m = \lambda_{II}$, $\lambda_i = \lambda_I$, $\lambda^{(2)} = \frac{\lambda_i}{\lambda_m} = \frac{\lambda_I}{\lambda_{II}} \ll 1$.

    —   as a composite with non-conductive rhombic inclusions with rounded corners (Fig. 5.43b).

At the same time, it is obvious that since both shown in Fig. 5.43a and Fig. 5.43b models describe the same composite, then from a physical point of view these structures are equivalent. Note that the effective parameters of λ_{ef} physically equivalent composite structures are

$$\lambda_{ef}^{(1)}(\lambda) = \lambda \, \lambda_{ef}^{(2)}(\lambda^{-1}). \qquad (5.146)$$

Relation (5.146) is satisfied, in particular:

(i)   for absolutely conductive circle (5.119) and non-conductive curvilinear rhombic (5.143) inclusions touching along a thin layer at the phase boundary;

(ii)   for non-conductive circle (5.123) and absolutely conductive curvilinear rhombic (5.140) inclusions in the presence of a thin layer at the phase boundary.

In addition, using relation (5.146) and previously found asymptotic representations of the effective parameters of composites with circle and curvilinear rhombic inclusions in cases of their extremely high (low) conductivity (formulas (5.99), (5.104), (5.130), (5.135)), generalizing them, one can obtain asymptotics for structures physically equivalent to them.

Asymptotic representations of the effective parameters of composites of various structures with extremely large (small) physical characteristics of the inclusions are given in Table 5.2. Analysis carried out based on given in Table 5.2 asymptotic expressions shows that all of them, chosen in pairs in an appropriate way, satisfy the so-called "complementary systems" formula obtained in [80], which can be interpreted as a generalization of the well-known Dykhne formula for composites with equally represented phases. Namely, if in a two-phase composite the conductivities of phases I and II are, respectively, λ_1 and λ_2, and their volume fractions c and $(1-c)$, and $c \neq \frac{1}{2}$, then when replacing $\lambda_1 \leftrightarrows \lambda_2$, or which is the same, replacing $c \leftrightarrows (1-c)$, effective parameters of structures $\lambda_{ef}(c)$ and $\lambda_{ef}(1-c)$ linked by the ratio

$$\lambda_{ef}(c)\,\lambda_{ef}(1-c) = \lambda_1\,\lambda_2. \tag{5.147}$$

Let us show the fulfillment of the Dykhne relation (5.147) for the given in Table 5.2 asymptotic formulas ($\lambda_{ci} = \lambda_1$, $\lambda_{cri} = \lambda_2$):

(i) relations (5.99) and (5.2):

$$\lambda_{ef}(c)\,\lambda_{ef}(1-c) = \left(\pi\sqrt{\frac{\lambda_1}{2\lambda_2}}\,\lambda_2\right)\left(\frac{1}{\pi}\sqrt{\frac{2\lambda_1}{\lambda_2}}\,\lambda_2\right) = \lambda_1\,\lambda_2;$$

(ii) relations (5.2) and (5.130):

$$\lambda_{ef}(c)\,\lambda_{ef}(1-c) = \left(\frac{2\lambda_1}{\pi\lambda_2}\,\lambda_2\right)\left(\frac{\pi}{2}\,\lambda_2\right) = \lambda_1\,\lambda_2;$$

(iii) relations (5.119) and (5.140):

$$\lambda_{ef}(c)\,\lambda_{ef}(1-c) = \left(\frac{\pi\,\lambda_1}{2\lambda_2\left(\ln\frac{\lambda_1}{\lambda_2} - \ln 2\right)}\,\lambda_2\right) \times$$

$$\times \left(\frac{2}{\pi}\left(\ln\frac{\lambda_1}{\lambda_2} - \ln 2\right)\lambda_2\right) = \lambda_1\,\lambda_2;$$

(iv) relations (5.104) and (5.2):

$$\lambda_{ef}(c)\,\lambda_{ef}(1-c) = \left(\frac{1}{\pi}\sqrt{\frac{2\lambda_1}{\lambda_2}}\,\lambda_2\right)\left(\pi\sqrt{\frac{\lambda_1}{2\lambda_2}}\,\lambda_2\right) = \lambda_1\,\lambda_2;$$

(v) relations (5.2) and (5.135):

$$\lambda_{ef}(c)\,\lambda_{ef}(1-c) = \left(\frac{\pi\lambda_1}{2\lambda_2}\,\lambda_2\right)\left(\frac{2}{\pi}\,\lambda_2\right) = \lambda_1\,\lambda_2;$$

(vi) relations (5.123) and (5.143):

$$\lambda_{ef}(c)\,\lambda_{ef}(1-c) = \left(\frac{2}{\pi}\frac{\lambda_1}{\lambda_2}\left(\ln\frac{\lambda_2}{\lambda_1} - \ln 2\right)\lambda_2\right)\left(\frac{\pi}{2}\frac{1}{\left(\ln\frac{\lambda_2}{\lambda_1} - \ln 2\right)}\,\lambda_2\right)$$

$$= \lambda_1\,\lambda_2.$$

Table 5.2

Asymptotic representations of the effective parameters of composites of various structures.

1. Circle absolutely conducting inclusions $(\lambda \to \infty)$	
	1.1. Contact by matrix material (curvilinear rhombus) $$\lambda = \frac{\lambda_{ci}}{\lambda_{cri}}$$ $$q = \pi\sqrt{\frac{\lambda}{2}}$$ relation (5.99)
	1.2. Contact by inclusion material (circles) $$\lambda = \frac{\lambda_{ci}}{\lambda_{cri}}$$ $$q = \frac{2\lambda}{\pi}$$ Relation (5.147) (derived from (5.146) and (5.135))
	1.3. Material contact with average matrix and inclusion conductivities $$\lambda = \frac{\lambda_{ci}}{\lambda_{cri}}$$ $$q = \frac{\pi\lambda}{2(\ln\lambda - \ln 2)}$$ relation (5.119)

(Continued on next page)

Table 5.2 (Continued)

2. Circle non-conductive inclusions $(\lambda \to \infty)$	
	2.1. Contact by matrix material (curvilinear rhombus) $$\lambda = \frac{\lambda_{ci}}{\lambda_{cri}}$$ $$q = \frac{\sqrt{2\lambda}}{\pi}$$ relation (5.104)
	2.2. Contact by inclusion material (circles) $$\lambda = \frac{\lambda_{ci}}{\lambda_{cri}}$$ $$q = \frac{\pi\lambda}{2}$$ Relation (5.148)(derived from (5.146) and (5.130))
	2.3. Material contact along interphase with average matrix and inclusion conductivities $$\lambda = \frac{\lambda_{ci}}{\lambda_{cri}}$$ $$q = \frac{2\lambda\left(\ln\lambda^{-1} - \ln 2\right)}{\pi}$$ Relation (5.123)

(Continued on next page)

Table 5.2 (Continued)

3. Curvilinear rhombic absolutely conducting inclusions $(\lambda \to \infty)$	
	3.1. Contact by matrix material (circles) $$\lambda = \frac{\lambda_{cri}}{\lambda_{ci}}$$ $$\boxed{q = \frac{\pi}{2}}$$ relation (5.130)
	3.2. Contact on the material of inclusions (curvilinear rhoms) $$\lambda = \frac{\lambda_{cri}}{\lambda_{ci}}$$ $$\boxed{q = \frac{\sqrt{2\lambda}}{\pi}}$$ Relation (5.149) (derived from (5.146) and (5.104))
	3.3. Material contact along interphase with average matrix and inclusion conductivities $$\lambda = \frac{\lambda_{cri}}{\lambda_{ci}}$$ $$\boxed{q = \frac{2(\ln\lambda - \ln 2)}{\pi}}$$ Relation (5.140)

(Continued on next page)

Table 5.2 (Continued)

4. Curvilinear rhombic non-conductive inclusions $(\lambda \to \infty)$	
	4.1. Contact by matrix material (circles) $$\lambda = \frac{\lambda_{cri}}{\lambda_{ci}}$$ $$q = \frac{2}{\pi}$$ relation (5.135)
	4.2. Contact by the material of inclusions (curvilinear rhombus) $$\lambda = \frac{\lambda_{cri}}{\lambda_{ci}}$$ $$q = \pi\sqrt{\frac{\lambda}{2}}$$ Relation (5.150) (obtained using (5.146) and (5.99))
	4.3. Material contact along interface with average matrix and inclusion conductivities $$\lambda = \frac{\lambda_{cri}}{\lambda_{ci}}$$ $$q = \frac{\pi}{2\left(\ln\lambda^{-1} - \ln 2\right)}$$ Relation (5.143)

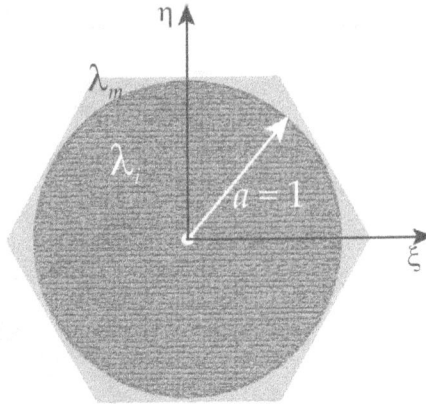

Figure 5.44 Characteristic cell of a composite array of hexagonal structure in the limiting case of inclusion sizes.

5.3.9 HEXAGONAL ARRAY OF CIRCLE INCLUSIONS

1) Using LA, the expressions for the effective thermal conductivity parameter of a composite with a hexagonal lattice of inclusions were found under the condition of their large sizes ($a \to 1$) and high conductivity ($\lambda \to \infty$). In the limiting case of inclusion sizes ($a = 1$) (Fig. 5.44), the main term of the resulting asymptotic expansion can be represented as follows:

$$q = \frac{\pi\sqrt{3}\,\frac{\tilde{\lambda}-1}{\tilde{\lambda}+1}}{\sqrt{1 - \frac{\tilde{\lambda}-1}{\tilde{\lambda}+1}}}. \tag{5.148}$$

Let us further consider a composite material of a hexagonal lattice of circle inclusions of high conductivity ($\lambda \gg 1$) and an extremely large size ($a \to 1$), which have a thin interlayer at the phase boundary, characterized by the conductivity $\lambda_{\text{int}} = \frac{\lambda_i + \lambda_m}{2}$ (Fig. 5.45). To find the effective parameter of the composite in the presence of a thin layer at the interface of its phases, we use the technique of non-smooth transformation of the argument, assuming that the conductivities of the inclusions and the matrix are presented in the form of analytical dependences (5.106), (5.107).

Let's evaluate the expression $\dfrac{2\,\frac{\tilde{\lambda}-1}{\tilde{\lambda}+1}}{\sqrt{1 - \frac{\tilde{\lambda}-1}{\tilde{\lambda}+1}}}$ from (5.148) at $\tilde{\lambda} \to 1$ in the follow-

ing way:

$$\frac{2\,\frac{\tilde{\lambda}-1}{\tilde{\lambda}+1}}{\sqrt{1 - \frac{\tilde{\lambda}-1}{\tilde{\lambda}+1}}} \sim \frac{2\left(\frac{\tilde{\lambda}}{2} + \frac{1}{2\tilde{\lambda}} - 1\right)^{\frac{3}{2}} \left(\frac{\tilde{\lambda}}{2}\right)^{\frac{1}{4}}}{\ln\left(\frac{\tilde{\lambda}}{2} + \frac{1}{2\tilde{\lambda}}\right) \left(\frac{\tilde{\lambda}}{2} + \frac{1}{2\tilde{\lambda}}\right)^{\frac{1}{2}} \left(\frac{\tilde{\lambda}}{2} + \frac{1}{2\tilde{\lambda}} + 1\right)^{\frac{1}{4}}} \tag{5.149}$$

at $\tilde{\lambda} \to 1$.

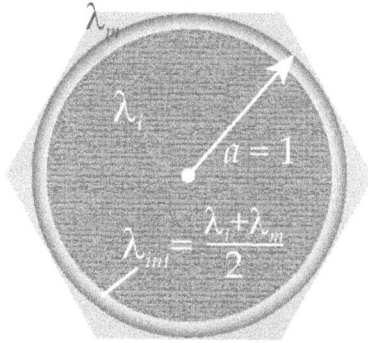

Figure 5.45 Composite of hexagonal structure with a thin layer at the phase boundary.

As in the case of a square lattice, for a hexagonal packing of inclusions, the approximating function $f_0\left(\tilde{\lambda}\right)$ on the right side of expression (5.149):

(i)   correctly describes the function $f_1\left(\tilde{\lambda}\right)$, standing on the left side of the equality, in a neighborhood of the point $\tilde{\lambda} = 1$; expansions into series of functions $f_1\left(\tilde{\lambda}\right)$, $f_0\left(\tilde{\lambda}\right)$ coincide up to terms of order $\left(\tilde{\lambda}-1\right)^4$ inclusive:

$$\frac{2\frac{\tilde{\lambda}-1}{\tilde{\lambda}+1}}{\sqrt{1-\frac{\tilde{\lambda}-1}{\tilde{\lambda}+1}}} = \tilde{\lambda}-1-\frac{1}{4}\left(\tilde{\lambda}-1\right)^2+\frac{3}{32}\left(\tilde{\lambda}-1\right)^3-\frac{5}{128}\left(\tilde{\lambda}-1\right)^4+$$

$$+\frac{35}{2048}\left(\tilde{\lambda}-1\right)^5+O\left(\tilde{\lambda}-1\right)^6;$$

$$\frac{2\left(\frac{\tilde{\lambda}}{2}+\frac{1}{2\tilde{\lambda}}-1\right)^{\frac{3}{2}}\left(\frac{\tilde{\lambda}}{2}\right)^{\frac{1}{4}}}{\ln\left(\frac{\tilde{\lambda}}{2}+\frac{1}{2\tilde{\lambda}}\right)\left(\frac{\tilde{\lambda}}{2}+\frac{1}{2\tilde{\lambda}}\right)^{\frac{1}{2}}\left(\frac{\tilde{\lambda}}{2}+\frac{1}{2\tilde{\lambda}}+1\right)^{\frac{1}{4}}} = \tilde{\lambda}-1-\frac{1}{4}\left(\tilde{\lambda}-1\right)^2+$$

$$+\frac{3}{32}\left(\tilde{\lambda}-1\right)^3-\frac{5}{128}\left(\tilde{\lambda}-1\right)^4+\frac{169}{6144}\left(\tilde{\lambda}-1\right)^5+O\left(\tilde{\lambda}-1\right)^6;$$

(ii)   a symmetrical saw-tooth function (Fig. 5.46, Fig. 5.47) is the same;

(iii) allows us to correctly perform the transition from the saw-tooth to the original smooth argument. Taking into account relations (5.116), (5.117) at $\lambda \to \infty$, we have

$$\frac{2\left(\frac{\tilde{\lambda}}{2}+\frac{1}{2\tilde{\lambda}}-1\right)^{\frac{3}{2}}\left(\frac{\tilde{\lambda}}{2}\right)^{\frac{1}{4}}}{\ln\left(\frac{\tilde{\lambda}}{2}+\frac{1}{2\tilde{\lambda}}\right)\left(\frac{\tilde{\lambda}}{2}+\frac{1}{2\tilde{\lambda}}\right)^{\frac{1}{2}}\left(\frac{\tilde{\lambda}}{2}+\frac{1}{2\tilde{\lambda}}+1\right)^{\frac{1}{4}}} \sim \frac{\lambda}{\ln\frac{\lambda}{2}}. \qquad (5.150)$$

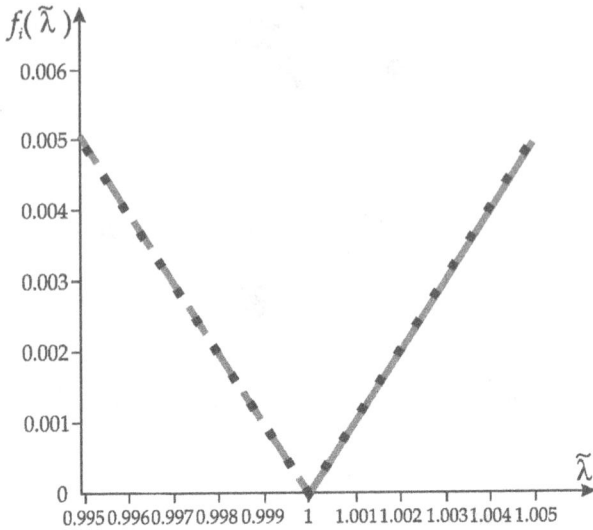

Figure 5.46 Graphs of asymptotically equivalent at $\tilde{\lambda} \to 1$ functions ($f_1(\tilde{\lambda})$ – solid line, $f_2(\tilde{\lambda}) = f_1(\tilde{\lambda}^{-1})$ – dashed line, $f_0(\tilde{\lambda})$ – dotted line).

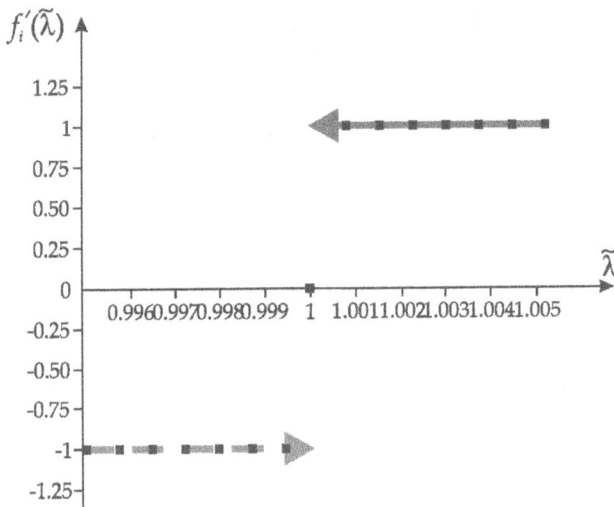

Figure 5.47 Graphs of derivatives of asymptotically equivalent at $\tilde{\lambda} \to 1$ functions ($f_1'(\tilde{\lambda})$ – solid line, $f_2'(\tilde{\lambda})$ – dashed line, $f_0'(\tilde{\lambda})$ – dotted line).

Thus, taking into account relations (5.148)–(5.150), the expression for the effective parameter of a composite of a hexagonal structure with absolutely conducting inclusions ($\lambda \to \infty$) of an extremely large size ($a = 1$) and a thin layer at the phase boundary is defined as

$$q_{hex}^{(\infty)} = \frac{\pi \lambda \sqrt{3}}{2 \left(\ln \lambda - \ln 2 \right)}. \tag{5.151}$$

The main part of the asymptotic formula (5.151) coincides with the result obtained in [187].

2) Let us show that for a hexagonal array an expression for the effective thermal conductivity parameter can be obtained in the case of extremely large sizes $(a = 1)$ of non-conductive inclusions $(\lambda \to 0)$ in the presence of a thin layer at the interface of the composite phases.

Using the expression for the effective parameter at $\lambda \ll 1$, obtained in [122], we represent the leading term of the asymptotic expansion in the form

$$q = \frac{\sqrt{1 - \frac{1-\lambda}{1+\lambda}}}{\pi\sqrt{3}\left(\frac{1-\lambda}{1+\lambda}\right)}. \tag{5.152}$$

Rewriting (5.152) using a non-smooth argument transformation and replacing the expression

$$\frac{2\left(\frac{1-\tilde{\lambda}}{1+\tilde{\lambda}}\right)}{\sqrt{1 - \frac{1-\tilde{\lambda}}{1+\tilde{\lambda}}}}$$

by an asymptotically equivalent function, we have

$$\frac{2\left(\frac{1-\tilde{\lambda}}{1+\tilde{\lambda}}\right)}{\sqrt{1 - \frac{1-\tilde{\lambda}}{1+\tilde{\lambda}}}} \sim \frac{2\left(\frac{\tilde{\lambda}}{2} + \frac{1}{2\tilde{\lambda}} - 1\right)^{\frac{3}{2}}}{\ln\left(\frac{\tilde{\lambda}}{2} + \frac{1}{2\tilde{\lambda}}\right)\left(\frac{\tilde{\lambda}}{2} + \frac{1}{2\tilde{\lambda}}\right)^{\frac{1}{2}}\left(\frac{\tilde{\lambda}}{2} + \frac{1}{2\tilde{\lambda}} + 1\right)^{\frac{1}{4}}\left(2\tilde{\lambda}\right)^{\frac{1}{4}}} \tag{5.153}$$

at $\tilde{\lambda} \to 1$.

Moreover, the expansions into series of functions in both parts of the asymptotic equality (5.153) coincide up to terms of the order $\left(1 - \tilde{\lambda}\right)^4$ inclusive:

$$\frac{2\left(\frac{1-\tilde{\lambda}}{1+\tilde{\lambda}}\right)}{\sqrt{1 - \frac{1-\tilde{\lambda}}{1+\tilde{\lambda}}}} = 1 - \tilde{\lambda} + \frac{3}{4}\left(1 - \tilde{\lambda}\right)^2 + \frac{19}{32}\left(1 - \tilde{\lambda}\right)^3 + \frac{63}{128}\left(1 - \tilde{\lambda}\right)^4 +$$

$$+ \frac{867}{2048}\left(1 - \tilde{\lambda}\right)^5 + O\left(1 - \tilde{\lambda}\right)^6;$$

$$\frac{2\left(\frac{\tilde{\lambda}}{2} + \frac{1}{2\tilde{\lambda}} - 1\right)^{\frac{3}{2}}}{\ln\left(\frac{\tilde{\lambda}}{2} + \frac{1}{2\tilde{\lambda}}\right)\left(\frac{\tilde{\lambda}}{2} + \frac{1}{2\tilde{\lambda}}\right)^{\frac{1}{2}}\left(\frac{\tilde{\lambda}}{2} + \frac{1}{2\tilde{\lambda}} + 1\right)^{\frac{1}{4}}\left(2\tilde{\lambda}\right)^{\frac{1}{4}}} = 1 - \tilde{\lambda} + \frac{3}{4}\left(1 \quad \tilde{\lambda}\right)^2 +$$

$$+ \frac{19}{32}\left(1 - \tilde{\lambda}\right)^3 + \frac{63}{128}\left(1 - \tilde{\lambda}\right)^4 + \frac{2665}{6144}\left(1 - \tilde{\lambda}\right)^5 + O\left(1 - \tilde{\lambda}\right)^6.$$

Using (5.116), (5.117) at $\lambda \to 0$, we get

$$\frac{2\left(\frac{\tilde{\lambda}}{2}+\frac{1}{2\lambda}-1\right)^{\frac{3}{2}}}{\ln\left(\frac{\tilde{\lambda}}{2}+\frac{1}{2\lambda}\right)\left(\frac{\tilde{\lambda}}{2}+\frac{1}{2\lambda}\right)^{\frac{1}{2}}\left(\frac{\tilde{\lambda}}{2}+\frac{1}{2\lambda}+1\right)^{\frac{1}{4}}\left(2\tilde{\lambda}\right)^{\frac{1}{4}}} \sim \frac{1}{\lambda\ln\frac{1}{2\lambda}}. \qquad (5.154)$$

Cosequently, the expression of the effective parameter (5.152) is transformed in the case of contact of non-conductive inclusions in the presence of a thin layer at the interface of the composite phases to the form:

$$q_{hex}^{(0)} = \frac{2\lambda\left(\ln\lambda^{-1}-\ln 2\right)}{\pi\sqrt{3}}. \qquad (5.155)$$

Relations (5.151), (5.155), obtained, respectively, for absolutely conductive and non-conductive inclusions of extremely large size $(a=1)$ with a thin interlayer at the phase boundary, satisfy Keller's theorem relation

$$q_{hex}^{(\infty)}(\lambda) = \frac{1}{q_{hex}^{(0)}(\lambda^{-1})}.$$

5.4 PERCOLATION THRESHOLD FOR ELASTIC PROBLEMS: SELF-CONSISTENT APPROACH AND PADÉ APPROXIMANTS

Mathematical models of composite materials can be quite complex due to the distribution and orientation of multiple inclusions within the matrix. The properties of the inclusions usually differ significantly from the properties of the matrix. If the distribution of inclusions is completely random, then with an increase in their concentration c_1, chains of contacting inclusions (clusters) are formed in the material. The critical value of $c_1 = c_p$, at which a cluster of infinite length is formed, is called the percolation threshold.

The properties of such composite materials cannot be described in terms of regular or quasi-regular models. In this case, it is necessary to use the theory of percolation, which has been intensively developed in recent decades [94, 95, 96, 216, 218, 222, 226]. The effective characteristics k_0 of the composite near the percolation threshold $(c_1 \to c_p)$ are determined by asymptotic relations of the form

$$k_0 \sim |c_1 - c_p|^t,$$

where c_p is the critical volume fraction of inclusions; t is the critical index of the corresponding physical property of the composite.

A review of various models of percolation media, the corresponding methods for calculating the percolation threshold, and critical exponents is presented in [222, 226]. At the same time, it should be noted that there are still certain discrepancies between the results of different authors, especially in the 3D case.

For conductivity problems, it has been shown that the Bruggemann formulas (self-consistent approximation) make it possible to qualitatively describe the percolation threshold [51, 216]. However, the accuracy of the Bruggemann formula is not high enough. In [216], a modification of the Maxwell formula based on AP is proposed, which makes it possible to give a qualitative explanation for the existence of the percolation threshold.

In this chapter, we analyze the use of the self-consistency theory and the AP apparatus for calculating the percolation threshold in the problem of elasticity theory. The percolation threshold depends on the shape of the inclusions. Let us consider the case of a three-dimensional medium with spherical inclusions.

5.4.1 SELF-CONSISTENCY THEORY APPROXIMATION

Within the framework of the self-consistency approach [62, 212], we arrive at the following equations for the effective shear modulus μ^*, bulk modulus K^*, and the Poisson's coefficient v^*:

$$\frac{c_1}{1+\alpha^*\left(\frac{K_1}{K^*}-1\right)}+\frac{c_2}{1+\alpha^*\left(\frac{K_2}{K^*}-1\right)}=1,$$

$$\frac{c_1}{1+\beta^*\left(\frac{\mu_1}{\mu^*}-1\right)}+\frac{c_2}{1+\beta^*\left(\frac{\mu_2}{\mu^*}-1\right)}=1, \qquad (5.156)$$

$$v^*=\frac{3K^*-2\mu^*}{6K^*+2\mu^*},$$

$$\alpha^*=\frac{1+v^*}{3\left(1-v^*\right)}, \quad \beta^*=\frac{2\left(4-5v^*\right)}{15\left(1-v^*\right)}, \qquad (5.157)$$

where K_1, μ_1 and K_2, μ_2 are the elastic constants of inclusions and matrices, respectively; c_1, c_2 are their volume fractions, $c_1+c_2=1$.

The system of equations (5.156), (5.157) admits an exact analytical solution (we do not present this solution here because of its cumbersomeness), which allows us to obtain an expression for the effective Young's modulus E^*.

Fig. 5.48 shows the graphs of the effective Young's modulus E^*, obtained as a result of solving the system of equations (5.156), (5.157), for various values of the elastic characteristics of the matrix K_2, μ_2 and inclusions K_1, μ_1. An analysis of these dependences shows that in the case of rigid inclusions, the elastic characteristics of which significantly exceed the values of the corresponding matrix parameters $(K_1 \gg K_2, \mu_1 \gg \mu_2)$, self-consistent solution qualitatively describes the percolation threshold. In particular, for the values of the elastic constants of the matrix $K_2=10^{10}$, $\mu_2=10^5$ and inclusions $K_1=10^{12}$, $\mu_1=10^{12}$ the percolation threshold obtained using the self-consistent approach agrees with the experimental data of [43], where it was shown that for a composite with the indicated elastic characteristics, the value of the percolation threshold lies between 0.40 and 0.41.

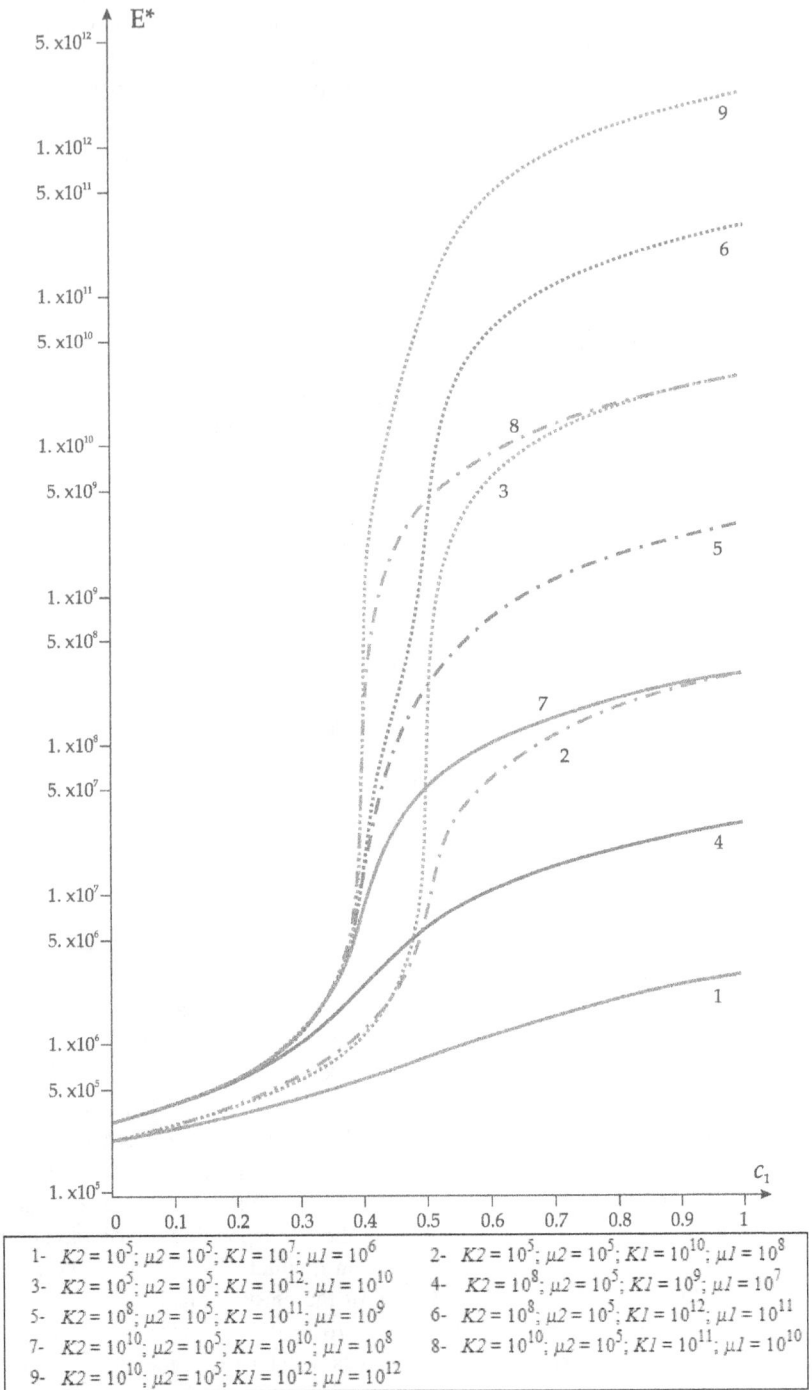

Figure 5.48 Graphs of the effective Young's modulus E^* for various values of the elastic characteristics of the matrix and inclusions.

5.4.2 AP FOR VIRIAL EXPANSIONS

Virial expansions for the effective shear modulus μ^* and the bulk modulus of elasticity K^* at low concentrations of inclusions can be written as follows [212]:

$$\mu^* = \left[1 + \frac{15(1-v_2)(\mu_1-\mu_2)c_1}{\mu_2(7-5v_2)+2\mu_1(4-5v_2)}\right]\mu_2, \qquad (5.158)$$

$$K^* = \left[1 + \frac{3(1-v_2)(K_1-K_2)c_1}{2K_2(1-2v_2)+K_1(1+v_2)}\right]K_2. \qquad (5.159)$$

Using known relations

$$E = \frac{9K\mu}{3K+\mu}, \qquad v = \frac{1}{2}\frac{3K-2\mu}{3K+\mu}$$

and formulas (5.158), (5.159), we obtain an expression for the effective Young's modulus E^*:

$$E^* = \left[9\mu_2\left(4\mu_2^2(2-5c_1)+\mu_1(3K_2(3-5_1)+4\mu_1(3+5_1))+\right.\right.$$
$$+5K_2\mu_1(2+3_1))\left(\mu_2(4K_2(1-c_1)+4K_1c_1)+\right.$$
$$+3K_2(-K_2c_1+K_1(1+c_1)))\right] / \left[16(2-5c_1)\mu_2^4+\right. \qquad (5.160)$$
$$+4(3K_2(11-13c_1)+3K_1(2+3c_1)+4\mu_1(3+5c_1))\mu_2^3+$$
$$+3(K_2(28\mu_1(2-c_1)+3K_1(11+15c_1))+4K_1\mu_1(3+17c_1))\mu_2^2+$$
$$+9K_2\left(K_2\left(4\mu_1(2-5c_1)+9(K_1(1+c_1)-K_2c_1)\right)+\right.$$
$$+K_1\mu_1(14+25c_1)\mu_2)+54K_2^2\mu_1(-K_2c_1+K_1(1+c_1))\right].$$

The area of applicability of the virial expansion method is limited by low concentrations of one of the components; therefore, relations (5.158)–(5.160) are inapplicable for large inclusions and do not describe the percolation threshold even at a qualitative level. To improve formula (5.160), we use AP. AP $[0/1]$ for E^*, and we get

$$E^*{}^{(0)}_{[0/1]}(c_1) = \frac{9K_2^2\mu_2\Delta_1^{(1)}}{K_2(3K_2+\mu_2)\Delta_1^{(1)}-(3K_2+4\mu_2)\Delta_2^{(1)}c_1} \qquad (5.161)$$

where

$$\Delta_1^{(1)} = (3K_1+4\mu_2)\left(6K_2\mu_1+9K_2\mu_2+12\mu_1\mu_2+8\mu_2^2\right),$$

$$\Delta_2^{(1)} = 45K_1K_2^2(\mu_1-\mu_2)+3K_1K_2\mu_2(2\mu_1+3\mu_3)+$$
$$+3K_2^2\mu_2(18\mu_1-23\mu_2)+4(K_1-K_2)\mu_2^2(3\mu_1+2\mu_2).$$

Similarly, we construct AP $[1/0]$ for the expression obtained from (5.160), taking into account the replacement: $K_{1(2)} \rightleftarrows K_{2(1)}, \mu_{1(2)} \rightleftarrows \mu_{2(1)}, c_1 \rightarrow c_2$. From a

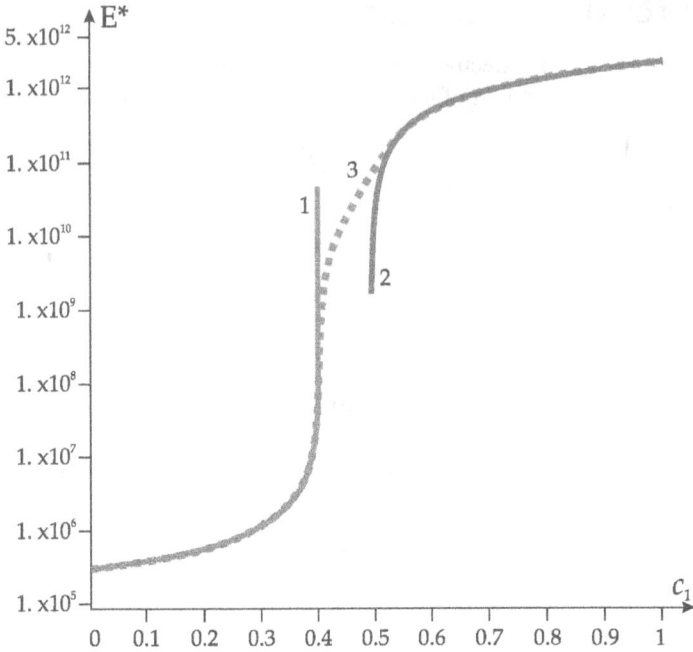

Figure 5.49 Comparison of the results of calculating the effective Young's modulus by the self-consistent approach and AP (1– AP[0/1], 2– AP[1/0], 3– self-consistent).

physical point of view, such a replacement means that we have changed the roles of the phases of the composite. In this case, the corresponding AP is governed by the formula

$$E^{*\,(0)}_{\,[1/0]}\,(c_2) = \frac{9K_1\mu_1}{3K_1+\mu_1} + \frac{9\mu_1\,(3K_1+4\mu_1)\,\Delta_2^{(2)}}{(3K_1+\mu_1)^2\Delta_1^{(2)}}\,c_2, \tag{5.162}$$

where $\Delta_i^{(2)} = \Delta_i^{(1)}$ $(i = 1, 2)$ subject to replacement: $K_{1(2)} \rightleftarrows K_{2(1)}, \mu_{1(2)} \rightleftarrows \mu_{2(1)}$.
Passing in equation (5.162) to the variable c_1 $(c_2 = 1 - c_1)$, we finally have

$$E^{*\,(1)}_{\,[1/0]}\,(c_1) = \tag{5.163}$$

$$= \frac{9\mu_1\left((3K_1+\mu_1)\left(K_1\Delta_1^{(2)}+\Delta_2^{(2)}\right)+\mu_1\Delta_2^{(2)} - (3K_1+4\mu_1)\,\Delta_2^{(2)}c_1\right)}{(3K_1+\mu_1)^2\Delta_1^{(2)}}.$$

Fig. 5.49 presents effective Young's modulus E^* for $K_1 = 10^{12}$, $\mu_1 = 10^{12}$, $K_2 = 10^{10}$, $\mu_2 = 10^5$, obtained by the self-consistency method (5.156), (5.157) and using AP (5.162), (5.163).

Thus, analyzing the calculation results, we can conclude that:

(i) AP makes it possible to significantly expand the range of applicability of the virial expansion;

(ii) comparison with the self-consistent solution shows that AP (5.161) correctly describes the effective parameter up to the percolation threshold;

(iii) the AP relation (5.163) can be used for large inclusions: the results almost coincide with the self-consistent solution.

5.4.3 ESTIMATION OF AP ACCURACY

In order to estimate the accuracy of the AP, we use the Hill equation [117], which is an exact relation that does not depend on the microstructure of the composite. This equation is valid for composites consisting of isotropic components having the same shear modulus. For a two-dimensional two component composite with $\mu_1 = \mu_2 = \mu$, this equation is written as follows:

$$\frac{1}{K^* + \mu} = \frac{c_1}{K_1 + \mu} + \frac{c_2}{K_2 + \mu}. \tag{5.164}$$

From (5.164) yields the expression for the effective bulk modulus of elasticity

$$K_H^* = \frac{(c_1 K_1 + c_2 K_2)\,\mu + K_1 K_2}{c_1 K_2 + c_2 K_1 + \mu}. \tag{5.165}$$

Comparison with the exact solution (5.165) of the expression

$$K^* = \left[1 + \frac{(3K_2 + 4\mu)\,(K_1 - K_2)\,c_1}{(3K_1 + 4\mu)\,K_2}\right] K_2, \tag{5.166}$$

obtained from (5.159) in the particular case $\mu_1 = \mu_2 = \mu$ indicates a very limited area of applicability of the latter (Fig. 5.50, dotted line).

We transform solution (5.166) into AP [0/1] for a low concentration of inclusions, $c_1 \to 0$:

$$K^{*\,(0)}_{[0/1]}(c_1) = \frac{(3K_1 + 4\mu)\,K_2^2}{(3K_1 + 4\mu)\,K_2 - (3K_2 + 4\mu)\,(K_1 - K_2)\,c_1}. \tag{5.167}$$

Similarly, we construct AP [1/0] for a high concentration of inclusions, $c_2 = 1 - c_1 \to 0$:

$$K^{*\,(1)}_{[1/0]}(c_1) = \tag{5.168}$$

$$= \frac{(3K_1 + 4\mu)\,K_2 - 3\,(K_1 - K_2)\,K_1 + (3K_1 + 4\mu)\,(K_1 - K_2)\,c_1}{3K_2 + 4\mu}.$$

Results based on AP (5.167), (5.168) are close to the exact solution (5.165) for small and large inclusions, respectively (Fig. 5.50).

Thus, as follows from the analysis of the calculated data, AP (5.161) correctly describes the effective parameter up to the beginning of the percolation threshold.

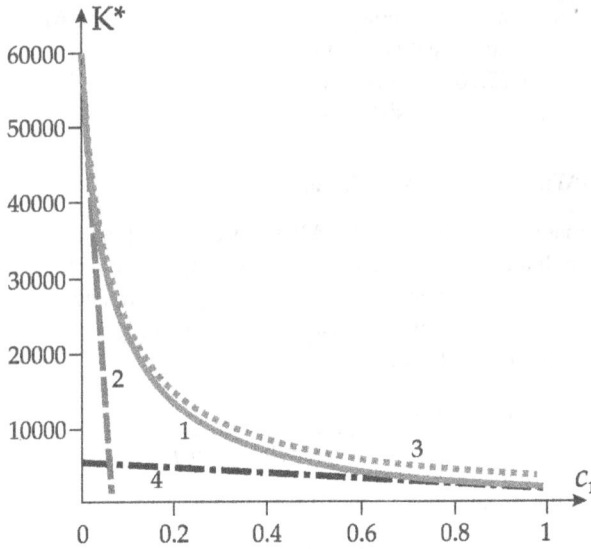

Figure 5.50 Comparison of the exact solution (5.165) with AP (5.167), (5.168) in the particular case of identical shear moduli of the composite components (1– (5.165) Hill, 2– (5.166), 3– AP[0/1], 4– AP[1/0], $K_1 = 2.4 \cdot 10^3$, $K_2 = 60 \cdot 10^3$, $\mu = 1.2 \cdot 10^3$).

6 Construction of Corrections to the Maxwell Formula

This chapter is devoted to construction of corrections to the Maxwell formula (MF). First, in section 6.1, a brief history to the problem is given, and the state of the art of the problem is described. Section 6.2 presents analysis of the MF formula based on a two phase composite model (2PhM). It includes the study of composites with parallelepiped inclusions, solution of the problem for a composite with small cylindrical inclusions, and analysis of the 2PhM solution for small inclusions. Section 6.3 is focused on refinement of the Maxwell formula for a fibrous composite with square cross section inclusions. The carried out consideration include: higher-order iterations of 2PhM, the first corrections to the MF, analysis of the Schwarz alternating method solution, inclusions with regard to heat insulators of large sizes, asymptotic estimation for the corrected MF, numerical results, and the generalization of the problem. Refinement of the MF for composites reinforced by circle cross-section fibers is investigated in section 6.4. It contains the following research steps: introduction to the problem, solving of the local problem using the Schwartz alternating method, construction of the generalized formulas for MF, analysis of the solution based on the N-iterative procedure of the Schwartz alternating method, MF refinement using the Padé approximations, as well as the combined Schwarz–Padé approach.

6.1 MAXWELL APPROXIMATION: APPLICATIONS, ESTIMATES, GENERALIZATIONS

The classical MF [154] is one of the most successful, formulas of the theory of composites, combining simplicity and elegance with high accuracy, evident physical meaning, and a wide range of practical applications. This approximation is also called the Maxwell Garnett, Maxwell-Odelevsky, Clausius-Mossotti, Lorenz-Lorentz, Landauer, Wiener-Wagner formula [140, 141, 144, 161, 188, 213, 226, 72, 147, 148, 150, 149, 180, 234].

Originally, MF was obtained for the conductivity of a dilute suspension of conducting spheres in a conducting matrix [154]. Subsequently, the Maxwell model found wide application in physics in determining the effective permittivity of matrix media [220], and the volume fraction of the inclusions is small. In [78], MF was generalized for the dynamic analysis of a composite dielectric random medium with small spherical inclusions. It has been found that the MF remains very accurate at high concentrations of scatterers, describes the effects of multiple scattering, as well as the weakening of the mean field. In [66], MF was used to derive the equation for the magnetic permeability of an isotropic material contained in a cluster of spherical homogeneous magnetic particles. The book [232] presents the effective medium

DOI: 10.1201/9781003391029-6

theory for continuous structures and discusses the applicability of the MF within the framework of this approach. In [131], MF was used to study the problem of obtaining the needed optical frequency characteristics of composite media containing conducting particles. The authors of the article [211] generalized MF for chiral materials and expressed the permittivity, magnetic permeability, and chirality of a two component mixture with spherical inclusions as functions of the parameters of the components. The effective properties of nanocomposite structures using MF are studied in [161]. In [87], MF was used for simulation of the dielectric properties of a binary mixture of a nematic liquid crystal and the carbon nanotubes.

In [131, 132, 143, 214, 220], the MF was generalized to the case of anisotropic inclusions. The modified MF for inclusions in the form of ellipsoids, infinite cylinders, and plates is given in [214, 220]. The authors of [143] used MF to calculate the effective dielectric property of a medium consisting of anisotropic inclusions placed in an isotropic matrix. This approximation has been used to analyze organic polymers consisting of randomly oriented polymer chains embedded in an isotropic matrix.

In the mechanics of composites, the use of the MF model is associated, first of all, with estimates of the effective parameters: it is included in the expression of the known Hashin–Shtrikman bounds [71, 110, 111, 161], which are the best ones that can be obtained without specifying the inclusion form. Namely, the MF represents the lower bound of the effective composite parameter if the ratio of the inclusion conductivity to the matrix conductivity is greater than 1, and the upper bound if this ratio is less than 1. In the book by Berdichevsky [49], it is shown that MF gives a very good approximation for the effective thermal conductivity coefficient of cubic lattices; corrections were found for various types of lattices: simple cubic, body-centered, and face-centered.

Generally speaking, MF was obtained under the assumption of a low concentration of inclusions c, and therefore, it is commonly believed [216, 218] that the MF approximation works well only for $c \ll 1$. For problems of thermal conductivity this means that for large sizes of inclusions of sufficiently high conductivity, the MF does not describe the formation of an infinite cluster in the composite and does not give estimates of the percolation treshold. However, in the general case of inclusions of an arbitrary shape, MF also works quite well for medium and large sizes of inclusions, except for high contrast composites [161, 226].

The analysis of the solution and comparison with the known asymptotic expansions showed that in the case of square inclusions, the expression MF quite accurately describes the effective thermal conductivity parameter over the entire range of the geometric size of inclusions a and their thermal conductivity λ, including limiting cases: $a \to 0$, $a \to 1$ and $\lambda \to 0$, $\lambda \to \infty$.

The refinement and generalizations of MF are still of interest and attract the attention of researchers, despite more than a century of history of its existence. In the monograph by Milton [161] for two- and three-dimensional composites, two-sided Hashin–Shtrikman estimates of the effective parameters are given and corrections to them are constructed, taking into account the shape, degree of inhomogeneity, and the nature of the distribution of inhomogeneities in the composite structure. In other words, relations are determined that make it possible to refine the MF solution for 2D and 3D composites, taking into account the geometry of the phase arrangement and the physical characteristics of the composites.

A generalization of MF, considering the arbitrary orientation of ellipsoidal inclusions is presented in [206]. A special case of a thin composite layer with a two-dimensional distribution of inclusions is also considered, and a simplified expression is derived. In [176], the Maxwell approach is used for circular disks; in this case, the final formulas can be written explicitly. The contrast parameter series for effective conductivity is truncated, and the second and third order terms are analyzed. It is shown that for macroscopically isotropic composites, the second order term does not depend on the location of the inclusions, while the third order term already depends. In [142], the Maxwell scheme was formulated in terms of the induced dipole moments of a representative volume element of the composite and a properly defined equivalent inclusion. The numerical study showed that the proposed version of the MF yields good estimation of the effective properties of composites with a periodic and random structure and with an accuracy comparable to that of the Rayleigh method.

Paper [164] is devoted to the problem of conductivity of doubly periodic composite materials with circular inclusions. An exact relation is obtained for the effective conductivity tensor, which reduces to MF in the case of small conductivity of inclusions.

In a number of publications, the Maxwell approach has been developed in combination with the cluster method. Thus, in [170], the generalization of Maxwell's approach from problems for single inclusions to problems with n inclusions was obtained. The MF, which is a special case of effective medium theory models [69], has a number of modifications, such as the Maxwell-Burger-Aiken model [84], the Maxwell-Garnett model [155], and their submodels [189, 201].

Analysis of the considered scientific publications allows us to draw the following conclusions:

(i) in the theory of composites, there are significant number of theoretical and practical problems for which it is expedient and natural to use relations of the MF type and/or its modifications;

(ii) the accuracy of MF is not always sufficient;

(iii) refinement of MF can be based on the use of certain physically justified hypotheses [142, 161], as well as on the construction of higher approximations of some asymptotic processes [5, 6, 21, 22].

In this chapter, an asymptotic approach is proposed and analytical relations are obtained that refine MF for composites reinforced with fibers of square and circle cross section.

6.2   ANALYSIS OF THE MAXWELL FORMULA BASED ON A TWO-PHASE COMPOSITE MODEL

6.2.1   COMPOSITES WITH PARALLELEPIPED INCLUSIONS

Composites with parallelepiped inclusions have been studied by many authors in relation to solving a wide range of problems in the theory of elasticity, thermal and electrical conductivity, vibrations, etc. Let us note only some of these works.

The monograph by Torquato [226] considers composites with different microstructures, including arrays of oriented rectangles or cylinders of arbitrary but constant cross section. Chaikin and Lubensky [65] studied idealized nematic liquid crystals using the schemes of oriented rectangles $(D = 2)$ and oriented rectangular parallelepipeds $(D = 3)$. Milton [161] investigated heat conduction problems for periodic microstructures that have square cross section inclusions.

Bakhvalov and Panasenko [38] analyzed frame structures that are intensively used in civil engineering. Such systems can be interpreted as composites with parallelepiped cavities. In book [231], it is noted that glass fibers are widely used in the creation of non-metallic structural composites – glass-reinforced plastics. The cross-sectional shapes of glass fibers can be, in particular, a square or a rectangle. The handbook on composite materials [151] describes the form of composite materials called structural sandwich structures. As a filler in such structures, metals, plastics, reinforced plastics, etc., are used. The cell shape can have a different configuration depending on the fillers of the composites, including square profiles.

A two-phase composite material consisting of a continuous matrix with cylindrical inclusions periodically located in it, the geometric sizes of which are small compared to the period of the structure, is considered. The proposed approach is illustrated by the example of determining the effective thermal conductivity of a composite with square fibers forming a simple square lattice (Fig. 6.1).

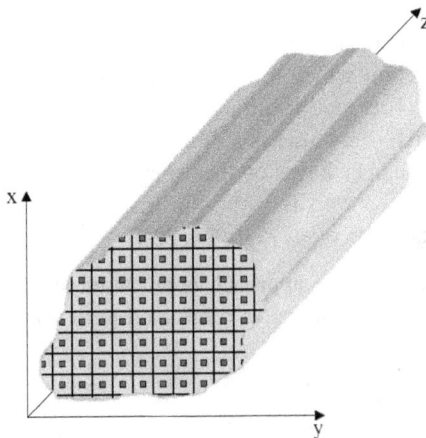

Figure 6.1   Composite structure with small square cylindrical inclusions.

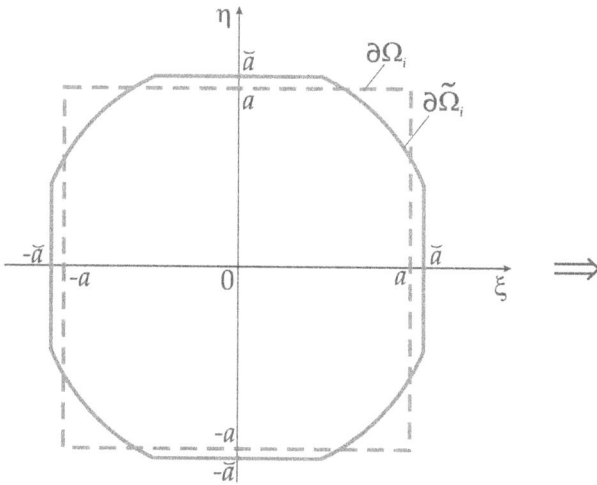

Figure 6.2 Square inclusion approximation: contour of square $\partial\Omega_i$ (dashed line), contour with rounded corners inclusion $\partial\tilde{\Omega}_i$ (solid line).

A cell problem, governed by the Laplace equation (3.105) with conditions at the phase boundary (3.106) and periodicity conditions at the outer contour of the cell (3.107), is considered. The phases of the composite array has different thermal conductivities – λ^+ and λ^-, respectively, in the matrix (region Ω_i^+) and inclusions (region Ω_i^-), where $\frac{\lambda^-}{\lambda^+} = \lambda$.

To solve the problem, the two-phase composite model (2PhM) is proposed. The physical essence of the idealization proposed in 2PhM is determined by the small geometric sizes of the inclusions ($a \ll 1$) and is based on the following assumptions:

(i) the shape of the inclusions does not fundamentally influence the homogenized conductivity;

(ii) periodicity conditions do not significantly affect the conductivity of the structure near the inclusion.

1) The construction of the 2PhM consists of two main stages and is carried out according to the following scheme:

(i) approximation of the square contour of the inclusion based on the PPB (Section 3.2.3);

(ii) transformation of the outer contour of the cell and the use of the Schwarz alternating method [160].

2) We pass from the original square inclusion Ω_i^- to a similar inclusion with rounded corners $\tilde{\Omega}_i^-$ (Fig. 6.2), defined in such a way that the equality of the areas of the original and transformed areas is preserved: $S_{\Omega_i^-} = S_{\tilde{\Omega}_i^-}$.

Then, the equation of the transformed inclusion contour with rounded corners in polar coordinates r, θ is written on the interval $\theta \in \left[0; \frac{\pi}{2}\right]$ in the

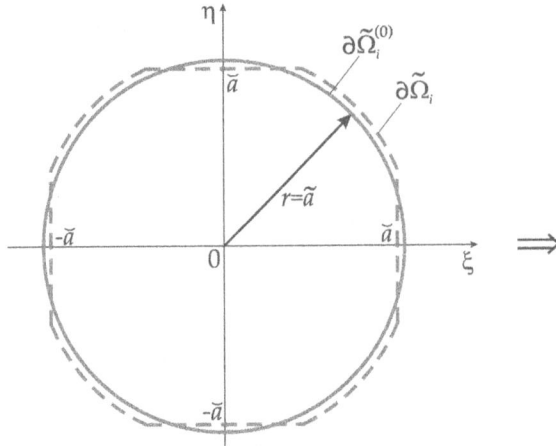

Figure 6.3 Contour in the zeroth approximation: contour with rounded corners inclusion $\partial \tilde{\Omega}_i$ (dashed line), contour of inclusion in zero-order approximation $\partial \tilde{\Omega}_i^{(0)}$ (solid line).

following way:

$$
r = \begin{cases}
\dfrac{\sqrt{4+2\sqrt{2}}}{\sqrt{\pi+2\sqrt{2}}} \dfrac{a}{\cos \theta} = \dfrac{\breve{a}}{\cos \theta} & \text{at} \quad \theta \in \left[0; \dfrac{\pi}{8}\right] \\[3mm]
\dfrac{2\sqrt{2}a}{\sqrt{\pi+2\sqrt{2}}} & \text{at} \quad \theta \in \left[\dfrac{\pi}{8}; \dfrac{3\pi}{8}\right] \\[3mm]
\dfrac{\sqrt{4+2\sqrt{2}}}{\sqrt{\pi+2\sqrt{2}}} \dfrac{a}{\sin \theta} = \dfrac{\breve{a}}{\sin \theta} & \text{at} \quad \theta \in \left[\dfrac{3\pi}{8}; \dfrac{\pi}{2}\right]
\end{cases}
\tag{6.1}
$$

Obviously, due to the symmetry of the region with respect to both coordinate axes ξ, η, the equation of the contour is obtained by relations (6.1) in an even way to the interval $\theta \in [0; 2\pi]$. The described transformation of the square contour makes it possible to use the PPB for solving the problem, according to which the inclusion contour equation can be represented in the form:

$$
r = r_0 + \varepsilon_1 f(\theta),
\tag{6.2}
$$

where $r_0 = const > 0$; $f(\theta)$ is the differentiable function characterizing the shape of the contour; ε_1 is the small parameter ($|\varepsilon_1| \ll 1$).
We represent relation (6.2) as follows:

$$
r = \tilde{a} \left(1 + \varepsilon_1 \cos 4\theta + \ldots\right).
\tag{6.3}
$$

The mathematical meaning of this approximation is that, in the zeroth approximation, the inclusion contour $\partial \tilde{\Omega}_i^{(0)}$ is a circle of radius $\tilde{a}$ (Fig. 6.3).

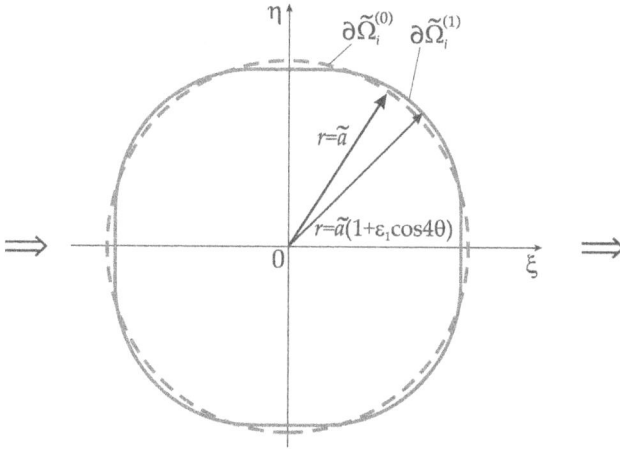

Figure 6.4 Comparison of approximations of the inclusion contour in the zeroth and first approximations: contour of inclusion in zero-order approximation $\partial\tilde{\Omega}_i^{(0)}$ (dashed line), contour of inclusion in first-order approximation $\partial\tilde{\Omega}_i^{(1)}$ (solid line).

Circle radius $\tilde{a}$ is found in such a way that it coincides with the zero term of the expansion in the Fourier series of the function $r(\theta)$, defined by relation (6.1), i.e. describing the transformed inclusion contour with rounded corners.

Then, we get

$$\tilde{a} = \frac{2}{\pi}\left[\int_0^{\frac{\pi}{8}} \frac{\sqrt{4+2\sqrt{2}}}{\sqrt{\pi+2\sqrt{2}}}\frac{a}{\cos\theta}\,d\theta + \int_{\frac{\pi}{8}}^{\frac{3\pi}{8}} \frac{2\sqrt{2}a}{\sqrt{\pi+2\sqrt{2}}}\,d\theta + \right.$$

$$\left. + \int_{\frac{3\pi}{8}}^{\frac{\pi}{2}} \frac{\sqrt{4+2\sqrt{2}}}{\sqrt{\pi+2\sqrt{2}}}\frac{a}{\sin\theta}\,d\theta\right] = \left(\frac{4}{\pi}\ln\frac{2+\sqrt{2-\sqrt{2}}}{\sqrt{2+\sqrt{2}}} + \frac{1}{\sqrt{2+\sqrt{2}}}\right) \times$$

$$\tag{6.4}$$

$$\times \frac{\sqrt{4+2\sqrt{2}}}{\sqrt{\pi+2\sqrt{2}}}a \approx 1.1278\,a \approx \frac{2a}{\sqrt{\pi}}.$$

A small parameter ε_1 is defined as follows:

$$\varepsilon_1 = \frac{\breve{a}-\tilde{a}}{\tilde{a}} = \frac{1 - \frac{4}{\pi}\ln\frac{2+\sqrt{2-\sqrt{2}}}{\sqrt{2+\sqrt{2}}} - \frac{1}{\sqrt{2+\sqrt{2}}}}{\frac{4}{\pi}\ln\frac{2+\sqrt{2-\sqrt{2}}}{\sqrt{2+\sqrt{2}}} + \frac{1}{\sqrt{2+\sqrt{2}}}} \approx -0.0517. \tag{6.5}$$

The inclusion contour $\partial\Omega_i^{(1)}$, taking into account the first term of the asymptotic expansion (6.3), is shown in Fig. 6.4. For comparison, Fig. 6.5

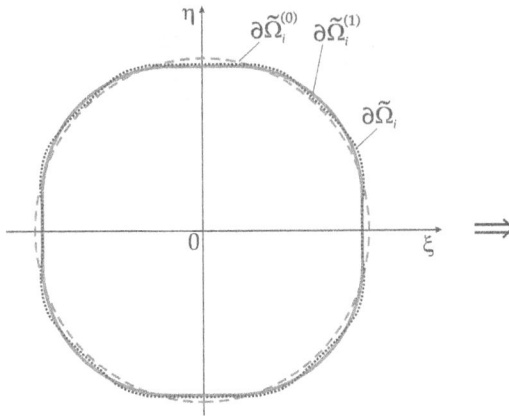

Figure 6.5 Transformed contour of a square inclusion and its approximations: contour with rouded corners inclusion $\partial\tilde{\Omega}_i$ (dotted line), contour of inclusion in zero-order approximation $\partial\tilde{\Omega}_i^{(0)}$ (dashed line), contour of inclusion in first-order approximation $\partial\tilde{\Omega}_i^{(1)}$ (solid line).

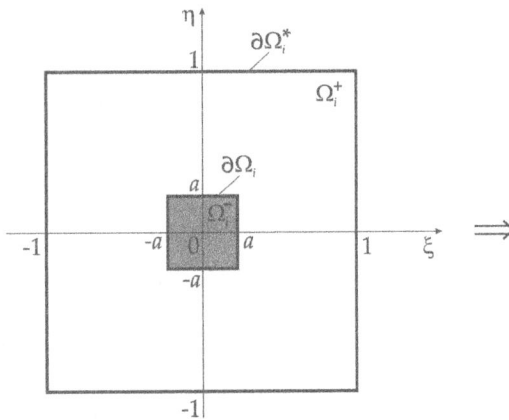

Figure 6.6 Cell with small square inclusion.

illustrates approximations of the inclusion contour with rounded corners $\partial\tilde{\Omega}_i$ taking into account zero $\partial\tilde{\Omega}_i^{(0)}$ and first $\partial\Omega_i^{(1)}$ terms of the asymptotic expansion (6.3).

3) Figures 6.2–6.5 illustrate the approximation of the contour; all the necessary numerical calculations and graphical constructions are performed, taking into account the relations (6.4), (6.5).

Consider the outer contour of the cell, which is a square with side 1, and apply the Schwarz alternating method [160] to the solution of the problem. Due to the small geometric dimensions of the inclusion (Fig. 6.6) in

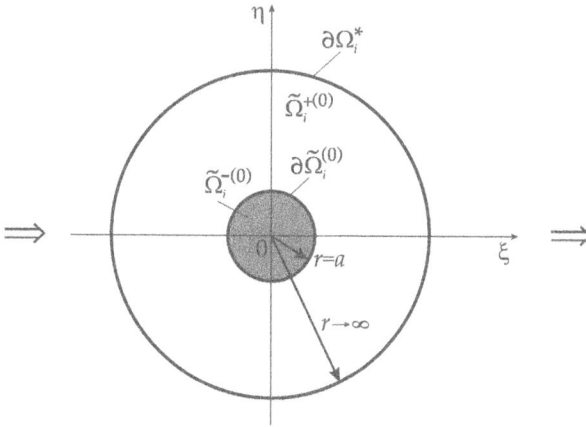

Figure 6.7 Circular inclusion in an infinite domain.

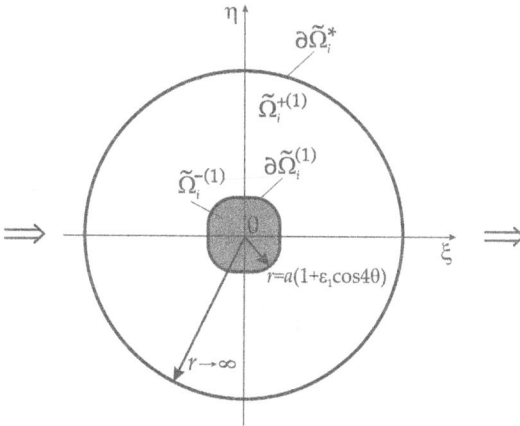

Figure 6.8 An inclusion taking into account the first approximation of its form in an infinite region.

comparison with the sizes of the cell, we pass in the first approximation to the problem of inclusion in an infinite region (Fig. 6.7 and Fig. 6.8). Mathematically, this means that, in the first approximation, the conditions of conjugation of the matrix and inclusion will be exactly satisfied, and the periodicity conditions should be replaced by the decaying conditions at the matrix at infinity.

The solution of the second approximation should be determined in a homogeneous square area of the matrix (Fig. 6.9) and the discrepancies on sides of the cell should be removed. Taking into account the above approximations of the contour, the construction of a two-phase model is carried out in accordance with the one shown in Figs. 6.6–6.9 scheme. Namely, the following problems are solved:

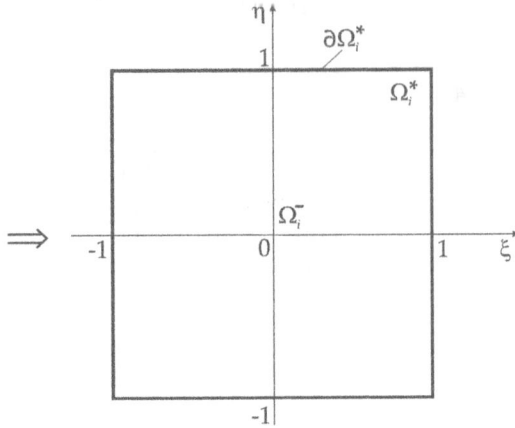

Figure 6.9 Cell without inclusion.

(i) problem for circle inclusion $\tilde{\Omega}_i^{-(0)}$ in an infinite domain $\tilde{\Omega}_i^{+(0)}$ (Fig. 6.7);

(ii) problem, considering the first approximation of contour $\tilde{\Omega}_i^{-(1)}$, in an infinite domain $\tilde{\Omega}_i^{+(1)}$ (Fig. 6.8);

(iii) problem in the area of a cell without inclusion (Fig. 6.9).

6.2.2 SOLUTION OF THE PROBLEM FOR A COMPOSITE WITH SMALL CYLINDRICAL INCLUSIONS

For the problem of thermal conductivity of a doubly periodic composite structure with small ($a \ll 1$) cylindrical inclusions of a square profile (a characteristic structural cell is shown in Fig. 6.6), the local problem is written as follows:

$$\frac{\partial^2 u_1^+}{\partial \xi^2} + \frac{\partial^2 u_1^+}{\partial \eta^2} = 0 \quad \text{in} \quad \Omega_i^+, \tag{6.6}$$

$$\frac{\partial^2 u_1^-}{\partial \xi^2} + \frac{\partial^2 u_1^-}{\partial \eta^2} = 0 \quad \text{in} \quad \Omega_i^-, \tag{6.7}$$

$$u_1^+ = u_1^-, \quad \frac{\partial u_1^+}{\partial \xi} - \lambda \frac{\partial u_1^-}{\partial \xi} = (\lambda - 1) \frac{\partial u_0}{\partial \xi} \quad \text{at} \quad \xi = \pm a,$$

$$u_1^+ = u_1^-, \quad \frac{\partial u_1^+}{\partial \eta} - \lambda \frac{\partial u_1^-}{\partial \eta} = (\lambda - 1) \frac{\partial u_0}{\partial \eta} \quad \text{at} \quad \eta = \pm a \tag{6.8}$$

$$u_1^+ \Big|_{\xi=1} = u_1^+ \Big|_{\xi=-1}, \quad \frac{\partial u_1^+}{\partial \xi} \Big|_{\xi=1} = \frac{\partial u_1^+}{\partial \xi} \Big|_{\xi=-1},$$

$$u_1^+ \Big|_{\eta=1} = u_1^+ \Big|_{\eta=-1}, \quad \frac{\partial u_1^+}{\partial \eta} \Big|_{\eta=1} = \frac{\partial u_1^+}{\partial \eta} \Big|_{\eta=-1}. \tag{6.9}$$

1) We obtain a solution to the problem in the (01) approximation (Fig. 6.7), i.e., the problem for a circle inclusion $\tilde{\Omega}_i^{-\,(0)}$ (zero approximation in representation (6.3) of the PPB) in an infinite domain $\tilde{\Omega}_i^{+\,(0)}$ (first approximation of the Schwarz alternating method).

In this case, the cell problem (6.6)–(6.9) in fast polar coordinates r, θ will be written as follows:

$$\frac{\partial^2 u_1^{-\,(01)}}{\partial r^2} + \frac{1}{r}\frac{\partial u_1^{-\,(01)}}{\partial r} + \frac{1}{r^2}\frac{\partial^2 u_1^{-\,(01)}}{\partial \theta^2} = 0 \quad \text{in} \quad \Omega_i^{-\,(0)}, \tag{6.10}$$

$$\frac{\partial^2 u_1^{+\,(01)}}{\partial r^2} + \frac{1}{r}\frac{\partial u_1^{+\,(01)}}{\partial r} + \frac{1}{r^2}\frac{\partial^2 u_1^{+\,(01)}}{\partial \theta^2} = 0 \quad \text{in} \quad \Omega_i^{+\,(0)}, \tag{6.11}$$

$$u_1^{+\,(01)} = u_1^{-\,(01)},$$

$$\frac{\partial u_1^{+\,(01)}}{\partial r} - \lambda \frac{\partial u_1^{-\,(01)}}{\partial r} = (\lambda - 1)\left(\frac{\partial u_0}{\partial x}\cos\theta + \frac{\partial u_0}{\partial y}\sin\theta\right) \quad \text{at} \quad r = \tilde{a} \tag{6.12}$$

$$u_1^{+\,(01)} \to 0, \quad \frac{\partial u_1^{+\,(01)}}{\partial r} \to 0 \quad \text{at} \quad r \to \infty. \tag{6.13}$$

The solution of the problem (6.10)–(6.13) has the following form

$$u_1^{-\,(01)} = A_1^{(01)} r\cos\theta + A_2^{(01)} r\sin\theta, \tag{6.14}$$

$$u_1^{+\,(01)} = \frac{B_1^{(01)}}{r}\cos\theta + \frac{B_2^{(01)}}{r}\sin\theta, \tag{6.15}$$

where $A_1^{(01)}$, $A_2^{(01)}$, $B_1^{(01)}$, $B_2^{(01)}$ are the arbitrary constants.

Note that the representation of the function $u_1^{-\,(01)}$ in the form (6.14) is written, taking into account, the boundedness of the temperature distribution function and its derivative $\frac{\partial u_1^{-\,(01)}}{\partial r}$ at $r = 0$, and the expression of function $u_1^{+\,(01)}$ (6.15) which satisfies the decaying conditions for these characteristics at $r \to \infty$ (6.13).

Relations (6.14), (6.15) include four arbitrary constants, which are determined from the conjugation conditions (6.12). Systems of equations for finding constants $A_1^{(01)}$, $B_1^{(01)}$ and $A_2^{(01)}$, $B_2^{(01)}$ are completely identical, so we present only one of them:

$$\begin{cases} A_1^{(01)}\tilde{a} = B_1^{(01)}\tilde{a}^{-1} \\ -B_1^{(01)}\tilde{a}^{-2} - \lambda A_1^{(01)} = \dfrac{\partial u_0}{\partial x}(\lambda - 1). \end{cases} \tag{6.16}$$

Solving the system of equations (6.16), we find the values of the constants in the form

$$\begin{cases} A_1^{(01)} = -\dfrac{\lambda - 1}{\lambda + 1} \dfrac{\partial u_0}{\partial x}, \\ B_1^{(01)} = -\dfrac{(\lambda - 1)\, \tilde{a}^2}{\lambda + 1} \dfrac{\partial u_0}{\partial x}. \end{cases} \tag{6.17}$$

Obviously, for constants $A_2^{(01)}$, $B_2^{(01)}$, we have

$$A_2^{(01)} = A_1^{(01)}, \qquad B_2^{(01)} = B_1^{(01)} \left(\frac{\partial u_0}{\partial x} \to \frac{\partial u_0}{\partial y} \right). \tag{6.18}$$

Then relations (6.17), (6.18) can be represented as follows:

$$A_1^{(01)} = \frac{\partial u_0}{\partial x} A^{(01)*}, \qquad A_2^{(01)} = \frac{\partial u_0}{\partial y} A^{(01)*},$$

$$B_1^{(01)} = \frac{\partial u_0}{\partial x} B^{(01)*}, \qquad B_2^{(01)} = \frac{\partial u_0}{\partial y} B^{(01)*}, \tag{6.19}$$

$$A^{(01)*} = -\frac{\lambda - 1}{\lambda + 1}, \qquad B^{(01)*} = -\frac{(\lambda - 1)\, \tilde{a}^2}{\lambda + 1}. \tag{6.20}$$

2) In solving the problem, we restrict ourselves to terms of the zero approximation in representation (6.3) and proceed to the construction of the second approximation of the Schwarz alternating method (Fig. 6.9).
We now satisfy the periodicity conditions (6.9) on opposite sides of the cell, neglecting the conditions (6.12). Function $u_1^{(02)}$ is defined as a solution to a boundary value problem of the following form

$$\Delta u_1^{(02)} = 0 \quad \text{in} \quad \Omega_i^*, \tag{6.21}$$

$$\left(u_1^{+\,(01)} + u_1^{(02)} \right) \Big|_{\xi = 1} = \left(u_1^{+\,(01)} + u_1^{(02)} \right) \Big|_{\xi = -1},$$

$$\frac{\partial \left(u_1^{+\,(01)} + u_1^{(02)} \right)}{\partial \xi} \Big|_{\xi = 1} = \frac{\partial \left(u_1^{+\,(01)} + u_1^{(02)} \right)}{\partial \xi} \Big|_{\xi = -1}, \tag{6.22}$$

$$\left(u_1^{+\,(01)} + u_1^{(02)} \right) \Big|_{\eta = 1} = \left(u_1^{+\,(01)} + u_1^{(02)} \right) \Big|_{\eta = -1},$$

$$\frac{\partial \left(u_1^{+\,(01)} + u_1^{(02)} \right)}{\partial \eta} \Big|_{\eta = 1} = \frac{\partial \left(u_1^{+\,(01)} + u_1^{(02)} \right)}{\partial \eta} \Big|_{\eta = -1}. \tag{6.23}$$

Consider

$$u_1^{(02)} = u_{11}^{(02)} + u_{12}^{(02)}, \tag{6.24}$$

where function $u_{11}^{(02)}$ satisfies inhomogeneous boundary conditions in terms of ξ and homogeneous in terms of η, i.e., is a solution to the following problem

$$\Delta u_{11}^{(02)} = 0 \quad \text{in} \quad \Omega_i^*, \tag{6.25}$$

$$\left(u_1^{+ (01)} + u_{11}^{(02)} \right) \Big|_{\xi=1} = \left(u_1^{+ (01)} + u_{11}^{(02)} \right) \Big|_{\xi=-1},$$

$$\frac{\partial \left(u_1^{+ (01)} + u_{11}^{(02)} \right)}{\partial \xi} \Big|_{\xi=1} = \frac{\partial \left(u_1^{+ (01)} + u_{11}^{(02)} \right)}{\partial \xi} \Big|_{\xi=-1}, \tag{6.26}$$

$$u_{11}^{(02)} \Big|_{\eta=1} = u_{11}^{(02)} \Big|_{\eta=-1}, \quad \frac{\partial u_{11}^{(02)}}{\partial \eta} \Big|_{\eta=1} = \frac{\partial u_{11}^{(02)}}{\partial \eta} \Big|_{\eta=-1}. \tag{6.27}$$

It is obvious that the function $u_{12}^{(02)}$ is defined in a similar way, taking into account the change: $\xi \leftrightarrows \eta$.

The general solution of equation (6.25) has the following form

$$u_{11}^{(02)} = A_0^{(02)} + B_0^{(02)} \xi + \sum_{n=1}^{\infty} \left[\left(A_n^{(02)} \cosh \pi n \xi + B_n^{(02)} \sinh \pi n \xi \right) \cos \pi n \eta + \right.$$

$$\left. + \left(C_n^{(02)} \cosh \pi n \xi + D_n^{(02)} \sinh \pi n \xi \right) \sin \pi n \eta \right], \tag{6.28}$$

where $A_0^{(02)}$, $B_0^{(02)}$, $A_n^{(02)}$, $B_n^{(02)}$, $C_n^{(02)}$, $D_n^{(02)}$ $(n = 1, 2, ...)$ are the arbitrary constants.

To satisfy conditions (6.26), we rewrite them, taking into account (6.15), (6.19), (6.20), in the following form

$$u_{11}^{(02)} \Big|_{\xi=1} - u_{11}^{(02)} \Big|_{\xi=-1} = \frac{\partial u_0}{\partial x} \frac{2\tilde{a}^2}{1+\eta^2} \frac{\lambda-1}{\lambda+1},$$

$$\frac{\partial u_{11}^{(02)}}{\partial \xi} \Big|_{\xi=1} - \frac{\partial u_{11}^{(02)}}{\partial \xi} \Big|_{\xi=-1} = -\frac{\partial u_0}{\partial y} \frac{4\tilde{a}^2 \eta}{(1+\eta^2)^2} \frac{\lambda-1}{\lambda+1}. \tag{6.29}$$

We expand the functions on the right-hand sides of relations (6.29) into the Fourier series, and we get

$$\frac{1}{1+\eta^2} = \frac{\pi}{4} + \sum_{n=1}^{\infty} \left[e^{-\pi n} \operatorname{Im} E_1 \left(-\pi n + i\pi n \right) - \right.$$

$$\left. - e^{\pi n} \operatorname{Im} E_1 \left(\pi n + i\pi n \right) + \pi e^{-\pi n} \right] \cos \pi n \eta$$

$$\frac{\eta}{(1+\eta^2)^2} = \frac{\pi}{2} \sum_{n=1}^{\infty} n \left[e^{\pi n} \text{ Im } E_1 \left(\pi n + i\pi n \right) - \right. \tag{6.30}$$

$$\left. - e^{\pi n} \text{ Im } E_1 \left(\pi n + i\pi n \right) + \pi e^{-\pi n} \right] \sin \pi n \eta,$$

where $i = \sqrt{-1}$; Im E_1 is the imaginary part of familiar exponential integral [2].

Equating, taking into account the expression (6.30), the corresponding coefficients in (6.29), we find the values of the integration constants

$$A_0^{(02)} = 0, \quad A_n^{(02)} = D_n^{(02)} = 0, \quad n = 1, 2, \dots,$$

$$B_0^{(02)} = \frac{\partial u_0}{\partial x} B_0^{(02)*}, \quad B_n^{(02)} = \frac{\partial u_0}{\partial x} B_n^{(02)*}, \quad n = 1, 2, \dots, \tag{6.31}$$

$$C_n^{(02)} = \frac{\partial u_0}{\partial y} C_n^{(02)*}, \quad n = 1, 2, \dots,$$

$$B_0^{(02)*} = \frac{\lambda - 1}{\lambda + 1} \frac{\pi \tilde{a}^2}{4}, \tag{6.32}$$

$$B_n^{(02)*} = -C_n^{(02)*} = \frac{\lambda - 1}{\lambda + 1} \tilde{a}^2 S_n \tag{6.33}$$

where

$$S_n = \frac{e^{-\pi n} \text{ Im } E_1 \left(-\pi n + i\pi n \right) - e^{\pi n} \text{ Im } E_1 \left(\pi n + i\pi n \right) + \pi e^{-\pi n}}{\sinh \pi n}. \tag{6.34}$$

Therefore, we finally get

$$u_{11}^{(02)} = \frac{\partial u_0}{\partial x} B_0^{(02)*} \xi +$$

$$+ \sum_{n=1}^{\infty} \left(\frac{\partial u_0}{\partial x} B_n^{(02)*} \sinh \pi n \xi \cos \pi n \eta - \frac{\partial u_0}{\partial y} C_n^{(02)*} \cosh \pi n \xi \sin \pi n \eta \right). \tag{6.35}$$

It is obvious that the expression of function $u_{12}^{(02)}$ is written similarly as a result of the following substitutions

$$u_{12}^{(02)} = u_{11}^{(02)} \left(\frac{\partial u_0}{\partial x} \rightleftarrows \frac{\partial u_0}{\partial y}; \ \xi \rightleftarrows \eta \right). \tag{6.36}$$

Thus, the final solution of the local problem, taking into account (01) and (02) approximations, has the form:

$$u_1^{\pm} = u_1^{\pm (01)} + u_{11}^{(02)} + u_{12}^{(02)} = u_1^{\pm (01)} + u_1^{(02)}, \tag{6.37}$$

where

$$u_1^{- (01)} = A^{(01)*} \left(\frac{\partial u_0}{\partial x} \xi + \frac{\partial u_0}{\partial y} \eta \right), \tag{6.38}$$

$$u_1^{+\,(01)} = B^{(01)\,*} \left(\frac{\partial u_0}{\partial x} \frac{\xi}{\xi^2 + \eta^2} + \frac{\partial u_0}{\partial y} \frac{\eta}{\xi^2 + \eta^2} \right), \tag{6.39}$$

$$u_1^{(02)} = B_0^{(02)\,*} \left(\frac{\partial u_0}{\partial x} \xi + \frac{\partial u_0}{\partial y} \eta \right) +$$

$$+ \sum_{n=1}^{\infty} B_n^{(02)\,*} \left[\frac{\partial u_0}{\partial x} \left(\sinh \pi n \xi \cos \pi n \eta - \cosh \pi n \eta \sin \pi n \xi \right) + \right. \tag{6.40}$$

$$\left. + \frac{\partial u_0}{\partial y} \left(\sinh \pi n \eta \cos \pi n \xi - \cosh \pi n \xi \sin \pi n \eta \right) \right],$$

and constants $A^{(01)\,*}$, $B^{(01)\,*}$, $B_n^{(02)\,*}$ $(n = 0, 1, 2, ...)$ are defined by expressions (6.19), (6.20) and (6.31)–(6.34).

The further solution of the problem is reduced in accordance with the homogenization theory to the construction of the homogenized equation.

We apply to the equation

$$\frac{\partial^2 u_0}{\partial x^2} + \frac{\partial^2 u_0}{\partial y^2} + 2 \frac{\partial^2 u_1^+}{\partial x \partial \xi} + 2 \frac{\partial^2 u_1^+}{\partial y \partial \eta} + \frac{\partial^2 u_2^+}{\partial \xi^2} + \frac{\partial^2 u_2^+}{\partial \eta^2} +$$

$$+ \lambda \left(\frac{\partial^2 u_0}{\partial x^2} + \frac{\partial^2 u_0}{\partial y^2} + 2 \frac{\partial^2 u_1^-}{\partial x \partial \xi} + 2 \frac{\partial^2 u_1^-}{\partial y \partial \eta} + \frac{\partial^2 u_2^-}{\partial \xi^2} + \frac{\partial^2 u_2^-}{\partial \eta^2} \right) = F$$

averaging operator (3.25). Then, taking into account the solutions (6.37)–(6.40) and expressions (6.4), (6.19), (6.20), (6.31)–(6.34), we obtain the homogenized equation:

$$q_{2PhM} \left(\frac{\partial^2 u_0}{\partial x^2} + \frac{\partial^2 u_0}{\partial y^2} \right) = F, \tag{6.41}$$

where q_{2PhM} stands for the effective coefficient of thermal conductivity, which, up to terms of the order a^2 inclusive, is defined as follows:

$$q_{2PhM} = 1 + \frac{2(\lambda - 1)}{\lambda + 1} a^2. \tag{6.42}$$

6.2.3 ANALYSIS OF THE 2PHM SOLUTION FOR SMALL INCLUSIONS

Let us analyze the relation (6.42).

1) The expression for the effective parameter asymptotically at $a \to 0$ satisfies the Keller's theorem. Indeed, at $a \to 0$ we have

$$q_{2PhM}^{-1} \left(\frac{1}{\lambda} \right) = \left[1 + \frac{2 \left(\frac{1}{\lambda} - 1 \right)}{\frac{1}{\lambda} + 1} a^2 \right]^{-1} = 1 + \frac{2 \left(\frac{\lambda - 1}{\lambda + 1} \right) a^2}{1 - 2 \left(\frac{\lambda - 1}{\lambda + 1} \right) a^2} \sim$$

$$\sim 1 + \frac{2(\lambda - 1)}{\lambda + 1} a^2 + O\left(a^4 \right) = q_{2PhM} + O\left(a^4 \right).$$

2) Let us consider case $a \to 0$ and write the asymptotic representation of the effective thermal conductivity found on the basis of ThPhM identically coinciding with the MF approximation. We have

$$q_{ThPhM} \equiv q_{2PhM} = \frac{1 - a^2 + \lambda \left(1 + a^2\right)}{1 + a^2 + \lambda \left(1 - a^2\right)} \sim$$

$$\sim 1 + \frac{2\left(\lambda - 1\right)}{\lambda + 1} a^2 + \frac{2\left(\lambda - 1\right)^2}{\left(\lambda + 1\right)^2} a^4 + O\left(a^6\right). \tag{6.43}$$

From expressions (6.42) and (6.43), it directly follows that the effective coefficient of thermal conductivity in the two-term 2PhM approximation coincides up to terms of the order of a^2 inclusive with the ratio found by ThPhM (MF). As known [71], Voigt averaging $\bar{q}$ gives an upper bound of the effective coefficient, Reuss averaging $\underline{q}$ gives lower bound. In the case under consideration, we have

$$\bar{q} = 1 - a^2 + \lambda a^2 = 1 + (\lambda - 1)\, a^2, \tag{6.44}$$

$$\underline{q} = \frac{\lambda}{a^2 + \lambda \left(1 - a^2\right)}. \tag{6.45}$$

Rearranging the expression (6.42) in AP, we get

$$q_{2PhM\;[0/2]}\left(a\right) = \frac{\lambda + 1}{\left(\lambda + 1\right) - 2\left(\lambda - 1\right) a^2},$$

and it is easy to verify that

$$q_{2PhM\;[0/2]}\left(a\right) - \underline{q} = \frac{\left(\lambda - 1\right)^2 a^2}{\left(1 + 2a^2 + \lambda \left(1 - 2a^2\right)\right)\left(a^2 + \lambda \left(1 - a^2\right)\right)} \geq 0 \tag{6.46}$$

for any values $0 \leq \lambda < \infty$ and $a \leq \frac{1}{\sqrt{2}}$. Thus, as follows from (6.44) to (6.46), for all values of $0 \leq \lambda < \infty$, one obtains

$$\underline{q} \leq q_{2PhM} \leq \bar{q}. \tag{6.47}$$

The left side of inequality (6.47) is satisfied even for "not very small" values of the inclusion size $(a \leq \frac{1}{\sqrt{2}} \approx 0.71)$, while the right side of relation (6.47) holds for any values $0 \leq a \leq 1$.

However, it should be noted that the coincidence of the ratios q_{2PhM} (6.42) and q_{MF} (6.43) holds only up to terms of order a^2 inclusive. And this, as follows from (6.43), means that at $1 \leq \lambda < \infty$ the two-term 2PhM approximation gives lower estimation of the effective parameter.

3) $\lambda \to 0$: inclusions of low conductivity, in the limit – a composite with insulators.

In this case, the asymptotics of the effective parameter, found using 2PhM and ThPhM, are

$$q_{2PhM} \sim 1 - 2a^2 + 4a^2\lambda - 4a^2\lambda^2, \tag{6.48}$$

$$q_{ThPhM} \equiv q_{MF} \sim \frac{1-a^2}{1+a^2} + \frac{4a^2}{(1+a^2)^2}\lambda - \frac{4a^2(1-a^2)}{(1+a^2)^3}\lambda^2. \tag{6.49}$$

Obviously, for small values of λ, relation (6.47) will be satisfied for any size of inclusions a, since the lower bound in this case is trivial: $q \to 0$. As follows from expressions (6.48), (6.49), the asymptotics of 2PhM and ThPhM (MF) solutions coincide (up to terms of order a^2 inclusive) only for small values of the parameter a:

$$q_{ThPhM} \sim (1-a^2)^2 + 4a^2(1-2a^2)\lambda - 4a^2(1-a^2)(1-3a^2)\lambda^2. \tag{6.50}$$

This conclusion corresponds to the physical essence of the problem: the sizes of the inclusions have the more significant effect on the effective conductivity of the composite, the more the conductivities of the matrix and inclusions differ.

4) $\lambda \to 1$: thermal conductivity coefficients of the matrix and inclusions of the same order.

The asymptotic representations of the effective parameter expressions obtained using 2PhM and ThPhM are defined, respectively, as

$$q_{2PhM} \sim 1 + a^2(\lambda - 1) - \frac{a^2}{2}(\lambda - 1)^2, \tag{6.51}$$

$$q_{ThPhM} \equiv q_{MF} \sim 1 + a^2(\lambda - 1) - \frac{a^2(1-a^2)}{2}(\lambda - 1)^2, \tag{6.52}$$

i.e., coincide up to terms of the order $\varepsilon = \lambda - 1$ inclusive.

For comparison, Voigt and Reuss averaging leads to the following results:

$$\bar{q} = 1 + a^2(\lambda - 1),$$

$$\underline{q} \sim 1 + a^2(\lambda - 1) - a^2(1 - a^2)(\lambda - 1)^2.$$

5) $\lambda \to \infty$: inclusions of high conductivity. Asymptotic representations of the effective thermal conductivity using 2PhM, ThPhM, and Reuss averaging are determined, respectively, by the following relations

$$q_{2PhM} \sim 1 + 2a^2 - 4a^2\frac{1}{\lambda} + 4a^2\frac{1}{\lambda^2}, \tag{6.53}$$

$$q_{ThPhM} \equiv q_{MF} \sim \frac{1+a^2}{1-a^2} - \frac{4a^2}{(1-a^2)^2}\frac{1}{\lambda} + \frac{4a^2(1+a^2)}{(1-a^2)^3}\frac{1}{\lambda^2}, \tag{6.54}$$

$$\underline{q} \sim \frac{1}{1-a^2} - \frac{a^2}{(1-a^2)^2}\frac{1}{\lambda} + \frac{a^4}{(1-a^2)^3}\frac{1}{\lambda^2}. \qquad (6.55)$$

In this case, Voigt bound is trivial: $\bar{q} \to \infty$.

For small values of a, relations (6.54), (6.55) are transformed, respectively, to the form

$$q_{ThPhM} \equiv q_{MF} \sim \left(1+a^2\right)^2 - 4a^2\left(1+2a^2\right)\frac{1}{\lambda} + \qquad (6.56)$$

$$+ 4a^2\left(1+a^2\right)\left(1+3a^2\right)\frac{1}{\lambda^2},$$

$$\underline{q} \sim 1 + a^2 - a^2\left(1+2a^2\right)\frac{1}{\lambda} + a^4\left(1+3a^2\right)\frac{1}{\lambda^2}. \qquad (6.57)$$

As follows from a comparison of expressions (6.48), (6.50) and (6.53), (6.56), the structure and qualitative nature of the asymptotics of 2PhM and ThPhM at $\lambda \to 0$ and $\lambda \to \infty$ are identical. Based on the latter observations, the following conclusions can be formulated:

(i) first, it corresponds to the essence of Keller's theorem [55];
(ii) secondly, it confirms the closeness of the results obtained for small inclusion sizes a using 2PhM and ThPhM, for all values of the inclusion conductivity.

In addition, analyzing the expressions of the effective parameter found by means of 2PhM and ThPhM, we come to a conclusion that is obvious from a physical point of view: the discrepancy between the results when using these approaches is the greater:

(i) the larger the inclusion size a;
(ii) the more significantly the conductivities of the matrix and inserts differ, i.e., at $\lambda \to 0$ and $\lambda \to \infty$.

In this regard, the following criterion can be proposed for which inclusion can be considered as a "small". If the maximum deviation in the values of the effective coefficient found from 2PhM and ThPhM is no more than 3%, what corresponds to an inclusion size not exceeding the value $a \approx 0.35$, then inclusion is understood as a "small" one.

Figures 6.10–6.13 illustrate the results of calculations of the effective thermal conductivity obtained using the 2PhM and ThPhM (MF), as well as the Voigt-Reuss bounds (in cases where they are not trivial).

Figures 6.10–6.12 present the effective parameter as a function of the size of inclusions a at their fixed conductivity λ. The choice of λ values is due to all possible limiting cases of the physical structure of the composite, which are listed below:

(i) $\lambda = 0$ (Fig. 6.10) corresponds to non-conductive inclusions;
(ii) $\lambda \to \infty$ determines the absolute conductivity of the inclusions (Fig. 6.11);

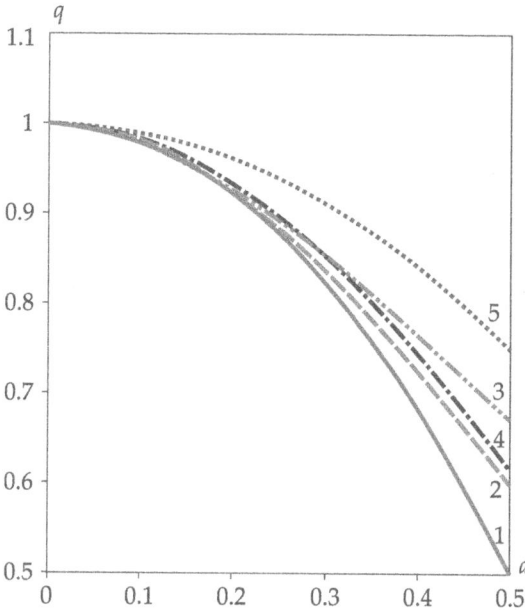

Figure 6.10 Dependence of the effective coefficient q on the size of inclusions a for non-conductive inclusions $\lambda = 0$ (1– 2PhM, 2– MF, 3– Bubnov-Galerkin method, 4– R-function method, 5– Voigt averaging).

(iii) for $\lambda = 0.5$ (Fig. 6.12a) and $\lambda = 2$ (Fig. 6.12b) we have the case when the conductivity of the inclusions and the matrix are of the same order;

(iv) Fig. 6.10 shows effective thermal conductivity calculated according to the two-term 2PhM approximation (6.42) for the case of non-conductive inclusions $\lambda = 0$, presented in comparison with known results, obtained using ThPhM (MF) [27, 28]; the Bubnov-Galerkin method [25]; Rvachev R-functions [26].

Figures 6.13a, Fig. 6.13b show graphs of the effective coefficient q depending on the conductivity of the inclusions with their fixed geometric size a. The choice of calculated values of the size of inclusions a is determined by the following case studies:

(i) $a = 0.15$ (Fig. 6.13a) – very small inclusion;
(ii) $a = 1/3$ (Fig. 6.13b) – small inclusion.

Qualitative picture of the dependence of the effective thermal conductivity coefficient q on two parameters: size of inclusions a and conductivity of inclusions λ, found using the two-term 2PhM approximation (6.42) as shown in Fig. 6.14.

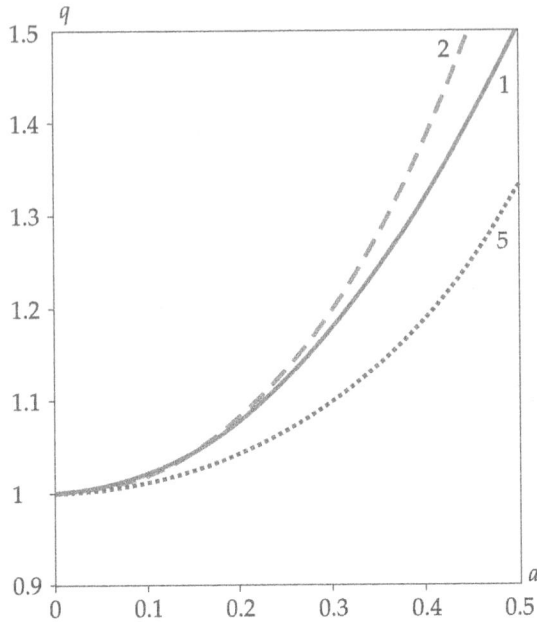

Figure 6.11 Dependence of the effective coefficient q on the size of inclusions a for absolutely conducting inclusions $\lambda \to \infty$ (1– 2PhM, 2– MF, 5– Reuss averaging).

6.3 REFINEMENT OF THE MAXWELL FORMULA FOR A FIBROUS COMPOSITE WITH SQUARE CROSS SECTION INCLUSIONS

6.3.1 HIGHER-ORDER ITERATIONS OF 2PHM

Let us perform one more iteration of the Schwarz alternating method, i.e., construct (03) and (04) approximations, limiting the series in expression (6.35) to terms of the zero approximation:

$$u_1^{(02.0)} = B_0^{(02)*} \left(\frac{\partial u_0}{\partial x} \xi + \frac{\partial u_0}{\partial y} \eta \right). \tag{6.58}$$

Let us represent the solution of the approximation (03) in the following form

$$u_1^{-(03.0)} = A_1^{(03.0)} r \cos \theta + A_2^{(03.0)} r \sin \theta, \tag{6.59}$$

$$u_1^{+(03.0)} = \frac{B_1^{(03.0)}}{r} \cos \theta + \frac{B_2^{(03.0)}}{r} \sin \theta. \tag{6.60}$$

Then from conditions similar to relations (6.16), we find

$$A_1^{(03.0)} = \frac{\partial u_0}{\partial x} A^{(03.0)*}, \qquad A_2^{(03.0)} = \frac{\partial u_0}{\partial y} A^{(03.0)*},$$

$$B_1^{(03.0)} = \frac{\partial u_0}{\partial x} B^{(03.0)*}, \qquad B_2^{(03.0)} = \frac{\partial u_0}{\partial y} B^{(03.0)*}, \tag{6.61}$$

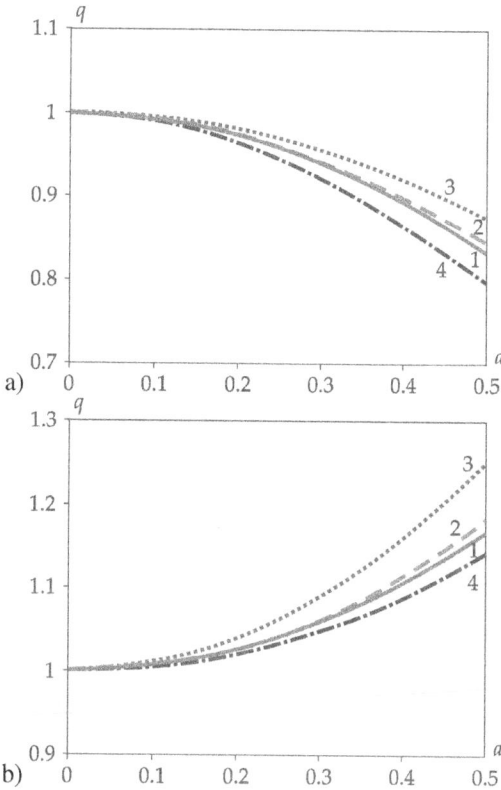

Figure 6.12 Dependence of the effective coefficient q on the size of inclusions a for the case of conductivity of inclusions and a matrix of the same order: a) $\lambda = 0.5$, b) $\lambda = 2$ (1– 2PhM, 2– MF, 3– Voigt averaging, 4– Reuss averaging).

$$A^{(03.0)*} = -\left(\frac{\lambda - 1}{\lambda + 1}\right)^2 \frac{\pi \tilde{a}^2}{4}, \qquad B^{(03.0)*} = -\left(\frac{\lambda - 1}{\lambda + 1}\right)^2 \frac{\pi \tilde{a}^4}{4}. \tag{6.62}$$

In the (04) approximation, we have

$$u_1^{(04.0)} = B_0^{(04.0)*}\left(\frac{\partial u_0}{\partial x}\xi + \frac{\partial u_0}{\partial y}\eta\right) +$$

$$+ \sum_{n=1}^{\infty} B_n^{(04.0)*}\left[\frac{\partial u_0}{\partial x}\left(\sinh \pi n\xi \cos \pi n\eta - \cosh \pi n\eta \sin \pi n\xi\right) + \right. \tag{6.63}$$

$$\left. + \frac{\partial u_0}{\partial y}\left(\sinh \pi n\eta \cos \pi n\xi - \cosh \pi n\xi \sin \pi n\eta\right)\right],$$

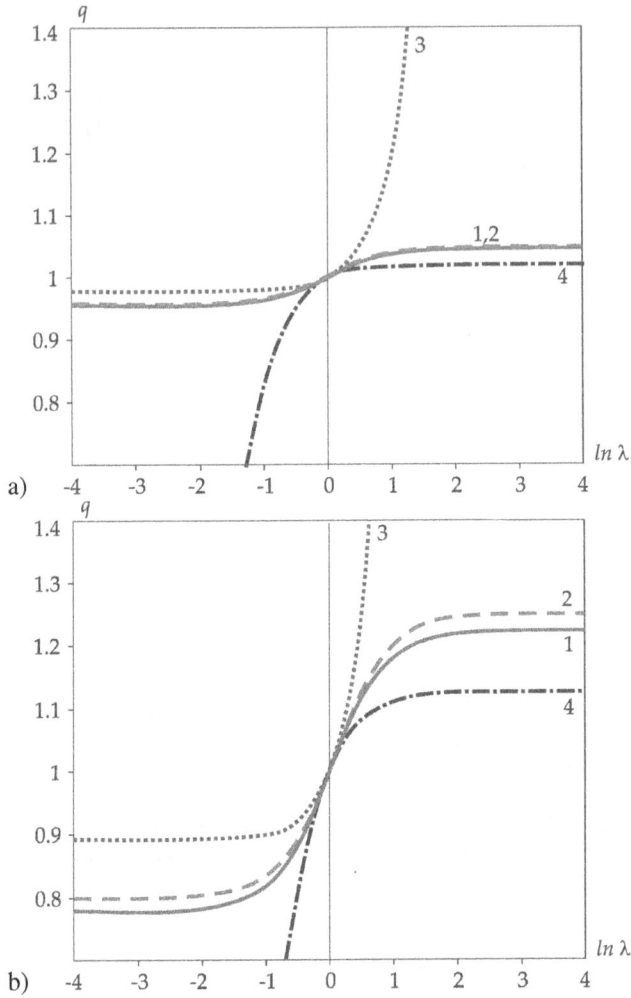

Figure 6.13 Dependence of the effective coefficient q on the conductivity of inclusions λ with size $a = 0.15$ (a) and $a = 1/3$ (b) (1– 2PhM, 2– MF, 3– Voigt averaging, 4– Reuss averaging).

where

$$B_0^{(04.0)^*} = \left(\frac{\lambda-1}{\lambda+1}\right)^2 \frac{\pi^2 \tilde{a}^4}{16}, \quad B_n^{(04.0)^*} = \left(\frac{\lambda-1}{\lambda+1}\right)^2 \frac{\pi \tilde{a}^4}{4} S_n, \quad n = 1, 2, \ldots$$

$$(6.64)$$

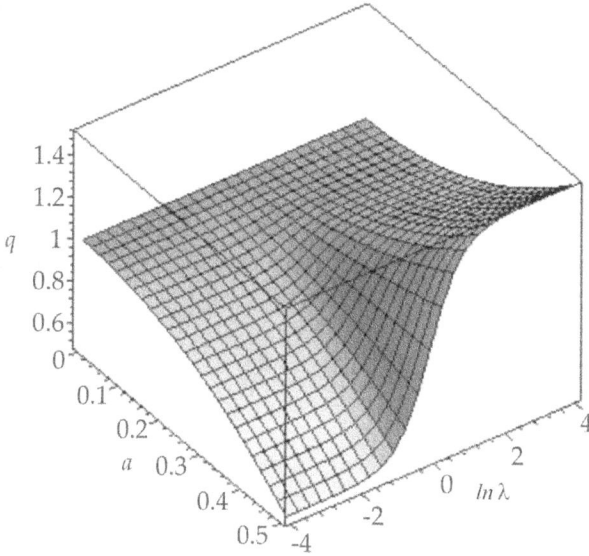

Figure 6.14 Effective thermal conductivity q as a function of two variables: inclusion size a and their conductivity λ.

Therefore, taking into account solutions (6.59)–(6.64), the homogenized equation has the following form

$$
(1 - a^2 + \lambda a^2) \left(\frac{\partial^2 u_0}{\partial x^2} + \frac{\partial^2 u_0}{\partial y^2} \right) + \frac{1}{|\Omega_i^*|} \left[\iint\limits_{\Omega_i^+} \left(\frac{\partial^2 u_1^{+\,(01)}}{\partial x \partial \xi} + \frac{\partial^2 u_1^{+\,(01)}}{\partial y \partial \eta} + \right. \right.
$$

$$
+ \frac{\partial^2 u_1^{(02)}}{\partial x \partial \xi} + \frac{\partial^2 u_1^{(02)}}{\partial y \partial \eta} + \frac{\partial^2 u_1^{+\,(03.0)}}{\partial x \partial \xi} + \frac{\partial^2 u_1^{+\,(03.0)}}{\partial y \partial \eta} + \frac{\partial^2 u_1^{(04.0)}}{\partial x \partial \xi} + \frac{\partial^2 u_1^{(04.0)}}{\partial y \partial \eta} \right) d\xi\, d\eta +
$$

$$
+ \lambda \iint\limits_{\Omega_i^-} \left(\frac{\partial^2 u_1^{-\,(01)}}{\partial x \partial \xi} + \frac{\partial^2 u_1^{-\,(01)}}{\partial y \partial \eta} + \frac{\partial^2 u_1^{(02)}}{\partial x \partial \xi} + \frac{\partial^2 u_1^{(02)}}{\partial y \partial \eta} + \right.
$$

$$
\left. + \frac{\partial^2 u_1^{-\,(03.0)}}{\partial x \partial \xi} + \frac{\partial^2 u_1^{-\,(03.0)}}{\partial y \partial \eta} + \frac{\partial^2 u_1^{(04.0)}}{\partial x \partial \xi} + \frac{\partial^2 u_1^{(04.0)}}{\partial y \partial \eta} \right) d\xi\, d\eta \right] =
$$

$$
= \left(1 - a^2 + \lambda a^2 - \lambda \frac{\lambda - 1}{\lambda + 1} a^2 + \frac{\lambda - 1}{\lambda + 1} a^2 (1 - a^2) + \lambda \frac{\lambda - 1}{\lambda + 1} a^4 - \right.
$$

$$
\left. - \lambda \left(\frac{\lambda - 1}{\lambda + 1} \right)^2 a^4 + \left(\frac{\lambda - 1}{\lambda + 1} \right)^2 a^4 \right) \left(\frac{\partial^2 u_0}{\partial x^2} + \frac{\partial^2 u_0}{\partial y^2} \right) = F.
$$

Thus, four approximations according to the Schwarz alternating method lead, up to terms of the order of a^4 inclusive, yield the following expression $q^{(04)}$ for the effective thermal conductivity

$$q^{(04)} = 1 + 2\frac{\lambda-1}{\lambda+1}a^2 + 2\left(\frac{\lambda-1}{\lambda+1}\right)^2 a^4. \tag{6.65}$$

It is easy to see that the continuation of the iterative process according to the Schwarz alternating method gives, after construction of the (05) and (06) approximations, the following effective coefficient in

$$q^{(06)} = 1 + 2\frac{\lambda-1}{\lambda+1}a^2 + 2\left(\frac{\lambda-1}{\lambda+1}\right)^2 a^4 + 2\left(\frac{\lambda-1}{\lambda+1}\right)^3 a^6. \tag{6.66}$$

Therefore, after performing $2n$ iterations, we have

$$q^{(2n)} = 1 + 2\frac{\lambda-1}{\lambda+1}a^2 + 2\left(\frac{\lambda-1}{\lambda+1}\right)^2 a^4 +$$
$$+ 2\left(\frac{\lambda-1}{\lambda+1}\right)^3 a^6 + \ldots + 2\left(\frac{\lambda-1}{\lambda+1}\right)^n a^{2n} + \ldots \tag{6.67}$$

Summing series (6.67) under the condition $a \ll 1$, we obtain the MF relation:

$$q_{MF} = \frac{1 - a^2 + \lambda\left(1 + a^2\right)}{1 + a^2 + \lambda\left(1 - a^2\right)}. \tag{6.68}$$

Indeed, we have

$$q_{MF} = \frac{1 - a^2 + \lambda\left(1 + a^2\right)}{1 + a^2 + \lambda\left(1 - a^2\right)} = \frac{(\lambda+1) + (\lambda-1)a^2}{(\lambda+1) - (\lambda-1)a^2} = \frac{1 + \frac{\lambda-1}{\lambda+1}a^2}{1 - \frac{\lambda-1}{\lambda+1}a^2} =$$

$$= \left(1 + \frac{\lambda-1}{\lambda+1}a^2\right)\left(1 + \frac{\lambda-1}{\lambda+1}a^2 + \left(\frac{\lambda-1}{\lambda+1}\right)^2 a^4 + \ldots\right) =$$

$$= 1 + 2\frac{\lambda-1}{\lambda+1}a^2 + 2\left(\frac{\lambda-1}{\lambda+1}\right)^2 a^4 + 2\left(\frac{\lambda-1}{\lambda+1}\right)^3 a^6 + \ldots$$

Thus, the application of the Schwarz alternating method allows us to analytically show that the main term of the asymptotic expansion of the effective thermal conductivity parameter in the case of square cylindrical inclusions of low concentration is the MF and coincides with upper Hashin-Shtrikman bound at $0 \leq \lambda \leq 1$ and with the Hashin-Shtrikman lower bound at $1 \leq \lambda < \infty$.

6.3.2 THE FIRST CORRECTION TO THE MF

To improve MF (6.68), we employ the following series

$$
\begin{aligned}
u_1^{(02.\Sigma)} = \sum_{n=1}^{\infty} B_n^{(02)*} \Bigg[& \frac{\partial u_0}{\partial x} (\sinh \pi n \xi \cos \pi n \eta - \cosh \pi n \eta \sin \pi n \xi) + \\
& + \frac{\partial u_0}{\partial y} (\sinh \pi n \eta \cos \pi n \xi - \cosh \pi n \xi \sin \pi n \eta) \Bigg],
\end{aligned}
\tag{6.69}
$$

where

$$
B_n^{(02)*} = \frac{\lambda - 1}{\lambda + 1} \tilde{a}^2 S_n.
\tag{6.70}
$$

To remove the discrepancies in the boundary conditions on the contour of radius $r = \tilde{a}$, we expand the expression $u_1^{(02.n)}$ into a series in polar coordinates r, θ for small values of r we have

$$
\begin{aligned}
\sum_{n=1}^{\infty} B_n^{(02)*} \Bigg[& \frac{\partial u_0}{\partial x} (\sinh \pi n \xi \cos \pi n \eta - \cosh \pi n \eta \sin \pi n \xi) + \\
& + \frac{\partial u_0}{\partial y} (\sinh \pi n \eta \cos \pi n \xi - \cosh \pi n \xi \sin \pi n \eta) \Bigg] = \\
= 2 \sum_{n=1}^{\infty} B_n^{(02)*} \Bigg[& \frac{\partial u_0}{\partial x} \sum_{k=1}^{\infty} \frac{(\pi n r)^{4k-1} \cos (4k - 1)\theta}{(4k - 1)!} + \\
& + \frac{\partial u_0}{\partial y} \sum_{k=1}^{\infty} \frac{(\pi n r)^{4k-1} \sin (4k - 1)\theta}{(4k - 1)!} \Bigg],
\end{aligned}
\tag{6.71}
$$

where the series on the right side of (6.71) converges for all values $0 \le r < \infty$.
The approximation of (03) order are written as follows:

$$
u_1^{-(03.\Sigma)} = \sum_{k=1}^{\infty} \left(A_{1k}^{(03.\Sigma)} r^{4k-1} \cos (4k - 1)\theta + A_{2k}^{(03.\Sigma)} r^{4k-1} \sin (4k - 1)\theta \right),
\tag{6.72}
$$

$$
u_1^{+(03.n)} = \sum_{k=1}^{\infty} \left(B_{1k}^{(03.n)} \frac{\cos (4k - 1)\theta}{r^{4k-1}} + B_{2k}^{(03.n)} \frac{\sin (4k - 1)\theta}{r^{4k-1}} \right),
\tag{6.73}
$$

where the constants

$$
A_{1k}^{(03.\Sigma)} = A_{1k}^{(03.\Sigma)*} \frac{\partial u_0}{\partial x}, \qquad B_{1k}^{(03.\Sigma)} = B_{1k}^{(03.\Sigma)*} \frac{\partial u_0}{\partial x}
$$

and

$$
A_{2k}^{(03.\Sigma)} = A_{2k}^{(03.\Sigma)*} \frac{\partial u_0}{\partial y}, \qquad B_{2k}^{(03.\Sigma)} = B_{2k}^{(03.\Sigma)*} \frac{\partial u_0}{\partial y}
$$

are determined from the solution of the system of equations

$$
\begin{cases}
A_{mk}^{(03.\Sigma)\,*}\,\tilde{a}^{4k-1} = \dfrac{B_{mk}^{(03.\Sigma)\,*}}{\tilde{a}^{4k-1}} \\[4mm]
-\dfrac{B_{mk}^{(03.\Sigma)\,*}}{\tilde{a}^{4k}} - \lambda\,\tilde{a}^{4k-2}A_{mk}^{(03.\Sigma)\,*} = \dfrac{2\,(\lambda-1)\,\pi^{4k-1}\tilde{a}^{4k-2}}{(4k-1)!}\,\sum_{n=1}^{\infty}B_n^{(02)\,*}\,n^{4k-1}.
\end{cases}
\tag{6.74}
$$

Finally, we obtain

$$
A_{mk}^{(03.\Sigma)\,*} = -\frac{2\,\pi^{4k-1}}{(4k-1)!}\,\frac{\lambda-1}{\lambda+1}\,\sum_{n=1}^{\infty}B_n^{(02)\,*}\,n^{4k-1},
\tag{6.75}
$$

$$
B_{mk}^{(03.\Sigma)\,*} = -\frac{2\,\pi^{4k-1}\tilde{a}^{8k-2}}{(4k-1)!}\cdot\frac{\lambda-1}{\lambda+1}\,\sum_{n=1}^{\infty}B_n^{(02)\,*}\,n^{4k-1}, \quad m=1,\,2.
\tag{6.76}
$$

Thus, taking into account expressions (6.75), (6.76), we get

$$
\bar{u}_1^{\,(03.\Sigma)} = \sum_{k=1}^{\infty}\left(-\frac{2\,\pi^{4k-1}}{(4k-1)!}\,\frac{\lambda-1}{\lambda+1}\,\frac{\partial u_0}{\partial x}\left(\sum_{n=1}^{\infty}B_n^{(02)\,*}\,n^{4k-1}\right)r^{4k-1}\cos(4k-1)\,\theta-\right.
$$

$$
\left.-\frac{2\,\pi^{4k-1}}{(4k-1)!}\,\frac{\lambda-1}{\lambda+1}\,\frac{\partial u_0}{\partial y}\left(\sum_{n=1}^{\infty}B_n^{(02)\,*}\,n^{4k-1}\right)r^{4k-1}\sin(4k-1)\,\theta\right) =
$$

$$
= -2\frac{\lambda-1}{\lambda+1}\,\sum_{n=1}^{\infty}B_n^{(02)\,*}\left(\frac{\partial u_0}{\partial x}\,\sum_{k=1}^{\infty}\frac{(\pi n r)^{4k-1}}{(4k-1)!}\cos(4k-1)\,\theta+\right.
\tag{6.77}
$$

$$
\left.+\frac{\partial u_0}{\partial y}\,\sum_{k=1}^{\infty}\frac{(\pi n r)^{4k-1}}{(4k-1)!}\sin(4k-1)\,\theta\right),
$$

$$
u_1^{+\,(03.\Sigma)} = -2\frac{\lambda-1}{\lambda+1}\,\sum_{n=1}^{\infty}B_n^{(02)\,*}\left(\frac{\partial u_0}{\partial x}\,\sum_{k=1}^{\infty}\frac{(\pi n\tilde{a}^2)^{4k-1}}{(4k-1)!}\,\frac{\cos(4k-1)\,\theta}{r^{4k-1}}+\right.
$$

$$
\left.+\frac{\partial u_0}{\partial y}\,\sum_{k=1}^{\infty}\frac{(\pi n\tilde{a}^2)^{4k-1}}{(4k-1)!}\,\frac{\sin(4k-1)\,\theta}{r^{4k-1}}\right).
\tag{6.78}
$$

Summing the series over k in expressions (6.77), (6.78), we obtain

$$
\bar{u}_1^{\,(03.\Sigma)} = -\frac{\lambda-1}{\lambda+1}\,\sum_{n=1}^{\infty}B_n^{(02)\,*}\left[\frac{\partial u_0}{\partial x}\,(\sinh\pi n\xi\,\cos\pi n\eta - \cosh\pi n\eta\,\sin\pi n\xi)+\right.
$$

$$
\left.+\frac{\partial u_0}{\partial y}\,(\sinh\pi n\eta\,\cos\pi n\xi - \cosh\pi n\xi\,\sin\pi n\eta)\right],
\tag{6.79}
$$

$$
u_1^{+\,(03.\Sigma)} = -\frac{\lambda-1}{\lambda+1}\sum_{n=1}^{\infty}B_n^{(02)\,*}\left[\frac{\partial u_0}{\partial x}\times\right.
$$

$$
\times\left(\sinh\frac{\pi n\tilde{a}^2\xi}{\xi^2+\eta^2}\cos\frac{\pi n\tilde{a}^2\eta}{\xi^2+\eta^2}-\cosh\frac{\pi n\tilde{a}^2\eta}{\xi^2+\eta^2}\sin\frac{\pi n\tilde{a}^2\xi}{\xi^2+\eta^2}\right)+ \tag{6.80}
$$

$$
\left.+\frac{\partial u_0}{\partial y}\left(\sinh\frac{\pi n\tilde{a}^2\eta}{\xi^2+\eta^2}\cos\frac{\pi n\tilde{a}^2\xi}{\xi^2+\eta^2}-\cosh\frac{\pi n\tilde{a}^2\xi}{\xi^2+\eta^2}\sin\frac{\pi n\tilde{a}^2\eta}{\xi^2+\eta^2}\right)\right].
$$

The discrepancies in the boundary conditions on the outer contour of the cell are removed in the same way as it was shown earlier. We introduce function $u_1^{(04.\Sigma)}$ as follows

$$
u_1^{(04.\Sigma)} = u_{11}^{(04.\Sigma)}+u_{12}^{(04.\Sigma)}, \tag{6.81}
$$

where

$$
u_{11}^{(04.\Sigma)} = A_0^{(04.\Sigma)}+B_0^{(04.\Sigma)}\xi+\sum_{m=1}^{\infty}\left[\left(A_m^{(04.\Sigma)}\cosh\pi m\xi+B_m^{(04.\Sigma)}\sinh\pi m\xi\right)\times\right.
$$

$$
\left.\times\cos\pi m\eta+\left(C_m^{(04.\Sigma)}\cosh\pi m\xi+D_m^{(04.\Sigma)}\sinh\pi m\xi\right)\sin\pi m\eta\right], \tag{6.82}
$$

$$
u_{12}^{(04.\Sigma)} = u_{11}^{(04.\Sigma)}\ (\xi\rightleftarrows\eta), \tag{6.83}
$$

$A_0^{(04.\Sigma)}$, $B_0^{(04.\Sigma)}$, $A_m^{(04.\Sigma)}$, $B_m^{(04.\Sigma)}$, $C_m^{(04.\Sigma)}$, $D_m^{(04.\Sigma)}$ ($m=1,\,2,\,...$) are the arbitrary constants.

Let us write the conditions on the contours $\xi=1$ and $\xi=-1$. We have

$$
\left(u_{11}^{(04.\Sigma)}+u_1^{+\,(03.\Sigma)}\right)\bigg|_{\xi=1} = \left(u_{11}^{(04.\Sigma)}+u_1^{+\,(03.\Sigma)}\right)\bigg|_{\xi=-1}, \tag{6.84}
$$

$$
\left(\frac{\partial u_{11}^{(04.\Sigma)}}{\partial\xi}+\frac{\partial u_1^{+\,(03.\Sigma)}}{\partial\xi}\right)\bigg|_{\xi=1} = \left(\frac{\partial u_{11}^{(04.\Sigma)}}{\partial\xi}+\frac{\partial u_1^{+\,(03.\Sigma)}}{\partial\xi}\right)\bigg|_{\xi=-1} \tag{6.85}
$$

as

$$
u_{11}^{(04.\Sigma)}\bigg|_{\xi-1}-u_{11}^{(04.\Sigma)}\bigg|_{\xi=-1} = \tag{6.86}
$$

$$
= \frac{\partial u_0}{\partial x}\cdot 2\frac{\lambda-1}{\lambda+1}\sum_{n=1}^{\infty}B_n^{(02)\,*}\left(\sinh\frac{\pi n\tilde{a}^2}{1+\eta^2}\cos\frac{\pi n\tilde{a}^2\eta}{1+\eta^2}-\cosh\frac{\pi n\tilde{a}^2\eta}{1+\eta^2}\sin\frac{\pi n\tilde{a}^2}{1+\eta^2}\right),
$$

$$
\frac{\partial u_{11}^{(04.\Sigma)}}{\partial\xi}\bigg|_{\xi=1}-\frac{\partial u_{11}^{(04.\Sigma)}}{\partial\xi}\bigg|_{\xi=-1} = \tag{6.87}
$$

$$
= -\frac{\partial u_0}{\partial y}\cdot 2\pi\tilde{a}^2\frac{\lambda-1}{\lambda+1}\sum_{n=1}^{\infty}B_n^{(02)\,*}n\left[\frac{1-\eta^2}{(1+\eta^2)^2}\sinh\frac{\pi n\tilde{a}^2}{1+\eta^2}\sin\frac{\pi n\tilde{a}^2\eta}{1+\eta^2}+\right.
$$

$$
+\frac{2\eta}{(1+\eta^2)^2}\cosh\frac{\pi n\tilde{a}^2}{1+\eta^2}\cos\frac{\pi n\tilde{a}^2\eta}{1+\eta^2}-\frac{1-\eta^2}{(1+\eta^2)^2}\sinh\frac{\pi n\tilde{a}^2\eta}{1+\eta^2}\sin\frac{\pi n\tilde{a}^2}{1+\eta^2}+
$$

$$
-\frac{2\eta}{(1+\eta^2)^2}\cosh\frac{\pi n\tilde{a}^2\eta}{1+\eta^2}\cos\frac{\pi n\tilde{a}^2}{1+\eta^2}\Bigg].
$$

Expanding the functions on the right-hand sides of relations (6.86), (6.87) into Fourier series and neglecting terms of order $\tilde{a}^{16}$ and higher, we have

$$
A_0^{(04.\Sigma)}=0, \quad A_m^{(04.\Sigma)}=D_m^{(04.\Sigma)}=0, \quad m=1,2,...,
$$

$$
B_0^{(04.\Sigma)}=\frac{\partial u_0}{\partial x}B_0^{(04.\Sigma)*}, \quad B_m^{(04.\Sigma)}=\frac{\partial u_0}{\partial x}B_m^{(04.\Sigma)*}, \quad C_m^{(04.\Sigma)}=\frac{\partial u_0}{\partial y}C_m^{(04.\Sigma)*}, \tag{6.88}
$$

$$
B_0^{(04.\Sigma)*}=\frac{\lambda-1}{\lambda+1}\frac{\pi^3\tilde{a}^6}{12}\sum_{n=1}^{\infty}B_n^{(02)*}n^3=\left(\frac{\lambda-1}{\lambda+1}\right)^2\frac{\pi^3\tilde{a}^8}{12}\sum_{n=1}^{\infty}S_n n^3, \tag{6.89}
$$

$$
B_m^{(04.\Sigma)*}=-C_m^{(04.\Sigma)*}=\frac{\lambda-1}{\lambda+1}\frac{\pi^3\tilde{a}^6}{6}\left(\pi^2 m^2 S_m+(-1)^m\right)\sum_{n=1}^{\infty}B_n^{(02)*}n^3=
$$

$$
=\left(\frac{\lambda-1}{\lambda+1}\right)^2\frac{\pi^3\tilde{a}^8}{6}\left(\pi^2 m^2 S_m+(-1)^m\right)\sum_{n=1}^{\infty}S_n n^3. \tag{6.90}
$$

After averaging, taking into account expressions (6.69), (6.70), (6.79), (6.80) and (6.81)–(6.83), (6.88)–(6.90), one obtains

$$
\tilde{q}=\frac{1}{|\Omega_i^*|}\Bigg[\iint_{\Omega_i^+}\left(\frac{\partial^2 u_1^+}{\partial x\partial\xi}^{(03.\Sigma)}+\frac{\partial^2 u_1^+}{\partial y\partial\eta}^{(03.\Sigma)}+\frac{\partial^2 u_1}{\partial x\partial\xi}^{(04.\Sigma)}+\frac{\partial^2 u_1}{\partial y\partial\eta}^{(04.\Sigma)}\right)d\xi\, d\eta+
$$

$$
+\lambda\iint_{\Omega_i^-}\left(\frac{\partial^2 u_1^-}{\partial x\partial\xi}^{(03.\Sigma)}+\frac{\partial^2 u_1^-}{\partial y\partial\eta}^{(03.\Sigma)}+\frac{\partial^2 u_1}{\partial x\partial\xi}^{(04.\Sigma)}+\frac{\partial^2 u_1}{\partial y\partial\eta}^{(04.\Sigma)}\right)d\xi\, d\eta\Bigg]=
$$

$$
=-\frac{\lambda-1}{\lambda+1}\frac{\pi^3\tilde{a}^6}{12}\sum_{n=1}^{\infty}B_n^{(02)*}n^3+B_0^{(04.\Sigma)*}+O\left(\tilde{a}^8\right),
$$

i.e., up to terms of order $\tilde{a}^8$ inclusive, the corrections to the MF (6.68) is equal to zero.

6.3.3 THE FIRST NON-ZERO CORRECTION TO MF

We continue the solution, taking into account in (04) the terms of the order $\tilde{a}^{12}$ and constructing relations (05)–(08) of iterations using the Schwarz alternating method. Omitting cumbersome calculations, we present the final results.

- (05) iteration:

$$u_1^{-\,(05.0)} = A^{(05.0)^*} r \left(\frac{\partial u_0}{\partial x} \cos\theta + \frac{\partial u_0}{\partial y} \sin\theta \right), \tag{6.91}$$

$$u_1^{+\,(05.0)} = \frac{B^{(05.0)^*}}{r} \left(\frac{\partial u_0}{\partial x} \cos\theta + \frac{\partial u_0}{\partial y} \sin\theta \right), \tag{6.92}$$

where

$$\begin{aligned}
A^{(05.0)^*} &= -\left(\frac{\lambda-1}{\lambda+1}\right)^3 \frac{\pi^3 \tilde{a}^8}{12} \left(1 + \frac{\lambda-1}{\lambda+1} \frac{\pi\tilde{a}^2}{4}\right) \sum_{n=1}^{\infty} S_n n^3, \\
B^{(05.0)^*} &= -\left(\frac{\lambda-1}{\lambda+1}\right)^3 \frac{\pi^3 \tilde{a}^{10}}{12} \left(1 + \frac{\lambda-1}{\lambda+1} \frac{\pi\tilde{a}^2}{4}\right) \sum_{n=1}^{\infty} S_n n^3,
\end{aligned} \tag{6.93}$$

S_n is determined by the formula (6.34);
- (06) iteration:

$$u_1^{(06.0)} = B_0^{(06.0)^*} \left(\frac{\partial u_0}{\partial x} \xi + \frac{\partial u_0}{\partial y} \eta \right), \tag{6.94}$$

where

$$B_0^{(06.0)^*} = \left(\frac{\lambda-1}{\lambda+1}\right)^3 \frac{\pi^4 \tilde{a}^{10}}{48} \left(1 + \frac{\lambda-1}{\lambda+1} \frac{\pi\tilde{a}^2}{4}\right) \sum_{n=1}^{\infty} S_n n^3; \tag{6.95}$$

- (07) iteration:

$$u_1^{-\,(07.0)} = A^{(07.0)^*} r \left(\frac{\partial u_0}{\partial x} \cos\theta + \frac{\partial u_0}{\partial y} \sin\theta \right), \tag{6.96}$$

$$u_1^{+\,(07.0)} = \frac{B^{(07.0)^*}}{r} \left(\frac{\partial u_0}{\partial x} \cos\theta + \frac{\partial u_0}{\partial y} \sin\theta \right), \tag{6.97}$$

where

$$\begin{aligned}
A^{(07.0)^*} &= -\left(\frac{\lambda-1}{\lambda+1}\right)^4 \frac{\pi^4 \tilde{a}^{10}}{48} \left(1 + \frac{\lambda-1}{\lambda+1} \cdot \frac{\pi\tilde{a}^2}{4}\right) \sum_{n=1}^{\infty} S_n n^3, \\
B^{(07.0)^*} &= -\left(\frac{\lambda-1}{\lambda+1}\right)^4 \frac{\pi^4 \tilde{a}^{12}}{48} \left(1 + \frac{\lambda-1}{\lambda+1} \cdot \frac{\pi\tilde{a}^2}{4}\right) \sum_{n=1}^{\infty} S_n n^3,
\end{aligned} \tag{6.98}$$

- (08) iteration:

$$u_1^{(08.0)} = B_0^{(08.0)^*} \left(\frac{\partial u_0}{\partial x} \xi + \frac{\partial u_0}{\partial y} \eta \right), \tag{6.99}$$

where

$$B_0^{(08.0)^*} = \left(\frac{\lambda-1}{\lambda+1}\right)^4 \frac{\pi^5 \tilde{a}^{12}}{192} \left(1 + \frac{\lambda-1}{\lambda+1} \frac{\pi\tilde{a}^2}{4}\right) \sum_{n=1}^{\infty} S_n n^3. \tag{6.100}$$

After averaging, taking into account expressions (6.81)–(6.83), (6.88)–(6.90), (6.91)–(19.3.10) and employ restrictions to terms of order $\tilde{a}^{14}$ inclusive, one obtains

$$\frac{1}{|\Omega_i^*|}\left[\iint\limits_{\Omega_i^+}\left(\frac{\partial^2}{\partial x\partial\xi}+\frac{\partial^2}{\partial y\partial\eta}\right)\left(u_1^{+\,(04.\Sigma)}+u_1^{+\,(05.\Sigma)}+u_1^{(06.\Sigma)}+\right.$$

$$+u_1^{+\,(07.\Sigma)}+u_1^{(08.\Sigma)}\right)d\xi\,d\eta+\lambda\iint\limits_{\Omega_i^+}\left(\frac{\partial^2}{\partial x\partial\xi}+\frac{\partial^2}{\partial y\partial\eta}\right)\left(u_1^{(04.\Sigma)}+u_1^{+\,(05.\Sigma)}+\right.$$

$$+u_1^{(06.n)}+u_1^{-\,(07.n)}+u_1^{(08.n)}\right)d\xi\,d\eta\Bigg]=\left(\frac{\lambda-1}{\lambda+1}\right)^2\frac{\pi^4\tilde{a}^{10}}{48}\left(1+\frac{\lambda-1}{\lambda+1}\frac{\pi\tilde{a}^2}{4}\right)\times$$

$$\times\sum_{n=1}^{\infty}S_nn^3\left(-1+\lambda-\lambda\frac{\lambda-1}{\lambda+1}+\frac{\lambda-1}{\lambda+1}\left(1-\frac{\pi\tilde{a}^2}{4}\right)+\right.$$

$$+\lambda\frac{\lambda-1}{\lambda+1}\frac{\pi\tilde{a}^2}{4}-\lambda\left(\frac{\lambda-1}{\lambda+1}\right)^2\frac{\pi\tilde{a}^2}{4}+\left(\frac{\lambda-1}{\lambda+1}\right)^2\frac{\pi\tilde{a}^2}{4}\right).$$

We obtain correction $\tilde{q}$ to the MF in the following form

$$\tilde{q}=\left(\frac{\lambda-1}{\lambda+1}\right)^3\frac{128a^{10}}{3\pi}\left(1+\frac{\lambda-1}{\lambda+1}a^2\right)^2\sum_{n=1}^{\infty}S_nn^3.\qquad(6.101)$$

Thus, the solution found under the assumption of small sizes of inclusions a using the procedure of the Schwarz alternating method has the form

$$q_{ASM}=\frac{1-a^2+\lambda\,(1+a^2)}{1+a^2+\lambda\,(1-a^2)}+$$
$$\left(\frac{\lambda-1}{\lambda+1}\right)^3\frac{128a^{10}}{3\pi}\left(1+\frac{\lambda-1}{\lambda+1}a^2\right)^2\sum_{n=1}^{\infty}S_nn^3.\qquad(6.102)$$

6.3.4 ANALYSIS OF THE SCHWARZ ALTERNATING METHOD SOLUTION

Analysis of the solution obtained by the Schwarz alternating method (6.102) yields the following results.

1) For small inclusion sizes a, the expression q_{ASM} asymptotically, up to terms of order a^{14} inclusive, satisfies the Keller's theorem

$$q\,(\lambda)=q^{-1}\left(\lambda^{-1}\right).\qquad(6.103)$$

Indeed, by presenting

$$q_{ASM}=\frac{1-a^2+\lambda\,(1+a^2)}{1+a^2+\lambda\,(1-a^2)}+\left(\frac{\lambda-1}{\lambda+1}\right)^3a^{10}S+$$

$$+2\left(\frac{\lambda-1}{\lambda+1}\right)^4 a^{12} S + \left(\frac{\lambda-1}{\lambda+1}\right)^5 a^{14} S,$$

where

$$S = \frac{128}{3\pi}\sum_{n=1}^{\infty} S_n n^3, \tag{6.104}$$

we get

$$q_{ASM} \sim 1 + 2\frac{\lambda-1}{\lambda+1}a^2 + 2\left(\frac{\lambda-1}{\lambda+1}\right)^2 a^4 + 2\left(\frac{\lambda-1}{\lambda+1}\right)^3 a^6 + 2\left(\frac{\lambda-1}{\lambda+1}\right)^4 a^8 +$$

$$+\left(\frac{\lambda-1}{\lambda+1}\right)^3\left(2\left(\frac{\lambda-1}{\lambda+1}\right)^2 + S\right) a^{10} + \tag{6.105}$$

$$+2\left(\frac{\lambda-1}{\lambda+1}\right)^4\left(\left(\frac{\lambda-1}{\lambda+1}\right)^2 + S\right) a^{12} +$$

$$+\left(\frac{\lambda-1}{\lambda+1}\right)^5\left(2\left(\frac{\lambda-1}{\lambda+1}\right)^2 + S\right) a^{14} + o\left(a^{14}\right),$$

$$q_{ASM}^{-1}\left(\lambda^{-1}\right) \sim 1 + 2\frac{\lambda-1}{\lambda+1}a^2 + 2\left(\frac{\lambda-1}{\lambda+1}\right)^2 a^4 + 2\left(\frac{\lambda-1}{\lambda+1}\right)^3 a^6 +$$

$$+2\left(\frac{\lambda-1}{\lambda+1}\right)^4 a^8 + \left(\frac{\lambda-1}{\lambda+1}\right)^3\left(2\left(\frac{\lambda-1}{\lambda+1}\right)^2 + S\right) a^{10} + \tag{6.106}$$

$$+2\left(\frac{\lambda-1}{\lambda+1}\right)^4\left(\left(\frac{\lambda-1}{\lambda+1}\right)^2 + S\right) a^{12} +$$

$$+\left(\frac{\lambda-1}{\lambda+1}\right)^5\left(2\left(\frac{\lambda-1}{\lambda+1}\right)^2 + S\right) a^{14} + o\left(a^{14}\right).$$

Expressions (6.105) and (6.106) coincide, which proves that relation (6.103) holds.

2) A series in an expression (6.104)

$$T = \sum_{n=1}^{\infty} S_n n^3$$

converges quickly:

$$\frac{T_{n+1}}{T_n} = \frac{S_{n+1}(n+1)^3}{S_n n^3} \sim e^{-\pi}.$$

3) Comparison q_{MF} (6.68) and q_{ASM} (6.94) show that for all values of the conductivity of the inclusions, including $\lambda \to \infty$, and for any sizes of inclusions the following inequalities hold

$$q_{ASM} \leq q_{MF} \quad \text{for} \quad 0 \leq \lambda \leq 1,$$

$$q_{ASM} \geq q_{MF} \quad \text{for} \quad \lambda \geq 1.$$

Therefore, the expression (6.102) found by the Schwarz alternating method gives for MF (6.68) lower bound at $0 \leq \lambda \leq 1$, and upper bound for $1 \leq \lambda < \infty$ and falls within the Hashin-Shtrikman bounds $\underline{q}_{HS}$, $\bar{q}_{HS}$:

at $0 \leq \lambda \leq 1$:

$$\lambda \frac{2 - a^2 + \lambda a^2}{a^2 + \lambda (2 - a^2)} = \underline{q}_{HS} \leq q_{ASM} \leq \bar{q}_{HS} \equiv q_{MF} = \frac{1 - a^2 + \lambda (1 + a^2)}{1 + a^2 + \lambda (1 - a^2)},$$

at $1 \leq \lambda < \infty$:

$$\frac{1 - a^2 + \lambda (1 + a^2)}{1 + a^2 + \lambda (1 - a^2)} = q_{MF} \equiv \underline{q}_{HS} \leq q_{ASM} \leq \bar{q}_{HS} = \lambda \frac{2 - a^2 + \lambda a^2}{a^2 + \lambda (2 - a^2)},$$

or

$$q_{MF} + \widehat{q} \leq q_{ASM} = q_{MF} + \tilde{q} \leq q_{MF} \quad \text{at} \quad 0 \leq \lambda \leq 1,$$

$$q_{MF} \leq q_{ASM} = q_{MF} + \tilde{q} \leq q_{MF} + \widehat{q} \quad \text{at} \quad 1 \leq \lambda < \infty,$$

where $\tilde{q}$ is defined by the expression (6.101);

$$\widehat{q} = \frac{a^2 (\lambda - 1)^3 (1 - a^2)}{(1 + a^2 + \lambda (1 - a^2)) (a^2 + \lambda (2 - a^2))}.$$

6.3.5 INCLUSIONS–HEAT INSULATORS OF LARGE SIZES

The solution by the Schwarz alternating method (6.102) was obtained for small inclusion sizes, i.e. under the assumption $a \ll 1$. It is interesting to analize the possibility of using expression (6.102) in the case of inclusions of the form of heat insulators ($\lambda = 0$) of large sizes ($a \to 1$).

In the absence of heat sources in the area of inclusions, the homogenized equation (6.102) takes the following form

$$q_{ASM} \left(\frac{\partial^2 u_0}{\partial x^2} + \frac{\partial^2 u_0}{\partial y^2} \right) = (1 - a^2) F, \tag{6.107}$$

where F stands for the density of heat sources in the area of the matrix.

In this case

$$q_{ef} = \frac{q_{ASM}}{1 - a^2},$$

and q_{ef} is the effective thermal conductivity. Then from (6.107), (6.102) at $\lambda = 0$, $a \to 1$ we have the following asymptotic expression for the effective parameter q_{ef}:

$$q_{ef} = \frac{1}{2} - \left(\frac{256}{3\pi} \sum_{n=1}^{\infty} S_n n^3 - \frac{1}{2}\right)(1-a) + o\,(1-a), \qquad (6.108)$$

i.e., the leading term of the asymptotics (6.108), $q_{ef}^{(0)} \sim 0.5$, coincides with the well-known result [38].

6.3.6 ASYMPTOTIC ESTIMATION FOR THE CORRECTED MF

Let us carry out an asymptotic analysis (up to a constant factor S (6.104)) of the corrected MF (6.101), for various values of the geometric and physical characteristics of the composite.

1) For small sizes of inclusions, $a \to 0$, we have

$$\left(\frac{\lambda-1}{\lambda+1}\right)^3 \left(1 + \frac{\lambda-1}{\lambda+1}a^2\right)^2 a^{10} = \frac{(\lambda-1)^3}{(\lambda+1)^3}a^{10} + \frac{2(\lambda-1)^4}{(\lambda+1)^4}a^{12} \sim O\,(a^{10})$$

$$\text{at} \quad a \to 0.$$

In particular:

1.1) small highly conducting inclusions ($a \to 0,\ \lambda \gg 1$):

$$\left(\frac{\lambda-1}{\lambda+1}\right)^3 \left(1 + \frac{\lambda-1}{\lambda+1}a^2\right)^2 a^{10} = (1+2a^2)\,a^{10} - 2\,(3+8a^2)\,\frac{a^{10}}{\lambda} + \ldots \sim$$

$$\sim O\left(a^{10}\left(\frac{1}{\lambda}\right)^0\right) \quad \text{at} \quad a \to 0, \quad \lambda \gg 1;$$

1.2) small inclusions whose conductivity is of the order of the matrix conductivity ($a \to 0,\ \lambda \sim 1$):

$$\left(\frac{\lambda-1}{\lambda+1}\right)^3 \left(1 + \frac{\lambda-1}{\lambda+1}a^2\right)^2 a^{10} = \frac{1}{8}a^{10}(\lambda-1)^3 - \frac{3-2a^2}{16}a^{10}(\lambda-1)^4 + \ldots \sim$$

$$\sim O\left(a^{10}(\lambda-1)^3\right) \quad \text{at} \quad a \to 0, \quad \lambda \sim 1;$$

1.3) small low conducting inclusions ($a \to 0,\ \lambda \ll 1$):

$$\left(\frac{\lambda-1}{\lambda+1}\right)^3 \left(1 + \frac{\lambda-1}{\lambda+1}a^2\right)^2 a^{10} = -\left(1-2a^2\right)a^{10} + 2\left(3-8a^2\right)a^{10}\lambda + \ldots \sim$$

$$\sim O\left(a^{10}\lambda^0\right) \quad \text{at} \quad a \to 0, \quad \lambda \ll 1.$$

2) Inclusions of large sizes close to the limit possible value ($a \to 1$):

$$\left(\frac{\lambda-1}{\lambda+1}\right)^3 \left(1 + \frac{\lambda-1}{\lambda+1}a^2\right)^2 a^{10} = \frac{4(\lambda-1)^3\lambda^2}{(\lambda+1)^5} -$$

$$-\frac{8(\lambda-1)^3\lambda(6\lambda-1)}{(\lambda+1)^5}(1-a) + \dots \sim O\left((1-a)^0\right) \quad \text{at} \quad a \to 1.$$

In special cases:

2.1) large highly conductive inclusions ($a \to 1$, $\lambda \gg 1$)

$$\left(\frac{\lambda-1}{\lambda+1}\right)^3 \left(1 + \frac{\lambda-1}{\lambda+1}a^2\right)^2 a^{10} = 4(1 - 12(1-a)) - 8(4 - 49(1-a))\frac{1}{\lambda} + \dots \sim$$

$$\sim O\left((1-a)^0\left(\frac{1}{\lambda}\right)^0\right) \quad \text{at} \quad a \to 1, \quad \lambda \gg 1;$$

2.2) large inclusions whose conductivity is of the order of the matrix conductivity ($a \to 1$, $\lambda \sim 1$):

$$\left(\frac{\lambda-1}{\lambda+1}\right)^3 \left(1 + \frac{\lambda-1}{\lambda+1}a^2\right)^2 a^{10} = \left(\frac{1}{8} - \frac{5}{4}(1-a)\right)(\lambda-1)^3 -$$

$$-\left(\frac{1}{16} - \frac{3}{8}(1-a)\right)(\lambda-1)^4 + \dots \sim O\left((1-a)^0(\lambda-1)^3\right) \quad \text{at} \quad a \to 1, \quad \lambda \sim 1;$$

2.3) large low conducting inclusions ($a \to 1$, $\lambda \ll 1$):

$$\left(\frac{\lambda-1}{\lambda+1}\right)^3 \left(1 + \frac{\lambda-1}{\lambda+1}a^2\right)^2 a^{10} = -8(1-a)\lambda - 4(1-28(1-a))\lambda^2 + \dots \sim$$

$$\sim O((1-a)\lambda) \quad \text{at} \quad a \to 0, \quad \lambda \ll 1.$$

3) Absolutely conducting inclusions ($\lambda \to \infty$):

$$\left(\frac{\lambda-1}{\lambda+1}\right)^3 \left(1 + \frac{\lambda-1}{\lambda+1}a^2\right)^2 a^{10} = (1+a^2)^2 a^{10} - 2(1+a^2)(3+5a^2)\frac{a^{10}}{\lambda} + \dots \sim$$

$$\sim O(\lambda^0) \quad \text{at} \quad \lambda \to \infty.$$

In particular cases, we have,

3.1) absolutely conducting inclusions of small sizes ($\lambda \to \infty$, $a \ll 1$):

$$\left(\frac{\lambda-1}{\lambda+1}\right)^3 \left(1 + \frac{\lambda-1}{\lambda+1}a^2\right)^2 a^{10} = \left(1 - \frac{6}{\lambda}\right)a^{10} + 2\left(1 - \frac{8}{\lambda}\right)a^{12} + \dots \sim$$

$$\sim O\left(\left(\frac{1}{\lambda}\right)^0 a^{10}\right) \quad \text{at} \quad \lambda \to \infty, \quad a \ll 1;$$

3.2) absolutely conducting inclusions of large sizes ($\lambda \to \infty$, $a \gg 0$):

$$\left(\frac{\lambda-1}{\lambda+1}\right)^3 \left(1+\frac{\lambda-1}{\lambda+1}a^2\right)^2 a^{10} = 4\left(1-\frac{8}{\lambda}\right) - 8\left(6-\frac{49}{\lambda}\right)(1-a)+...\sim$$

$$\sim O\left(\left(\frac{1}{\lambda}\right)^0 (1-a)^0\right) \quad \text{at} \quad \lambda \to \infty, \quad a \gg 0.$$

4) Conductivities of the matrix and inclusions of the same order ($\lambda \sim 1$):

$$\left(\frac{\lambda-1}{\lambda+1}\right)^3 \left(1+\frac{\lambda-1}{\lambda+1}a^2\right)^2 a^{10} = \frac{a^{10}}{8}(\lambda-1)^3 - \frac{a^{10}(3-2a^2)}{16}(\lambda-1)^4+...\sim$$

$$\sim O\left((\lambda-1)^3\right) \quad \text{at} \quad \lambda \sim 1.$$

In particular cases, we have,

4.1) small sizes inclusions with a conductivity of the order of conductivity of a matrix ($\lambda \sim 1$, $a \ll 1$):

$$\left(\frac{\lambda-1}{\lambda+1}\right)^3 \left(1+\frac{\lambda-1}{\lambda+1}a^2\right)^2 a^{10} = \frac{1}{8}\left(1-\frac{3}{2}(\lambda-1)\right)(\lambda-1)^3 a^{10}+\frac{1}{8}(\lambda-1)^4 a^{12}\sim$$

$$\sim O\left((\lambda-1)^3 a^{10}\right) \quad \text{at} \quad \lambda \sim 1, \quad a \ll 1;$$

4.2) inclusions of large sizes with a conductivity of the order of the conductivity of a matrix ($\lambda \sim 1$, $a \gg 0$):

$$\left(\frac{\lambda-1}{\lambda+1}\right)^3 \left(1+\frac{\lambda-1}{\lambda+1}a^2\right)^2 a^{10} = \frac{1}{8}\left(1-\frac{1}{2}(\lambda-1)\right)(\lambda-1)^3-$$

$$-\frac{1}{4}\left(5-\frac{3}{2}(\lambda-1)\right)(\lambda-1)^3(1-a)+...\sim O\left((\lambda-1)^3(1-a)^0\right)$$

$$\text{at} \quad \lambda \sim 1, \quad a \gg 0.$$

5) Non-conductive inclusions ($\lambda \to 0$):

$$\left(\frac{\lambda-1}{\lambda+1}\right)^3 \left(1+\frac{\lambda-1}{\lambda+1}a^2\right)^2 a^{10} = -(1-a^2)^2 a^{10}+$$

$$+2(1-a^2)(3-5a^2) a^{10}\lambda+...\sim O(\lambda^0) \quad \text{at} \quad \lambda \to 0.$$

In particular:

5.1) non-conductive inclusions of small sizes ($\lambda \to 0$, $a \ll 1$):

$$\left(\frac{\lambda-1}{\lambda+1}\right)^3 \left(1+\frac{\lambda-1}{\lambda+1}a^2\right)^2 a^{10} = -(1-6\lambda)a^{10}+2(1-8\lambda)a^{12}+...\sim$$

$$\sim O(\lambda^0 a^{10}) \quad \text{at} \quad \lambda \to 0, \quad a \ll 1;$$

5.2) large non-conductive inclusions ($\lambda \to 0$, $a \gg 0$):

$$\left(\frac{\lambda - 1}{\lambda + 1}\right)^3 \left(1 + \frac{\lambda - 1}{\lambda + 1} a^2\right)^2 a^{10} = -8\lambda\,(1 - a) - 4\,(1 - 31\lambda)\,(1 - a)^2 + ... \sim$$

$$\sim O\,(\lambda\,(1 - a)) \quad \text{at} \quad \lambda \to 0, \quad a \gg 0.$$

From a comparison of the above relations, it follows that for small inclusions, the conductivity of which does not differ significantly from the conductivity of the matrix, the asymptotic expansions of the correction to the MF will have a higher order of smallness than in the case of high-contrast inclusions of large sizes.

6.3.7 NUMERICAL RESULTS

The results of calculations of the reduced coefficient of thermal conductivity calculated for various values of the geometric and physical characteristics of the composite according to the Schwarz alternating method q_{ASM} (6.102) and MF q_{MF} (6.68), illustrated graphically in Fig. 6.15–6.16 for the following cases:

   (i) non-conductive inclusions (Fig. 6.15a);
  (ii) inclusions of low conductivity (Fig. 6.15b);
 (iii) inclusions of high conductivity (Fig. 6.15c);
 (iv) absolutely conducting inclusions (Fig. 6.15d);
  (v) small inclusions (Fig. 6.16a);
 (vi) medium size inclusions (Fig. 6.16b);
(vii) large inclusions (Fig. 6.16c).

In Table 6.1, the results of the calculation of the effective coefficients by the Schwarz alternating method (6.102) are compared with the solution by MF (6.68).

The results shown in Figs. 6.15–6.16 and calculated data of Table 6.1 imply that solution by the Schwarz alternating method gives an adequate estimate not only for small inclusions. Formula q_{ASM} (6.102) can be used for inclusions of any conductivity for all inclusion sizes, even at $a \to 1$ and inclusions-thermal insulators ($\lambda \to 0$), as well as for ideally conducting inclusions ($\lambda \to \infty$).

A relatively large discrepancy between the results obtained by the MF and the refined solution q_{ASM} (6.102) at large ($a \gg 0$) and close to the limit possible ($a \to 1$) inclusion sizes and significantly different phase conductivities ($\lambda \gg 1$ or $\lambda \ll 1$) is explained by the fact that correction to the MF has a more significant effect on the effective conductivity in this case.

6.3.8 POSSIBLE GENERALIZATIONS

From a mathematical point of view, the proposed method can be interpreted as an asymptotic approach that combines the homogenization technique with the PPB and the Schwarz alternating method.

The following important results have been obtained:

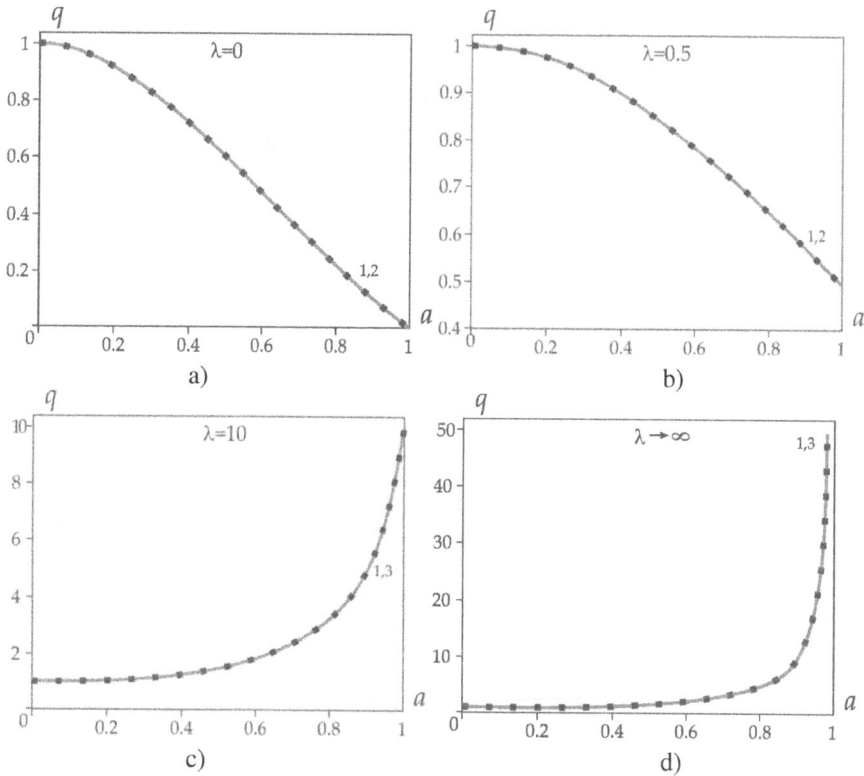

Figure 6.15 The effective coefficients calculated by the Schwarz alternating method (q_{ASM}) and MF (q_{MF}) for: (a) non-conductive inclusions ($\lambda = 0$); (b) inclusions of low conductivity ($\lambda = 0.5$); (c) inclusions of high conductivity ($\lambda = 10$); (d) absolutely conducting inclusions ($\lambda \to \infty$) ($1 - q_{ASM}$, $2 - q_{MF} \equiv \bar{q}_{HS}$, $3 - q_{MF} \equiv \underline{q}_{HS}$).

1. Based on 2PhM, a simple analytical expression for the thermal conductivity parameter was obtained, which gives reliable results for small sizes of inclusions of any conductivities, including limiting cases: non-conductive inserts; perfectly conducting inclusions.
2. Using 2PhM and the Schwarz alternating method, the effective thermal conductivity of composite with small-sized cylindrical inclusions of a square profile was estimated.
3. It is analytically shown that the main term of the asymptotic expansion of the obtained effective thermal conductivity parameter coincides for inclusions of low concentration MF.
4. An analytical expression (6.101) for the first nonzero correction to the MF was derived.
5. It was shown that for large-sized high-contrast inclusions, the correction to MF has a more significant effect on the effective conductivity of the composite than in the case of their small sizes and low contrast.

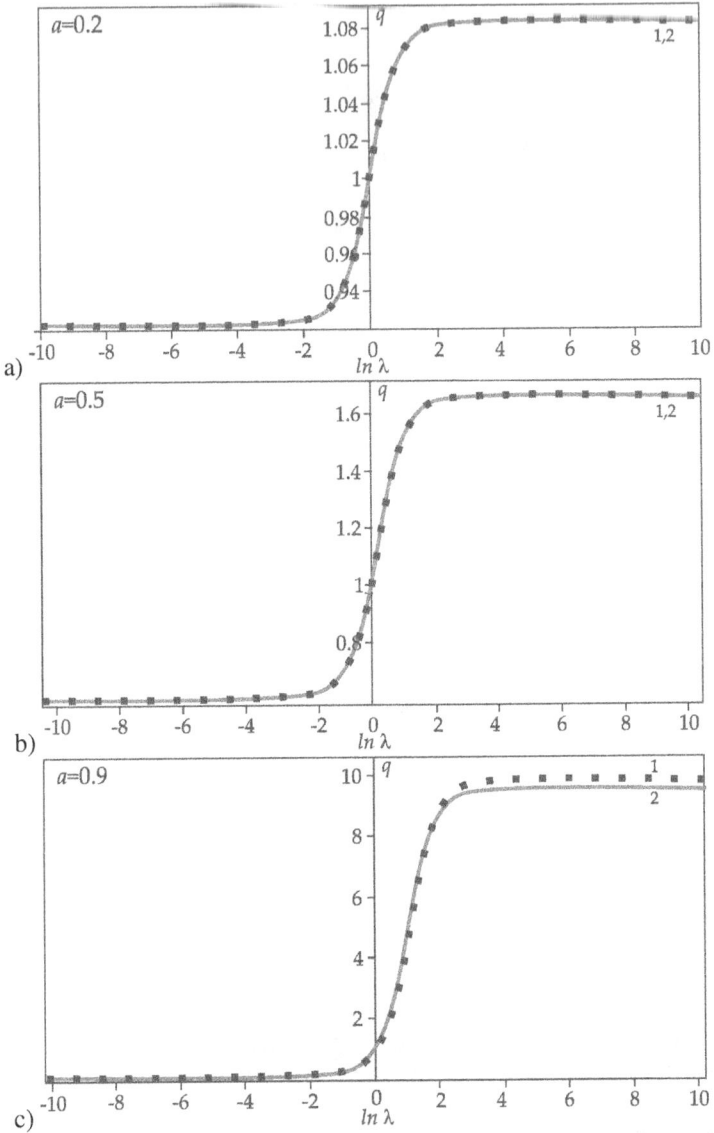

Figure 6.16 The effective coefficients calculated by the Schwarz alternating method (q_{ASM}) and MF (q_{MF}) for: (a) small inclusions ($a = 0.2$); (b) medium size inclusions ($a = 0.5$); (c) large inclusions ($a = 0.9$) ($1 - q_{ASM}$, $2 - q_{MF} \equiv q_{HS}$).

The correction to the MF was obtained, taking into account the shape of the inclusions.

6. It was shown that the solution based on the Schwarz alternating method: satisfies Keller's theorem; falls within Hashin–Shtrikman bounds; can be used for inclusions of any conductivity in the entire range of their size,

including asymptotics of extremely large size ($a \to 1$) for absolutely conductive inclusions ($\lambda \to \infty$) and thermal insulators ($\lambda = 0$).

7. The correctness of the results obtained was confirmed by using the asymptotic expansions of the effective thermal conductivity; using the comparative of results obtained by various methods; using analysis of the physical nature of the problem.

Let us briefly refer to the term "high-contrast composites". This term is intuitively understandable, but difficult to formalize concepts. The definition usually found in articles is, for example, as follows: "We are particularly interested in the case of

---

Table 6.1

Comparison of the results of the calculation of the effective coefficients calculated by the Schwarz alternating method is compared with the solution by MF.

Inclusion size a	Error $\dfrac{q_{MF} - q_{ASM}}{q_{MF}} \cdot 100\%$	Inclusion size a	Error $\dfrac{q_{MF} - q_{ASM}}{q_{MF}} \cdot 100\%$
	$\lambda = 0$		$\lambda = 0.5$
0.0	0.0000%	0.0	0.0000%
0.1	$0.2588 \cdot 10^{-8}\%$	0.1	$0.9585 \cdot 10^{-10}\%$
0.2	$0.2646 \cdot 10^{-5}\%$	0.2	$0.9814 \cdot 10^{-7}\%$
0.3	$0.1516 \cdot 10^{-3}\%$	0.3	$0.5655 \cdot 10^{-5}\%$
0.4	$0.2644 \cdot 10^{-2}\%$	0.4	$0.1002 \cdot 10^{-3}\%$
0.5	$0.2369 \cdot 10^{-1}\%$	0.5	$0.9296 \cdot 10^{-3}\%$
0.6	0.1362%	0.6	$0.5712 \cdot 10^{-2}\%$
0.7	0.5555%	0.7	$0.2635 \cdot 10^{-1}\%$
0.8	1.6406%	0.8	$0.9824 \cdot 10^{-1}\%$
0.9	3.1033%	0.9	0.3099%
1.0	0.0000%	1.0	0.8520%
	$\lambda = 10$		$\lambda \to \infty$
0.0	0.0000%	0.0	0.0000%
0.1	$-0.1417 \cdot 10^{-8}\%$	0.1	$-0.2588 \cdot 10^{-8}\%$
0.2	$-0.1450 \cdot 10^{-5}\%$	0.2	$-0.2646 \cdot 10^{-5}\%$
0.3	$0.8325 \cdot 10^{-4}\%$	0.3	$-0.1516 \cdot 10^{-3}\%$
0.4	$-0.1461 \cdot 10^{-2}\%$	0.4	$-0.2644 \cdot 10^{-2}\%$
0.5	$-0.1326 \cdot 10^{-1}\%$	0.5	$-0.2369 \cdot 10^{-1}\%$
0.6	$-0.7827 \cdot 10^{-1}\%$	0.6	-0.1362%
0.7	-0.3360%	0.7	-0.5555%
0.8	-1.1047%	0.8	-1.6406%
0.9	-2.7717%	0.9	-3.1033%
1.0	$-4,6859\%$	1.0	0.0000%

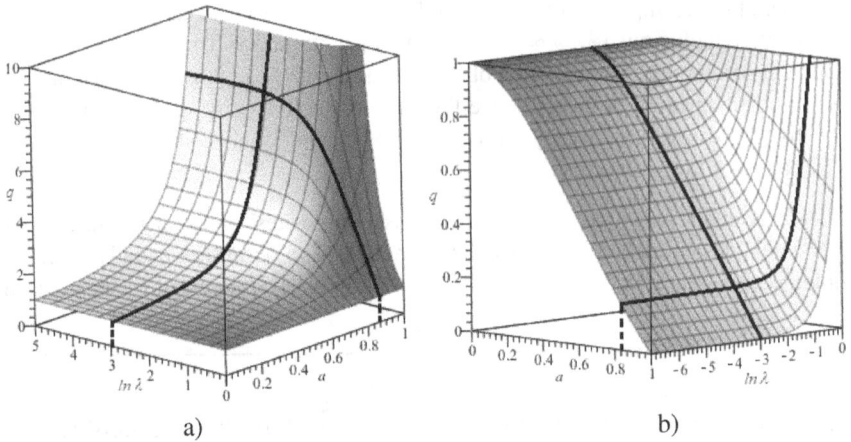

Figure 6.17 Dependence of the effective thermal conductivity parameter q on the size of square inclusions a and their conductivity λ: a) $1 \leq \lambda < \infty$, b) $0 \leq \lambda \leq 1$.

the high contrast composites, that is, when the conductivity of the inclusions is much larger than the conductivity of the hosting medium". Such a qualitatively (physically) justified definition does not quantitatively explain the situation.

The results of the study allow us to make some quantitative estimates. In particular, the calculated data show that for inclusions of any size $(0 \leq a \leq 1)$ an increase in their conductivity over $\lambda = 10^3$ does not significantly affect the effective conductivity of the composite (Fig. 6.17a).

Therefore, in engineering applications, the inclusions of conductivity $\lambda \geq 10^3$ with sufficient accuracy for practical purposes can be considered absolutely conductive.

A similar conclusion takes place in the case of inclusions of low conductivity: at $\lambda \leq 10^{-3}$ (Fig. 6.17b) the inclusions are practically non-conductive.

6.4 REFINEMENT OF THE MAXWELL FORMULA FOR COMPOSITES REINFORCED BY CIRCLE CROSS SECTION FIBERS

6.4.1 INTRODUCTORY REMARKS

In the case of a composite structure with periodically arranged cylindrical inclusions of a circle profile forming a square lattice, the expression for the effective thermal conductivity parameter obtained on the basis of a ThPhM leads to a relation that coincides with the MF:

$$q = \frac{\lambda \left(1 + \frac{\pi a^2}{4}\right) + 1 - \frac{\pi a^2}{4}}{\lambda \left(1 - \frac{\pi a^2}{4}\right) + 1 + \frac{\pi a^2}{4}}. \tag{6.109}$$

Analysis of the MF shows that

(i) relation (6.109) reliably describes the effective conductivity of the composite in the case of a low concentration of inclusions ($a \ll 1$) for any values of their conductivity λ;

(ii) for large sizes of inclusions ($a \to 1$) the ThPhM gives acceptable results only in the case of conductivity $\lambda \sim 1$;

(iii) ThPhM cannot be used, even for qualitative estimations, for large inclusion sizes ($a \to 1$) and extremely large ($\lambda \to \infty$) or extremely low ($\lambda \to 0$) conductivity. The asymptotic expression obtained from (6.109) at $\lambda \to \infty$, $a \to 1$ does not coincide with known asymptotic representation [55]:

$$q_{as}^{(\infty)} = \frac{\pi}{\sqrt{1-a^2}} - \pi + 1 \quad \text{at} \quad \lambda \to \infty, \quad a \to 1. \tag{6.110}$$

A similar conclusion takes place in the case of extremely low conductivity of inclusions ($\lambda \to 0$) of large sizes ($a \to 1$), where, by virtue of the Keller's theorem and formula (6.110), we have the formula

$$q_{as}^{(0)} = \frac{\sqrt{1-a^2}}{\pi - (\pi - 1)\sqrt{1-a^2}} \quad \text{at} \quad \lambda \to 0, \quad a \to 1. \tag{6.111}$$

The asymptotic formula (6.110) was refined in [46] using the method of functional equations:

$$q_{as}^{(\infty)} = \frac{2.78125}{\sqrt{\frac{\pi}{4}-c}} - 1.84934 = \frac{3.13830}{\sqrt{1-a^2}} - 1.84934 \tag{6.112}$$

$$\text{at} \quad \lambda \to \infty, \quad c \to c_{cr},$$

where $c = \frac{\pi a^2}{4}$, $c_{cr} = \frac{\pi}{4}$.

Let us briefly describe the following matter: why do we need solutions for a regular periodic lattice of inclusions if the real structure of composites has imperfections compared to the considered ideal case?

It was shown in [56, 57, 135, 237] that a regular lattice has extreme properties in some sense. Therefore, the solution for a perfectly regular lattice can be considered as an upper or lower bound on the corresponding effective coefficient.

Besides, improved bounds for effective transport properties of a random non-percolated composite can be developed by means of the security-spheres approach [12, 226]. The key point of the method is to use a solution for a regular composite valid for all values of the volume fractions and properties of the components.

6.4.2 SOLVING A LOCAL PROBLEM USING THE SCHWARZ ALTERNATING METHOD

For the problem of thermal conductivity of a doubly periodic composite structure with small cylindrical inclusions of a circle profile ($a \ll 1$) constituting a square lattice, the local problem is written in the form of (3.30)–(3.32).

1) Using the ThPhM model, we obtain the solution of the cell problem in (01) approximation of the Schwarz alternating method, i.e., the problem for one circular inclusion $\Omega_i^{-(0)}$ in a the plane $\Omega_i^{+(0)}$.

Note that the convergence of the Schwarz alternating method under fairly general assumptions was proved in [160]. In the same book, it is shown that the method also can be used for a multiply connected domain, provided that the inclusions are far enough apart. The convergence of the Schwarz alternating method in the general case of an arbitrary multiply connected domain was proved in the publications summarized in Chapter 3 of [94].

Let us define the cell problem (3.30)–(3.32) in fast polar co-ordinates r, θ in the following way:

$$\frac{\partial^2 u_1^{-(01)}}{\partial r^2} + \frac{1}{r}\frac{\partial u_1^{-(01)}}{\partial r} + \frac{1}{r^2}\frac{\partial^2 u_1^{-(01)}}{\partial \theta^2} = 0 \quad \text{in} \quad \Omega_i^{-(0)}, \quad (6.113)$$

$$\frac{\partial^2 u_1^{+(01)}}{\partial r^2} + \frac{1}{r}\frac{\partial u_1^{+(01)}}{\partial r} + \frac{1}{r^2}\frac{\partial^2 u_1^{+(01)}}{\partial \theta^2} = 0 \quad \text{in} \quad \Omega_i^{+(0)}, \quad (6.114)$$

$$u_1^{+(01)} = u_1^{-(01)}, \quad (6.115)$$

$$\frac{\partial u_1^{+(01)}}{\partial r} - \lambda\frac{\partial u_1^{-(01)}}{\partial r} = (\lambda - 1)\left(\frac{\partial u_0}{\partial x}\cos\theta + \frac{\partial u_0}{\partial y}\sin\theta\right) \quad \text{at} \quad r = a,$$

$$u_1^{+(01)} \to 0, \quad \frac{\partial u_1^{+(01)}}{\partial r} \to 0 \quad \text{at} \quad r \to \infty. \quad (6.116)$$

The solution of the problem (6.113)–(6.116) has the following form

$$u_1^{-(01)} = A_1^{(01)} r\cos\theta + A_2^{(01)} r\sin\theta, \quad (6.117)$$

$$u_1^{+(01)} = \frac{B_1^{(01)}}{r}\cos\theta + \frac{B_2^{(01)}}{r}\sin\theta, \quad (6.118)$$

where $A_1^{(01)}$, $A_2^{(01)}$, $B_1^{(01)}$, $B_2^{(01)}$ are the arbitrary constants.

The representation of the function $u_1^{-(01)}$ in the form (6.117) is written, taking into account the boundedness of the temperature distribution function and its derivative $\frac{\partial u_1^{-(01)}}{\partial r}$ (heat flux in the radial direction) at $r = 0$, and the function $u_1^{+(01)}$ (6.118) which satisfies the decaying conditions for at $r \to \infty$ (6.116).

Relations (6.117), (6.118) include four arbitrary constants, which are determined from the conditions (6.115). Systems of equations for constants

$A_1^{(01)}$, $B_1^{(01)}$ and $A_2^{(01)}$, $B_2^{(01)}$ are completely identical, so we present only one of them:

$$\begin{cases} A_1^{(01)} a = B_1^{(01)} a^{-1}, \\ -B_1^{(01)} a^{-2} - \lambda A_1^{(01)} = \dfrac{\partial u_0}{\partial x} (\lambda - 1). \end{cases} \tag{6.119}$$

Solving the system of equations (6.75), we find the values of the constants in the following form

$$\begin{cases} A_1^{(01)} = -\dfrac{\lambda - 1}{\lambda + 1} \dfrac{\partial u_0}{\partial x} = \dfrac{\partial u_0}{\partial x} A^{(01)*}, \\ B_1^{(01)} = -\dfrac{(\lambda - 1) a^2}{\lambda + 1} \dfrac{\partial u_0}{\partial x} = \dfrac{\partial u_0}{\partial x} B^{(01)*}, \end{cases} \tag{6.120}$$

where

$$A^{(01)*} = -\frac{\lambda - 1}{\lambda + 1}, \qquad B^{(01)*} = -\frac{(\lambda - 1) a^2}{\lambda + 1}. \tag{6.121}$$

Obviously, for arbitrary constants $A_2^{(01)}$, $B_2^{(01)}$, we have

$$A_2^{(01)} = A_1^{(01)}, \qquad B_2^{(01)} = B_1^{(01)} \left(\frac{\partial u_0}{\partial x} \to \frac{\partial u_0}{\partial y} \right). \tag{6.122}$$

Thus, solution of the (01) approximation has the form

$$\begin{aligned} u_1^{-\,(01)} &= -\frac{\partial u_0}{\partial x} \frac{\lambda - 1}{\lambda + 1} r \cos\theta - \frac{\partial u_0}{\partial y} \frac{\lambda - 1}{\lambda + 1} r \sin\theta, \\ u_1^{+\,(01)} &= -\frac{\partial u_0}{\partial x} \frac{\lambda - 1}{\lambda + 1} a^2 \frac{\cos\theta}{r} - \frac{\partial u_0}{\partial y} \frac{\lambda - 1}{\lambda + 1} a^2 \frac{\sin\theta}{r}, \end{aligned} \tag{6.123}$$

or

$$\begin{aligned} u_1^{-\,(01)} &= -\frac{\lambda - 1}{\lambda + 1} \left(\frac{\partial u_0}{\partial x} \xi + \frac{\partial u_0}{\partial y} \eta \right), \\ u_1^{+\,(01)} &= -\frac{\partial u_0}{\partial x} \frac{\lambda - 1}{\lambda + 1} a^2 \left(\frac{\partial u_0}{\partial x} \frac{\xi}{\xi^2 + \eta^2} + \frac{\partial u_0}{\partial y} \frac{\eta}{\xi^2 + \eta^2} \right). \end{aligned} \tag{6.124}$$

2) Let us proceed to the construction (02) approximation of the Schwarz alternating method.

We now satisfy the periodicity condition (3.32) on sides of the cell, neglecting the conditions (3.31). Function $u_1^{(02)}$ must remove discrepancies on the sides of the cell and is defined as a solution to a boundary value problem of the following form

$$\Delta u_1^{(02)} = 0 \quad \text{in} \quad \Omega_i^*, \tag{6.125}$$

$$\left(u_1^{+\,(01)} + u_1^{(02)}\right)\bigg|_{\xi=1} = \left(u_1^{+\,(01)} + u_1^{(02)}\right)\bigg|_{\xi=-1},$$

$$\frac{\partial\left(u_1^{+\,(01)} + u_1^{(02)}\right)}{\partial\xi}\bigg|_{\xi=1} = \frac{\partial\left(u_1^{+\,(01)} + u_1^{(02)}\right)}{\partial\xi}\bigg|_{\xi=-1}, \tag{6.126}$$

$$\left(u_1^{+\,(01)} + u_1^{(02)}\right)\bigg|_{\eta=1} = \left(u_1^{+\,(01)} + u_1^{(02)}\right)\bigg|_{\eta=-1},$$

$$\frac{\partial\left(u_1^{+\,(01)} + u_1^{(02)}\right)}{\partial\eta}\bigg|_{\eta=1} = \frac{\partial\left(u_1^{+\,(01)} + u_1^{(02)}\right)}{\partial\eta}\bigg|_{\eta=-1}. \tag{6.127}$$

Suppose that

$$u_1^{(02)} = u_{11}^{(02)} + u_{12}^{(02)}, \tag{6.128}$$

where function $u_{11}^{(02)}$ satisfies inhomogeneous (homogeneous) boundary conditions in terms of $\xi(\eta)$ and stands for a solution to the following problem

$$\Delta u_{11}^{(02)} = 0 \quad \text{in} \quad \Omega_i^*, \tag{6.129}$$

$$\left(u_1^{+\,(01)} + u_{11}^{(02)}\right)\bigg|_{\xi=1} = \left(u_1^{+\,(01)} + u_{11}^{(02)}\right)\bigg|_{\xi=-1},$$

$$\frac{\partial\left(u_1^{+\,(01)} + u_{11}^{(02)}\right)}{\partial\xi}\bigg|_{\xi=1} = \frac{\partial\left(u_1^{+\,(01)} + u_{11}^{(02)}\right)}{\partial\xi}\bigg|_{\xi=-1}, \tag{6.130}$$

$$u_{11}^{(02)}\bigg|_{\eta=1} = u_{11}^{(02)}\bigg|_{\eta=-1}, \quad \frac{\partial u_{11}^{(02)}}{\partial\eta}\bigg|_{\eta=1} = \frac{\partial u_{11}^{(02)}}{\partial\eta}\bigg|_{\eta=-1}. \tag{6.131}$$

Obviously, the function $u_{12}^{(02)}$ is defined in a similar way, taking into account the change $\xi \leftrightarrows \eta$. The general solution of equation (6.129) takes the following form

$$u_{11}^{(02)} = A_0^{(02)} + B_0^{(02)}\xi + \sum_{n=1}^{\infty}\left[\left(A_n^{(02)} \cosh \pi n\xi + B_n^{(02)} \sinh \pi n\xi\right) \cos \pi n\eta + \right.$$

$$\left. + \left(C_n^{(02)} \cosh \pi n\xi + D_n^{(02)} \sinh \pi n\xi\right) \sin \pi n\eta\right], \tag{6.132}$$

where $A_0^{(02)}$, $B_0^{(02)}$, $A_n^{(02)}$, $B_n^{(02)}$, $C_n^{(02)}$, $D_n^{(02)}$ ($n = 1, 2, \ldots$) are the arbitrary constants.

To satisfy conditions (6.130), we rewrite them, taking into account the solution (6.124), in the following form

$$u_{11}^{(02)}\Big|_{\xi=1} - u_{11}^{(02)}\Big|_{\xi=-1} = \frac{\partial u_0}{\partial x}\frac{2a^2}{1+\eta^2}\frac{\lambda-1}{\lambda+1},$$

$$\frac{\partial u_{11}^{(02)}}{\partial \xi}\Big|_{\xi=1} - \frac{\partial u_{11}^{(02)}}{\partial \xi}\Big|_{\xi=-1} = -\frac{\partial u_0}{\partial y}\frac{4a^2\eta}{(1+\eta^2)^2}\frac{\lambda-1}{\lambda+1}. \tag{6.133}$$

We expand the functions on the right-hand sides of relations (6.133) into the following Fourier series

$$\frac{1}{1+\eta^2} = \frac{\pi}{4} + \sum_{n=1}^{\infty} \left[e^{-\pi n}\,\mathrm{Im}\,E_1\left(-\pi n + i\pi n\right) - \right.$$

$$\left. -e^{\pi n}\,\mathrm{Im}\,E_1\left(\pi n + i\pi n\right) + \pi e^{-\pi n}\right]\cos \pi n\eta, \tag{6.134}$$

$$\frac{\eta}{(1+\eta^2)^2} = \frac{\pi}{2}\sum_{n=1}^{\infty} n\left[e^{\pi n}\,\mathrm{Im}\,E_1\left(\pi n + i\pi n\right) - \right.$$

$$\left. -e^{\pi n}\,\mathrm{Im}\,E_1\left(\pi n + i\pi n\right) + \pi e^{-\pi n}\right]\sin \pi n\eta,$$

where $i = \sqrt{-1}$; $\mathrm{Im}\,E_1$ is the imaginary part of the familiar integral exponential function [2].

Equating, taking into account the expression (6.134), the corresponding coefficients in (6.133), we find the values of the constants of the (02) approximation:

$$A_0^{(02)} = 0, \quad A_n^{(02)} = D_n^{(02)} = 0, \quad n = 1, 2, \ldots, \tag{6.135}$$

$$B_0^{(02)} = \frac{\partial u_0}{\partial x}B_0^{(02)*}, \quad B_n^{(02)} = \frac{\partial u_0}{\partial x}B_n^{(02)*}, \quad C_n^{(02)} = \frac{\partial u_0}{\partial y}C_n^{(02)*},$$

$$n = 1, 2, \ldots,$$

$$B_0^{(02)*} = \frac{\lambda-1}{\lambda+1}\frac{\pi a^2}{4}, \tag{6.136}$$

$$B_n^{(02)*} = -C_n^{(02)*} = \frac{\lambda-1}{\lambda+1}a^2 S_n, \tag{6.137}$$

where

$$S_n = \frac{e^{-\pi n}\,\mathrm{Im}\,E_1\left(-\pi n + i\pi n\right) - e^{\pi n}\,\mathrm{Im}\,E_1\left(\pi n + i\pi n\right) + \pi e^{-\pi n}}{\sinh \pi n}. \tag{6.138}$$

Therefore, we get

$$
u_{11}^{(02)} = \frac{\partial u_0}{\partial x} B_0^{(02)*} \xi + \sum_{n=1}^{\infty} \left(\frac{\partial u_0}{\partial x} B_n^{(02)*} \sinh \pi n \xi \cos \pi n \eta - \right.
$$
$$
\left. - \frac{\partial u_0}{\partial y} C_n^{(02)*} \cosh \pi n \xi \sin \pi n \eta \right).
$$

(6.139)

It is evident that

$$
u_{12}^{(02)} = u_{11}^{(02)} \left(\frac{\partial u_0}{\partial x} \rightleftarrows \frac{\partial u_0}{\partial y}; \ \xi \rightleftarrows \eta \right).
$$

(6.140)

Therefore, in the (02) approximation, we obtain

$$
u_1^{(02)} = \frac{\lambda - 1}{\lambda + 1} \frac{\pi a^2}{4} \left(\frac{\partial u_0}{\partial x} \xi + \frac{\partial u_0}{\partial y} \eta \right) +
$$

$$
+ \frac{\lambda - 1}{\lambda + 1} a^2 \sum_{n=1}^{\infty} S_n \left[\frac{\partial u_0}{\partial x} (\sinh \pi n \xi \cos \pi n \eta - \right.
$$

(6.141)

$$
\left. - \cosh \pi n \eta \sin \pi n \xi) + \frac{\partial u_0}{\partial y} (\sinh \pi n \eta \cos \pi n \xi - \cosh \pi n \xi \sin \pi n \eta) \right].
$$

3) In the (03) approximation, it is necessary to remove the discrepancies given by the function $u_1^{(02)}$ on a circular contour of the inclusion of radius $r = a$. To do this, we expand the expression $u_1^{(02)}$ in a series in polar coordinates r, θ for small values of the inclusion radius r:

$$
B_0^{(02)*} \left(\frac{\partial u_0}{\partial x} \xi + \frac{\partial u_0}{\partial y} \eta \right) + \sum_{n=1}^{\infty} B_n^{(02)*} \left[\frac{\partial u_0}{\partial x} (\sinh \pi n \xi \cos \pi n \eta - \right.
$$

$$
\left. - \cosh \pi n \eta \sin \pi n \xi) + \frac{\partial u_0}{\partial y} (\sinh \pi n \eta \cos \pi n \xi - \cosh \pi n \xi \sin \pi n \eta) \right] =
$$

$$
= B_0^{(02)*} \left(\frac{\partial u_0}{\partial x} r \cos \theta + \frac{\partial u_0}{\partial y} r \sin \theta \right) +
$$

$$
+ 2 \sum_{n=1}^{\infty} B_n^{(02)*} \left[\frac{\partial u_0}{\partial x} \sum_{k=1}^{\infty} \frac{(\pi n r)^{4k-1} \cos (4k-1) \theta}{(4k-1)!} + \right.
$$

$$
\left. + \frac{\partial u_0}{\partial y} \sum_{k=1}^{\infty} \frac{(\pi n r)^{4k-1} \sin (4k-1) \theta}{(4k-1)!} \right],
$$

or

$$
B_0^{(02)*} \left(\frac{\partial u_0}{\partial x} \xi + \frac{\partial u_0}{\partial y} \eta \right) + \sum_{n=1}^{\infty} B_n^{(02)*} \left[\frac{\partial u_0}{\partial x} (\sinh \pi n \xi \cos \pi n \eta - \right.
$$

$$-\cosh \pi n\eta \, \sin \pi n\xi) + \frac{\partial u_0}{\partial y} (\sinh \pi n\eta \, \cos \pi n\xi - \cosh \pi n\xi \, \sin \pi n\eta) \bigg] =$$

$$= B_0^{(02)*} \left(\frac{\partial u_0}{\partial x} r\cos\theta + \frac{\partial u_0}{\partial y} r\sin\theta \right) + \tag{6.142}$$

$$+2 \sum_{k=1}^{\infty} \frac{\pi^{4k-1}}{(4k-1)!} \left(\sum_{n=1}^{\infty} B_n^{(02)*} n^{4k-1} \right) \times$$

$$\times \left(\frac{\partial u_0}{\partial x} r^{4k-1} \cos (4k-1)\,\theta + \frac{\partial u_0}{\partial y} r^{4k-1} \sin (4k-1)\,\theta \right),$$

where the series on the right side of (6.142) converges for all values $0 \le r < \infty$.

Solution of the (03) approximation is written as follows:

$$u_1^{-\,(03)} = A_{10}^{(03)} r\cos\theta + A_{20}^{(03)} r\sin\theta +$$

$$+ \sum_{k=1}^{\infty} \left(A_{1k}^{(03)} r^{4k-1} \cos (4k-1)\,\theta + A_{2k}^{(03)} r^{4k-1} \sin (4k-1)\,\theta \right), \tag{6.143}$$

$$u_1^{+\,(03)} = \frac{B_{10}^{(03)}}{r} \cos\theta + \frac{B_{20}^{(03)}}{r} \sin\theta +$$

$$+ \sum_{k=1}^{\infty} \left(B_{1k}^{(03)} \frac{\cos (4k-1)\,\theta}{r^{4k-1}} + B_{2k}^{(03)} \frac{\sin (4k-1)\,\theta}{r^{4k-1}} \right). \tag{6.144}$$

The constants

$$A_{10}^{(03)} = A_{10}^{(03)*} \frac{\partial u_0}{\partial x}, \qquad B_{10}^{(03)} = B_{10}^{(03)*} \frac{\partial u_0}{\partial x},$$

$$A_{20}^{(03)} = A_{20}^{(03)*} \frac{\partial u_0}{\partial y}, \qquad B_{20}^{(03)} = B_{20}^{(03)*} \frac{\partial u_0}{\partial y},$$

and

$$A_{1k}^{(03.n)} = A_{1k}^{(03.n)*} \frac{\partial u_0}{\partial x}, \qquad B_{1k}^{(03.n)} = B_{1k}^{(03.n)*} \frac{\partial u_0}{\partial x},$$

$$A_{2k}^{(03.n)} = A_{2k}^{(03.n)*} \frac{\partial u_0}{\partial y}, \qquad B_{2k}^{(03.n)} = B_{2k}^{(03.n)*} \frac{\partial u_0}{\partial y}$$

are determined respectively from the solution of the systems of equations

$$\begin{cases} A_{m0}^{(03)*} a - \dfrac{B_{m0}^{(03)*}}{a}, \\[3mm] -\dfrac{B_{m0}^{(03)*}}{a^2} - \lambda A_{m0}^{(03)*} = (\lambda - 1) B_0^{(02)*} \end{cases}$$

and

$$
\begin{cases}
A_{mk}^{(03.n)^*}\, a^{4k-1} = \dfrac{B_{mk}^{(03.n)^*}}{a^{4k-1}}, \\[2mm]
-\dfrac{B_{mk}^{(03.n)^*}}{\tilde{a}^{4k}} - \lambda\, a^{4k-2} A_{mk}^{(03.n)^*} = \dfrac{2\,(\lambda-1)\,\pi^{4k-1}a^{4k-2}}{(4k-1)!} \sum_{n=1}^{\infty} B_n^{(02)^*}\, n^{4k-1}.
\end{cases}
$$

Finally, one obtains

$$
A_{m0}^{(03)^*} = -\frac{\lambda-1}{\lambda+1} B_0^{(02)^*} = -\left(\frac{\lambda-1}{\lambda+1}\right)^2 \frac{\pi a^2}{4},
$$
$$
B_{m0}^{(03)^*} = -\frac{\lambda-1}{\lambda+1} a^2 B_0^{(02)^*} = -\left(\frac{\lambda-1}{\lambda+1}\right)^2 \frac{\pi a^4}{4},
\tag{6.145}
$$

$$
A_{mk}^{(03)^*} = -\frac{2\,\pi^{4k-1}}{(4k-1)!}\frac{\lambda-1}{\lambda+1} \sum_{n=1}^{\infty} B_n^{(02)^*}\, n^{4k-1} =
$$
$$
= -\frac{2\,\pi^{4k-1}a^2}{(4k-1)!}\left(\frac{\lambda-1}{\lambda+1}\right)^2 \sum_{n=1}^{\infty} S_n n^{4k-1}
\tag{6.146}
$$

$$
B_{mk}^{(03)^*} = -\frac{2\,\pi^{4k-1}a^{8k-2}}{(4k-1)!}\frac{\lambda-1}{\lambda+1} \sum_{n=1}^{\infty} B_n^{(02)^*}\, n^{4k-1} =
$$
$$
= -\frac{2\,\pi^{4k-1}a^{8k}}{(4k-1)!}\left(\frac{\lambda-1}{\lambda+1}\right)^2 \sum_{n=1}^{\infty} S_n n^{4k-1}, \quad m=1,\,2.
\tag{6.147}
$$

Thus, taking into account expressions (6.145)–(6.147)), we get

$$
u_1^{-\,(03)} = -\frac{\lambda-1}{\lambda+1} B_0^{(02)^*}\, r\left(\frac{\partial u_0}{\partial x}\cos\theta + \frac{\partial u_0}{\partial y}\sin\theta\right) -
$$
$$
\frac{\lambda-1}{\lambda+1}\, 2\left(\sum_{n=1}^{\infty} B_n^{(02)^*}\, n^{4k-1}\right) \times
\tag{6.148}
$$
$$
\times \sum_{k=1}^{\infty} \frac{\pi^{4k-1}a^{4k-1}}{(4k-1)!}\left(\frac{\partial u_0}{\partial x}\cos(4k-1)\,\theta + \frac{\partial u_0}{\partial y}\sin(4k-1)\,\theta\right),
$$

$$
u_1^{+\,(03)} = -\frac{\lambda-1}{\lambda+1}\frac{B_0^{(02)^*}}{r}\left(\frac{\partial u_0}{\partial x}\cos\theta + \frac{\partial u_0}{\partial y}\sin\theta\right) -
$$
$$
\frac{\lambda-1}{\lambda+1}\, 2\left(\sum_{n=1}^{\infty} B_n^{(02)^*}\, n^{4k-1}\right) \times
\tag{6.149}
$$
$$
\times \sum_{k=1}^{\infty} \frac{\pi^{4k-1}a^{8k-2}}{(4k-1)!\, r^{4k-1}}\left(\frac{\partial u_0}{\partial x}\cos(4k-1)\,\theta + \frac{\partial u_0}{\partial y}\sin(4k-1)\,\theta\right).
$$

Summing the series in k in expressions (6.148), (6.149), yields

$$u_1^{-(03)} = -\frac{\lambda-1}{\lambda+1} B_0^{(02)*} \left(\frac{\partial u_0}{\partial x}\xi + \frac{\partial u_0}{\partial y}\eta\right) -$$

$$-\frac{\lambda-1}{\lambda+1}\sum_{n=1}^{\infty} B_n^{(02)*}\left[\frac{\partial u_0}{\partial x}(\sinh \pi n\xi \cos \pi n\eta - \cosh \pi n\eta \sin \pi n\xi) +\right.$$

$$\left. +\frac{\partial u_0}{\partial y}(\sinh \pi n\eta \cos \pi n\xi - \cosh \pi n\xi \sin \pi n\eta)\right],$$

$$(6.150)$$

$$u_1^{+(03)} = -\left(\frac{\lambda-1}{\lambda+1}\right)^2 \frac{\pi a^2}{4}\left(\frac{\partial u_0}{\partial x}\frac{\xi}{\xi^2+\eta^2} + \frac{\partial u_0}{\partial y}\frac{\eta}{\xi^2+\eta^2}\right) - \frac{\lambda-1}{\lambda+1}\times$$

$$\times\sum_{n=1}^{\infty} B_n^{(02)*}\left[\frac{\partial u_0}{\partial x}\left(\sinh\frac{\pi na^2\xi}{\xi^2+\eta^2}\cos\frac{\pi na^2\eta}{\xi^2+\eta^2} - \cosh\frac{\pi na^2\eta}{\xi^2+\eta^2}\sin\frac{\pi na^2\xi}{\xi^2+\eta^2}\right) +\right.$$

$$\left. +\frac{\partial u_0}{\partial y}\left(\sinh\frac{\pi na^2\eta}{\xi^2+\eta^2}\cos\frac{\pi na^2\xi}{\xi^2+\eta^2} - \cosh\frac{\pi na^2\xi}{\xi^2+\eta^2}\sin\frac{\pi na^2\eta}{\xi^2+\eta^2}\right)\right],$$

$$(6.151)$$

i.e.

$$u_1^{-(03)} = -\left(\frac{\lambda-1}{\lambda+1}\right)^2 \frac{\pi a^2}{4}\left(\frac{\partial u_0}{\partial x}\xi + \frac{\partial u_0}{\partial y}\eta\right) -$$

$$-\left(\frac{\lambda-1}{\lambda+1}\right)^2 a^2 \sum_{n=1}^{\infty} S_n\left[\frac{\partial u_0}{\partial x}(\sinh \pi n\xi \cos \pi n\eta - \cosh \pi n\eta \sin \pi n\xi) +\right.$$

$$\left. +\frac{\partial u_0}{\partial y}(\sinh \pi n\eta \cos \pi n\xi - \cosh \pi n\xi \sin \pi n\eta)\right],$$

$$(6.152)$$

$$u_1^{+(03)} = \left(\frac{\lambda-1}{\lambda+1}\right)^2 \frac{\pi a^2}{4}\left(\frac{\partial u_0}{\partial x}\frac{\xi}{\xi^2+\eta^2} + \frac{\partial u_0}{\partial y}\frac{\eta}{\xi^2+\eta^2}\right) - \left(\frac{\lambda-1}{\lambda+1}\right)^2 a^2 \times$$

$$\times\sum_{n=1}^{\infty} S_n\left[\frac{\partial u_0}{\partial x}\left(\sinh\frac{\pi na^2\xi}{\xi^2+\eta^2}\cos\frac{\pi na^2\eta}{\xi^2+\eta^2} - \cosh\frac{\pi na^2\eta}{\xi^2+\eta^2}\sin\frac{\pi na^2\xi}{\xi^2+\eta^2}\right) +\right.$$

$$\left. +\frac{\partial u_0}{\partial y}\left(\sinh\frac{\pi na^2\eta}{\xi^2+\eta^2}\cos\frac{\pi na^2\xi}{\xi^2+\eta^2} - \cosh\frac{\pi na^2\xi}{\xi^2+\eta^2}\sin\frac{\pi na^2\eta}{\xi^2+\eta^2}\right)\right].$$

$$(6.153)$$

4) In the (04) approximation, according to a scheme similar to the one given earlier, we remove the discrepancies on the outer contour of the cell. We write

$$u_{11}^{(04)} = A_0^{(04)} + B_0^{(04)} \xi + \sum_{\ell=1}^{\infty} \left[\left(A_\ell^{(04)} \cosh \pi\ell\xi + B_\ell^{(04)} \sinh \pi\ell\xi \right) \cos \pi\ell\eta + \right.$$

$$\left. + \left(C_\ell^{(02)} \cosh \pi\ell\xi + D_\ell^{(02)} \sinh \pi\ell\xi \right) \sin \pi\ell\eta \right]$$

conditions on the contours $\xi = 1$ and $\xi = -1$ in the following form

$$u_{11}^{(04)} \bigg|_{\xi=1} - u_{11}^{(04)} \bigg|_{\xi=-1} = \frac{\partial u_0}{\partial x} \frac{\lambda-1}{\lambda+1} B_0^{(02)*} \frac{2}{1+\eta^2} + \qquad (6.154)$$

$$+ \frac{\partial u_0}{\partial x} \frac{\lambda-1}{\lambda+1} 2 \sum_{n=1}^{\infty} B_n^{(02)*} \left(\sinh \frac{\pi na^2}{1+\eta^2} \cos \frac{\pi na^2 \eta}{1+\eta^2} - \cosh \frac{\pi na^2 \eta}{1+\eta^2} \sin \frac{\pi na^2}{1+\eta^2} \right),$$

$$\frac{\partial u_{11}^{(04)}}{\partial \xi} \bigg|_{\xi=1} - \frac{\partial u_{11}^{(04)}}{\partial \xi} \bigg|_{\xi=-1} = -\frac{\partial u_0}{\partial y} \frac{\lambda-1}{\lambda+1} B_0^{(02)*} \frac{4\eta}{(1+\eta^2)^2} - \qquad (6.155)$$

$$- \frac{\partial u_0}{\partial y} \frac{\lambda-1}{\lambda+1} 2\pi a^2 \sum_{n=1}^{\infty} B_n^{(02)*} n \left[\frac{2\eta}{(1+\eta^2)^2} \cosh \frac{\pi na^2 \eta}{1+\eta^2} \cos \frac{\pi na^2}{1+\eta^2} - \right.$$

$$- \frac{1-\eta^2}{(1+\eta^2)^2} \sinh \frac{\pi na^2 \eta}{1+\eta^2} \sin \frac{\pi na^2}{1+\eta^2} - \frac{1-\eta^2}{(1+\eta^2)^2} \sinh \frac{\pi na^2}{1+\eta^2} \sin \frac{\pi na^2 \eta}{1+\eta^2} -$$

$$\left. - \frac{2\eta}{(1+\eta^2)^2} \cosh \frac{\pi na^2}{1+\eta^2} \cos \frac{\pi na^2 \eta}{1+\eta^2} \right].$$

Expanding the functions on the right-hand sides of relations (6.154), (6.155) into Fourier series, we find the values of the coefficients

$$A_0^{(04)} = 0, \qquad A_\ell^{(04)} = D_\ell^{(04)} = 0, \qquad (6.156)$$

$$B_0^{(04)} = \frac{\partial u_0}{\partial x} B_0^{(04)*}, \qquad B_\ell^{(04)} = \frac{\partial u_0}{\partial x} B_\ell^{(04)*}, \qquad C_\ell^{(04)} = \frac{\partial u_0}{\partial y} C_\ell^{(04)*}, \qquad (6.157)$$

$$\ell = 0, 1, 2, \ldots,$$

$$B_0^{(04)*} = \frac{\lambda-1}{\lambda+1} \frac{\pi a^2}{4} B_0^{(02)*} + \frac{\lambda-1}{\lambda+1} \sum_{n=1}^{\infty} B_n^{(02)*} \sum_{m=1}^{\infty} \frac{(-1)^{m+1} \left(\pi na^2\right)^{4m-1}}{2^{2m-1}(2m-1)(4m-1)!} =$$

$$= \left(\frac{\lambda-1}{\lambda+1}\right)^2 \frac{\pi^2 a^4}{16} \left(1 + \frac{8}{\pi} \sum_{n=1}^{\infty} S_n n \sum_{m=1}^{\infty} \frac{(-1)^{m+1} (\pi n)^{4m-2} a^{8m-4}}{2^{2m-2} (2m-1)(4m-1)!}\right),$$

$$(6.158)$$

$$B_\ell^{(04)*} = -C_\ell^{(04)*} = \frac{\lambda-1}{\lambda+1} a^2 B_0^{(02)*} S_\ell +$$

$$+ \frac{\lambda-1}{\lambda+1} \sum_{n=1}^{\infty} B_n^{(02)*} \sum_{m=1}^{\infty} \frac{(-1)^{m+1} (\pi n a^2)^{4m-1}}{(2m-1)(4m-1)!(4m-3)!} \times$$

$$\times \left((\pi\ell)^{4m-2} S_\ell + \sum_{k=1}^{m} (-1)^{\ell+k+1} \frac{(4k-3)!}{2^{2k-2}} \frac{(\pi\ell)^{4m-4k}}{sh\, \pi\ell}\right) = \qquad (6.159)$$

$$= \left(\frac{\lambda-1}{\lambda+1}\right)^2 \frac{\pi a^4}{4} \left[S_\ell + 4 \sum_{n=1}^{\infty} S_n n \sum_{m=1}^{\infty} \frac{(-1)^{m+1} (\pi n)^{4m-2} a^{8m-4}}{(2m-1)(4m-1)!(4m-3)!} \times\right.$$

$$\left. \times \left((\pi\ell)^{4m-2} S_\ell + \sum_{k=1}^{m} (-1)^{\ell+k+1} \frac{(4k-3)!}{2^{2k-2}} \frac{(\pi\ell)^{4m-4k}}{\sinh \pi\ell}\right)\right].$$

Consequently, in (04) approximation, we have

$$u_1^{(04)} = \left(\frac{\lambda-1}{\lambda+1}\right)^2 \frac{\pi^2 a^4}{16} \left(1 + \frac{8}{\pi} \sum_{n=1}^{\infty} S_n n \sum_{m=1}^{\infty} \frac{(-1)^{m+1} (\pi n)^{4m-2} a^{8m-4}}{2^{2m-2} (2m-1)(4m-1)!}\right) \times$$

$$\times \left(\frac{\partial u_0}{\partial x} \xi + \frac{\partial u_0}{\partial y} \eta\right) + \left(\frac{\lambda-1}{\lambda+1}\right)^2 \frac{\pi a^4}{4} \left[\sum_{\ell=1}^{\infty} S_\ell +\right.$$

$$+ 4 \sum_{\ell=1}^{\infty} S_\ell \sum_{n=1}^{\infty} S_n n \sum_{m=1}^{\infty} \frac{(-1)^{m+1} (\pi^2 n\ell)^{4m-2} a^{8m-4}}{(2m-1)(4m-1)!(4m-3)!} +$$

$$+ 4 \sum_{\ell=1}^{\infty} \frac{1}{\sinh \pi\ell} \sum_{n=1}^{\infty} S_n n \sum_{m=1}^{\infty} \frac{(-1)^{m+1} (\pi n)^{4m-2} a^{8m-4}}{(2m-1)(4m-1)!(4m-3)!} \times$$

$$\times \sum_{k=1}^{m} \frac{(-1)^{\ell+k+1} (4k-3)! (\pi\ell)^{4m-4k}}{2^{2k-2}}\right] \times \qquad (6.160)$$

$$\times \left[\frac{\partial u_0}{\partial x} (\sinh \pi\ell\xi \cos \pi\ell\eta - \cosh \pi\ell\eta \sin \pi\ell\xi) +\right.$$

$$\left. + \frac{\partial u_0}{\partial y} (\sinh \pi\ell\eta \cos \pi\ell\xi - \cosh \pi\ell\xi \sin \pi\ell\eta)\right].$$

6.4.3   CONSTRUCTION OF GENERALIZING RELATIONS FOR MF

Further continuation of the iterative process is carried out according to a similar scheme and leads to the following equation

$$
\left(1 - \frac{\pi a^2}{4} + \lambda \frac{\pi a^2}{4}\right)\left(\frac{\partial^2 u_0}{\partial x^2} + \frac{\partial^2 u_0}{\partial y^2}\right) +
$$

$$
+ \frac{1}{|\Omega_i^*|}\left[\iint\limits_{\Omega_i^+}\left(\frac{\partial^2 u_1^{+\ (01)}}{\partial x \partial \xi} + \frac{\partial^2 u_1^{+\ (01)}}{\partial y \partial \eta} + \frac{\partial^2 u_1^{(02)}}{\partial x \partial \xi} + \frac{\partial^2 u_1^{(02)}}{\partial y \partial \eta} + \right.
$$

$$
+ \frac{\partial^2 u_1^{+\ (03)}}{\partial x \partial \xi} + \frac{\partial^2 u_1^{+\ (03)}}{\partial y \partial \eta} + \frac{\partial^2 u_1^{(04)}}{\partial x \partial \xi} + \frac{\partial^2 u_1^{(04)}}{\partial y \partial \eta} + \frac{\partial^2 u_1^{+\ (05)}}{\partial x \partial \xi} + \frac{\partial^2 u_1^{+\ (05)}}{\partial y \partial \eta} +
$$

$$
\left. + \frac{\partial^2 u_1^{(06)}}{\partial x \partial \xi} + \frac{\partial^2 u_1^{(06)}}{\partial y \partial \eta} + ...\right) d\xi\, d\eta + \lambda \iint\limits_{\Omega_i^-}\left(\frac{\partial^2 u_1^{-\ (01)}}{\partial x \partial \xi} + \frac{\partial^2 u_1^{-\ (01)}}{\partial y \partial \eta} +\right.
$$

$$
+ \frac{\partial^2 u_1^{(02)}}{\partial x \partial \xi} + \frac{\partial^2 u_1^{(02)}}{\partial y \partial \eta} + \frac{\partial^2 u_1^{-\ (03)}}{\partial x \partial \xi} + \frac{\partial^2 u_1^{-\ (03)}}{\partial y \partial \eta} + \frac{\partial^2 u_1^{(04)}}{\partial x \partial \xi} + \frac{\partial^2 u_1^{(04)}}{\partial y \partial \eta} +
$$

$$
\left.\left. + \frac{\partial^2 u_1^{-\ (05)}}{\partial x \partial \xi} + \frac{\partial^2 u_1^{-\ (05)}}{\partial y \partial \eta} + \frac{\partial^2 u_1^{(06)}}{\partial x \partial \xi} + \frac{\partial^2 u_1^{(06)}}{\partial y \partial \eta} + ...\right) d\xi\, d\eta\right] = F.
$$

The effective thermal conductivity

$$
q = 1 + 2\frac{\lambda - 1}{\lambda + 1}\frac{\pi a^2}{4} + 2\left(\frac{\lambda - 1}{\lambda + 1}\frac{\pi a^2}{4}\right)^2 + 2\left(\frac{\lambda - 1}{\lambda + 1}\frac{\pi a^2}{4}\right)^3 + ... +
$$

$$
+ \left(\frac{\lambda - 1}{\lambda + 1}\right)^3\frac{\pi^2 a^6}{4}\left(1 + \frac{\lambda - 1}{\lambda + 1}\frac{\pi a^2}{4} + ...\right)^2 \sum_{\ell=1}^{\infty} S_{\ell\ell} \sum_{n=1}^{\infty}\frac{(-1)^{n+1}(\pi\ell)^{4n-2} a^{8n-4}}{2^{2n-1}(2n-1)(4n-1)!} +
$$

$$
+ \left(\frac{\lambda - 1}{\lambda + 1}\right)^4\frac{\pi^3 a^8}{16}\left(1 + \frac{\lambda - 1}{\lambda + 1}\frac{\pi a^2}{4} + ...\right)^2 \times \qquad (6.161)
$$

$$
\times \sum_{\ell=1}^{\infty}\ell\sum_{j=1}^{\infty} S_{jj}\sum_{m=1}^{\infty}\frac{(-1)^{m+1}(\pi j)^{4m-2} a^{8m-4}}{(2m-1)(4m-1)!(4m-3)!} \times
$$

$$
\times \left((\pi\ell)^{4m-2} S_\ell + \sum_{k=1}^{m}(-1)^{\ell+k+1}\frac{(4k-3)!}{2^{2k-2}}\frac{(\pi\ell)^{4m-4k}}{\sinh \pi\ell}\right) \times
$$

$$
\times \sum_{n=1}^{\infty}\frac{(-1)^{n+1}(\pi\ell)^{4n-2} a^{8n-4}}{2^{2n-1}(2n-1)(4n-1)!}.
$$

Summing up series in (6.161)

$$1+2\frac{\lambda-1}{\lambda+1}\frac{\pi a^2}{4}+2\left(\frac{\lambda-1}{\lambda+1}\frac{\pi a^2}{4}\right)^2+2\left(\frac{\lambda-1}{\lambda+1}\frac{\pi a^2}{4}\right)^3+...+$$

we get the MF

$$q_{MF}=\frac{\lambda\left(1+\frac{\pi a^2}{4}\right)+1-\frac{\pi a^2}{4}}{\lambda\left(1-\frac{\pi a^2}{4}\right)+1+\frac{\pi a^2}{4}}. \tag{6.162}$$

Indeed,

$$q_{MF}=\frac{\lambda\left(1+\frac{\pi a^2}{4}\right)+1-\frac{\pi a^2}{4}}{\lambda\left(1-\frac{\pi a^2}{4}\right)+1+\frac{\pi a^2}{4}}=\frac{(\lambda+1)+(\lambda-1)\frac{\pi a^2}{4}}{(\lambda+1)-(\lambda-1)\frac{\pi a^2}{4}}=\frac{1+\frac{\lambda-1}{\lambda+1}\frac{\pi a^2}{4}}{1-\frac{\lambda-1}{\lambda+1}\frac{\pi a^2}{4}}=$$

$$=\left(1+\frac{\lambda-1}{\lambda+1}\frac{\pi a^2}{4}\right)\left(1+\frac{\lambda-1}{\lambda+1}\frac{\pi a^2}{4}+\left(\frac{\lambda-1}{\lambda+1}\frac{\pi a^2}{4}\right)^2+...\right)=$$

$$=1+2\frac{\lambda-1}{\lambda+1}\frac{\pi a^2}{4}+2\left(\frac{\lambda-1}{\lambda+1}\frac{\pi a^2}{4}\right)^2+2\left(\frac{\lambda-1}{\lambda+1}\frac{\pi a^2}{4}\right)^3+...$$

Thus, the application of the Schwarz alternating method shows that the main term of the effective thermal conductivity parameter in the case of small inclusions is the MF and coincides with Hashin-Shtrikman upper bound at $0\leq\lambda\leq1$, and with the Hashin-Shtrikman lower bound at $1\leq\lambda<\infty$.

The first two non-zero corrections to MF, taking into account the obvious relation

$$1+\frac{\lambda-1}{\lambda+1}\frac{\pi a^2}{4}+\left(\frac{\lambda-1}{\lambda+1}\frac{\pi a^2}{4}\right)^2+...=$$

$$=\frac{1}{1-\frac{\lambda-1}{\lambda+1}\frac{\pi a^2}{4}}=\frac{\lambda+1}{\lambda\left(1-\frac{\pi a^2}{4}\right)+1+\frac{\pi a^2}{4}}\quad\text{at}\quad a\to0$$

and restricting ourselves to the approximation (6.161), we obtain in the following form

$$\Delta=\left(\frac{\lambda-1}{\lambda+1}\right)^3\frac{\pi^2 a^6}{4}\frac{(\lambda+1)^2}{\left(\lambda\left(1-\frac{\pi a^2}{4}\right)+1+\frac{\pi a^2}{4}\right)^2}\times$$

$$\times\sum_{\ell=1}^{\infty}S_{\ell\ell}\sum_{n=1}^{\infty}\frac{(-1)^{n+1}(\pi\ell)^{4n-2}a^{8n-4}}{2^{2n-2}(2n-1)(4n-1)!}+\left(\frac{\lambda-1}{\lambda+1}\right)^4\frac{\pi^3 a^8}{16}\times \tag{6.163}$$

$$\times\frac{(\lambda+1)^2}{\left(\lambda\left(1-\frac{\pi a^2}{4}\right)+1+\frac{\pi a^2}{4}\right)^2}\sum_{\ell=1}^{\infty}\ell\sum_{j=1}^{\infty}S_{jj}\sum_{m=1}^{\infty}\frac{(-1)^{m+1}(\pi j)^{4m-2}a^{8m-4}}{(2m-1)(4m-1)!(4m-3)!}\times$$

$$\times \left((\pi\ell)^{4m-2} S_\ell + \sum_{k=1}^{m} (-1)^{\ell+k+1} \frac{(4k-3)!}{2^{2k-2}} \frac{(\pi\ell)^{4m-4k}}{\sinh \pi\ell} \right) \times$$

$$\times \sum_{n=1}^{\infty} \frac{(-1)^{n+1} (\pi\ell)^{4n-2} a^{8n-4}}{2^{2n-1} (2n-1)(4n-1)!}.$$

Thus, the finally transformed expression for the effective thermal conductivity, found as a generalized (N-iterative) solution of the problem by the Schwarz alternating method, is reduced to the form

$$q_{ASM(N)} = \frac{\lambda\left(1+\frac{\pi a^2}{4}\right)+1-\frac{\pi a^2}{4}}{\lambda\left(1-\frac{\pi a^2}{4}\right)+1+\frac{\pi a^2}{4}} + \left(\frac{\lambda-1}{\lambda+1}\right)^3 \frac{\pi^2 a^6}{4} \times$$

$$\times \frac{(\lambda+1)^2}{\left(\lambda\left(1-\frac{\pi a^2}{4}\right)+1+\frac{\pi a^2}{4}\right)^2} \times \left(\Delta_1 + \frac{\lambda-1}{\lambda+1}\frac{\pi a^2}{4}\Delta_2\right) = \qquad (6.164)$$

$$= q_{MF} + \frac{\lambda-1}{\lambda+1}\frac{\pi^2 a^6}{4} \frac{(\lambda-1)^2}{\left(\lambda\left(1-\frac{\pi a^2}{4}\right)+1+\frac{\pi a^2}{4}\right)^2} \left(\Delta_1 + \frac{\lambda-1}{\lambda+1}\frac{\pi a^2}{4}\Delta_2\right),$$

where Δ_1, Δ_2 follow

$$\Delta_1 = \sum_{\ell=1}^{\infty} S_{\ell}\ell \sum_{n=1}^{\infty} \frac{(-1)^{n+1}(\pi\ell)^{4n-2} a^{8n-4}}{2^{2n-2}(2n-1)(4n-1)!}, \qquad (6.165)$$

$$\Delta_2 = \sum_{\ell=1}^{\infty} \ell \sum_{j=1}^{\infty} S_j j \sum_{m=1}^{\infty} \frac{(-1)^{m+1}(\pi j)^{4m-2} a^{8m-4}}{(2m-1)(4m-1)!(4m-3)!} \times \left((\pi\ell)^{4m-2} S_\ell + \right.$$

$$\left. + \sum_{k=1}^{m} (-1)^{\ell+k+1} \frac{(4k-3)!}{2^{2k-2}} \frac{(\pi\ell)^{4m-4k}}{\sinh \pi\ell} \right) \sum_{n=1}^{\infty} \frac{(-1)^{n+1}(\pi\ell)^{4n-2} a^{8n-4}}{2^{2n-1}(2n-1)(4n-1)!}. \qquad (6.166)$$

Observe that the series in expression (6.163) of the form

$$T^{(i)} = \sum_{n=1}^{\infty} S_n n^i$$

quickly converge

$$\frac{T_{n+1}^{(i)}}{T_n^{(i)}} = \frac{S_{n+1}^{(i)}(n+1)^i}{S_n^{(i)} n^i} \sim e^{-\pi}.$$

For small inclusion sizes, the resulting expression of the effective parameter $q_{ASM(N)}$ (6.164), up to terms of order a^{14} inclusive, satisfies Keller's theorem, because

$$q_{ASM(N)}(\lambda) = q_{ASM(N)}^{-1}(\lambda^{-1}).$$

In order to show this, we transform expressions (6.165), (6.166) to the following form

$$\Delta_1 = \sum_{\ell=1}^{\infty} S_{\ell}\ell \sum_{n=1}^{\infty} \frac{(-1)^{n+1} 2^{6n-2} \ell^{4n-2}}{(2n-1)(4n-1)!} \left(\frac{\pi a^2}{4}\right)^{4n-2} = \sum_{n=1}^{\infty} \delta_1^{(n)} \left(\frac{\pi a^2}{4}\right)^{4n-2},$$

(6.167)

$$\Delta_2 = \sum_{\ell=1}^{\infty} S_{\ell}\ell \sum_{j=1}^{\infty} S_j j \sum_{m=1}^{\infty} \frac{(-1)^{m+1}(4\pi j\ell)^{4m-2}}{(2m-1)(4m-1)!(4m-3)!} \left(\frac{\pi a^2}{4}\right)^{4m-2} \times$$

$$\times \sum_{n=1}^{\infty} \frac{(-1)^{n+1} 2^{6n-3} \ell^{4n-2}}{(2n-1)(4n-1)!} \left(\frac{\pi a^2}{4}\right)^{4n-2} +$$

(6.168)

$$+ \sum_{\ell=1}^{\infty} \frac{\ell}{sh\,\pi\ell} \sum_{j=1}^{\infty} S_j j \sum_{m=1}^{\infty} \frac{(-1)^{m+1}(4j)^{4m-2}}{(2m-1)(4m-1)!(4m-3)!} \left(\frac{\pi a^2}{4}\right)^{4m-2} \times$$

$$\times \sum_{k=1}^{m} \frac{(-1)^{\ell+k+1}(4k-3)!(\pi\ell)^{4m-4k}}{2^{2k-2}} \sum_{n=1}^{\infty} \frac{2^{6n-3}(-1)^{n+1}\ell^{4n-2}}{(2n-1)(4n-1)!} \left(\frac{\pi a^2}{4}\right)^{4n-2} =$$

$$= \sum_{m=1}^{\infty} \frac{(-1)^{m+1} 2^{8m-4}}{(2m-1)(4m-1)!(4m-3)!} \sum_{n=1}^{\infty} \frac{(-1)^{n+1} 2^{6n-3}}{(2n-1)(4n-1)!} \sum_{j=1}^{\infty} S_j j^{4m-1} \times$$

$$\times \sum_{\ell=1}^{\infty} \ell^{4n-1} \left((\pi\ell)^{4m-2} S_{\ell} + \sum_{k=1}^{m} (-1)^{\ell+k+1} \frac{(4k-3)!}{2^{2k-2}} \frac{(\pi\ell)^{4m-4k}}{\sinh\pi\ell}\right) \left(\frac{\pi a^2}{4}\right)^{4(m+n-1)} =$$

$$= \sum_{m=1}^{\infty} \sum_{n=1}^{\infty} \frac{(-1)^{m+n+2} 2^{8m+6n-7}}{(2m-1)(4m-1)!(4m-3)!(2n-1)(4n-1)!} \sum_{j=1}^{\infty} S_j j^{4m-1} \times$$

$$\times \sum_{\ell=1}^{\infty} \ell^{4n-1} \left((\pi\ell)^{4m-2} S_{\ell} + \sum_{k=1}^{m} (-1)^{\ell+k+1} \frac{(4k-3)!}{2^{2k-2}} \frac{(\pi\ell)^{4m-4k}}{\sinh\pi\ell}\right) \left(\frac{\pi a^2}{4}\right)^{4(m+n-1)} =$$

$$= \sum_{m=1}^{\infty} \sum_{n=1}^{\infty} \frac{(-1)^{m+n+2} \pi^{4(m+n-1)} a^{8(m+n-1)}}{(2m-1)(4m-1)!(4m-3)!2^{2n-1}(2n-1)(4n-1)!} \times$$

$$\times \sum_{j=1}^{\infty} S_j j^{4m-1} \sum_{\ell=1}^{\infty} \ell^{4n-1} \left((\pi\ell)^{4m-2} S_{\ell} + \sum_{k=1}^{m} (-1)^{\ell+k+1} \frac{(4k-3)!}{2^{2k-2}} \frac{(\pi\ell)^{4m-4k}}{\sinh\pi\ell}\right) =$$

$$= \sum_{m=1}^{\infty} \sum_{n=1}^{\infty} \delta_2^{(mn)} \left(\frac{\pi a^2}{4}\right)^{4(m+n-1)},$$

where

$$\delta_1^{(n)} = \sum_{\ell=1}^{\infty} S_{\ell}\ell^{4n-1} \frac{(-1)^{n+1} 2^{6n-2}}{(2n-1)(4n-1)!},$$

(6.169)

$$\delta_2^{(mn)} = \frac{(-1)^{m+1} 2^{8m-4}}{(2m-1)(4m-1)!(4m-3)!} \frac{(-1)^{n+1} 2^{6n-3}}{(2n-1)(4n-1)!} \sum_{j=1}^{\infty} S_j j^{4m-1} \times$$

$$\times \sum_{\ell=1}^{\infty} \ell^{4n-1} \left((\pi\ell)^{4m-2} S_\ell + \sum_{k=1}^{m} (-1)^{\ell+k+1} \frac{(4k-3)!}{2^{2k-2}} \frac{(\pi\ell)^{4m-4k}}{sh\,\pi\ell} \right). \quad (6.170)$$

Then, taking into account (6.167)–(6.170), we obtain

$$q_{ASM(N)}(\lambda) = \frac{1 + \frac{\lambda-1}{\lambda+1}\frac{\pi a^2}{4}}{1 - \frac{\lambda-1}{\lambda+1}\frac{\pi a^2}{4}} + \frac{\left(\frac{\lambda-1}{\lambda+1}\right)^3}{\left(1 - \frac{\lambda-1}{\lambda+1}\frac{\pi a^2}{4}\right)^2} \frac{\pi^2 a^6}{4} \left[\delta_1^{(1)} \frac{\pi^2 a^4}{4^2} + \right.$$

$$+ \delta_1^{(2)} \frac{\pi^6 a^{12}}{4^6} + \ldots + \frac{\lambda-1}{\lambda+1}\frac{\pi a^2}{4} \left(\delta_2^{(11)} \frac{\pi^4 a^8}{4^4} + \left(\delta_2^{(12)} + \delta_2^{(21)} \right) \frac{\pi^8 a^{16}}{4^8} + \ldots \right) \Big] \sim$$

$$\sim 1 + 2\frac{\lambda-1}{\lambda+1}\frac{\pi}{4} a^2 + 2\left(\frac{\lambda-1}{\lambda+1}\frac{\pi}{4}\right)^2 a^4 + 2\left(\frac{\lambda-1}{\lambda+1}\frac{\pi}{4}\right)^3 a^6 +$$

$$+ 2\left(\frac{\lambda-1}{\lambda+1}\frac{\pi a^2}{4}\right)^4 a^8 + \left(2\left(\frac{\lambda-1}{\lambda+1}\frac{\pi}{4}\right)^5 + \left(\frac{\lambda-1}{\lambda+1}\right)^2 \frac{\pi^4}{4^3} \delta_1^{(1)} \right) a^{10} +$$

$$+ \left(2\left(\frac{\lambda-1}{\lambda+1}\frac{\pi}{4}\right)^6 + 2\left(\frac{\lambda-1}{\lambda+1}\right)^4 \frac{\pi^5}{4^4} \delta_1^{(1)} \right) a^{12} + \quad (6.171)$$

$$+ \left(2\left(\frac{\lambda-1}{\lambda+1}\frac{\pi}{4}\right)^7 + 3\left(\frac{\lambda-1}{\lambda+1}\right)^5 \frac{\pi^6}{4^5} \delta_1^{(1)} \right) a^{14} +$$

$$+ \left(2\left(\frac{\lambda-1}{\lambda+1}\frac{\pi}{4}\right)^8 + 4\left(\frac{\lambda-1}{\lambda+1}\right)^6 \frac{\pi^7}{4^6} \delta_1^{(1)} + \left(\frac{\lambda-1}{\lambda+1}\right)^4 \frac{\pi^7}{4^6} \delta_2^{(11)} \right) a^{16} + o\left(a^{16}\right)$$

at $a \to 0$,

$$q_{ASM(N)}^{-1}(\lambda^{-1}) = \left\{ \frac{1 - \frac{\lambda-1}{\lambda+1}\frac{\pi a^2}{4}}{1 + \frac{\lambda-1}{\lambda+1}\frac{\pi a^2}{4}} - \frac{\left(\frac{\lambda-1}{\lambda+1}\right)^3}{\left(1 + \frac{\lambda-1}{\lambda+1}\frac{\pi a^2}{4}\right)^2} \frac{\pi^2 a^6}{4} \left[\delta_1^{(1)} \frac{\pi^2 a^4}{4^2} + \right. \right.$$

$$+ \delta_1^{(2)} \frac{\pi^6 a^{12}}{4^6} + \ldots - \frac{\lambda-1}{\lambda+1}\frac{\pi a^2}{4} \left(\delta_2^{(11)} \frac{\pi^4 a^8}{4^4} + \left(\delta_2^{(12)} + \delta_2^{(21)} \right) \frac{\pi^8 a^{16}}{4^8} + \ldots \right) \Big] \Big\}^{-1} \sim$$

$$\sim 1 + 2\frac{\lambda-1}{\lambda+1}\frac{\pi}{4} a^2 + 2\left(\frac{\lambda-1}{\lambda+1}\frac{\pi}{4}\right)^2 a^4 + 2\left(\frac{\lambda-1}{\lambda+1}\frac{\pi}{4}\right)^3 a^6 +$$

$$+ 2\left(\frac{\lambda-1}{\lambda+1}\frac{\pi a^2}{4}\right)^4 a^8 + \left(2\left(\frac{\lambda-1}{\lambda+1}\frac{\pi}{4}\right)^5 + \left(\frac{\lambda-1}{\lambda+1}\right)^2 \frac{\pi^4}{4^3} \delta_1^{(1)} \right) a^{10} +$$

$$+\left(2\left(\frac{\lambda-1}{\lambda+1}\frac{\pi}{4}\right)^6+2\left(\frac{\lambda-1}{\lambda+1}\right)^4\frac{\pi^5}{4^4}\delta_1^{(1)}\right)a^{12}+ \qquad (6.172)$$

$$+\left(2\left(\frac{\lambda-1}{\lambda+1}\frac{\pi}{4}\right)^7+3\left(\frac{\lambda-1}{\lambda+1}\right)^5\frac{\pi^6}{4^5}\delta_1^{(1)}\right)a^{14}+$$

$$+\left(2\left(\frac{\lambda-1}{\lambda+1}\frac{\pi}{4}\right)^8+4\left(\frac{\lambda-1}{\lambda+1}\right)^6\frac{\pi^7}{4^6}\delta_1^{(1)}-\left(\frac{\lambda-1}{\lambda+1}\right)^4\frac{\pi^7}{4^6}\delta_2^{(11)}\right)a^{16}+o\left(a^{16}\right)$$

at $a\to 0$.

From a comparison of relations (6.171), (6.172), an essential conclusion follows: in the solution (6.164)–(6.166) based on the Schwarz alternating method, the sum of the main part of the asymptotic representation q_{MF} and the first correction to it, Δ_1, i.e.

$$q_{ASM(N)}^{(1)}=\frac{\lambda\left(1+\frac{\pi a^2}{4}\right)+1-\frac{\pi a^2}{4}}{\lambda\left(1-\frac{\pi a^2}{4}\right)+1+\frac{\pi a^2}{4}}+$$

$$+\left(\frac{\lambda-1}{\lambda+1}\right)^3\frac{\pi^2 a^6}{4}\frac{(\lambda+1)^2}{\left(\lambda\left(1-\frac{\pi a^2}{4}\right)+1+\frac{\pi a^2}{4}\right)^2}\Delta_1 \qquad (6.173)$$

satisfies Keller's theorem up to order terms a^{2n} inclusive for any values n. The error is already introduced by the second correction Δ_2, whose order for small values of a is a^{16}.

6.4.4 ANALYSIS OF THE SOLUTION BASED ON THE N-ITERATIVE PROCEDURE OF THE SCHWARZ ALTERNATING METHOD

In relation (6.164), the MF (6.162) stands for the main part of the effective parameter, Δ_1 and Δ_2 are corrections of order a^{10} and higher which is determined by series. For $n=1$ the minimum order of correction is a^{10}. This conclusion qualitatively coincides with the result obtained in [165] in the particular case of non-conductive inclusions (cavities). Indeed, the asymptotic solution obtained in [165] can be written as follows

$$q_{as\;s}^{(0)}=\frac{1-c}{1+c}-\frac{59.5808}{\pi^4}\frac{c^5}{(1+c)^2} \qquad at \quad c\to 0, \qquad (6.174)$$

where $c=\frac{\pi a^2}{4}$.

Then, taking into account relations (6.167) for Δ_1 and (6.169) for $\delta_1^{(n)}$, expression (6.173) at $\lambda=0$ is

$$q_{ASM(N)}^{(1)}=\frac{1-c}{1+c}-\frac{128}{3\pi}\sum_{\ell=1}^{\infty}S_\ell\ell^3\frac{c^5}{(1+c)^2}. \qquad (6.175)$$

From (6.174), (6.175) it follows that

$$\frac{c^5}{(1+c)^2}=c^5-2c^6+3c^7+O\left(c^8\right) \quad at \quad c\to 0.$$

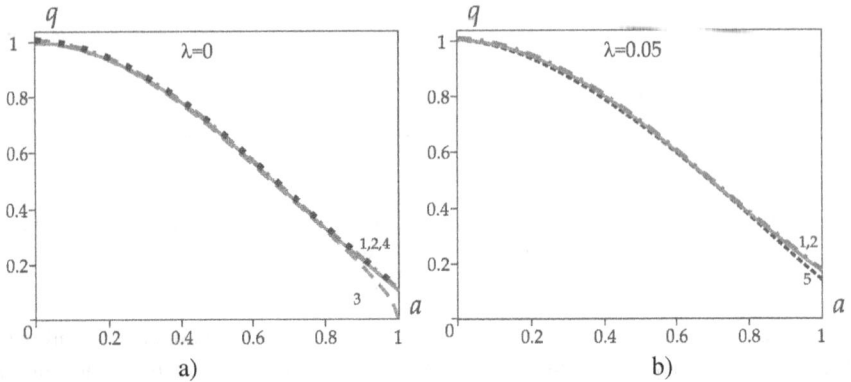

Figure 6.18 The effective thermal conductivity for inclusions of low conductivity: a) $\lambda = 0$, b) $\lambda = 0.05$ ($1-q_{ASM\,(N)}$, $2-q_{MF}$, $3-q_{as}^{(0)}$, $4-q_{as\,s}^{(0)}$, $5-q_{as}$).

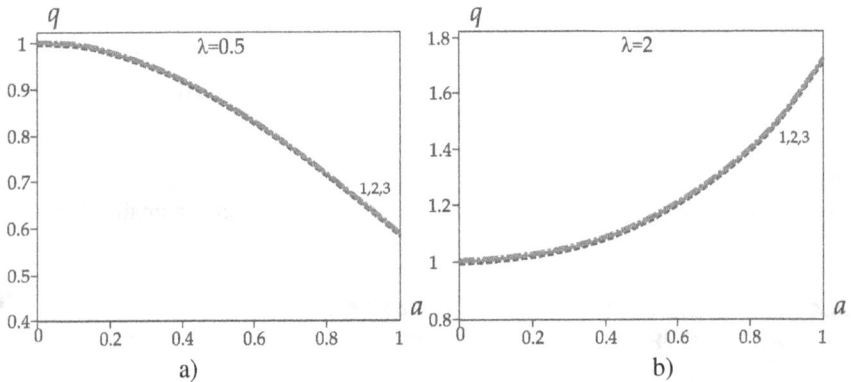

Figure 6.19 The effective thermal conductivity for the conductivity of inclusions of the order of the conductivity of the matrix: a) $\lambda \sim 1$: $\lambda = 0.5$, b) $\lambda \sim 1$: $\lambda = 2$ ($1-q_{ASM\,(N)}$, $2-q_{MF}$, $3-q_{as}$).

Thus, the first non-zero correction to MF has the order a^{10} – by the size of the inclusions, or c^5 – by their concentration.

Puting in expressions (6.165), (6.166) $n = 2$ and resticting it to the terms of order a^{18} inclusive, we calculate by formulas (6.164)–(6.166) of the effective thermal conductivity for various sizes and conductivity of inclusions (Figs. 6.18–6.22). For comparison, we also present the results obtained by other authors.

The following notation is used in Figures: 6.18–6.22: q_{MF} – MF (6.162); $q_{ASM\,(N)}$ – generalized Schwarz approximation (6.164)–(6.166); $q_{as}^{(0)}$ – asymptotic solution (6.111) [158], transformed using Keller's theorem; $q_{as\,s}^{(0)}$ – asymptotic solution (6.174) [165]; q_{as} – asymptotics [194], (formula (14)); $q_{as}^{(\infty)}$ – asymptotic solution (6.110) [158]; $q_{as\,s}^{(\infty)}$ – asymptotic solution (6.174) [165] transformed using

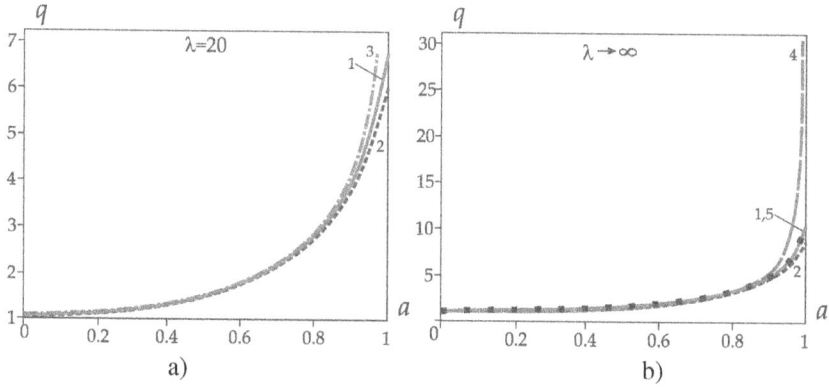

Figure 6.20 The effective thermal conductivity for inclusions of high conductivity: a) $\lambda \gg 1$: $\lambda = 20$, b) $\lambda \to \infty$ $(1 - q_{ASM\,(N)}, 2 - q_{MF}, 3 - q_{as}, 4 - q_{as}^{(\infty)}, 5 - q_{as\,s}^{(\infty)})$.

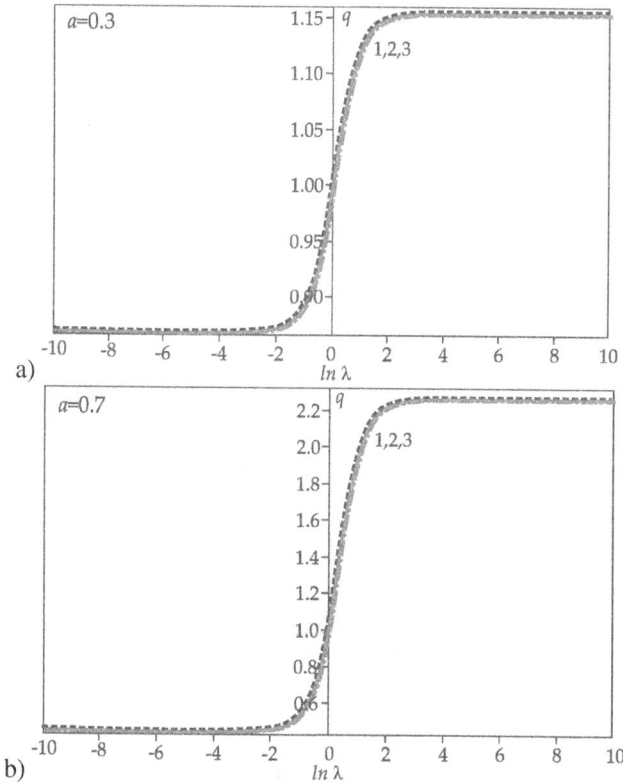

Figure 6.21 The effective thermal conductivity for inclusions of: a) small size $(a = 0.3)$, b) medium size $(a = 0.7)$ $(1 - q_{ASM\,(N)}, 2 - q_{MF}, 3 - q_{as})$.

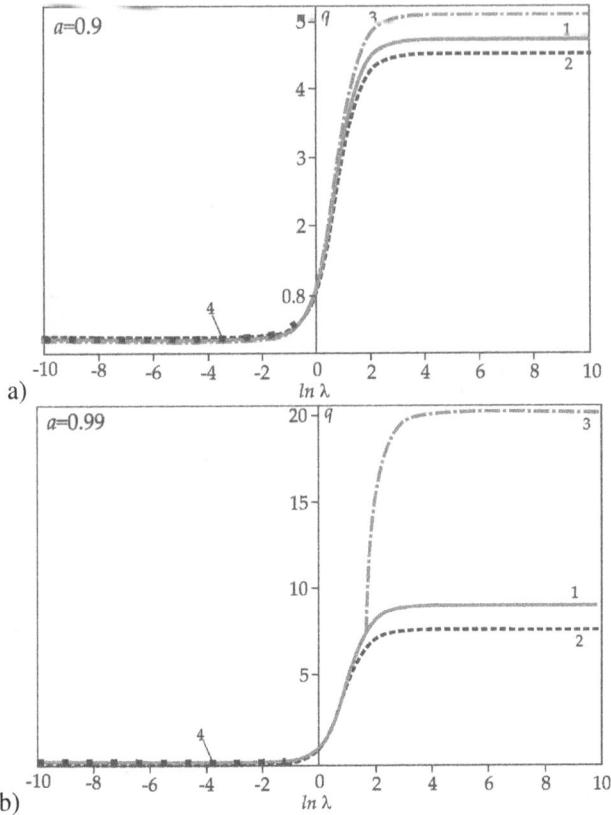

Figure 6.22 The effective thermal conductivity for: a) large inclusions ($a = 0.9$), b) size of inclusions close to the limit possible value ($a = 0.99$) (1– $q_{ASM\,(N)}$, 2– q_{MF}, 3– $q_{as\,f}^{(\infty)}$, 4– $q_{as\,f}^{(0)}$).

Keller's theorem; $q_{as\,f}^{(\infty)}$ – asymptotic formula (49) from [158]; $q_{as\,f}^{(0)}$ – asymptotic formula (49) from [158], transformed using the Keller's theorem.

Analysis of the presented graphical results yields the following conclusions.

1) The MF well describes the effective parameter for small and medium inclusion sizes (up to $a \approx 0.7$) of any conductivity, including limiting cases ($\lambda = 0$ and $\lambda \to \infty$). In this case, the results of calculations for MF and the generalized Schwarz alternating method practically coincide (Table 6.2).

2) For conductivities of the matrix and inclusions of the same order (from $\lambda \approx 0.5$ up to $\lambda \approx 2$), the results by MF and the Schwarz alternating method are quite close (Table 6.3) over the entire range of inclusion sizes ($0 \le a \le 1$).

3) For large inclusions ($a > 0.8$) and large ($\lambda > 10$) or small ($\lambda < 0.1$) conductivity, the corrections obtained by the Schwarz alternating method (Table 6.4) are sufficient.

Table 6.2

The effective thermal conductivity coefficient for small and medium sizes of inclusions.

Conductivity of inclusions $\lambda = 0$			
Inclusion size a	Maxwell Formula q_{MF}	Generalized Schwarz approximation $q_{ASM(N)}$	Error $\dfrac{q_{MF} - q_{ASM(N)}}{q_{MF}} \cdot 100\%$
0.1	0.9844	0.9844	$0.7735 \cdot 10^{-9}$
0.2	0.9391	0.9391	$0.7928 \cdot 10^{-6}$
0.3	0.8680	0.8680	$0.4589 \cdot 10^{-4}$
0.4	0.7767	0.7767	$0.8238 \cdot 10^{-3}$
0.5	0.6718	0.6717	$0.7847 \cdot 10^{-2}$
0.6	0.5592	0.5589	$0.5065 \cdot 10^{-1}$
0.7	0.4442	0.4431	0.2541
Conductivity of inclusions $\lambda = 0.05$			
0.1	0.9859	0.9859	$0.5729 \cdot 10^{-9}$
0.2	0.9447	0.9447	$0.5870 \cdot 10^{-6}$
0.3	0.8798	0.8798	$0.3396 \cdot 10^{-4}$
0.4	0.7958	0.7958	$0.6083 \cdot 10^{-3}$
0.5	0.6983	0.6983	$0.5770 \cdot 10^{-2}$
0.6	0.5926	0.5924	$0.3694 \cdot 10^{-1}$
0.7	0.4835	0.4826	0.1826
Conductivity of inclusions $\lambda = 20$			
0.1	1.0143	1.0143	$-0.5729 \cdot 10^{-9}$
0.2	1.0585	1.0585	$-0.5870 \cdot 10^{-6}$
0.3	1.1366	1.1366	$-0.3397 \cdot 10^{-4}$
0.4	1.2566	1.2566	$-0.6087 \cdot 10^{-3}$
0.5	1.4321	1.4321	$-0.5782 \cdot 10^{-2}$
0.6	1.6875	1.6881	$-0.3717 \cdot 10^{-1}$
0.7	2.0684	2.0722	-0.1855
Conductivity of inclusions $\lambda \to \infty$			
0.1	1.0158	1.0158	$-0.7735 \cdot 10^{-9}$
0.2	1.0649	1.0649	$0.7928 \cdot 10^{-6}$
0.3	1.1521	1.1521	$-0.4590 \cdot 10^{-4}$
0.4	1.2874	1.2875	$-0.8243 \cdot 10^{-3}$
0.5	1.4886	1.4888	$-0.7865 \cdot 10^{-2}$
0.6	1.7884	1.7893	$-0.5100 \cdot 10^{-1}$
0.7	2.2512	2.2570	-0.2586

4) For large sizes of inclusions close to the limiting $(a \to 1)$ and extremely large $(\lambda \to \infty)$ or extremely small $(\lambda \to 0)$ conductivity, neither the MF, nor the N-iterative solution by the Schwarz alternating method give a correct value of the effective parameter.

Table 6.3

The effective thermal conductivity for conductivities of the matrix and inclusions of the same order.

Conductivity of inclusions $\lambda = 0.5$			
Inclusion size a	Maxwell Formula q_{MF}	Generalized Schwarz approximation $q_{ASM(N)}$	Error $\frac{q_{MF} - q_{ASM(N)}}{q_{MF}} \cdot 100\%$
0.1	0.9948	0.9948	$0.2865 \cdot 10^{-10}$
0.2	0.9793	0.9793	$0.2934 \cdot 10^{-7}$
0.3	0.9540	0.9540	$0.1692 \cdot 10^{-5}$
0.4	0.9196	0.9196	$0.3009 \cdot 10^{-4}$
0.5	0.8771	0.8771	$0.2808 \cdot 10^{-3}$
0.6	0.8277	0.8277	$0.1745 \cdot 10^{-2}$
0.7	0.7726	0.7725	$0.8200 \cdot 10^{-2}$
0.8	0.7130	0.7128	$0.3140 \cdot 10^{-1}$
0.9	0.6501	0.6494	0.1029
1.0	0.5850	0.5833	0.2978
Conductivity of inclusions $\lambda = 2$			
0.1	1.0052	1.0052	$-0.2865 \cdot 10^{-10}$
0.2	1.0212	1.0212	$-0.2934 \cdot 10^{-7}$
0.3	1.0483	1.0483	$-0.1692 \cdot 10^{-5}$
0.4	1.0874	1.0874	$-0.3009 \cdot 10^{-4}$
0.5	1.1401	1.1401	$-0.2810 \cdot 10^{-3}$
0.6	1.2081	1.2081	$-0.1749 \cdot 10^{-2}$
0.7	1.2943	1.2944	$-0.8247 \cdot 10^{-2}$
0.8	1.4026	1.4030	$-0.3181 \cdot 10^{-1}$
0.9	1.5383	1.5399	-0.1056
1.0	1.7093	1.7146	-0.3129

In addition, the following general results can be formulated.

1. For small inclusion sizes ($a << 1$), the main part of the asymptotic expression for the effective thermal conductivity parameter coincides with MF.
2. The first nonzero correction to MF has the order a^{10}.
3. The resulting solution satisfies Keller's theorem up to terms of order a^{14} inclusive.

4. An analysis of the numerical results, carried out taking into account terms of order a^{18} inclusive, showed:

   (i) in the case of small and medium sizes inclusions and their arbitrary conductivity ($0 \le \lambda < \infty$, $\lambda \to \infty$), the MF correctly describes the effective parameter;

Table 6.4

The effective thermal conductivity coefficient for large sizes of inclusions of high or low conductivity.

Inclusion size a	Maxwell formula q_{MF}	Generalized Schwarz approximation $q_{ASM(N)}$	Error $\dfrac{q_{MF} - q_{ASM(N)}}{q_{MF}} \cdot 100\%$
Conductivity of inclusions $\lambda = 0.05$			
0.80	0.3748	0.3719	0.7610
0.85	0.3215	0.3167	1.4899
0.90	0.2694	0.2616	2.8720
0.95	0.2185	0.2065	5.5130
1.00	0.1692	0.1510	10.7180
Conductivity of inclusions $\lambda = 20$			
0.80	2.6683	2.6893	−0.7881
0.85	3.1102	3.1590	−1.5670
0.90	3.7124	3.8268	−3.0837
0.95	4.5759	4.8543	−6.0841
1.00	5.9108	6.6355	−12.2612

(ii) in the case of composites with the conductivity of inclusions close to the conductivity of the matrix ($\lambda \sim 1$), the MF and the Schwarz alternating method are in good agreement for any size of inclusions ($0 \leq a \leq 1$);

(iii) in the case of composites with large inclusion sizes ($a \gg 0$) large ($\lambda \gg 1$) or small ($\lambda \ll 1$) conductivity, one should take into account the correction based on the Schwarz alternating method N-iteration solution;

(iv) in the case of large sizes of inclusions close to the limit possible value ($a \rightarrow 1$), extremely large ($\lambda \rightarrow \infty$) or extremely small ($\lambda \rightarrow 0$) conductivity, MF or the Schwarz alternating method cannot be used.

5. In this regard, it becomes obvious that the Schwarz alternating method, despite its many advantages, does not allow to obtain a solution that correctly describes the effective characteristics of the composite for any values of its physical and geometric parameters.

Therefore, the solution obtained on the basis of the Schwarz alternating method requires modification. The authors of monograph [79] have already pointed out a similar problem. To eliminate some shortcomings of the Schwarz alternating method, they carried out a rigorous analysis based on infinite series. The proposed further modification of the Schwarz alternating method is based on the AP.

Table 6.5

The effective coefficient of thermal conductivity for inclusion sizes close to limiting, extremely large or extremely small conductivity.

				Conductivity of inclusions $\lambda = 0$	
Inclusion	MF	General Schwarz	Asympt. solution	Error	Error
size a	q_{MF}	appr. $q_{ASM(N)}$	$q_{as}^{(0)}$	$\dfrac{q_{as}^{(0)} - q_{MF}}{q_{as}^{(0)}} \cdot 100\%$	$\dfrac{q_{as}^{(0)} - q_{ASM(N)}}{q_{as}^{(0)}} \cdot 100\%$
0.80	0.3310	0.3274	0.3232	−2.4176	−1.3028
0.85	0.2760	0.2699	0.2616	−5.4749	−3.1765
0.90	0.2224	0.2127	0.1974	−12.6438	−7.7561
0.95	0.1704	0.1555	0.1263	−34.9469	−23.1268
0.99	0.1301	0.1095	0.0497	−161.8549	−120.3139
				Conductivity of inclusions $\lambda \to \infty$	
Inclusion	MF	General Schwarz	Asympt. solution	Error	Error
size a	q_{MF}	appr. $q_{ASM(N)}$	$q_{as}^{(0)}$	$\dfrac{q_{as}^{(0)} - q_{MF}}{q_{as}^{(0)}} \cdot 100\%$	$\dfrac{q_{as}^{(0)} - q_{ASM(N)}}{q_{as}^{(0)}} \cdot 100\%$
0.80	3.0214	3.0555	3.0944	2.3605	1.2558
0.85	3.6237	3.7072	3.8221	5.1908	3.0064
0.90	4.4971	4.7082	5.0657	11.2246	7.0576
0.95	5.8686	6.4419	7.9196	25.8968	18.6587
0.99	7.6869	9.0895	20.1286	61.8019	54.8427

6.4.5 MF REFINEMENT USING THE PADÉ APPROXIMANTS

Let us consider the expression for the effective thermal conductivity parameter obtained by the N-iterative Schwarz alternating method (6.161), restricting ourselves to the (04) approximation:

$$q = 1 + 2\frac{\lambda - 1}{\lambda + 1}\frac{\pi a^2}{4} + 2\left(\frac{\lambda - 1}{\lambda + 1}\frac{\pi a^2}{4}\right)^2 + 2\left(\frac{\lambda - 1}{\lambda + 1}\frac{\pi a^2}{4}\right)^3 + ... +$$

$$+ \left(\frac{\lambda - 1}{\lambda + 1}\right)^3 \frac{\pi^2 a^6}{4}\left(1 + \frac{\lambda - 1}{\lambda + 1}\frac{\pi a^2}{4}\right)^2 \sum_{\ell=1}^{\infty} S_{\ell\ell} \sum_{n=1}^{\infty} \frac{(-1)^{n+1}(\pi\ell)^{4n-2}a^{8n-4}}{2^{2n-2}(2n-1)(4n-1)!} +$$

$$+ \left(\frac{\lambda - 1}{\lambda + 1}\right)^3 \frac{\pi^3 a^8}{16}\left(1 + \frac{\lambda - 1}{\lambda + 1}\frac{\pi a^2}{4}\right)^2 \sum_{j=1}^{\infty} S_{jj} \sum_{m=1}^{\infty} \frac{(-1)^{m+1}(\pi j)^{4m-2}a^{8m-4}}{(2m-1)(4m-1)!(4m-3)!} \times$$

$$\times \sum_{\ell=1}^{\infty} \ell \left((\pi\ell)^{4m-2} S_\ell + \sum_{k=1}^{m} (-1)^{\ell+k+1} \frac{(4k-3)!}{2^{2k-2}} \frac{(\pi\ell)^{4m-4k}}{\sinh \pi\ell} \right) \times \qquad (6.176)$$

$$\times \sum_{n=1}^{\infty} \frac{(-1)^{n+1}(\pi\ell)^{4n-2} a^{8n-4}}{2^{2n-1}(2n-1)(4n-1)!}.$$

In expansion (6.176), we take into account the terms corresponding to $n=2$, $m=1$ in the series (6.167)–(6.170), the order of which does not exceed a^{18}:

$$q = 1 + 2\sum_{j=1}^{9} \left(\frac{\lambda-1}{\lambda+1}\frac{\pi}{4}\right)^j a^{2j} + \left(\frac{\lambda-1}{\lambda+1}\right)^3 \frac{\pi^2 a^6}{4} \left(1 + \frac{\lambda-1}{\lambda+1}\frac{\pi a^2}{4}\right)^2 \times$$

$$\times \left(\delta_1^{(1)} \frac{\pi^2 a^4}{4^2} + \delta_1^{(2)} \frac{\pi^6 a^{12}}{4^6} + \frac{\lambda-1}{\lambda+1}\frac{\pi a^2}{4} \delta_2^{(11)} \frac{\pi^4 a^8}{4^4}\right) = \qquad (6.177)$$

$$= 1 + 2\sum_{j=1}^{9} \left(\frac{\lambda-1}{\lambda+1}\frac{\pi a^2}{4}\right)^j + \left(\frac{\lambda-1}{\lambda+1}\right)^3 \frac{\pi^4 a^{10}}{4^3} \left(1 + \frac{\lambda-1}{\lambda+1}\frac{\pi a^2}{4}\right)^2 \times$$

$$\times \left(\delta_1^{(1)} + \delta_1^{(2)} \frac{\pi^4 a^8}{4^4} + \frac{\lambda-1}{\lambda+1}\frac{\pi^3 a^6}{4^3} \delta_2^{(11)}\right).$$

Now we the solution (6.177) as the sum of two components.

1) To do this, let us transform the (6.177) in AP [0/18] according to the size of the inclusion a:

$$q_{[0/18]}(\lambda, a) = \left[1 - 2\frac{\lambda-1}{\lambda+1}\frac{\pi}{4}a^2 + 2\left(\frac{\lambda-1}{\lambda+1}\frac{\pi}{4}\right)^2 a^4 - 2\left(\frac{\lambda-1}{\lambda+1}\frac{\pi}{4}\right)^3 a^6 + \right.$$

$$+2\left(\frac{\lambda-1}{\lambda+1}\frac{\pi a^2}{4}\right)^4 a^8 - \left(2\left(\frac{\lambda-1}{\lambda+1}\frac{\pi}{4}\right)^5 + \left(\frac{\lambda-1}{\lambda+1}\right)^3 \frac{\pi^4}{4^3}\delta_1^{(1)}\right) a^{10} +$$

$$+ \left(2\left(\frac{\lambda-1}{\lambda+1}\frac{\pi}{4}\right)^6 + 2\left(\frac{\lambda-1}{\lambda+1}\right)^4 \frac{\pi^5}{4^4}\delta_1^{(1)}\right) a^{12} - \left(2\left(\frac{\lambda-1}{\lambda+1}\frac{\pi}{4}\right)^7 + \right.$$

$$\qquad (6.178)$$

$$+ \left(\frac{\lambda-1}{\lambda+1}\right)^5 \frac{\pi^6}{4^5}\delta_1^{(1)}\right) a^{14} + \left(2\left(\frac{\lambda-1}{\lambda+1}\frac{\pi}{4}\right)^8 - \left(\frac{\lambda-1}{\lambda+1}\right)^4 \frac{\pi^7}{4^6}\delta_2^{(11)}\right) a^{16} -$$

$$- \left(2\left(\frac{\lambda-1}{\lambda+1}\frac{\pi}{4}\right)^9 + \left(\frac{\lambda-1}{\lambda+1}\right)^3 \frac{\pi^8}{4^7}\delta_1^{(2)} - 2\left(\frac{\lambda-1}{\lambda+1}\right)^5 \frac{\pi^8}{4^7}\delta_2^{(11)}\right) a^{18}\right]^{-1} =$$

$$= \left[1 + 2\sum_{j=1}^{9} \left(-\frac{\lambda-1}{\lambda+1}\frac{\pi a^2}{4}\right)^j - \left(\frac{\lambda-1}{\lambda+1}\right)^3 \frac{\pi^4 a^{10}}{4^3}\left(1 - \frac{\lambda-1}{\lambda+1}\frac{\pi a^2}{4}\right)^2 \delta_1^{(1)} - \right.$$

$$- \left(\frac{\lambda-1}{\lambda+1}\right)^4 \frac{\pi^7 a^{16}}{4^6}\left(1 - 2\frac{\lambda-1}{\lambda+1}\frac{\pi a^2}{4}\right)\delta_2^{(11)} - \left(\frac{\lambda-1}{\lambda+1}\right)^3 \frac{\pi^8 a^{18}}{4^7}\delta_1^{(2)}\right]^{-1}.$$

2) Let us transform the (6.177) in AP [18/0] according to $q^{-1}_{[18/0]} \left(\lambda^{-1}, a \right)$:

$$q^{-1}\left(\lambda^{-1}, a \right) = \left[1 + 2 \sum_{j=1}^{9} \left(-\frac{\lambda-1}{\lambda+1} \frac{\pi a^2}{4} \right)^j - \left(\frac{\lambda-1}{\lambda+1} \right)^3 \frac{\pi^4 a^{10}}{4^3} \times \right.$$

$$\left. \times \left(1 - \frac{\lambda-1}{\lambda+1} \frac{\pi a^2}{4} \right)^2 \left(\delta_1^{(1)} + \delta_1^{(2)} \frac{\pi^4 a^8}{4^4} - \frac{\lambda-1}{\lambda+1} \frac{\pi^3 a^6}{4^3} \delta_2^{(11)} \right) \right]^{-1}.$$

Therefore, we have

$$q^{-1}_{[18/0]} \left(\lambda^{-1}, a \right) = 1 + 2\frac{\lambda-1}{\lambda+1} \frac{\pi}{4} a^2 + 2\left(\frac{\lambda-1}{\lambda+1} \frac{\pi}{4} \right)^2 a^4 +$$

$$+ 2\left(\frac{\lambda-1}{\lambda+1} \frac{\pi}{4} \right)^3 a^6 + 2\left(\frac{\lambda-1}{\lambda+1} \frac{\pi a^2}{4} \right)^4 a^8 +$$

$$+ \left(2\left(\frac{\lambda-1}{\lambda+1} \frac{\pi}{4} \right)^5 + \left(\frac{\lambda-1}{\lambda+1} \right)^3 \frac{\pi^4}{4^3} \delta_1^{(1)} \right) a^{10} + \tag{6.179}$$

$$+ \left(2\left(\frac{\lambda-1}{\lambda+1} \frac{\pi}{4} \right)^6 + 2\left(\frac{\lambda-1}{\lambda+1} \right)^4 \frac{\pi^5}{4^4} \delta_1^{(1)} \right) a^{12} + \left(2\left(\frac{\lambda-1}{\lambda+1} \frac{\pi}{4} \right)^7 + \right.$$

$$\left. + \left(\frac{\lambda-1}{\lambda+1} \right)^5 \frac{\pi^6}{4^5} \delta_1^{(1)} \right) a^{14} + \left(2\left(\frac{\lambda-1}{\lambda+1} \frac{\pi}{4} \right)^8 - \left(\frac{\lambda-1}{\lambda+1} \right)^4 \frac{\pi^7}{4^6} \delta_2^{(11)} \right) a^{16} +$$

$$+ \left(2\left(\frac{\lambda-1}{\lambda+1} \frac{\pi}{4} \right)^9 + \left(\frac{\lambda-1}{\lambda+1} \right)^3 \frac{\pi^8}{4^7} \delta_1^{(2)} - 2\left(\frac{\lambda-1}{\lambda+1} \right)^4 \frac{\pi^8}{4^7} \delta_2^{(11)} \right) a^{18} =$$

$$= 1 + 2 \sum_{j=1}^{9} \left(\frac{\lambda-1}{\lambda+1} \frac{\pi a^2}{4} \right)^j + \left(\frac{\lambda-1}{\lambda+1} \right)^3 \frac{\pi^4 a^{10}}{4^3} \left(1 + \frac{\lambda-1}{\lambda+1} \frac{\pi a^2}{4} \right)^2 \delta_1^{(1)} -$$

$$- \left(\frac{\lambda-1}{\lambda+1} \right)^4 \frac{\pi^7 a^{16}}{4^6} \left(1 + 2\frac{\lambda-1}{\lambda+1} \frac{\pi}{4} a^2 \right) \delta_2^{(11)} + \left(\frac{\lambda-1}{\lambda+1} \right)^3 \frac{\pi^8 a^{18}}{4^7} \delta_1^{(2)}.$$

Now we match function $q_{[0/18]} \left(\lambda, a \right)$ (6.178) and $q^{-1}_{[18/0]} \left(\lambda^{-1}, a \right)$ (6.179) in the following way

$$q\left(a, \lambda \right) = \frac{1}{\lambda+1} q^{-1}_{[18/0]} \left(\lambda^{-1}, a \right) + \frac{\lambda}{\lambda+1} q_{[0/18]} \left(\lambda, a \right). \tag{6.180}$$

Relation (6.180) can be treated as a three-point AP, since
(i) at $\lambda = 0$ decomposition

$$q(a,\lambda)\bigg|_{\lambda=0} = q_{[18/0]}^{-1}(\lambda^{-1},a)\bigg|_{\lambda=0} = 1 + 2\sum_{j=1}^{9}\left(-\frac{\pi a^2}{4}\right)^j -$$

$$-\frac{\pi^4 a^{10}}{4^3}\left(1-\frac{\pi a^2}{4}\right)^2\delta_1^{(1)} - \frac{\pi^7 a^{16}}{4^6}\left(1-\frac{\pi a^2}{2}\right)\delta_2^{(11)} - \frac{\pi^8 a^{18}}{4^7}\delta_1^{(2)}$$

coincides with expression (6.177) for $\lambda = 0$ up to terms a^{14};
(ii) at $\lambda = 1$ we have

$$q(a,\lambda)\bigg|_{\lambda=1} = \frac{1}{2}q_{[18/0]}^{-1}(\lambda^{-1},a)\bigg|_{\lambda=1} + \frac{1}{2}q_{[0/18]}(\lambda,a)\bigg|_{\lambda=1} \equiv 1;$$

(iii) at $\lambda \to \infty$ expansion into a series of expressions

$$q(a,\lambda)\bigg|_{\lambda\to\infty} = q_{[0/18]}(\lambda,a)\bigg|_{\lambda\to\infty} =$$

$$= \frac{1}{1 + 2\sum_{j=1}^{9}\left(-\frac{\pi a^2}{4}\right)^j - \frac{\pi^4 a^{10}}{4^3}\left(1-\frac{\pi a^2}{4}\right)^2\delta_1^{(1)} - \frac{\pi^7 a^{16}}{4^6}\left(1-\frac{\pi a^2}{2}\right)\delta_2^{(11)} - \frac{\pi^8 a^{18}}{4^7}\delta_1^{(2)}} \sim$$

$$\sim 1 + 2\sum_{j=1}^{9}\frac{\pi a^{2j}}{4} + \frac{\pi^4 a^{10}}{4^3}\left(1+\frac{\pi a^2}{4}\right)^2\delta_1^{(1)} + \frac{\pi^7 a^{16}}{4^6}\left(1+\frac{\pi a^2}{2}\right)\delta_2^{(11)} + \frac{\pi^8 a^{18}}{4^7}\delta_1^{(2)}$$

coincides with relation (6.177) at $\lambda \to \infty$ up to terms of the order a^{18}.

Thus, taking into account the described transformations, we finally have the analytical expression for the effective thermal conductivity parameter

$$q_{ASM-AP} = \frac{1}{\lambda+1}\left[1 + 2\sum_{j=1}^{9}\left(\frac{\lambda-1}{\lambda+1}\frac{\pi a^2}{4}\right)^j +\right.$$

$$+\left(\frac{\lambda-1}{\lambda+1}\right)^3\frac{\pi^4 a^{10}}{4^3}\left(1+\frac{\lambda-1}{\lambda+1}\frac{\pi a^2}{4}\right)^2\delta_1^{(1)} - \qquad (6.181)$$

$$-\left(\frac{\lambda-1}{\lambda+1}\right)^4\frac{\pi^7 a^{16}}{4^6}\left(1+2\frac{\lambda-1}{\lambda+1}\frac{\pi a^2}{4}\right)\delta_2^{(11)} + \left(\frac{\lambda-1}{\lambda+1}\right)^3\frac{\pi^8 a^{18}}{4^7}\delta_1^{(2)}\right] +$$

$$+\frac{\lambda}{\lambda+1}\left[1 + 2\sum_{j=1}^{9}\left(-\frac{\lambda-1}{\lambda+1}\frac{\pi a^2}{4}\right)^j - \left(\frac{\lambda-1}{\lambda+1}\right)^3\frac{\pi^4 a^{10}}{4^3}\left(1-\frac{\lambda-1}{\lambda+1}\frac{\pi a^2}{4}\right)^2\delta_1^{(1)} -\right.$$

$$\left.-\left(\frac{\lambda-1}{\lambda+1}\right)^4\frac{\pi^7 a^{16}}{4^6}\left(1-2\frac{\lambda-1}{\lambda+1}\frac{\pi a^2}{4}\right)\delta_2^{(11)} - \left(\frac{\lambda-1}{\lambda+1}\right)^3\frac{\pi^8 a^{18}}{4^7}\delta_1^{(2)}\right]^{-1},$$

where expressions $\delta_1^{(1)}$, $\delta_1^{(2)}$, $\delta_2^{(11)}$ are defined by formulas (6.169), (6.170) as follows

$$\delta_1^{(1)} = \frac{8}{3} \sum_{\ell=1}^{\infty} S_\ell \ell^3, \qquad \delta_1^{(2)} = -\frac{64}{945} \sum_{\ell=1}^{\infty} S_\ell \ell^7, \qquad (6.182)$$

$$\delta_2^{(11)} = \frac{32}{9} \sum_{\ell=1}^{\infty} \left((\pi\ell)^2 S_\ell + \frac{(-1)^{\ell+2}}{\sinh \pi\ell} \right) \ell^3 \sum_{j=1}^{\infty} S_j j^3. \qquad (6.183)$$

6.4.6 SCHWARZ–PADÉ APPROACH

Let us analyze the expression for the effective thermal conductivity q_{ASM-AP} (6.181), (6.182), obtained using the Schwarz alternating method and the AP.

1) It is significant that the expansion at $a \to 0$ of expression (6.181) coincides for any values of λ with the expansion (6.177) of the effective coefficient up to terms of the order a^{14} inclusive:

$$q_{ASM-AP} \sim 1 - 2\frac{\lambda-1}{\lambda+1}\frac{\pi}{4}a^2 + 2\left(\frac{\lambda-1}{\lambda+1}\frac{\pi}{4}\right)^2 a^4 - 2\left(\frac{\lambda-1}{\lambda+1}\frac{\pi}{4}\right)^3 a^6 +$$

$$+2\left(\frac{\lambda-1}{\lambda+1}\frac{\pi a^2}{4}\right)^4 a^8 + \left(2\left(\frac{\lambda-1}{\lambda+1}\frac{\pi}{4}\right)^5 + \left(\frac{\lambda-1}{\lambda+1}\right)^3\frac{\pi^4}{4^3}\delta_1^{(1)}\right)a^{10} +$$

$$+\left(2\left(\frac{\lambda-1}{\lambda+1}\frac{\pi}{4}\right)^6 + 2\left(\frac{\lambda-1}{\lambda+1}\right)^4\frac{\pi^5}{4^4}\delta_1^{(1)}\right)a^{12} + \left(2\left(\frac{\lambda-1}{\lambda+1}\frac{\pi}{4}\right)^7 +\right.$$

$$(6.184)$$

$$\left.+\left(\frac{\lambda-1}{\lambda+1}\right)^5\frac{\pi^6}{4^5}\delta_1^{(1)}\right)a^{14} + \left(2\left(\frac{\lambda-1}{\lambda+1}\frac{\pi}{4}\right)^8 + \left(\frac{\lambda-1}{\lambda+1}\right)^5\frac{\pi^7}{4^6}\delta_2^{(11)}\right)a^{16}$$

$$+O\left(a^{18}\right)$$

$$\text{at} \quad a \to 0;$$

$$q \sim 1 - 2\frac{\lambda-1}{\lambda+1}\frac{\pi}{4}a^2 + 2\left(\frac{\lambda-1}{\lambda+1}\frac{\pi}{4}\right)^2 a^4 - 2\left(\frac{\lambda-1}{\lambda+1}\frac{\pi}{4}\right)^3 a^6 +$$

$$+2\left(\frac{\lambda-1}{\lambda+1}\frac{\pi a^2}{4}\right)^4 a^8 + \left(2\left(\frac{\lambda-1}{\lambda+1}\frac{\pi}{4}\right)^5 + \left(\frac{\lambda-1}{\lambda+1}\right)^3\frac{\pi^4}{4^3}\delta_1^{(1)}\right)a^{10} +$$

$$+\left(2\left(\frac{\lambda-1}{\lambda+1}\frac{\pi}{4}\right)^6 + 2\left(\frac{\lambda-1}{\lambda+1}\right)^4\frac{\pi^5}{4^4}\delta_1^{(1)}\right)a^{12} + \left(2\left(\frac{\lambda-1}{\lambda+1}\frac{\pi}{4}\right)^7 +\right.$$

$$(6.185)$$

$$+\left(\frac{\lambda-1}{\lambda+1}\right)^5 \frac{\pi^6}{4^5}\delta_1^{(1)}\right) a^{14} + \left(2\left(\frac{\lambda-1}{\lambda+1}\frac{\pi}{4}\right)^8 + \left(\frac{\lambda-1}{\lambda+1}\right)^4 \frac{\pi^7}{4^6}\delta_2^{(11)}\right) a^{16}$$

$$+O\left(a^{18}\right)$$

at $a \to 0$.

2) Expression q_{ASM-AP} (6.181) satisfies the Keller's theorem:

$$q_{ASM-AP}\left(\lambda, a\right) \sim q_{ASM-AP}^{-1}\left(\lambda^{-1}, a\right) + O\left(a^{16}\right) \quad \text{at} \quad a \to 0.$$

3) Solution q_{ASM-AP} (6.181) falls into Hashin–Shtrikman bounds:

(i) at $1 \le \lambda < \infty$, we have

$$\frac{1 - \frac{\pi a^2}{4} + \lambda\left(1 + \frac{\pi a^2}{4}\right)}{1 + \frac{\pi a^2}{4} + \lambda\left(1 - \frac{\pi a^2}{4}\right)} = \underline{q}_{HS} \le q_{ASM-AP} \le \bar{q}_{HS} =$$

$$= \frac{\lambda\left(2 - \frac{\pi a^2}{4} + \lambda\frac{\pi a^2}{4}\right)}{\frac{\pi a^2}{4} + \lambda\left(2 - \frac{\pi a^2}{4}\right)}.$$

The validity of the left hand side of this inequality follows from a comparison of the asymptotic expansions q_{ASM-AP} (6.184) and $\underline{q}_{HS}$:

$$\underline{q}_{HS} \sim 1 - 2\frac{\lambda-1}{\lambda+1}\frac{\pi}{4}a^2 + 2\left(\frac{\lambda-1}{\lambda+1}\frac{\pi}{4}\right)^2 a^4 - 2\left(\frac{\lambda-1}{\lambda+1}\frac{\pi}{4}\right)^3 a^6 +$$

$$+2\left(\frac{\lambda-1}{\lambda+1}\frac{\pi a^2}{4}\right)^4 a^8 + 2\left(\frac{\lambda-1}{\lambda+1}\frac{\pi}{4}\right)^5 a^{10} + o\left(a^{10}\right) \qquad (6.186)$$

at $a \to 0$,

$$q_{ASM-AP} \sim 1 - 2\frac{\lambda-1}{\lambda+1}\frac{\pi}{4}a^2 + 2\left(\frac{\lambda-1}{\lambda+1}\frac{\pi}{4}\right)^2 a^4 -$$

$$-2\left(\frac{\lambda-1}{\lambda+1}\frac{\pi}{4}\right)^3 a^6 + 2\left(\frac{\lambda-1}{\lambda+1}\frac{\pi a^2}{4}\right)^4 a^8 + \qquad (6.187)$$

$$+\left(2\left(\frac{\lambda-1}{\lambda+1}\frac{\pi}{4}\right)^5 + \left(\frac{\lambda-1}{\lambda+1}\right)^3 \frac{\pi^4}{4^3}\delta_1^{(1)}\right) a^{10} + o\left(a^{10}\right) \quad \text{at} \quad a \to 0,$$

where $\lambda \ge 1$, $\delta_1^{(1)} > 0$;

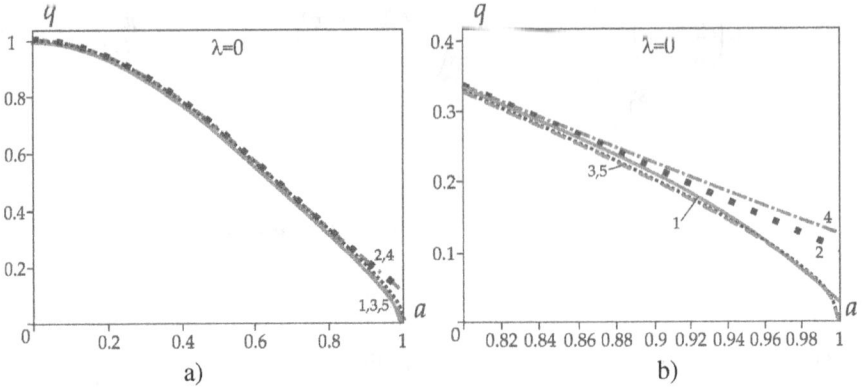

Figure 6.23 Effective thermal conductivity coefficient for non-conductive inclusions ($\lambda = 0$) ($1- q_{ASM-AP}$, $2- q_{ASM\,(N)}$, $3- q_{as\,s}^{(0)}$, $4- q_{MF}\,(\bar{q}_{HS})$, $5- q_{as}^{(0)}$).

(ii) if $0 \leq \lambda \leq 1$, then the estimate

$$\frac{\lambda \left(2 - \frac{\pi a^2}{4} + \lambda \frac{\pi a^2}{4}\right)}{\frac{\pi a^2}{4} + \lambda \left(2 - \frac{\pi a^2}{4}\right)} = \underline{q}_{HS} \leq q_{ASM-AP} \leq \bar{q}_{HS} =$$

$$= \frac{1 - \frac{\pi a^2}{4} + \lambda \left(1 + \frac{\pi a^2}{4}\right)}{1 + \frac{\pi a^2}{4} + \lambda \left(1 - \frac{\pi a^2}{4}\right)}$$

follows from the expressions (6.186), (6.187) considered at $\lambda \leq 1$.

Figures 6.23–6.27 present the effective thermal conductivity coefficient q_{ASM-AP} described by formulas (6.181), (6.182), in comparison with asymptotic solutions, Schwarz alternating method and MF.

The following notation has been used in Figs. 6.23–6.27: q_{ASM-AP} – Schwarz–Padé approach (6.181), (6.182); $q_{ASM\,(N)}$ – generalized Schwarz approximation (6.164)–(6.166); $q_{MF}\,(\bar{q}_{HS})$ – MF (Hashin–Shtrikman upper bound) (6.162); $q_{MF}\left(\underline{q}_{HS}\right)$ – MF (Hashin–Shtrikman lower bound) (6.162); $q_{as}^{(\infty)}$ – asymptotic solution (6.110) [158]; $q_{as}^{(0)}$ – asymptotic solution (6.111) [158] transformed using Keller's theorem [127]; $q_{as\,s}^{(0)}$ – asymptotic solution (6.6.61) from [94]; $q_{as\,s}^{(0)}$ – asymptotic solution (6.6.61) from [94], transformed using Keller's theorem [127]; $q_{as\,f}^{(\infty)}$ – asymptotic formula (49) from [158]; $q_{as\,f}^{(0)}$ – asymptotic formula (49) from [158], transformed using the Keller's theorem [127]; q_f – formula (6.8.90) from [94]; q_{num} – numerical solution [194].

For inclusions of medium and large sizes, Table 6.6 shows the values of the effective thermal conductivity coefficient calculated in [94, 119, 120, 121, 158, 194] in

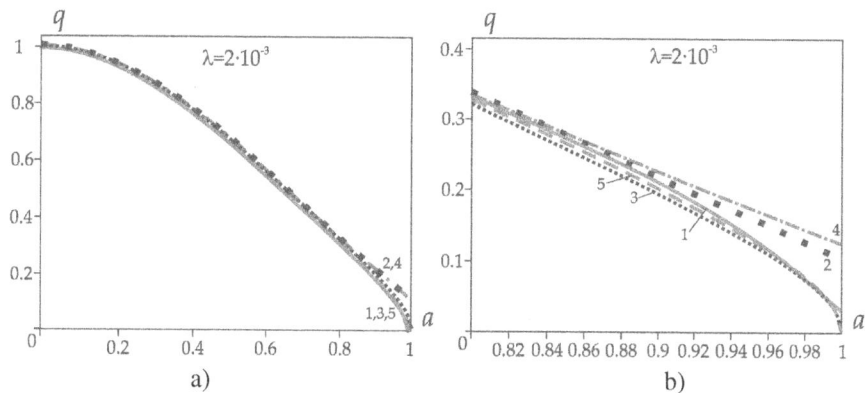

Figure 6.24 Effective thermal conductivity coefficient for inclusions of low conductivity ($\lambda = 2 \cdot 10^{-3}$) ($1- q_{ASM-AP}, 2- q_{ASM\,(N)}, 3- q_f, 4- q_{MF}(\bar{q}_{HS}), 5- q_{as\,f}^{(0)}$).

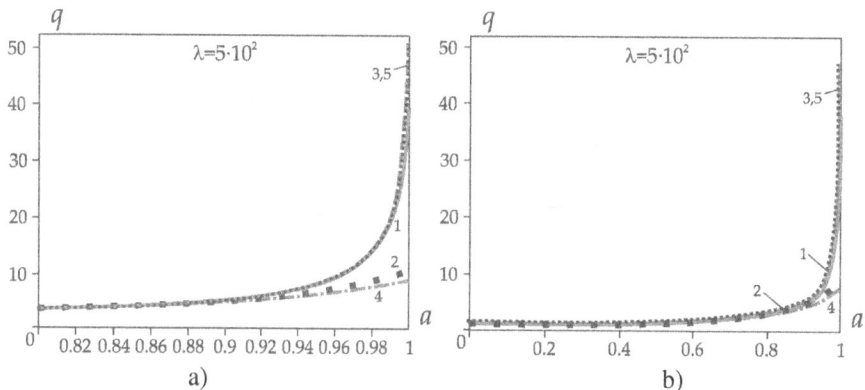

Figure 6.25 Effective thermal conductivity coefficient for inclusions of high conductivity ($\lambda = 5 \cdot 10^{2}$) ($1- q_{ASM-AP}, 2- q_{ASM\,(N)}, 3- q_f, 4- q_{MF}(\underline{q}_{HS}), 5- q_{as\,f}^{(\infty)}$).

various ways for absolutely conducting inclusions ($\lambda \to \infty$) and compared with the calculation data using the Schwarz–Padé method.

Table 6.7 reports the data of calculations of the effective thermal conductivity using formulas (6.181), (6.182) in comparison with asymptotic solutions [13, 94, 120, 121, 158] for inclusions of large sizes close to the limit value ($a \to 1$) and high conductivity $\lambda \gg 1$ (including the limiting case $\lambda \to \infty$).

Thus, the analysis of the above analytical and calculated results yields the following conclusions:

1. Using AP to the solution obtained by the Schwarz alternating method does not change the structure of the expression for effective thermal conductivity parameter, namely:

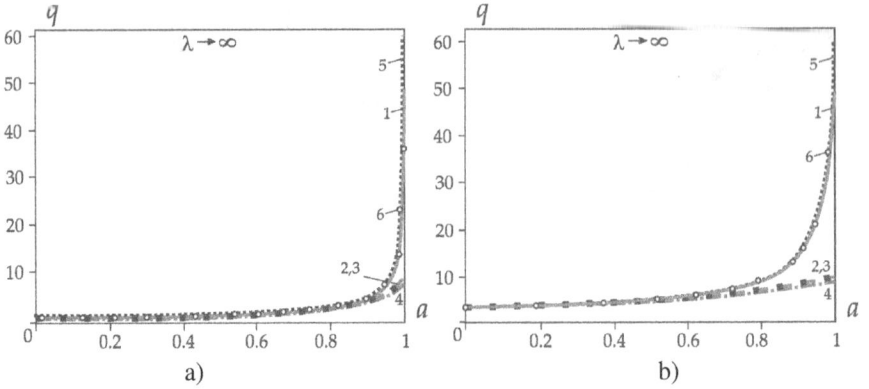

Figure 6.26   Effective thermal conductivity coefficient for absolutely conductive inclusions $(\lambda \to \infty)$ $(1- q_{ASM-AP}, 2- q_{ASM\,(N)}, 3- q_{as\,s}^{(\infty)}, 4- q_{MF}(q_{HS}), 5- q_{as}^{(\infty)}, 6- q_{num})$.

Table 6.6

Comparison of the results of calculation of the effective thermal conductivity coefficient for absolutely conductive inclusions by different methods.

Inclusion concentration c	Inclusion size a	Decomposition (3.9) [119]	AP (3.12) [119]	Formula (3.13) [119]	Numerical results [194]	Asymptotic formula [194]	Asymptotic solution (6.6.61) from [94]	MThPhM [13]	ThPhM and AP [121]	Schwarz–Padé decomposition (6.181), (6.182)
0.1	0.3568	1.210	1.247	1.223	1.2222	1.2214	1.2222	1.2234	1.2222	1.2222
0.2	0.5046	1.470	1.544	1.506	1.5003	1.4973	1.5003	1.5065	1.5000	1.5001
0.3	0.6180	1.811	1.918	1.879	1.8602	1.8546	1.8602	1.8790	1.8572	1.8582
0.4	0.7136	2.306	2.417	2.395	2.3510	2.3432	2.3510	2.3955	2.3341	2.3394
0.5	0.7979	3.270	3.145	3.172	3.0802	3.0700	3.0802	3.1720	3.0118	3.0301
0.6	0.8740	7.106	4.386	4.517	4.3418	4.3245	4.3418	4.5175	4.1247	4.1771
0.7	0.9441	–	7.409	7.769	7.4327	7.3857	7.4327	7.7695	6.9814	7.1213
0.74	0.7136	–	10.91	11.46	11.0062	10.9254	11.0065	11.4624	10.7726	10.9935
0.76	0.7979	–	15.29	15.99	15.4412	15.3284	15.4411	15.9902	15.7958	16.0741
0.77	0.8740	–	20.18	21.04	20.4334	20.2952	20.4317	21.0488	21.2061	21.4902
0.78	0.9441	–	35.01	36.60	35.934	35.7525	35.9216	36.6519	33.4056	33.4999

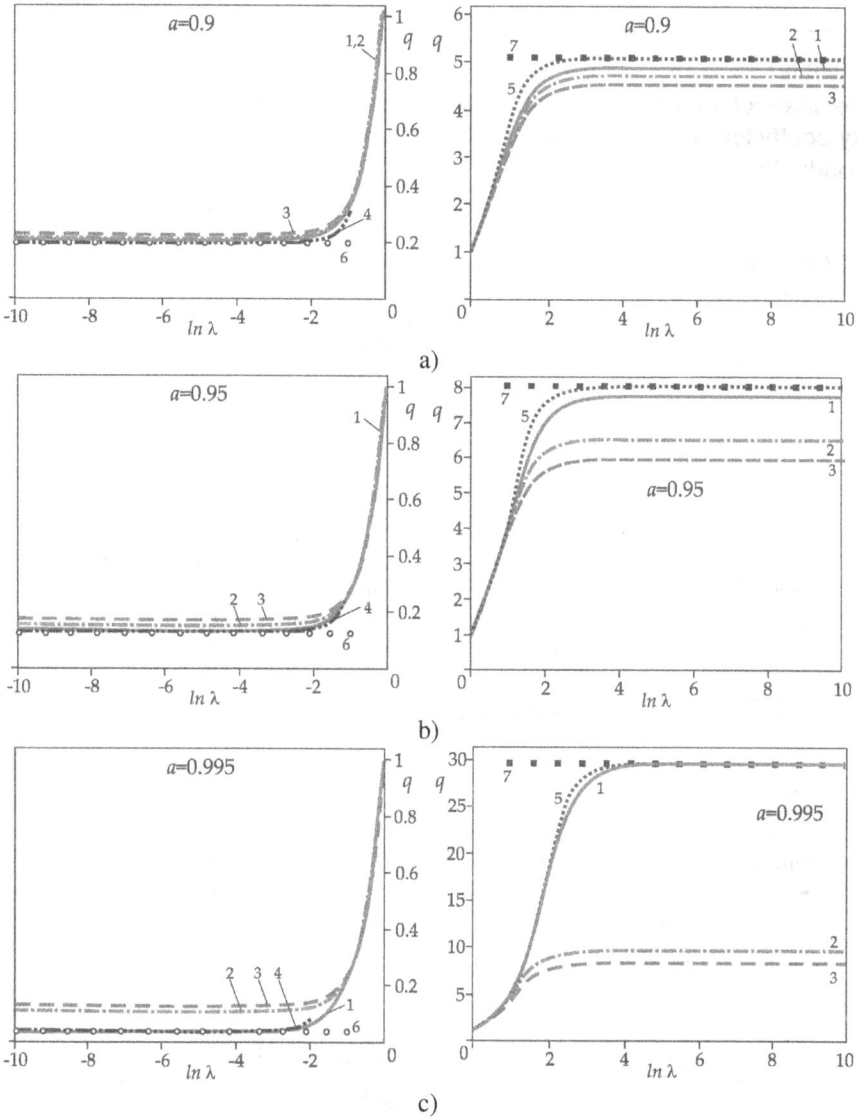

Figure 6.27 Effective thermal conductivity coefficient for: a) large sizes of inclusions ($a = 0.9$), b) very large inclusions ($a = 0.95$), c) extremely large sizes of inclusions ($a = 0.995$) ($1- q_{ASM-AP}$, $2- q_{ASM\,(N)}$, $3- q_{MF}(q_{HS})$, $4- q_{as\,f}^{(0)}$, $5- q_{as\,f}^{(\infty)}$, $6- q_{as}^{(0)}$, $7- q_{as}^{(\infty)}$).

1.1 the series expansion of the Schwarz–Padé relation for small inclusion sizes coincides with the expansion of the effective coefficient up to terms of the order a^{14} inclusive;

1.2 asymptotically, up to terms of order a^{14} inclusive, Keller's theorem holds;

Table 6.7

Comparison of the results of calculation of the effective thermal conductivity coefficient by different methods for inclusions of large sizes with high conductivity.

	Conductivity of inclusions $\lambda = 10^2$				
Inclusion size a	Asymptotic solution [158]	MThPhM [13]	ThPhM and AP [121]	Schwarz–Padé decomposition (6.181), (6.182)	Formula (6.8.90) from [94]
0.9	4.9108	5.0044	4.5270	4.5926	5.0255
0.91	5.2473	5.3449	4.8118	4.8861	5.3756
0.92	5.6413	5.7411	5.1482	5.2325	5.7840
0.93	6.1112	6.2097	5.5546	5.6500	6.2683
0.94	6.6841	6.7753	6.0584	6.1665	6.8542
0.95	7.4037	7.4765	6.7043	6.8268	7.5818
0.96	8.3438	8.3770	7.5684	7.7068	8.5175
0.97	9.6432	9.5934	8.7920	8.9479	9.7815
0.98	11.5936	11.3680	10.6703	10.8431	11.6229
0.99	14.8171	14.3260	13.9399	14.1167	14.6791
0.991	15.2200	14.7431	14.4058	14.5808	15.1086
0.992	15.6149	15.1940	14.9093	15.0816	15.5725
0.993	15.9721	15.6836	15.4549	15.6237	16.0759
0.994	16.2297	16.2179	16.0484	16.2124	16.6246
0.995	16.2513	16.8041	16.6962	16.8541	17.2260
0.996	–	–	17.4063	17.5561	17.8893
	Conductivity of inclusions $\lambda = 5 \cdot 10^2$				
Inclusion size a	Asymptotic solution [158]	MThPhM [13]	ThPhM and AP [121]	Schwarz–Padé decomposition (6.181), (6.182)	Formula (6.8.90) from [94]
0.9	5.0347	5.2456	4.7283	4.8007	5.2733
0.91	5.3980	5.6282	5.0517	5.1345	5.6675
0.92	5.8277	6.0792	5.4404	5.5353	6.1333
0.93	6.3467	6.6213	5.9197	6.0287	6.6945
0.94	6.9901	7.2893	6.5296	6.6555	7.3875
0.95	7.8164	8.1407	7.3380	7.4842	8.2719
0.96	8.9315	9.2776	8.4687	8.6398	9.4535
0.97	10.5536	10.9067	10.1733	10.3755	11.1447
0.98	13.2351	13.5334	13.0554	13.2939	13.8619
0.99	19.0663	18.9367	19.0093	19.0093	19.4089
0.991	20.1058	19.8518	19.9743	20.2269	20.3435
0.992	21.3187	20.8999	21.0549	21.3008	21.4123
0.993	22.7595	22.1169	22.2732	22.5087	22.6517
0.994	24.5101	23.5539	23.6571	23.8773	24.1128
0.995	26.7012	25.2868	25.2431	25.4411	25.8720
0.996	29.5560	–	27.0788	27.2450	28.0480

(Continued on next page)

Table 6.7 (Continued)

	Conductivity of inclusions $\lambda = 10^3$				
Inclusion size a	Asymptotic solution [158]	MThPhM [13]	ThPhM and AP [121]	Schwarz–Padé decomposition (6.181), (6.182)	Formula (6.8.90) from [94]
0.9	5.0502	5.2776	4.7552	4.8286	5.3064
0.91	5.4168	5.6661	5.0841	5.1681	5.7067
0.92	5.8510	6.1249	5.4803	5.5767	6.1805
0.93	6.3761	6.6774	5.9701	6.0812	6.7527
0.94	7.0283	7.3605	6.5958	6.7243	7.4613
0.95	7.8680	8.2343	7.4290	7.5789	8.3692
0.96	9.0049	9.4080	8.6017	8.7781	9.5890
0.97	10.6674	11.1050	10.3864	10.5962	11.3503
0.98	13.4403	13.8868	13.4495	13.6995	14.2264
0.99	19.5974	19.8425	19.9647	20.2348	20.3340
0.991	20.7166	20.8903	21.0452	21.3105	21.4026
0.992	22.0317	22.1070	22.2634	22.5206	22.6416
0.993	23.6079	23.5437	23.6475	23.8918	24,1024
0.994	25.5452	25.2763	25.2337	25.4588	25,8613
0.995	28.0074	27.4232	27.0699	27.2668	28,0371
0.996	31.2868	–	29.2203	29.3759	30,8282

	Conductivity of inclusions $\lambda = 10^5$			
Inclusion size a	Asymptotic solution [158]	MThPhM [13]	Schwarz–Padé decomposition (6.181), (6.182)	Formula (6.8.90) from [94]
0.9	5.0656	5.3099	4.8541	5.3395
0.91	5.4355	5.7043	5.1989	5.7461
0.92	5.8741	6.1709	5.6147	6.2282
0.93	6.4053	6.7342	6.1295	6.8116
0.94	7.0662	7.4325	6.7884	7.5362
0.95	7.9190	8.3297	7.6668	8.4683
0.96	9.0776	9.5418	8.9076	9.7279
0.97	10.7801	11.3109	10.8051	11.5637
0.98	13.6435	14.2620	14.0902	14.6131
0.99	20.1233	20.8685	21.2048	21.3804
0.991	21.3212	22.0828	22.4021	22.6170
0.992	22.7376	23.5164	23.7580	24.0747
0.993	24.4478	25.2448	25.3063	25.8294
0.994	26.5699	27.3860	27.0910	27.9994
0.995	29.3006	30.1366	29.1710	30.7820
0.996	32.8446	–	31.6260	34.5375

(Continued on next page)

Table 6.7 (Continued)

	Conductivity of inclusions $\lambda = \infty$			
Inclusion size a	Asymptotic solution [158]	MThPhM [13]	Schwarz–Padé decomposition (6.181), (6.182)	Formula (6.8.90) from [94]
0.9	5.0657	4.7826	4.8569	5.0893
0.91	5.4357	5.1171	5.2023	5.4627
0.92	5.8743	5.5210	5.6190	5.9057
0.93	6.4056	6.0218	6.1349	6.4424
0.94	7.0666	6.6639	6.7962	7.1103
0.95	7.9200	7.5230	7.6768	7.9721
0.96	9.0784	8.7403	8.9222	9.1425
0.97	10.7812	10.6109	10.8289	10.8610
0.98	13.6455	13.8728	14.1353	13.7475
0.99	20.1286	21.0343	21.3190	20.2645
0.991	21.3273	22.2524	22.5311	21.4676
0.992	22.7447	23.6364	23.9047	22.8897
0.993	24.4563	25.2227	25.4748	24.6062
0.994	26.5802	27.0590	27.2865	26.7353
0.995	29.3137	29.2098	29.4004	29.4742
0.996	33.0177	31.7631	31.8990	33.1837

1.3 the results obtained using Schwarz–Padé approximation falls into the Hashin–Shtrikman bounds.

2. Comparison of the calculation results with the known data of other authors showed that the transformation of the N-iterative solution according to the Schwarz alternating method using AP significantly expands the area of its applications:

2.1 the Schwarz–Padé method (6.181), (6.182) adequately describes the effective parameter for the limiting values of the geometric and physical parameters and gives quantitative estimates close to asymptotic when the inclusion sizes approaches 1 fairly close;

2.2 the Schwarz–Padé approximation can be used with practical purposes accuracy for inclusions of any conductivity ($0 \leq \lambda < \infty$, $\lambda \to \infty$) and size ($0 \leq a \leq 0.996$).

References

1. Aboudi, J., Arnold, S.M., Bednarcyk, B.A. (2013) *Micromechanics of Composite Materials. A Generalized Multiscale Analysis Approach.* Elsevier: Amsterdam.
2. Abramowitz, M., Stegun, I.A. (Eds.) (1965) *Handbook of Mathematical Functions, with Formulas, Graphs, and Mathematical Tables.* Dover Publications: New York.
3. Achenbach, J.D., Zhu, H. (1990) Effect of interphases on micro and macromechanical behavior of hexagonal-array fiber composites. *J. Appl. Mech.* **57**(4):956–963.
4. Adams, R.D., Harris, J.A. (1987) The influence of local geometry on the strength of adhesive joints. *Int. J. Adh. Adh.* **7**(2):69–80.
5. Andrianov, I.I., Awrejcewicz, J., Starushenko, G.A., Gabrinets, V.A. (2020) Refinement of the Maxwell formula for composite reinforced by circular cross-section fibers. Part I: using the Schwarz alternating method. *Acta Mech.* **231**:4971–4990.
6. Andrianov, I.I., Awrejcewicz, J., Starushenko, G.A., Gabrinets, V.A. (2020) Refinement of the Maxwell formula for composite reinforced by circular cross-section fibers. Part II: using Padé approximants. *Acta Mech.* **231**(12):5145–5157.
7. Andrianov, I.V. (2021) Mathematical models in pure and applied mathematics. In: Abramyan, A., Andrianov, I., Gaiko. V. (Eds.) *Nonlinear Dynamics of Discrete and Continuous Systems.* Springer Nature: Basingstoke, pp. 15–30.
8. Andrianov, I.V., Awrejcewicz, J. (2001) New trends in asymptotic approaches: Summation and interpolation methods. *Appl. Mech. Rev.* **54**:69–92.
9. Andrianov, I.V., Awrejcewicz, J. (2000) Numbers or understanding: analytical and numerical methods in the theory of plates and shells. *Facta Univ.* **2**(10):1319–1327.
10. Andrianov, I.V., Awrejcewicz, J. (2003) Homo analyticus or homo computicus? *Facta Univ.* **3**(13):765–770.
11. Andrianov, I.V., Awrejcewicz, J., Danishevskyy, V.V., Ivankov, A.O. (2014) *Asymptotic Methods in the Theory of Plates with Mixed Boundary Conditions.* Wiley: Chichester.
12. Andrianov, I.V., Awrejcewicz, J., Danishevskyy, V.V. (2018) *Asymptotical Mechanics of Composites. Modelling Composites without FEM.* Springer Nature: Cham.
13. Andrianov, I.V., Awrejcewicz, J., Starushenko, G.A. (2013) Application of an improved three-phase model to calculate effective characteristics for a composite with cylindrical inclusions. *Lat. Am. J. Sol. Struct.* **10**(1):197–222.
14. Andrianov, I. V., Awrejcewicz, J., Starushenko, G.A. (2015) Asymptotic analysis of the Maxwell Garnett formula using the two-phase composite model. *Int. J. Appl. Mech.* **7**(2):1550025-1–1550025-27.
15. Andrianov, I.V., Awrejcewicz, J., Starushenko, G.A. (2017) Asymptotic models and transport properties of densely packed, high contrast fibre composites. Part I: Square lattice of circular inclusions. *Compos. Struct.* **179**:617–627.
16. Andrianov, I.V., Awrejcewicz, J., Starushenko, G.A. (2017) Asymptotic models for transport properties of densely packed, high contrast fibre composites. Part II: Square lattices of rhombic inclusions and hexagonal lattices of circular inclusions. *Compos. Struct.* **180**:351–359.
17. Andrianov, I.V., Danishevs'kyy V.V., Kalamkarov A.L. (2010) Analysis of the effective conductivity of composite materials in the entire range of volume fractions of inclusions up to the percolation threshold. *Compos. Part B* **41**(6):503–507.

18. Andrianov, I.V., Danishevs'kyy V.V., Kalamkarov, A.L. (2012) Asymptotic analysis of perforated plates and membranes. Part 1: Static problems for small holes. *Int. J. Sol. Struct.* **49**(2):298–310.

19. Andrianov, I., Danishevskyi, V., Tokarzewski, S. (1996) Two-point quasifractional approximates for effective conductivity of a simple cubic lattice of spheres. *Int. J. Heat Mass Trans.* **39**(11):2349–2352.

20. Andrianov, I., Danishevsky, V., Tokarzewski, S. (2000) Quasifractional approximants in the theory of composite materials. *Acta Appl. Math.* **61**(1-3):29–35.

21. Andrianov, I.V., Kalamkarov, A.L., Starushenko, G.A. (2013) Analytical expressions for effective thermal conductivity of composite materials with inclusions of square cross-section. *Compos. Part B* **50**:44–53.

22. Andrianov, I.V., Kalamkarov, A.L., Starushenko, G.A. (2013) Three-phase model for a fiber-reinforced composite material. *Compos. Struct.* **95**:95–104.

23. Andrianov, I.V., Manevitch, L.I. (2002) *Asymptotology: Ideas, Methods, and Applications.* Kluwer Academic Publishers: Dordrecht.

24. Andrianov, I., Mityushev, V. (2018) Exact and "exact" formulae in the theory of composites. In: Drygaś, P., Rogosin, S. (Eds.) *Modern Problems in Applied Analysis.* Birkhauser: Cham, pp. 15–34.

25. Andrianov, I.V., Starushenko, G.A. (1995) Asymptotic methods in the theory of perforated membranes of nonhomogeneous structure. *Eng. Transact.* **43**(1-2):5–18.

26. Andrianov, I.V., Starushenko, G.A. (1987) Homogenization method for problems of mechanics in a multiply connected domain. *Prob. Mashinostroy.* **27**:48–54 (in Russian).

27. Andrianov, I.V., Starushenko, G.A., Danishevs'kyy, V.V., Tokarzewski, S. (1999) Homogenization procedure and Padé approximants for effective heat conductivity of composite materials with cylindrical inclusions having square cross-section. *Proc. Roy. Soc. Math. Phys. Eng. Sci.* **455**(1989):3401–3413.

28. Andrianov, I.V., Starushenko, G.A., Tokarzewski, S. (1998) Homogenization procedure and Padé approximations in the theory of composite materials with parallelepiped inclusions. *Int. J. Heat Mass Transf.* **41**(1):175–181.

29. Andrianov, I.V., Starushenko, G.A. (1988) Application of the averaging method for the calculation of perforated plates. *Sov. Appl. Mech.* **24**(4):410–415.

30. Baker, G.A. (1975) *Essential of Padé Approximants.* Academic Press: New York.

31. Baker, G.A.Jr., Graves-Morris, P. (1996) *Padé Approximants.* 2nd ed. Cambridge University Press: Cambridge.

32. Bakhvalov, N.S., Églit, M.É. (1994) An estimate of the error of averaging the dynamics of small perturbations of very inhomogeneous mixtures. *Comput. Math. Math. Phys.* **34**(3):333–349.

33. Bakhvalov, N.S., Églit, M.É. (1993) Averaging of the equations of the dynamics of composites of slightly compressible elastic components. *Comput. Math. Math. Phys.* **33**(7):939–952.

34. Bakhvalov, N.S., Églit, M.É. (1995) The limiting behaviour of periodic media with soft-modular inclusions. *Comput. Math. Math. Phys.* **35**(6):719–729.

35. Bakhvalov, N.S., Églit, M.É. (1992) Variational properties of averaged equations for periodic media. *Proc. Steklov Inst. Math.* **192**:3–18.

36. Bakhvalov, N.S., Églit, M.É. (1998) Effective moduli of composites reinforced by systems of plates and bars. *Comput. Math. Math. Phys.* **38**(5):783–804.

37. Bakhvalov, N.S., Knyazev, A.V. (1991) Effective computation of averaged characteristics of composites of periodic structure that consist of essentially different materials.
Dokl. Math. **42**(1):57–62.

38. Bakhvalov, N., Panasenko, G. (1989) *Averaging Processes in Periodic Media. Mathematical Problems in Mechanics of Composite Materials.* Kluwer: Dordrecht.

39. Balagurov, B.Ya. (2001) Effective electrical characteristics of a two-dimensional three-
component doubly-periodic system with circular inclusions. *J. Exp. Theor. Phys.*
92(1):123–134.

40. Balagurov, B.Ya., Kashin, V.A. (2005) Analytic properties of the effective dielectric
constant of a two-dimensional Rayleigh mode. *J. Exp. Theor. Phys.* **100**(4):731–741.

41. Batchelor, G.K. (1974) Transport properties of two-phase materials with random structure. *Ann. Rev. Fluid Mech.* **6**:227–255.

42. Batchelor, G.K., O'Brien, R.W. (1977) Thermal or electrical conduction through a
granular material. *Proc. Roy. Soc. London Ser. A, Math. Phys. Sci.* **355**(1682):313–333.

43. Benguigui, L., Ron, P. (1995) A direct comparison of the exponents of superconductivity and superelasticity near the percolation threshold. *J. Phys. I, EDP Sci.* **5**(4):451–
453.

44. Bensoussan, A., Lions, J.-L., Papanicolaou, G. (1978) *Asymptotic Analysis for Periodic
Structures.* North-Holland Publishing Company: Amsterdam.

45. Benveniste, Y., Miloh, T. (2001) Imperfect soft and stiff interfaces in two-dimensional
elasticity. *Mech. Mater.* **33**(6):309–324.

46. Benveniste, Y., Milton, G.W. (2010) The effective medium and the average field approximations vis-a-vis the Hashin–Shtrikman bounds. I. The self-consistent scheme in
matrix-based composites. *J. Mech. Phys. Sol.* **58**(7):1026–1038.

47. Benveniste, Y., Milton, G.W. (2010) The effective medium and the average field
approximations vis-a-vis the Hashin–Shtrikman bounds. II. The generalized self-
consistent scheme in matrix-based composites. *J. Mech. Phys. Sol.* **58**(7):1039–1056.

48. Berdichevsky, V.L. (1985) Heat conduction of checkerboard structures. *Moscow Univ.
Mech. Bull.* **40**:15–25.

49. Berdichevsky, V.L. (1983) Variational Principles of the Continuum Mechanics. Nauka:
Moscow (in Russian).

50. Bergman, D.J. (1978) The dielectric constants of a composite material—a problem in
classical physics. *Phys. Rep. C* **43**:377–407.

51. Bergman, D.J. (2007) The self-consistent effective medium approximation (SEMA):
New tricks from an old dog. *Phys. B* **394**(2):344–350.

52. Bergman, D.J., Dunn, K.-J. (1992) Bulk effective dielectric constant of a composite
with a periodic microgeometry. *Phys. Rev. B* **45**(23):13262–13271.

53. Berlin, A.A., Wolfson, S.A., Oshmyan, V.G., Enikolopyan, N.S. (1990) *Principles for
Polymer Composites Design.* Chemistry: Moscow (in Russian).

54. Berlyand, L., Kolpakov, A.G., Novikov, A. (2013) *Introduction to the Network Approximation.* Cambridge University Press: New York.

55. Berlyand, L., Kolpakov, A. (2001) Network approximation in the limit of small inter-
particle distance of the effective properties of a high contrast random dispersed composite. *Arch. Ratio. Mech. Anal.* **159**(3):179–227.

56. Berlyand, L., Mityushev, V. (2001) Generalized Clausius-Mossotti formula for random
composite with circular fibers. *J. Stat. Phys.* **102**(1–2):115–145.

57. Berlyand, L., Mityushev, V. (2005) Increase and decrease of the effective conductivity
of two phase composites due to polydispersity. *J. Stat. Phys.* **118**(3-4):481–509.

58. Berlyand, L., Novikov, A. (2002) Error of the network approximation for densely packed composites with irregular geometry. *SIAM J. Math. Anal.* **34**(2):385–408.

59. Bolotin, V.V., Novichkov, Yu.N. (1980) *Mechanics of Multilayer Structures.* Mashinostroyenie: Moscow (in Russian).

60. Bourgat, J.F. (1979) Numerical experiments of the homogenization method for operators with periodic coefficients. *Lect. Not. Math.* **704**:330–356.

61. Bruggeman, D.A.G. (1935) Berechnung verschiedener physikalischer Konstanten von heterogenen Substanzen. I. Dielektrizitätskonstanten und Leitfähigkeiten der Mischkörper aus isotropen Substanzen. *Ann. Phys.* **416**(7):636–664; **416**(8):665–679; (1936) II. Dielecktrizitätskonstanten und Leitfähigkeiten von Vielkristallen der nichtregulären Systeme. *Ibid* **417**(7):645–672; (1937) III. Die elastischen Konstanten der quasiisotropen Mischkörper aus isotropen Substanzen. *Ibid* **421**(2), 160–178. **416**(8):665–679; (1936) II. Dielecktrizitätskonstanten und Leitfähigkeiten von Vielkristallen der nichtregulären Systeme. *Ibid* **417**(7):645–672; (1937) III. Die elastischen Konstanten der quasiisotropen Mischkörper aus isotropen Substanzen. *Ibid* **421**(2), 160–178.

62. Budiansky, B. (1965) On the elastic moduli of some heterogeneous materials. *J. Mech. Phys. Sol.* **13**(4):223–227.

63. Buryachenko, V.A. (2001) Multiparticle effective field and related methods in micromechanics of composite materials. *Appl. Mech. Rev.* **54**(1):1–47.

64. Carbon-fiber epoxy honeycombs mimic the material performance of balsa wood. URL: https://www.seas.harvard.edu/news/2014/06/carbon-fiber-epoxy-honeycombs-mimic-material-performance-of-balsa-wood.

65. Chaikin, P.M., Lubensky, T.C. (1995) *Principles of Condensed Matter Physics.* Cambridge University Press: New York.

66. Chang, H.-M., Liao, C. (2011) A parallel derivation to the Maxwell-Garnett formula for the magnetic permeability of mixed materials. *World J. Cond. Matt. Phys.* **1**(2):55–58.

67. Chen, H.S., Acrivos, A. (1978) The effective elastic moduli of composite materials containing spherical inclusions at non-dilute contractions. *Int. J. Sol. Struct.* **14**:349–364.

68. Chen, X., Liu, Y. (2001) Multiple-cell modelling of fiber reinforced composites with the presence of interphases using the boundary element method. *Comput. Mater. Sci.* **21**(1):86–94.

69. Choy, T.C. (2016) *Effective Medium Theory: Principles and Applications.* Second Edition. Oxford University Press: Oxford.

70. Christensen, R.M., Lo, K.H. (1979) Solutions for effective shear properties in three phase sphere and cylinder models. *J. Mech. Phys. Sol.* **27**(4):315–330.

71. Christensen, R.M. (2005) *Mechanics of Composite Materials.* Dover Publications: Mineola, New York.

72. Clausius, R. (1879) *Die mechanische Behandlung der Elektrizität.* Friedrich Vieweg und Sohn: Braunschweig.

73. Cole, K.S., Li, Ch., Bak, A.F. (1969) Electrical analogues for tissues. *Exper. Neuro.* **24**(3):459–473.

74. Craster, R.V., Obnosov, Yu.V. (2001) Checkerboard composites with separated phases. *J. Math. Phys.* **42**(11):5379–5388.

75. Craster, R.V., Obnosov, Yu.V. (2001) Four-phase checkerboard composites. *SIAM J. Appl. Math.* **61**:1839–1856.

76. Craster, R.V., Obnosov, Yu.V. (2004) A three-phase tessellation: solution and effective properties. *Proc. Roy. Soc. London A* **460**:1017–1037.

77. Crighton, D.G. (1994) Asymptotics-an indispensable complement to thought, computation and experiment in applied mathematical modeling. In: Fasano, A., Primicerio, M. (Eds.) *Proceedings of the 7th European Conference on Mathematics in Industry.* B.G.Teubner: Stuttgart, pp. 3–19.

78. Da Silva, L.F.M., Das Neves, P.J.C., Adams, R.D., Spelt, J.K., Mallet, P., Guerin, C.A., Sentenac, A. (2005) Maxwell-Garnett mixing rule in the presence of multiple scattering: Derivation and accuracy. *Phys. Rev. B* **72**(1):014205.

79. Drygas, P., Gluzman, S., Mityushev, V., Nawalaniec, W. (2020) *Applied Analysis of Composite Media: Analytical and Computational Approaches for Materials Scientists and Engineers.* Elsevier: New York.

80. Dykhne, A.M. (1970) Conductivity of a two-dimensional system. *Sov. Phys. JETP* **32**:63–65.

81. Einstein, A. (1906) Eine neue Bestimmung der Moleküldimensionen. *Ann. Phys.* **19**(2):289–306; (1911) Berichtigung zu meiner Arbeit: "Eine neue Bestimmung der Moleküldimensionen". *Ibid* **339**(3):591–592.

82. Einstein, A. (1994) About Religion. In: Seelig, C. (Ed.) *Ideas and Opinions.* Modern Library: New York, pp. 39–57.

83. Eshelby, J.D. (2006) *Collected Works of J.D. Eshelby. The Mechanics of Defects and Inhomogeneities.* Springer: New York.

84. Eucken, A. (1940) Allgemeine Gesetzmäßigkeiten für das Wärmeleitvermögen verschiedener Stoffarten und Aggregatzustände. *Forschung im Ingenieurwesen.* **11**(1):6–20.

85. Faraday, M. (1839) *Experimental Researches in Electricity.* London : Richard and John Edward Taylor: London.

86. Fokin, A.G. (1996) Macroscopic conductivity of random inhomogeneous media. Calculation methods. *Phys. Usp.* **39**:1009–1032.

87. Fuh, A.Y.-G., Lee, W., Huang, K.Y.-C. (2013) Derivation of extended Maxwell Garnett formula for carbon-nanotube-doped nematic liquid crystal. *Liq. Cryst.* **40**(6):745–755.

88. Gai, M.I., Manevitch, L.I., Oshmyan, V.G. (1990) Percolation effects in the mechanics of composite materials. *Mech. Compos. Mater.* **26**(3):310–314.

89. Gai, M.I., Zelenskii, E.S., Manevitch, L.I., Oshmyan, V.G., Sochnev, V.I., Sulyaeva, Z.P., Turusov, R.A. (1987) Elastic characteristics of randomly nonuniform composites. *Mech. Compos. Mater.* **23**(2):172–178.

90. Garboczi, E.J., Bentz, D.P. (1997) Analytical formulas for interfacial transition zone properties. *Adv. Cem. Bas. Mater.* **6**(3-4):99–108.

91. Garboczi, E.J., Berryman, J.G. (2000) New effective medium theory for the diffusivity or conductivity of a multi scale concrete microstructure model. *Conc. Sci. Eng. J.* **2**(6):88–96.

92. Gluzman, S., Mityushev, V. (2015) Series, index and threshold for random 2D composite. *Arch. Mech.* **67**(1):75–93.

93. Gluzman, S., Mityushev, V., Nawalaniec, W. (2014) Crossproperties of the effective conductivity of the regular array of ideal conductors. *Arch. Mech.* **66**(4):287–301.

94. Gluzman, S., Mityushev, V., Nawalaniec, W. (2018) *Computational Analysis of Structured Media.* Elsevier: London.

95. Gluzman, S., Mityushev, V., Nawalaniec, W., Sokal, G. (2016) Random composite: stirred or shaken? *Arch. Mech.* **68**:229–241.

96. Gluzman, S., Mityushev, V., Nawalaniec, W., Starushenko, G. (2016) Effective conductivity and critical properties of a hexagonal array of superconducting cylinders. In: Pardalos, P.M., Rassias, T.M. (Eds.) Contributions in Mathematics and Engineering. In Honor of Constantin Carathéodory. Springer, pp. 255–297.

97. Gorbachev, V.I. (2016) The Bakhvalov–Pobedrya homogenization method in composite mechanics. *Moscow Univ. Mech. Bull.* **71**(6):137–141.

98. Grigolyuk, E.I., Filshtinskii, L.A. (1966) Elastic equilibrium of an isotropic plane with a doubly periodic system of inclusions. *Sov. Appl. Mech.* **2**:1–5.

99. Grigolyuk, E.I., Filshtinsky, L.A. (1970) *Perforated Plates and Shells*. Nauka: Moscow (in Russian).

100. Grigolyuk, E.I., Filshtinskii, L.A. (1992) *Periodical Piece–Homogeneous Elastic Structures*. Nauka: Moscow (in Russian).

101. Grigolyuk, E.I., Filshtinskii, L.A. (1994) *Regular Piece-Homogeneous Structures with Defects*. Fiziko-Matematicheskaja Literatura: Moscow (in Russian).

102. Gringauz, M.G., Filshtinskii, L.A. (1975) Theory of an elastic linearly reinforced composite. *J. Appl. Math. Mech.* **39**(3):510–519.

103. Guckenheimer, J. (1998) Computer simulation and beyond – for the 21st century. *Not. Am. Math. Soc.* **45**(9):1120–1123.

104. Guinovart-Diaz, R., Bravo-Castillero, J., Rodriguez-Ramos, R., Sabina, F.J. (2001) Closed-form expressions for the effective coefficients of a fiber-reinforced composite with transversely-isotropic constituents - I. Elastic and hexagonal symmetry. *J. Mech. Phys. Sol.* **49**:1445–1462.

105. Guz, A.N., Nemish, Yu.N. (1987) Perturbation of boundary shape in continuum mechanics (review). *Sov. Appl. Mech.* **23**(9):799–822.

106. Guz, A.N., Golovchan, V.T., Kokhanenko, Yu.V., Kush, V.I. (1993) *Statics of Materials*. Naukova Dumka: Kiev (in Russian).

107. Hamming, R.W. (1973) *Numerical Methods for Scientists and Engineers*. 2nd ed. McGraw Hill: New York.

108. Hashin, Z. (1962) The elastic moduli of heterogeneous materials. *J. Appl. Mech.* **29**(1):143–150.

109. Hashin, Z., Rosen, B.W. (1964) The elastic moduli of fiber reinforced materials. *J. Appl. Mech.* **31**(2):223–232.

110. Hashin, Z., Shtrikman, S. (1962) A variational approach to the theory of the effective magnetic permeability of multiphase materials. *J. Appl. Phys.* **33**(10):3125–3131.

111. Hashin, Z., Shtrikman, S. (1963) A variational approach to the theory of the elastic behaviour of multiphase materials. *J. Mech. Phys. Sol.* **11**(2):127–140.

112. Helsing, J. (1991) Transport properties of two-dimensional tilings with corners. *Phys. Rev. B* **44**(21):11677–11682.

113. Helsing, J. (1995) An integral equation method for electrostatics of anisotropic composites. *Proc. Roy. Soc. Math. Phys. Sci.* **450**(1939):343–350.

114. Hershey, A.V. (1954) The elasticity of an isotropic aggregate of anisotropic cubic crystals. *J. Appl. Mech.* **21**(3):236–240.

115. Hill, R. (1952) The elastic behavior of a crystalline aggregate. *Proc. Phys. Soc. A* **65**:349–354.

116. Hill, R. (1963) New derivations of some elastic extremum principles. In: *Progress in Applied Mathematics: The Prager Anniversary Volume*. Macmillan: New York, pp. 99–106.

117. Hill, R. (1965) A self-consistent mechanics of composite materials. *J. Mech. Phys. Sol.* **13**(4):213–222.

118. Hill, R. (1964) Theory of mechanical properties of fibre-strengthened materials: I. Elastic behavior. *J. Mech. Phys. Sol.* **12**:199–212.

119. Kalamkarov, A.L., Andrianov, I.V., Danishevs'kyy, V.V. (2009) Asymptotic homogenization of composite materials and structures. *Appl. Mech. Rev.* **62**(3):030802-1–030802-20.

120. Kalamkarov, A.L., Andrianov, I.V., Starushenko, G.A. (2014) Three-phase model for a composite material with cylindrical circular inclusions. Part I: Application of the boundary shape perturbation method. *Int. J. Eng. Sci.* **78**:154–177.

121. Kalamkarov, A.L., Andrianov, I.V., Starushenko, G.A. (2014) Three-phase model for a composite material with cylindrical circular inclusions. Part II: Application of Padé approximants. *Int. J. Eng. Sci.* **78**:178–191.

122. Kalamkarov, A.L., Andrianov, I.V., Pacheco, P.M.C.L., Savi, M.A., Starushenko, G.A. (2016) Asymptotic analysis of fiber-reinforced composites of hexagonal structure. *J. Multiscale Model.* **7**(3):1650006-1–1650006-32.

123. Katz, H.S., Milewski, V. (Ed.) (1978) *Handbooks of Fillers and Reinforcements for Plastics.* Van Nostrand Reinhold Company: New York.

124. Kanaun, S.K., Levin, V.M. (2008) *Self-Consistent Methods for Composites. Vol. 1: Static Problems.* Springer: New York.

125. Kantorovich, L.V., Krylov, V.I. (1958) *Approximate Methods of Higher Analysis.* Groningen: Noordhoff.

126. Keller, J.B. (1963) Conductivity of a medium containing a dense array of perfectly conducting spheres or cylinders or nonconducting cylinders. *J. Appl. Phys.* **34**:991–993.

127. Keller, J.B. (1964) A theorem on the conductivity of a composite medium. *J. Math. Phys.* **5**(4):548–549.

128. Keller, J.B. (1993) Stresses in narrow regions. *Trans. ASME J. Appl. Mech.* **60**:1054–1056.

129. Kerner, E.H. (1956) The elastic and thermos-elastic properties of composite media. *Proc. Phys. Soc. Sec. B* **69**(8):808–813.

130. Knunyants, N.N., Lyapunova, M.A., Manevitch, L.I., Oshmyan, V.G., Shaulov, A.Yu. (1986) Modeling the effect of a nonideal adhesive bond on the elastic properties of a dispersively filled composite. *Mech. Compos. Mater.* **22**(2):162–165.

131. Koledintseva, M.Y., DuBroff, R.E., Schwartz, R.W. (2006) Maxwell Garnett model for dielectric mixtures containing particles optical frequencies. *Prog. Electromag. Res.* **63**:223–242.

132. Koledintseva, M.Y., DuBroff, R.E., Schwartz, R.W. (2009) Maxwell Garnett rule for dielectric mixtures with statistically distributed orientations of inclusions. *Prog. Electromag. Res.* **99**:131–148.

133. Kolpakov, A.A., Kolpakov, A.G. (2009) *Capacity and Transport in Contrast Composite Structures: Asymptotic Analysis and Applications.* CRC Press: Boca Raton.

134. Kopysov, S.P., Sagdeeva, Yu.A. (2007) The application of the wavelet transform in the numeric averaging of differential equations with quickly oscillating coefficients and in calculation of effective characteristics. *Russ. Math. (Izv. VUZ).* **51**(7):76–79.

135. Kozlov, G.M. (1989) Geometrical aspects of averaging. *Russ. Math. Surv.* **44**:91–144.

136. Krieger, I.M. (1972) Rheology of monodisperse lattices. *Adv. Colloid Interface Sci.* **3**(2):111–136.

137. Kröner, E. (1958) *Kontinuumstheorie der Versetzungen und Eigenspannungen.* Springer: Berlin.

138. Kushch, V.I. (2020) *Micromechanics of Composites. Multipole Expansion Approach.* Butterworth-Heinemann.

139. Lagzdins, A.Zh., Tamuzh, V.P., Teters, G.A., Kregers, A.F. (1992) *Orientational Averaging in Mechanics of Solids.* Longman Scientific and Technical: NewYork.

140. Landauer, R.C. (1978) Electrical conductivity in inhomogeneous media. In: Garland, J.C., Tanner, D.B. (Eds). *Electrical Transport and Optical Properties of Inhomogeneous Media.* AIP Conference Proceedings, 40, pp. 2–45.

141. Landauer, R. (1952) The electrical resistance of binary metallic mixture. *J. Appl. Phys.* **23**(7):779–784.

142. Levin, V., Kanaun, S., Markov, M. (2012) Generalized Maxwell's scheme for homogenization of poroelastic composites. *Int. J. Eng. Sci.* **61**:75–86.

143. Levy, O., Stroud, D. (1997) Maxwell Garnett theory for mixtures of anisotropic inclusions: Application to conducting polymers. *Phys. Rev. B.* **56**(13):8035–8046.

144. Lichtenecker, K. (1926) Die Dielektrizitätskonstante natürlicher und künstlicher Mischkorper. *Phys. Zeits.* **27**(4,5):115–158 .

145. Lions, J.-L. (1982) On some homogenisation problem. *ZAMM* **62**(5):251–262.

146. Lipatov, Yu.S. (1995) *Polymer Reinforcement.* ChemTec Publ.: Ontario.

147. Lorenz, L.V. (1870) Experimentale og theoretiske undersøgelser over legemernes brydningsforhold. *K. Dan. Vidensk. Selsk. Skrrift. Ser. 5, Naturvid. og Mathem.* **8**(1):203–248; (1875) *Ibid* **10**(2):483–518.

148. Lorenz, L. (1880) Über die Refraktionskonstante. *Ann. Phys. Chemie* **247**(9):70–103.

149. Lorentz, H.A. (1880) Über die Beziehung zwischen der Fortpflanzungsgeschwindigkeit des Lichtes der Körperdichte. *Ann. Phys. Chemie* **245**(4): 641–665.

150. Lorentz, H.A. (1909) *The Theory of Electrons and it's Applications to the Phenomena of Light and Radiant Heat.* B.G. Teubner: Leipzig.

151. Lubin, G. (Ed.) (1982) *Handbook of Composites.* Van Nostrand Reinhold Co.: New York.

152. Martin, P., Baker, G.A. Jr. (1991) Two-point quasifractional approximant in physics. Truncation error. *J. Math. Phys.* **32**:313–328.

153. Matheron, G. (1967) *Elements pour une Théorie des Milieux Poreux.* Masson: Paris .

154. Maxwell, J.C. (1873) *Treatise on Electricity and Magnetism.* Clarendon Press: Oxford.

155. Maxwell Garnett, J.C. (1904) Colours in metal glasses and in metallic films. *Phil. Transact. Roy. Soc. London* A. **203**:385–420.

156. Manevitch, L.I., Andrianov, I.V., Oshmyan, V.O. (2002) *Mechanics of Periodically Heterogeneous Structures.* Springer-Verlag: Berlin.

157. McPhedran, R.C., McKenzie, D.R. (1980) Electrostatic and optical resonances of arrays of cylinders. *Appl. Phys. A* **23**(3):223–235.

158. McPhedran, R.C., Poladian, L., Milton, G.W. (1988) Asymptotic studies of closely spaced, highly conducting cylinders. *Proc. Roy. Soc. London. Ser. A Math. Phys. Sci.* **415**(1848):185–196.

159. Mendelson, K.S. (1975) A theorem on the effective conductivity of a two-dimensional heterogeneous mediu. *J. Appl. Phys.* **46**:4740–4741.

160. Mikhlin, S.G. (1964) *Integral Equations and Their Applications to Certain Problems in Mechanics, Mathematical Physics, and Technology.* Oxford: Pergamon Press.

161. Milton, G.W. (2002) *The Theory of Composites.* University Press: Cambridge.

162. Milton, G.W., McPhedran, R.C., McKenzie, D.R. (1981) Transport properties of array of intersection cylinders. *Appl. Phys. A* **25**(1):23–30.

163. Mityushev, V. (1993) Plane problem for the steady heat conduction of material with circular inclusions. *Arch. Mech.* **45**:211–215.

164. Mityushev, V. (1997) Transport properties of double-periodic arrays of circular cylinders. *ZAMM* **77**(2):115–120.

165. Mityushev, V. (1998) Steady heat conduction of a material with an array of cylindrical holes in nonlinear case. *IMA J. Appl. Math.* **61**(1):91–102.

166. Mityushev, V. (2001) Transport properties of doubly periodic arrays of circular cylinders and optimal design problems. *Appl. Math. Opt.* **44**:17–31.

167. Mityushev, V. (2005) R-linear problem on torus and its application to composites. *Complex Var.* **50**:621–630.

168. Mityushev, V. (2007) Exact solution of the R-linear problem for a disk in a class of doubly periodic functions. *J. Appl. Funct. Anal.* **2**:115–127.

169. Mityushev, V. (2015) Random 2D composites and the generalized method of Schwarz. *Adv. Math. Phys.* **6**:1–15.

170. Mityushev, V. (2018) Cluster method in composites and its convergence. *Appl. Math. Lett.* **77**:44–48.

171. Mityushev, V., Andrianov, I., Gluzman, S. (2022) L.A. Filshtinsky's contribution to Applied Mathematics and Mechanics of Solids. In: Andrianov, I., Gluzman, S., Mityushev, V. (eds.) *Asymptotic and Integral Equations Methods of Leonid Filshtinsky.* London: Academic Press, pp. 1–40.

172. Mityushev, V., Drygas, P. (2019) Effective properties of fibrous composites and cluster convergence. *SIAM J. Multisc. Model. Simul.* **17**(2):696–715.

173. Mityushev, V., Nawalaniec, W. (2019) Effective conductivity of a random suspension of highly conducting spherical particles. *Appl. Math. Model.* **72**:230–246.

174. Mityushev, V., Obnosov, Yu., Pesetskaya, E., Rogosin, S. (2008) Analytical methods for heat conduction in composites. *Math. Model. Anal.* **13**(1):67–78.

175. Mityushev, V., Rylko, N. (2012) Optimal distribution of the nonoverlapping conducting disks. *Multisc. Model. Simul.* **10**(1):180–190.

176. Mityushev, V., Rylko, N. (2013) Maxwells approach to effective conductivity and its limitations. *Quart. J. Mech. Appl. Math.* **66**(2):241–251.

177. Mokryakov, V.V. (2014) Strength analysis of an elastic plane containing a square lattice of circular holes under mechanical loading. *Mech. Sol.* **49**(5):568–577.

178. Mokryakov, V.V. (2010) Study of the depence of effective compliences of a plane with an array of circular holes on array parameters. *Comput. Mech. Contin. Media* **3**(3):90–101 (in Russian).

179. Mol'kov, V.A., Pobedrya, B.E. (1984) Effective elastic moduli of a unidirectional fiber composite. *Sov. Phys. Dokl.* **29**:195–196.

180. Mossotti, O.F. (1850) Sobre las fuerzas que rigen la constitución de los cuerpos. Memorie di Matematica e di Fisica della Società. *Italiana delle Scienze Residente in Modena* **24**(2):49–74.

181. Movchan, A.B., Movchan, N.V., Poulton, C.G. (2002) *Asymptotic Models of Fields in Dilute and Densely Packed Composites.* Imperial College Press: London.

182. Nayfeh, A.H. (1981) *Introduction to Perturbation Techniques.* John Wiley & Sons: New York.

183. Nayfeh, A.H. (2000) *Perturbation Methods.* John Wiley & Sons: New York.

184. Nemish, Yu.N. (1989) *Elements of the Mechanics of Piecewise Homogeneous Bodies with Non-canonical Interfaces.* Naukova Dumka: Kiev (in Russian).

185. Nitham, B.W., Sammut, R.A. (1976) Refractive index of array of spheres and cylinders. *J. Theor. Biol.* **56**:125–149.

186. Novikov, V.V. (1985) Determination of the effective elastic moduli of inhomogeneous materials. *J. Appl. Mech. Tech. Phys.* **26**:739–746.

187. O'Brien, R.W. (1977) *Properties of Suspensions of Interacting Particles: Doctoral Thesis.* University of Cambridge: Cambridge.

188. Odelevskii, V.I. (1951) Calculation of generalized conductivity of heterogeneous systems. *J. Tech. Phys.* **21**(6):667–685 (in Russian).

189. Olives, R., Mauran, S. (2001) A highly conductive porous medium for solid-gas reactions: Effect of the dispersed phase on the thermal tortuosity. *Trans. Por. Med.* **43**(2):377–394.

190. Obnosov, Yu.V. (1996) Exact solution of a boundary-value problem for a rectangular checkerboard field. *Proc. Roy. Soc. Lond A* **452**:2423–2442.

191. Panasenko, G.P. (2005) *Multi-Scale Modeling for Structures and Composites.* Springer-Verlag: Berlin.

192. Panasenko, G.P. (1993) Asymptotic solutions of the system of elasticity theory for rod and frame structures. *Russ. Acad. Sci. Sb. Math.* **75**(1):85–110.

193. Panasenko, G.P. (2003) Method of asymptotic partial decomposition of domain and partial homogenization. *Proc. Steklov Inst. Math.* **1**:161–167.

194. Perrins, W.T., McKenzie, D. R., McPhedran, R.C. (1979) Transport properties of regular arrays of cylinders. *Proc. Roy. Soc. London. Ser. A Math. Phys. Sci.* **369**(1737):207–225.

195. Pilipchuk, V.N. (2010) *Nonlinear Dynamics: Between Linear and Impact Limits.* Springer: Berlin.

196. Pilipchuk, V.N., Andrianov, I.V., Markert, B. (2016) Analysis of micro-structural effects on phononic waves in layered elastic media with periodic nonsmooth coordinates. *Wave Mot.* **63**:149–169.

197. Pilipchuk, V.N., Starushenko, G.A. (1997). A version of non-smooth transformations of variables for one-dimensional elastic systems of periodic structures. *J. Appl. Math. Mech.* **61**(2):265–274.

198. Pobedrya, B.Ye. (1984) *Mechanics of Composite Materials.* MGU: Moscow (in Russian).

199. Pobedrya, B.Ye. (1983) On the theory of viscoelasticity of structurally inhomogeneous media. *J. Appl. Math. Mech.* **47**:103–109.

200. Poisson, S.D. (1826) Mémoires sur la théorie du magnétisme. *Mémoires de l'Académie (royale) des sciences de l'Institut (imperial) de France* **5**: 247–338.

201. Poon, Y.M., Shin, F.G. (2004) A simple explicit formula for the effective dielectric constant of binary 0-3 composites. *J. Mater. Sci.* **39**(4):1277–1281.

202. Rayleigh, L. (1887) On the maintenance of vibrations by forces of double frequency, and on the propagation of waves through a medium endowed with a periodic structure. *Phil. Mag.* Ser. 5, **24**(147):145–159.

203. Rayleigh, L. (1892) On the influence of obstacles arranged in rectangular order upon the properties of a medium. *Phil. Mag.* Ser. 5, **34**(211):481–502.

204. Reuss, A. (1929) Berechnung der Fließgrenze von Mischkristallen auf Grund der Plastizitätsbedingung für Einkristalle. *ZAMM*, **9**:49–58.

205. Sahimi, M. (2003) *Heterogeneous Materials I. Linear Transport and Optical Properties.* Springer: New York.

206. Salski, B. (2012) The extension of the Maxwell Garnett mixing rule for dielectric composites with nonuniform orientation of ellipsoidal inclusions. *Prog. Electromag. Res. Lett.* **30**:173–184.

207. Sanchez-Palencia, E. (1980) *Non-homogeneous Media and Vibrations Theory.* Springer: Berlin.

208. Sendeckyj, G.P. (Ed.) (2013) *Mechanics of Composite Materials: Composite Materials.* Vol. 2. Academic Press.

209. Sandrakov, G.V. (1991) Averaging principles for eguations with rapidly oscillatory co-efficients. *Math. USSR-Sb.* **68**(2):503–553.

210. Sevostianov, I., Kachanov, M. (2008) Connections between elastic and conductive properties of heterogeneous materials. *Adv. Appl. Mech.* **42**:69–251.

211. Sihvola, A.H., Lindell, I.V. (1990) Chiral Maxwell-Garnet mixing formula. *Electron. Lett.* **26**(2):118–119.

212. Shermergor, T.D. (1977) *The Theory of Elasticity of Microinhomogeneous Media.* Nauka: Moscow (in Russian).

213. Shvidler, M.I. (1985) *Statistical Hydrodynamics of Porous Media.* Nedra: Moscow (in Russian).

214. Skryabin, I.L., Radchik, A.V., Moses, P., Smith, G.B. (1997) The consistent application of Maxwell-Garnett effective medium theory to 635 anisotropic composites. *Appl. Phys. Lett.* **70**(17):2221–2223.

215. Slepyan, L.I., Yakovlev, Yu.S. (1980) *Integral Transforms in the Nonstationary Problems of Mechanics.* Sudostroyenie: Leningrad (in Russian).

216. Snarskii, A.A. (2007). Did Maxwell know about the percolation threshold? (on the 15th anniversary of percolation theory). *Physics-Uspekhi* **50**(12):1239–1242.

217. Snarskii, A.A. (2004) Effective conductivity of 2D macroscopic heterogeneous self-dual media. *Laser Phys.* **14**(3):1–7.

218. Snarskii, A.A., Bezsudnov, I.V., Sevryukov. V.A., Morozovskiy, A., Malinsky, J. (2016) *Transport Processes in Macroscopically Disordered Media (from Medium Field Theory to Percolation).* Springer.

219. Sokolkin, Yu.V., Tashkinov, A.A. (1984) *Mechanics of Deformation and Fracture of Structurally Inhomogeneous Bodies.* Nauka: Moscow (in Russian).

220. Spanier, J.E., Herman, I.P. (2000) Use of hybrid phenomenological and statistical effective-medium theories of dielectric functions to model the infrared reflectance of porous SiC films. *Phys. Rev. B* **61**(15):10437–10450.

221. Starushenko, G., Krulik, N., Tokarzewski, S. (2002) Employment of non-symmetrical saw-tooth argument transformation method in the elasticity theory for layered composites. *Int. J. Heat Mass Trans.* **45**(14):3055–3060.

222. Stauffer, D., Aharony, A. (1994) *Introduction to Percolation Theory.* 2nd ed. Taylor and Francis: London.

223. Tao, R., Chen, Z., Sheng, P. (1990) First-principles Fourier approach for the calculation of the effective dielectric constant of periodic composites. *Phys. Rev. B* **41**(4):2417–2420.

224. Tarnopolskii, Yu.M., Zhigun, I.G., Polyakov, V.A. (1992) *Spatially Reinforced Composites.* CRC Press.

225. Tokarzewski, S. (2013) Multipoint matrix Padé approximant bounds on effective anisotropic transport coefficients of two-phase media. *ZAMP* **4**(1):167–178.

226. Torquato, S. (2002) *Random Heterogeneous Materials. Microstructure and Macroscopic Properties.* Springer: New York.

227. Van der Poel, C. (1958) On the rheology of concentrated dispersions. *Rheolog. Acta* 1(2-3):198–205.

228. Van Fo Fy, G.A. (1971) *Theory of Reinforced Materials with Coatings*. Naukova Dumka: Kiev (in Russian).

229. Vanin, G.A. (1985) *Micromechanics of Composite Materials*. Naukova Dumka: Kiev: (in Russian).

230. Vasiliev, V.V., Morozov, E.V. (2018) *Advanced Mechanics of Composite Materials and Structures*. Fourth Edition. Elsevier.

231. Vasiliev, V.V., Tarnopolsky, Yu.M. (Eds.) (1990) *Composite Materials: Handbook*. Mashinostroyeniye: Moscow (in Russian).

232. Vinogradov, A.P. (2001) *Electrodynamics of Composite Materials*. URSS: Moscow (in Russian).

233. Voigt, W. (1889) Ueber die Beziehung zwischen den beiden Elasticitätsconstanten isotroper Körper. *Ann. Phys.* 274:573–587.

234. Wagner, K.W. (1914) Erklärung der dielektrischen Nachwirkungsvorgänge auf Grund Maxwellscher Vorstellungen. *Arch. Elektrotech.* 2:371–387.

235. Wiener, O. (1912) Die Theorie des Mischkörpers für das Feld der stationären Strömung. Erste Abhandlung die Mittelwertsätze für Kraft, Polarisation und Energie. *Der Abhandlungen der Mathematisch-Physischen Klasse der Königlich Sächsischen Gesellschaft der Wissenschaften* 32:509–604.

236. Wojnar, R. (1997) On elastic moduli of a two dimensional two phase system. *ZAMM* 77(2):S469–S472.

237. Zhikov, V.V. (1991) Estimates for the averaged matrix and the averaged tensor. *Russ. Math. Surv.* 46(3):65–136

238. Zhikov, V.V., Kozlov, S.M., Oleinik. O.A., Há Tien Ngoan (1979) Averaging and G-convergence of differential operators. *Russ. Math. Surv.* 34(5):69–147.

Index

Note:- Page numbers with *italics* represent the figure and **bold** for table.

For Product Safety Concerns and Information please contact our EU
representative GPSR@taylorandfrancis.com
Taylor & Francis Verlag GmbH, Kaufingerstraße 24, 80331 München, Germany